Northwest Lands, Northwest Peoples

Northwest Lands, Northwest Peoples

READINGS IN ENVIRONMENTAL HISTORY

EDITED BY

Dale D. Goble and Paul W. Hirt

University of Washington Press

SEATTLE AND LONDON

This book has been published
with the assistance of a grant from
the Sherman and Mabel Smith Pettyjohn Fund,
Washington State University.

Library of Congress Cataloging-in-Publication Data
Northwest lands, northwest peoples : readings in environmental history
/ edited by Dale D. Goble and Paul W. Hirt.
p. cm.
Includes index.
ISBN 0-295-97839-2 (alk. paper). — ISBN 0-295-97838-4 (pbk. : alk. paper)
1. Northwest, Pacific—Environmental conditions.
2. Human ecology—Northwest, Pacific.
I. Goble, Dale. II. Hirt, Paul W., 1954– .
G155.N75N66 1999 99-27118
333.7′13′09795—DC21 CIP

The paper used in this publication is acid-free and recycled from 10 percent post-consumer and at least 50 percent pre-consumer waste. It meets the minimum requirements of American National Standard for Information Sciences—Permanence of Paper for Printed Library Materials, ANSI Z39.48–1984.

Contents

Preface

PAUL W. HIRT AND DALE D. GOBLE

A half century ago, Aldo Leopold wrote of the need for a "land ethic," a perspective that recognized the artificiality of distinctions between "nature" and "culture." "Man is," he wrote, "only a member of a biotic team."[1] Despite Leopold's insight, most academic writing continues to treat the study of the human community—labeled sociology, economics, or history—as distinct from the study of its place in the natural community—denominated geology, biology, or ecology. There are, however, signs of change. When humanists, natural scientists, and social scientists get together in the same room, a discussion of environment and history produces a rich, fresh, and complex view of the world. The book you hold in your hands is the result of such a discussion.

It is also part of a growing discourse on human cultures in the context of their natural communities. Since the 1970s, concern for "the environment" has led to the creation of environmental subfields in many of the sciences and humanities. There are now recognized disciplinary specializations in environmental history, environmental sociology, resource economics, and environmental law. Ecocriticism and environmental ethics reflect the humanistic interest in culture and nature. In the physical sciences as well, interdisciplinary programs in environmental studies or natural resources sciences have recently flourished, with faculties made up of people trained in the hard sciences alongside colleagues trained in sociology, political science, economics, and ethics, as well as specialists in liminal fields such as recreation management and land-use planning.

A fundamental tenet of this discourse is that cultures evolve in places—

that societies develop within a geographical setting, filling an ecological niche. Imagine how different life would be in the Northwest if there were no Cascade Mountains or Columbia River; imagine Seattle or Tacoma without Puget Sound. The role of place in shaping human societies is sometimes apparent: geography influences the location of cities, transportation corridors, and the flow of commerce. At other times, the role of place in human affairs is subtler: climate and soil conditions largely determine vegetation patterns on the landscape, and vegetation patterns determine where and how we cut timber, graze livestock, and grow crops. Climate, soil, and water supply also limit the amount of each resource—trees, grass, or wheat—we can produce over time. Climatic fluctuations demonstrate the role of nature in shaping culture because even subtle changes can have profound ramifications on our economy, our society, and even our politics. Rivers provide hydropower, allow inexpensive transportation of goods, support fisheries and recreation, dilute and disperse wastes, and irrigate farms. A modest decline in winter snowpack brings conflict among hydropower, navigation, fisheries, and irrigation interests over access to scarce water. But too much snow and a warm chinook wind can cause floods and mudslides like those that plagued the Northwest in the springs of 1996 and 1997.

Just as surely as physical places and the plants and animals that inhabit them influence human societies, social systems in turn reshape the natural community. The Columbia River has been transformed into a series of slackwater lakes that provide hydropower, barge transportation, and flood control, but they also have transmogrified the river from salmon habitat into carp and squawfish habitat.

The boundaries between culture and nature are blurred—understandably so, since humans are undeniably part of nature. Although daily life often conspires to obscure this fact—living as we do in climate-controlled houses, driving self-contained vehicles, acquiring food without regard for the seasons, banishing darkness with the flip of a switch, disposing of our wastes with the push of a lever—our connectedness to nature is inescapable: our houses are constructed from trees, our cars originate in iron ores and run on fossil fuels, our groceries come from farms, our heat and light derive from falling water or the burning of coal or gas, and our wastes, whether flushed or hauled away, wind up back on the land. Humans are animals, filling an ecological niche, consuming resources, modifying the environment to suit their needs, cooperating with some species and competing against others.

But the relationship between nature and culture is more tangled still. The two dance a complex composition: the spring floods and landslides in

1996 and 1997 were exacerbated by excessive logging and road building, and so we could write an "unnatural history" of this "natural disaster."[2] Having transformed the Columbia to produce electricity and having created, as a byproduct, carp habitat, we seek to compensate by barging immature salmon downriver and placing a bounty on squawfish—action, reaction, and consequences.

The essays in this volume address questions of how humans have adapted to and modified nature over time in the Pacific Northwest of North America, and how changing ecological conditions have in turn affected human economies, laws, values, and social order. They do so from a variety of perspectives. The collection includes contributions from historians, anthropologists, ethnoecologists, a paleoecologist, a botanist, geographers, biologists, law professors, and a journalist. The collection evolved out of a symposium on Northwest environmental history that the editors organized in Pullman, Washington, in August 1996. Most of the papers were presented at the symposium, but we also solicited additional essays. We think it is an impressive, provocative collection by some of the most talented scholars in the region. It is a "sampler" of cutting-edge scholarship in environmental history and its related fields. Many of the authors—Blumm, Fiege, Hirt, Hunn, Kruckeberg, Langston, Morrissey, and Robbins—have written books on the topics they address here.

In compiling this collection, we struggled to balance competing aims: we wanted broad topical coverage but we also wanted to offer more than one perspective on each topic. Multiple points of view provoke analytical thinking and display the variety of scholarly approaches to a subject. For example, instead of one essay on American Indian land use, we solicited several covering both coastal and plateau tribes from different perspectives and time frames. Instead of one essay on Northwest forest history, we have three, each offering a different geographical and analytical focus. As a consequence, it was necessary to trade maximum coverage for multiple perspectives. Nonetheless, there are essays on more than a dozen distinct topics.

One topic we regret not covering is the region's romance with nuclear weapons production and nuclear energy. Hanford in Washington and the Idaho Nuclear Engineering Laboratory near Idaho Falls were important weapons research and production facilities. While this book was in production, however, the University of Washington Press simultaneously produced a collection of essays on the subject, *The Atomic West, 1942–1992*, edited by Bruce Hevly and John Findlay. We also regret not having additional essays on urban environmental history and the growing field of environmental justice.

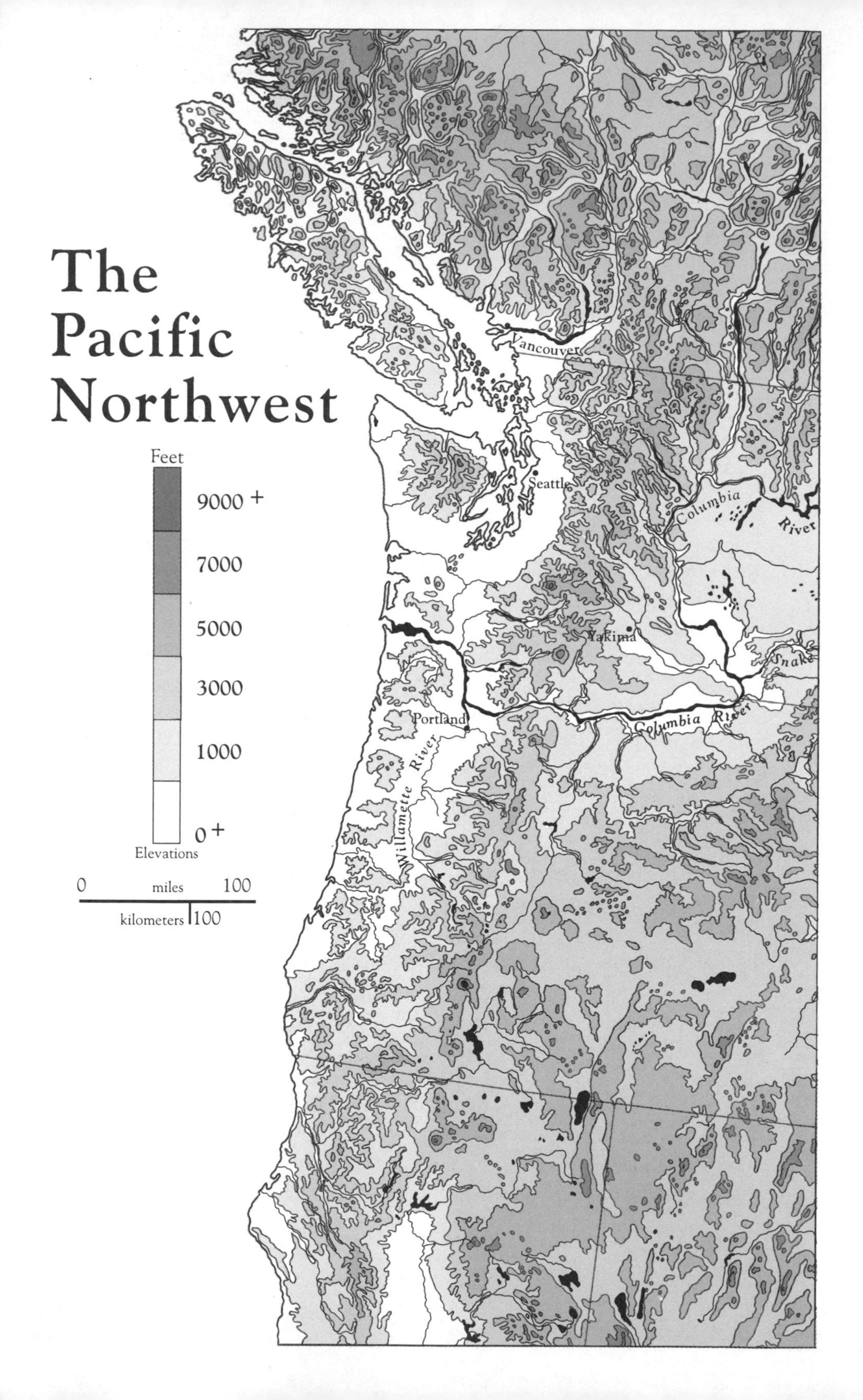
The Pacific Northwest
Feet
9000 +
7000
5000
3000
1000
0 +
Elevations
0
miles
100
kilometers
100
Vancouver
Seattle
Columbia
River
Yakima
Snake
Portland
Columbia
River
Willamette
River

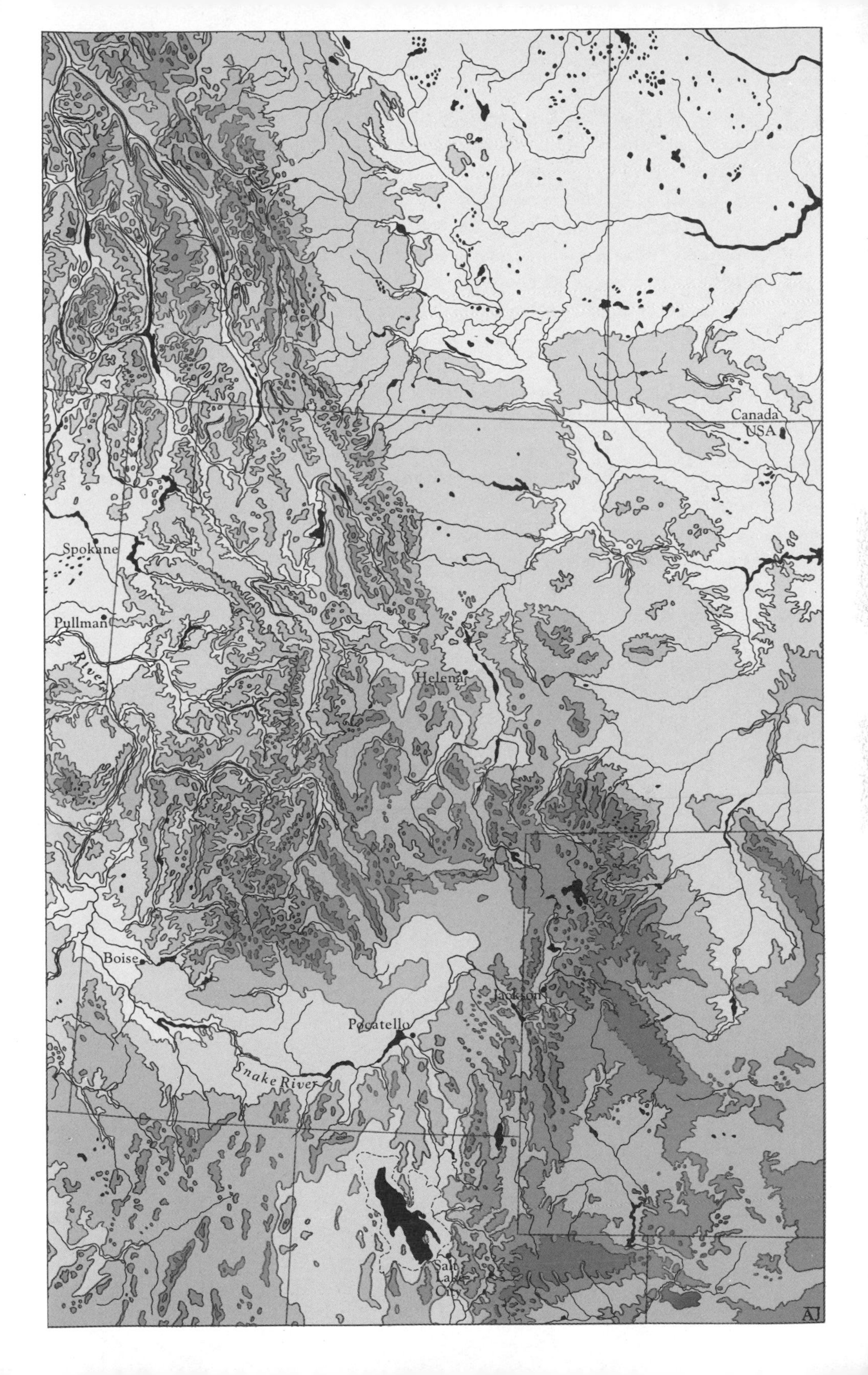

Canada
USA
Spokane
Pullman
River
Helena
Boise
Jackson
Pocatello
Snake River
Salt Lake City
AJ

Their absence suggests that the opportunity exists for a future compilation focusing more narrowly on the settled and humanized environments in which we live.

This collection offers a survey of subjects and approaches that is meant to serve as supplemental reading for college courses in Northwest history, geography, natural resource management, and environmental studies. It should also be useful for courses in American studies, environmental sociology, environmental policy, and environmental law. It offers a multidisciplinary approach to understanding nature and society in the Pacific Northwest. Although it is only a preliminary step, the effort to tell the history of humans and nature in one narrative is the goal that Leopold identified 50 years ago.

Finally, we want to mention that for researchers working in any of the environmental history subject areas covered in this volume, and for teachers planning writing assignments for their students on environmental history topics, we have compiled an extensive bibliography of published books and essays, Internet resources, films, and photographic archives on Northwest environmental history. The URL is http://www.wsu.edu:8080/~forrest/.

For those interested in environment and social justice issues, scholars at the University of California at Santa Cruz have compiled an extensive bibliography at http://www.cruzio.com/~meuser/ejwww.html.

NOTES

1. Aldo Leopold, *A Sand County Almanac and Sketches Here and There* (New York: Oxford University Press, 1949), p. 205.

2. Theodore Steinberg, "Do-It-Yourself Deathscape: The Unnatural History of Natural Disaster in South Florida," *Environmental History* 2 (October 1997): 414–38.

Northwest Lands,
Northwest Peoples

1 / Setting the Pacific Northwest Stage

The Influence of the Natural Environment

ERIC C. EWERT

In no other part of North America is so much physical geographic diversity compressed into such a small area as in the three states of Washington, Oregon, and Idaho. And in perhaps no other part of North America has the natural environment had such a profound effect on human activities. The Pacific Northwest's 250,000 square miles—6.8 percent of the United States' total—hold towering volcanic peaks, raging whitewater rivers, sagebrush-studded volcanic plains, deeply incised canyons, wave-battered rocky coastlines, lush agricultural valleys, and nearly impenetrable coniferous forests. In a half-day's drive from Spokane to Seattle, La Grande to Portland, or Boise to Coeur d'Alene, one experiences tremendous environmental diversity. The climate, landforms, flora, and fauna change repeatedly and remarkably in a mere two or three hundred miles.

On this varied natural stage we humans have settled the land. In the process, we dramatically changed it. We cut the trees, dammed the rivers, paved the pathways, plowed the prairies, dug up the minerals, and turned loose the cows. Our efforts made deserts verdant, turned rivers into reservoirs, reduced mountains to hills, erected cities and towns, and generally "tamed" the wilderness. American history is full of heroic accounts of people "conquering nature," "opening the frontier," and "developing natural resources." Testifying before Congress in 1828, unabashed Oregon booster Hall Jackson Kelley described the territory as "the most valuable of all the unoccupied parts of the earth." The remaking of nature was our manifest destiny—and certainly the Pacific Northwest fits this frontier model.

The diaries of explorers such as Captain James Cook, Meriwether Lewis, and William Clark are rich with details of the inspiring and challenging landscape. In the 1830s, an adventurous army captain named Benjamin L. E. Bonneville graphically described his encounter with Hells Canyon: "Nothing we had ever gazed upon in any other region could for a moment capture the wild majesty and impressive sternness with the series of scenes which here at every turn astonish our senses and filled us with awe and delight."[1] Coastal, Columbia Plateau, and Great Basin Native Americans, Oregon Trail pioneers, steamship pilots, cannery workers, lumberjacks, stump farmers, and dam builders all shared a common response to the Northwest's natural landscapes: a sense of awe.

Much of this awe derives from the power involved in the shaping of this dramatic landscape. On May 18, 1980, Mount St. Helens reminded us of the natural forces that created the Northwest. On that morning the mountain spewed hundreds of thousands of tons of ash into the air, loosed great debris floods, leveled millions of trees, took 60 human lives, and turned the day to night. Some 6,600 years ago, Mount Mazama—now Crater Lake—did the same thing, only with 50 times the fury. You can find Mazama ash on the East Coast. Indian folklore still recounts the event.

At the end of the Pleistocene—the epoch of the last Ice Age that ended some 12,000 years ago—unimaginable floods surged across what is today northern Idaho and central Washington. These "Missoula Floods"—named for great ice dams and lakes that formed near that present-day Montana town—periodically sent torrents racing into the Columbia River basin when the dams were breached. Coulees, potholes, channeled scablands, and other features of the landscape bear witness to these cataclysmic deluges.

Long before these events, great floods of fiery lava poured forth from fissures in the earth's crust, consuming everything in their paths. Huge portions of central Washington, northeastern Oregon, and southern Idaho were buried beneath hundreds of feet of molten rock. The flood basalts, cinder cones, and lava tubes of the Columbia Plateau and Snake River Plain are solidified testimony to this genesis of some of the Northwest's more striking landscapes.

Even today, we can stand near the floodgates of one of the Columbia or Snake River dams and sense the power of the natural, unharnessed Northwest. Witnessing ocean waves pounding a rocky headland or gale force winds buffeting a Cascade peak evokes a similar appreciation. Northwesterners are imbued with this sense of awe toward the natural environment.

Awe is not necessarily reverence, however. Euro-American immigrants

FIG. 1.1. Mount St. Helens erupts, May 18, 1980.

quickly busied themselves with altering the landscapes of the Northwest to fit their needs, desires, and visions of what the land *should* look like and what it *should* provide. Reservoirs, farmland, neatly platted towns, channelized rivers, and straight-line transportation routes were far more useful than the rough, "unfinished," natural Northwest. As we near the end of the second century of this transformation, contemporary residents wonder what happened to the Northwest of expansive forests, free-flowing rivers, uncountable salmon, and wide open spaces; they question whether the transformation has been an unmitigated success—whether what we have gained has always been worth the costs.

Much of this temporal, perceptual journey reflects the continued evolution of our relationship to the natural environment. Initially, the relationship was one of fear because most of the environment was "wild" and unknown. It was the haunt of beasts and bandits, the frightening land between civilizations, and the subject of terror in fairy tales. Later, the environment became a challenge, a place to be overcome and organized, site of a competition between nature and humans that we must not lose. These changes in our perceptions resulted in our treating nature as a commodity with a price,

FIG. 1.2. A transformed, modern Northwest landscape: Grande Coulee Dam on the Columbia River (photo by Paul Hirt).

something that could be bought or sold. Natural landscapes became board feet of timber, tons of coal, kilowatts of power, head of cattle, acre feet of irrigation, pounds of pelts, and bars of gold. The value of this natural bounty was determined only by the labor needed to transform it into commodities; in its natural state, it was thought to be valueless.

Recently, a new "value" is being assigned to the Northwest's natural places. This new approach views worth as intrinsic, incalculable in dollar amounts. Not everyone, however, believes that unspoiled views, free-flowing rivers, and abundant wildernesses are valuable for their own sake. The usual course of action is piecemeal protection: listing endangered species, imposing pollution controls, or establishing parks and monuments. These actions are tangible, unlike the altruistic or metaphysical reasons for protection. It is the job of environmental geographers and historians to document and illuminate this journey of changing environmental relations. Numerous contributions to this collection endeavor to do that.

Today, the natural environment is still very much a part of contemporary Pacific Northwest life; it is in part how northwesterners define themselves.

FIG. 1.3. Mount Rainier, view from Goat Rocks Wilderness, Gifford Pinchot National Forest (photo by Paul Hirt).

Several essays in this collection address how northwesterners develop a "sense of place." Indeed, much of their lifestyle centers on direct and indirect linkages to the natural environment—even in the region's largest cities. A friend in Seattle said of Mount Rainier, "It's my reference point, my compass bearing. Even on a cloudy day, I know it's there." Portlanders feel the same way about Mount Hood. Others locate themselves by the Columbia and Snake Rivers, Lake Coeur d'Alene, or the Sawtooth and Bitterroot Mountains. Even locked in traffic in one of the Northwest's large cities, residents can look just beyond the car windows to forested mountains, meandering rivers, undulating prairies, or smooth, gray saltwater bays.

In the Northwest, people's icons and contentious issues are typically environmental as well. Natural landscapes are often at the core of their political, economic, and social debates. Salmon, spotted owls, old-growth forests, wilderness, public land, Columbia and Snake River dams, and air and water pollution dominate the discussion; limiting growth, establishing greenbelts, enacting bottle bills, and channeling development are common issues.

This regional environmental connection is less ubiquitously strong elsewhere in the nation. Residents of the Midwest don't get as excited over the fate of their prairies, nor do south Atlantic Coast denizens over the condition of their Piedmont. Residents of other parts of the country *do* perceive the Northwest in these terms, however. The Northwest is less well known for memorable cities or notable cultural contributions than for national parks, outdoor recreation, fish, tall trees, and volcanic peaks—and perhaps coffee. These are the attributes northwesterners represent and relate to. They are *the* Northwest, even for Washington, Oregon, and Idaho's most recent residents.

Even as the traditional extractive industries of timber, mining, fishing, and ranching lose economic dominance, people's connection to the environment remains strong. Newcomers surveyed for the reasons they chose to move to the Northwest always rate the natural environment near the top; they even select this at the expense of a well-paying job. The newer "high technology" businesses—Sony in Eugene, Intel in Portland, Microsoft in Seattle, and Micron in Boise—report that quality-of-life reasons attracted them to the region. It is no accident that outdoor gear purveyors such as REI and Eddie Bauer chose to headquarter in Seattle. Most northwesterners would rather visit the mountains than a museum and would choose the ocean over an opera.

NATURAL DIVERSITY

Fundamental to understanding the role of the environment in the Pacific Northwest is a firm grounding in the region's physical geography. The Northwest's landscapes are exceptionally diverse. It is important to realize that this diversity exists by all measures. There are striking differences in precipitation, temperature, landforms, vegetation, rock types, and soils. Of these, the most important controlling factors are *climate* and *topography*. Together these two influence the remainder of the environmental attributes as well as much of the human experience in the Northwest. The remarkable contrast between eastern and western Oregon and Washington or between northern and southern Idaho is traceable primarily to the shape of the land and the long-term influence of weather. Native American and recent northwestern economic activities, settlement patterns, and transportation routes also reflect these natural controls.

Imagine running your hand over a raised-relief model of the Northwest from west to east. Your fingertips sense a wide variety of surfaces and sensa-

FIG. 1.4. The rocky, battered coast of Cape Blanco, southern Oregon (photo by Eric Ewert).

tions. After leaving the cold waters of the Pacific Ocean, your fingers quickly gain elevation as you climb the Coast Range. Unlike the south Atlantic coast—where the rolling Piedmont borders an expansive coastal plain—the Northwest leaps from the sea. In many places, the Coast Range mountains rise directly from the ocean, shouldering up in a jagged and abrupt edge of headlands, embayments, sea stacks, and islands. What beaches do exist tend to be narrow and subjected to intense wave activity that may coat them with gravel or cobbles instead of the sand of the East and Gulf coasts. This relationship between sea and mountains precluded the development of large cities on either Oregon's or Washington's coasts. Although Indian and Euro-American fishing villages and towns were and still are quite common, they have never grown to great size.

The journey up the Coast Range feels carpetlike as your fingers rub over a thick cloak of evergreen trees: fir, spruce, cedar, and hemlock. The sensation, however, is like that of a three-dimensional checkerboard, because many clear-cut logging swaths now riddle these forests. Some are old-growth forests, with trees that are 1,000 years old and home to the threatened spotted owl. In the far northwestern portion of Washington, the Olympic Mountains greet your hand with 8,000-foot summits, a temperate rainforest with an annual rainfall of well over 100 inches, and many small glaciers. In southwestern Oregon, the Coast Range gives way to the irregularly folded and faulted Klamath-Siskiyou mountain system, which spills into northern California.

After reaching only three to four thousand feet in most of the Coast Range, your fingers drop back down into an extensive inland valley. Stretching from Eugene in the south to Bellingham in the north, the Willamette Valley–Puget Sound lowland is the site of verdant agriculture and sprawling settlement. It is also the most densely populated part of the Pacific Northwest. These lowlands house the largest cities, the most abundant transportation routes, two state capitols, and the greatest concentration of industry. Here the rainfall is high enough, the seasons mild enough, and the soils deep enough for extensive cultivation. The rich farmland, especially along the Willamette River, attracted most of the region's first immigrants and was instrumental in Oregon's achieving statehood in 1859, a full 30 years before Washington and Idaho. Much of this river valley would be a mixed deciduous and evergreen forest if not for the intensive cultivation and historic Native American practice of clearing with fire.

The stunning juxtaposition of mountain and water, forest and farm, so famed in the Northwest, is most dramatic around the Pacific Northwest's inland sea, Puget Sound. Confined by the Olympic Mountains to the west and the Cascade Mountains to the east, great lobes of glacial ice rasped their way south through this lowland. The glaciers gouged channels hundreds of feet below sea level and piled the debris as islands and broad alluvial plains. Dense settlement from Olympia in the south to Vancouver, British Columbia, in the north has long focused on the splendid ports, deep-water channels, and fertile farmland the glaciers left as they retreated. One of the finest inland waterways in the world, Puget Sound is connected to the open Pacific by the Straight of Juan de Fuca, which passes between the Olympics and Vancouver Island.

Soon after leaving the valley and sound, your fingers climb rapidly again, up the slopes of the Cascade Mountains. This, the great backbone of the Northwest, stretches from California into British Columbia. It is broken only in two places along its entire length: by the Columbia River along the Washington-Oregon border and by the Klamath River in southern Oregon and northern California. Here, too, great forests cover the hills, nurtured by abundant precipitation. Clear-cutting is common again and displays an odd

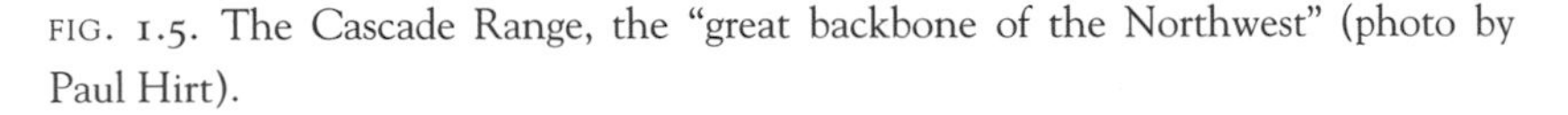

FIG. 1.5. The Cascade Range, the "great backbone of the Northwest" (photo by Paul Hirt).

geometry when one flies into Portland or Seattle. As you explore the range north to south, isolated peaks catch your fingertips. These are the great volcanic summits of the Cascades, the highest of which, Mount Rainier, looms 14,000 feet above Seattle. Its lofty slopes hold more than 25 active glaciers fed by remarkable snow accumulations, sometimes in excess of 100 feet in a single winter. Mount Hood similarly dominates the skyline above Portland. All of the Cascade peaks are dormant volcanoes. Despite Mount St. Helens' warning, northwesterners tend to treat them as extinct mammoths and erect ski resorts, vacation cottages, and recreation facilities on their precipitous slopes. People cluster in valleys in the mountains' shadows and along rivers that race down their margins.

As your exploration continues, the steep journey down the east side of the Cascades begins the most unexpected transition in northwestern geography: the change from the moist west-side forests to the arid east-side grasslands and sagebrush steppe. Swept along by the dominant westerly winds, water-laden air masses from the Pacific Ocean climb the Coast Range and the Cascades and drop most of their moisture as the lower air pressure and cooler surround-

FIG. 1.6. Clearcutting in the Colville National Forest, Washington (photo by Paul Hirt).

ings chill the air below its dew point, causing condensation and the formation of great clouds. As the cooling continues, the clouds deliver copious quantities of precipitation to western Oregon and Washington. On the east side of the Cascade summits, the process is reversed. The air descends in elevation and warms in temperature, and the clouds disappear. This is the *rain-shadow effect,* which influences two-thirds of the Northwest.

The aridity catches the uninitiated visitor by surprise. The scrubby plains, barren plateaus, and rocky ridges of southern Idaho and eastern Washington and Oregon exist in sharp contrast to the typical northwestern stereotype of Douglas firs and abundant streams. As they gaze at the sagebrush and bare rock, visitors to the Columbia River and Grand Coulee Dam in central Washington often wonder, "Where is all of this water coming from?" The answer lies in distant mountains that squeeze more moisture from the eastward-traveling air masses. These mountains—as far away as Wyoming, Montana, and southeastern British Columbia—catch more rain and snow and give birth to the Columbia River and its major tributary, the Snake River. These green mountains are remote from the stark desert on

FIG. 1.7. Scrubby plains and the Lemhi Range, southern Idaho (photo by Eric Ewert).

FIG. 1.8. The Grand Tetons, the Snake River, and terraces, as viewed from Ansel Adams Overlook, Grand Teton National Park, Wyoming (photo by Eric Ewert).

either side of the river at Grand Coulee, which in places receives a scant seven inches of annual precipitation.

The Columbia River is the great artery and lifeline of the Pacific Northwest. Along its circuitous course, Native Americans settled, explorers journeyed, and pioneers immigrated. Today, the most important east-west highways, railroads, utility corridors, pipelines, and communication facilities parallel at least part of its path. The river has always been the gateway through the Cascades, a rendezvous point for east and west. Native Americans from both sides of the mountains met at The Dalles and Celilo Falls at the east end of the Columbia Gorge to trade the products of their very different environments. Shellfish, cedar bark baskets, fish oil, and shell beads from the coast were swapped for buffalo tools, bison robes, and precious stones from the inland areas; this was the most important contact point between the Coast and Plateau Indian tribes.

Historically, the river delivered 10–15 million salmon each year to the interior of all three states. This abundant protein source nourished great populations. Today, the river and its tributaries have been engineered into a series of reservoirs for navigation, irrigation, recreation, electrical generation, and flood control. Because of the dams, Lewiston—400 miles inland and 700 feet above sea level—is effectively a Pacific Ocean port. The desert that would have greeted your exploring fingers on the Columbia Plateau a century ago is now in part a vast agricultural region made possible by dam, diversion, and ditch as a result of Great Depression and World War II–era government projects. Inexpensive irrigation and hydropower from Bonneville Dam and dozens of other projects have tempted population growth and electricity-hungry businesses such as the Boeing Corporation of Seattle.

Folk singer Woody Guthrie immortalized the great river with his verse, "Roll on Columbia, roll on." Except for a 50-mile stretch in southern Washington called the Hanford Reach, very little of the Columbia rolls freely today. Salmon no longer spawn in the river in any appreciable numbers. The mighty river of the Northwest, so critical to the region for so many reasons, remains at the heart of a number of its most contentious environmental issues.

Much of eastern Washington and Oregon and southern Idaho is arid, rocky, and lightly vegetated. Our tactile investigation meets sharp lava, prickly desert plants, jagged hills, cliffs, and gullies. Local mountain ranges such as the Blues, Ochocos, and Wallowas break the arid stranglehold, but only temporarily. This is a land of sparse settlement that serves primarily as open range for livestock and wildlife. Southern Idaho's Snake River Plain is

FIG. 1.9. The Columbia River and Interstate 90 near Vantage, Washington (photo by Paul Hirt).

equally arid, but there, as in the Columbia Basin, farmers tricked nature. The voluminous Snake River is diverted onto adjacent fields or pumped from its deep aquifers to form a rich agricultural arc that stretches from the Grand Teton Mountains in Wyoming to the Oregon border. Idaho's "famous potatoes," as well as a host of other crops, come mostly from this fertile plain, despite its meager 10 inches of annual rain.

Not until your fingers reach the Panhandle and central portion of Idaho does adequate water come naturally again. The Rocky Mountains spill into Idaho from Montana and form a highly varied landscape of steep mountains and deeply incised canyons. They house scattered veins of precious metals that drew hordes of prospectors to the area in the nineteenth century. Interesting to the touch but no easy place for settlement, these mountains sequester the largest protected wilderness areas outside of Alaska. Only Highway 12 between Lewiston and Missoula, paralleling Lewis and Clark's difficult trail, finds a route across this rugged and wild landscape. Reintroduction of endangered wolves and grizzlies is a violently contested issue here.

FIG. 1.10. The main stem of the Salmon River, Frank Church River of No Return Wilderness, Idaho (photo by Paul Hirt).

Tributaries to the Snake River, such as the Salmon and the Clearwater, etch contorted gorges through this mountainous country. The Snake itself carved a great defile that forms the Idaho-Oregon borderland; Hells Canyon—nearly 8,000 feet deep—has more vertical relief than the Grand Canyon. Its confused whitewater confounded early navigation and, together with the nearby mountains, effectively isolated the northern and southern halves of Idaho. So great was the topographic rift that during the petition for statehood, many northern Idahoans leaned toward their stronger affiliation with Washington. Even today, a landslide or washout on Highway 95 severs Idaho in two.

This portion of the Rockies supports forests that are generally drier than those on the west side of Oregon and Washington. White and ponderosa pines are much more common than cedar, spruce, and fir. These forests, too, have been extensively logged. They are also colder than the woodlands of the Cascades. West of the Cascade crest, a mild marine climate dominates, but to the east the pattern is more continental. Greater temperature extremes characterize the inland weather pattern: hot sunny summers and cold snowy

FIG. 1.11. Palouse landscape, Highway 12, southwest of Pullman, Washington (photo by Paul Hirt).

winters prevail in the Rockies and surrounding arid regions. One particularly fertile and mild area lies on the western flank of the Panhandle Rockies. The Palouse country south of Spokane supports some of the most productive dry-land wheat, pea, and lentil farming in the world. In extreme northern Idaho, glaciers excavated and later filled with their meltwater some of the West's largest natural lakes. The most prominent—Priest Lake, Lake Pend Oreille, and Lake Coeur d'Alene—became the center of a burgeoning tourism and retirement area. They are yet another example of the Northwest's highly variable natural landscapes.

SETTLEMENT: THE GEOGRAPHY OF LOCATION

Pacific Northwest settlements are location-bound: they were founded, platted, and populated in specific places because of what the natural environment around them had to offer. One has only to compare a map that carefully details the physical environment with one that shows the distribution of the Northwest's population to realize the relationship. In the

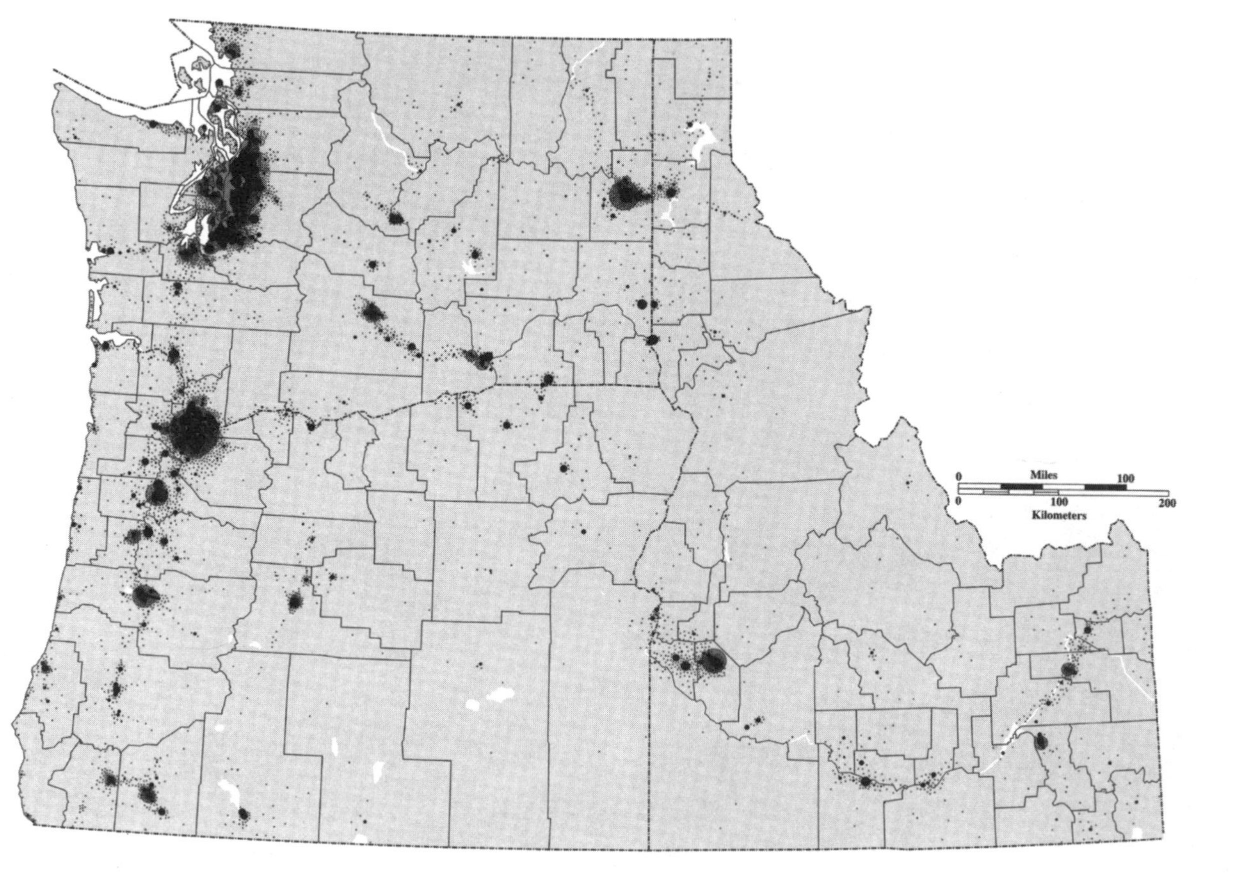

FIG. 1.12. Population distribution in the Northwest (source: *Atlas of the Pacific Northwest*, edited by P. L. Jackson and A. J. Kimerling [Corvallis: Oregon State University Press, 1993]).

midsections of North America—uniform in climate, soil fertility, and transportation potential—settlements are evenly spaced and graduated in size. In contrast, the Northwest is far from homogeneous. Its terrain, soils, water supplies, potential for transportation links, and occurrence of basic resources are widely scattered and differ markedly throughout the region. Thus, in the Pacific Northwest, the physical environment controls the potential for human habitation.

Geographers refer to the environmental characteristic of a particular location as *site*. Euro-American pioneers—like the Native Americans before them—gravitated to places on the land where they could make a living. Historically, these were locales where water, game, shelter, building supplies, transportation, and other raw materials of basic subsistence were available. Indians primarily located near rich coastal waters or along river cataracts where salmon could be netted. The first European explorers and colonizers sought out places where fur-bearing animals thrived: sea otters along the coasts of Oregon and Washington and beavers in the interiors of both states and Idaho. Trapping and fur trading centers sprang up where the abundant animals and their luxuriant pelts could be collected and shipped to distant markets. Native Americans and Euro-American trappers would return seasonally to these sites to sell, trade, and barter for the next foray's supplies. Settlements grew to extract the "soft gold" and meet the needs of the fur hunters. In this way the fur trade founded Astoria and Champoeg, Oregon; Vancouver, Washington; Fort Hall, Idaho; and Victoria, British Columbia.

Later, mineral strikes brought droves of prospectors in search of gold, silver, and other valuable ores. Settlements in Washington, Oregon, Idaho, and much of the mountainous West exploded into ramshackle tent villages overnight and bustling urban centers in a matter of months. Cries of "Eureka, gold!" sparked the imaginations not only of countless miners but of attendant entrepreneurs hell-bent on relieving the prospectors of their newfound wealth. Merchants, bankers, capitalists, shopkeepers, hoteliers, saloonists, prostitutes, shysters, racketeers, con artists, and bandits quickly gravitated to set up shop at the latest strike. Nearly vacant and wild landscapes rapidly became towns. Mining was the genesis for Jacksonville, Canyon City, and Baker, Oregon; Roslyn, Cle Elum, and Black Diamond, Washington; and Coeur d'Alene, Idaho City, and Silver City, Idaho.

Furs and mineral wealth were not the only resources to be wrested from the Pacific Northwest's bountiful land. The vast woodlands of the Pacific coast ranges, Cascades, and mountainous interior regions became valuable sources of fuel to power steamships and locomotives. Mills sawed raw timber into lumber to shore up mining excavations and build sluice boxes, to construct the burgeoning towns, and later to facilitate the expanding web of the railroad. Each devastating town fire, new mineral discovery, or railroad expansion demanded more trees, built another sawmill, created more jobs, and erected still more towns. Springfield and Bend, Oregon, Port Gamble and Longview, Washington, and Sandpoint and McCall, Idaho, were typical sawmill towns.

The job of feeding the Northwest's (as well as California's and the East's) expanding population also depended on the natural environment. Agriculture and ranching began as predominantly subsistence activities on tiny plots along the arid Northwest's few perennial freshwater courses or where west-side rains were plentiful. With population growth, however, the need for food pushed waves of farmers, sheep herders, and cattle ranchers onto other land and into surplus production for a commercial market.

Agricultural growth, like much before it, was facilitated by changing technologies and government investment. Dams, diversion canals, aquifer harvesting, and mechanization turned hundreds of thousands of Washington's, Oregon's, and Idaho's marginal arid acres into food machines. Villages grew up to plant, harvest, process, shear, or butcher the bounty and ship it to distant areas of demand. Virtually all of Oregon's Willamette Valley and Idaho's Snake River Plain towns began as farming communities. The Grand Coulee Dam and other irrigation projects enabled central Washington towns such as Moses Lake and Yakima to flourish. Diversion of the Boise River was instrumental in favoring Boise over other burgeoning agricultural villages in Idaho. Lakeview, Burns, Pendleton, and Hermiston, Oregon, all got their start in sheep and cattle ranching, as did Ellensburg, Washington.

The natural wealth was by no means confined to the terrestrial part of the region. The adjacent ocean waters and rich river systems, especially of Oregon and Washington, teemed with life. Towns emerged where fish and shellfish could be easily gathered. Sheltered natural harbors housed ports, docking facilities, and canneries. Settlements grew around these as workers and their wages supported attendant businesses and services. Astoria, Bellingham, and Aberdeen were early cannery centers. The lower Columbia and especially Puget Sound developed rapidly as sites of safe harbor and easy water transport. They supplied fish, timber, and agricultural products to the West Coast and later to the Pacific Rim. Gateway to the Sound, Port Townsend was the largest city in the Northwest until the late nineteenth century.

While the specific attributes of *site* were critical, the geography of the surrounding region also played an important role. Geographers refer to the relationship and connections between a settlement and its hinterland as *situation*. The geographic business of populating and supplying the Northwest's far-flung places founded its own share of towns. With each innovation in transportation, new townships arose, even if located on unfavorable sites. Often simply being in a strategic place was sufficient for town founding: wagon trains needed to resupply every few hundred miles; the insatiable boilers of steamboats required wood at regular intervals; and steam locomotives had to be quenched every 60 miles or so.

Initially, overland transportation by foot, horse, or wagon demanded settlements at strategic points for supplies, medical care, or trading with the area's earliest inhabitants, the Indians. Fort Hall, Fort Boise, Fort Walla Walla, and Fort Spokane were early wagon trail protectors, barter centers, and suppliers. They and other towns served as critical waysides for the Oregon Trail and Mullan Road travelers, the stagecoach lines, and the overland mail system. Later, steamship travel deemed other locations essential fuel stops, entrepôts, or portage points, and towns grew up to meet the demand. At Oregon City, Cascade Locks, and Dalles City—treacherous spots along the important Columbia and Willamette River corridors—remarkable efforts were undertaken to bypass the rapids that in part fostered settlement. These included building parallel rail systems, dams, and locks, and dredging, channelizing, and removing obstacles to navigation. Wallula, near the Columbia-Snake confluence, and Lewiston upriver both grew to provide firewood and transfer raw materials from their expanding mineral and agricultural hinterlands to the steamboats.

Whereas stagecoach and steamboat pioneers were compelled to follow natural transportation routes, the great railroads forged entirely new paths across the Northwest's natural landscapes. The Northern Pacific Railway—completed in 1883 as the first northern transcontinental railroad—topped the mountains, bridged the rivers, and built trestles over the canyons in defiance of the physical environment. The Northern Pacific's tunnel *under* Stampede Pass—finished in 1887, the second longest tunnel in the United States—favored Puget Sound ports over those of the lower Columbia River. Seattle began to outpace Portland. The coming of the great railroads necessitated still more settlements, whether to construct the lines, water the steam engines, or serve as maintenance and marshaling yards. Competition for rail service was intense; the steel rails provided a community's link to the entire country. Tacoma temporarily beat out Seattle when it was chosen as the terminus of the Northern Pacific. Seattle soon rebounded when James Hill selected the city as the Pacific terminus for his Great Northern Railroad. Union, Oregon, never recovered when the rails were routed 10 miles west through La Grande. Yakima residents moved their entire town four miles north to join the newly finished rail line. Pocatello, Idaho, grew but for no other reason than as an important rail crossroads.

Still later, the arrival of the automobile, highways, and especially the Interstate system proved a further catalyst for town-building or the strengthening of existing settlements. Where location along a rail line was essential to the growth of late-nineteenth-century and early-twentieth-century population centers, the expanding highway web of the mid-twentieth century brought all places into contact. Taxpayer-financed roads, the rise of the trucking industry, and the affordability of the personal automobile allowed formerly remote locations to participate fully in the regional economy. The corridors along Interstates 5, 90, 82, and 84 have witnessed the most explosive growth. Freeway off-ramps are now far more important than railway stations. Where several modes of transportation converge—as road, rail, water, and air travel do in Portland and Seattle—growth is most spectacular. Today, high technology "transportation"—telephone, Internet, facsimile, satellite, and fiber optics—brings yet another rearrangement of the Northwest's settlement *situation*.

ENVIRONMENTAL PRECARIOUSNESS AND THE RISE OF A "NEW ECONOMY"

The commonality shared by all of the explorers, trappers, traders, farmers, ranchers, loggers, steamship companies, railroad corporations, town founders, and city builders was a reliance on the natural landscapes of the Pacific Northwest. Often the entire economy of a population center depended on a single natural resource such as salmon or timber. So dominant was a particular industry in many settlements that they were referred to simply as "company towns." The economy of the Northwest was measured according to what could be extracted from the natural environment. Unfortunately for the stability of the region's settlements, most of these commodities were subject to the whims of the market or the exhaustibility of the resource. When the vein of gold ran out, the salmon stopped running, the trees were all logged, the soil became exhausted, or beaver felt hats went out of vogue, the reason for a town's existence all too often disappeared. Similarly, a change in technology—such as that from steamships to diesel locomotives—might render a town redundant or unnecessary. Just as dramatically as a boom began, it often collapsed. As fast as the frenzied growth had occurred, the evacuation precipitated. What had been a fair-sized city supporting a flush payroll, numerous services, and goods dealers could no longer keep anyone in business. And thus many northwestern towns went bust.

For some places, the boom might rebound. New technologies such as cyanide gold heap-leaching or steep-slope tree harvesting equipment might rejuvenate a stagnant mining or timber town. An expanded irrigation project or rising crop prices might temporarily reinvigorate a dying farm town. But predictably, when the latest innovation ran its course or the market swung again, bust usually returned to an extractive resource-based industry town. Until recently, this was the repetitive scenario in much of the Northwest. Settlements lost population, barely hung on, or simply disappeared from the landscape. Many survive today only as "ghost towns": Shaniko, Granite, Greenhorn, Bourne, and Blitzen, Oregon, are all but expunged from our maps and our collective consciousness.

This boom-and-bust cycle has changed however, with the advent and evolution of a different economy in the "New Northwest," an economy that doesn't extract in the traditional sense. Although it is still intimately tied to the land, it doesn't require digging up the rocks, cutting down the trees, turning the cows loose, or plowing the prairie. This new northwestern economy finds intrinsic, marketable values in the landscapes of Idaho, Oregon, and Washington. Lofty mountains, canyon views, mild climates, sprawling forests, lonely trails, fished-filled streams, rugged coastlines, wide vistas, and unhurried hamlets have recently become valuable. Former Northwest boom towns often possess these amenities in abundance. Interestingly, many of these new amenity assets were not so long ago regarded as liabilities. Historically, settlers judged mountains too steep to farm and too hard to push transportation routes through. Whitewater rivers were unnavigable. Small towns lacked the excitement, goods, and lure of big cities. Wild animals (predators) threatened livestock. Expansive deserts and rocky canyon complexes were devoid of life and deemed useless. Rocky coasts made poor harbors. Thick forests were forbidding and required taming. Now these impediments are the attractions.

What was once considered a difficult physical landscape to conquer is now valued for skiing, whitewater rafting, hunting and fishing, outdoor recreating, retiring, or simply living "close to nature." What were once considered inadequately endowed little towns are now revered as bucolic villages without the congestion, pollution, crime, and squalor of big cities. This newfound appreciation is not completely new, but increasingly, large numbers of tourists, hunters, anglers, recreationists, retirees, campers, lifestyle refugees, commuters, and high technology entrepreneurs are choosing the towns and cities of the Northwest as destinations. The wild, untamed, and formerly unproductive landscapes around them have become the "places to be." Pacific Northwest cities routinely make national "most liveable city" lists based on their environmental amenities. Many of the dying or stagnant towns are booming again—and this time the subsequent bust may not be so imminent.

CHALLENGES

The Pacific Northwest's new economic and demographic boom has a host of attendant concerns. The armies of recent migrants and visitors come with expectations that impact the environment. Although their expectations are embodied in very different uses of the land, often the results are equally damaging. Crowds can be far more damaging than cows. Strip malls might be more unsightly than strip mines.

This scenario is playing out in numerous Washington, Oregon, and Idaho locations. It challenges residents, migrants, and developers in ways remarkably similar to those experienced during the previous booms. The sprawl, the contamination, the lack of planning, the trammeling, and the get-rich-quick mentality descend once again on northwestern population concentrations. Recent migrants are often pitted against long-time residents, and communities and landscapes often suffer. The arrival of franchise America, an upwardly spiraling cost of living, and unrestrained growth betray many of the very things that were attractive about a place. Some wish to throw open the development gate; others would like to close it behind them. Developers trade quality for quantity in the pursuit of economic gain, and quality of life diminishes. The economic hierarchy is remarkably similar to that of past mining or timber towns: a few wealthy owners and many service-industry laborers.

The Pacific Northwest displays numerous examples of towns and cities going through this transition from historic periods of extraction to contemporary times of attraction. Some have just commenced their recent boom, whereas others have grown for quite some time. From Sandpoint to McCall, Idaho, Bend to Hood River, Oregon, and Port Townsend to Leavenworth, Washington, the geography of the "New West" replaces that of the "Old West." Logging roads become mountain bike trails, and farm fields become suburbs. One might term it "mining the view" or "selling the scenery," but the exponential growth and transformation is happening throughout the Northwest. The environmental ramifications of such expansion—pollution, spoiled views, habitat reduction, endangered species, overcrowding—are just now being realized, causing many to worry.

Our deep reliance on the natural environment continues to evolve in the Pacific Northwest. The environment is still the preeminent defining characteristic of the region and its residents. Careful extraction of its abundant natural resources sustained Native Americans for millennia and spurred Euro-American exploration, conquest, and settlement for two centuries. The deleterious legacy of 200 years of development sprawls everywhere around us. It manifests itself in virtually every northwestern landscape, from seacoast to mountaintop to desert plain. Culturally we have changed, and we now expect the Northwest's environments to sustain us in new but still consumptive ways. The question is, will we learn from historical relationships with our varied physical geography and create a future that is far more sustainable, or are we doomed to repeat past environmental transgressions?

NOTE

1. Quoted in Tim Palmer, *Endangered Rivers and the Conservation Movement* (Berkeley: University of California Press, 1986), p. 105.

PART I / THIS PLACE

What is this place commonly called the "Pacific Northwest"? We might begin by noting that it is the "Northwest" only from the Atlantic coast of the United States. From a Pacific perspective, the region is the "Pacific Northeast." Much depends, to paraphrase William Lang, "upon where you are standing."

The essays in part one examine "this place" from a variety of standpoints. Three environmental historians (Flores, Lang, and Robbins), a botanist (Kruckeberg), and an urban historian (Abbott) examine the Northwest and the ways in which humans have understood, lived upon, loved, feared, and altered it.

Dan Flores makes a provocative plea for a new kind of history: one that views humans as components of ecosystems and treats human history as deeply embedded in the physical landscapes and biological communities that help define those landscapes—a "bioregional" history. Drawing on examples from the northern Rockies and the contributions of others who have pioneered in this approach, Flores offers a road map for those who would pursue bioregional history.

Arthur Kruckeberg's essay on the natural history of the Puget Sound basin is an example of bioregional history. It is an analysis with deep roots in what Kruckeberg calls "regional natural history." Kruckeberg offers more than a traditional "natural history," however. Interlaced in his discussion of the physical forces that have shaped Puget Sound is a discussion of the human-produced changes that have also profoundly shaped the contemporary landscape. Kruckeberg's analysis exemplifies the interdisciplinary nature of environmental history. Just as humanists must become conversant in the natural sciences to write effective environmental histories, so scientists must be familiar with the humanities and social sciences to write about the complex, interconnected world of nature and humanity. The two cannot be separated.

William Lang also examines the concept of "place"—but from a perspective different from that offered by Flores. Lang emphasizes cultural perceptions of the landscape, focusing more on the "sense" of place than on the physical bioregion. As he states, "Place is a puzzle that to be understood requires delving into cognition, belief, emotion, and imagination." Thus, while Flores and Kruckeberg lean toward the natural science end of the environmental history continuum, Lang leans toward the humanities end. Like Flores, Lang assesses the scholarship of pioneers in the field, providing a comprehensive review of the literature on "sense of place"—a remarkably interdisciplinary literature that has grown exponentially in the 1980s and 1990s. Lang summarizes the arguments of novelists, poets, historians, legal scholars, cultural geographers, ecologists, psychologists, philosophers, anthropologists, and many others concerning the meaning of place in human society and the human psyche.

William Robbins provides an example of a cultural history of a place—the Willamette Valley. He demonstrates that nature and culture are "intertwined in an intimate choreography involving memory, experience, and understanding." Indians and settlers projected onto the landscape of the Willamette Valley a divergent set of perceptions and expectations; these changed over time in response to environmental and cultural changes in a complex, shifting dance. From the "Eden at the end of the trail" mythology to tales of gloomy, damp isolation, Robbins traces the varied representations of the Willamette landscape and explores their significance. Traditional narratives of the valley tell the story of environmental change as a grand tale of progressive achievements. Other, often more recent narratives argue that the past 150 years of history in the valley tell a sad tale of conquest, exploitation, and loss. Robbins juxtaposes the development of the valley against the cultural stories we have told about that development, concluding that people's perceptions are not always in line with reality. The legacy of our history there is much more ambiguous than the triumphal narratives would admit.

The final essay in part one, by Carl Abbott, provides a crucial urban perspective on this place we call the Northwest. Most Americans live in cities surrounded by largely "artificial" environments; from a demographic perspective, the Northwest has long been primarily an "urban" region. An environmental history of this "unnatural" landscape should, therefore, be a prominent part of any attempt to understand our history. Abbott focuses on this task in his essay on the "urban imprint" in the Northwest. Reflecting themes seen in the other chapters of this section, Abbott indicates how the landscape has helped shape patterns of urban growth, particularly in the first few generations of Euro-American settlement. He also examines how northwesterners have increasingly sought to reshape their environment—sometimes through massive earth-moving projects—to serve human purposes. Abbott concludes, interestingly, that the urban impact on the Northwest environment is relatively light in comparison with most other parts of the United States.

Abbott's essay also reflects another theme. Environmental historians have often focused on the interaction between humans and nature, as if the two were distinct. More recently, scholars have challenged the assumption that the two can be separated. The human-nature distinction is particularly ambiguous in the modern urban-industrial era of extensive landscape modification, pollution, and climate change. Even in remote rural areas of the Northwest, people live in places that are only ambiguously "natural" because of pervasive farming, grazing, logging, hunting, fishing, and fire suppression, as well as the introduction of non-native plants and animals.

Many of the themes introduced in this section appear again in later chapters. For example, several of the essays in part two, "First Peoples," discuss the natural history and human use of the plants and animals of the Northwest. Similarly, the later chapters by Fiege and Langston discuss the ambiguous distinction between "natural" and "human" landscapes. Zeisler-Vralstad and Morrissey emphasize the importance of cultural perceptions of landscapes, while the essays by Dwire, McIntosh, and Kauffman and by Wyckoff and Hansen emphasize physical conditions of the landscape and the impacts of development. Goble, Hirt, and Aiken assess the social and political implications of environmental deterioration. Each offers some additional perspective on this place called the Northwest.

2 / Place

An Argument for Bioregional History

DAN FLORES

As he told the story, Walter Prescott Webb—widely accepted among American environmental historians as one of the founding fathers of our discipline—began to conceptualize his most famous work, *The Great Plains: A Study in Institutions and Environment*, at the age of five. Raised as a young child in Panola County, Texas, deep in the heart of southern culture, where thickly timbered rolling hills screened the horizon and even the overhead sky was only partially visible through the soaring loblolly pines, Webb had not yet started school when his family moved to central Texas. In the western cross timbers province at the edge of the Great Plains, the future thinker of big ideas found himself stimulated by another world. Here no loblollies blocked the skies, and across the grasslands the horizon was miles distant, visible like the encircling rim of a bowl in every direction. Here king cotton and backwoods truck-garden farms gave way to fenced spreads enclosing the sacred cow. Young Walter Webb was fascinated by the differences between those two worlds. He remained fascinated as an adult and with his books and articles went on to stimulate reading Americans into pondering the peculiar dialogue between the western environmental setting and human technological adaptations to it.[1]

Like Frederick Jackson Turner (whom he claimed not to have read prior to writing *The Great Plains*) and his frontier thesis, Webb and his interpretation have taken their hits over the years. Fred Shannon's 200-page savaging of Webb's book in 1940 was the first of those critiques, made all the more famous by Webb's laconic refusal to acknowledge Shannon's points; his response to Shannon's critique was that he had conceived and written *The Great Plains* not as history but as art![2] In our own time, even his sympathizers acknowledge that Webb's interpretations were an exercise in rank environmental determinism, an approach that social scientists in geography and anthropology had long since abandoned. Somewhat in the manner of an intellectual Ulysses S. Grant, it has been asserted, Webb was moved to write big idea books like *The Great Plains* and later *The Great Frontier* (1950) essentially because he lacked the education to know better.[3]

I disagree. I think Webb was so moved because as a child he paid attention to what his senses told him about the difference between the piney woods and the western cross timbers in Texas. Just as William Cronon has properly credited Turner's Darwinian frontier construction with residing at the structural core of much contemporary American environmental history, so too Webb's intellectual vision continues to influence environmental historians, especially environmental historians of the American West.[4] Donald Worster, for example, has pointed out that Webb's insistence that aridity is a defining characteristic of the American West has taught us correctly that "the West" was not a process but was and is a *region* whose perimeter can be sensed on the ground and marked out on a map.[5] Beyond that, Webb's approach remains valuable in environmental history, I would assert, because of the attention he has forced us to pay to the confluence between specific ecological realities and specific human adaptations (such as, in the relatively simple terms of *The Great Plains*, the use of windmills and barbed wire in semiarid, open grasslands) that are a part of the evolution of cultures in place.

Having struggled with terms such as *region* and *place* in writing my own book about the Llano Estacado country of the American Southwest, I think I would argue that the intuitive foundation that Webb built—if refined and more surgically applied—can remain central to the modern arc of environmental history as a field.[6] Indeed, it seems to me that the particularism of distinctive *places* fashioned by human culture's peculiar and fascinating interpenetration with all the vagaries of topography, climate, and evolving ecology that define landscapes—and the continuing existence of such places despite the homogenizing forces of the modern world—ought to cause environmental historians to realize that one of their most crucial tasks is to write well what might be called *bioregional* histories.[7]

Toward the end of one of Patricia Limerick's essays in *Trails: Toward a New Western History*, she recounts a conversation after one of her public lectures with a man who said something like "I enjoyed your speech, but since I'm not a western historian, everything you said was obvious to me."[8] Here at the University of Montana, the graduate students in history and environmental studies tend to be young, bright nonresidents seeking exposure to the mountain West. Many of them take advantage of their location to travel widely, from Montana south to Texas and sometimes northward through Canada to Alaska. I suspect this is why they are puzzled when they read historical essays positing various reasons why the American West comprises a distinctive, singular region much as the American South appears to do. Their questions often follow these lines:

"So, if aridity is the defining characteristic of the West, then Alaska doesn't belong, and the Pacific Northwest doesn't either? And you mean that those high, wet lifezones disqualify the Rocky Mountains?"

"Kind of ironic, isn't it, that Texas has produced some of the most potent western symbols but isn't really a part of the West because it lacks the defining system of federal landownership?"

"To me, Colorado seems so different from Utah. But you're telling us that they were shaped into a consistent form by the same forces of global economic integration that forged the rest of the West?"

Uncorrupted by an impulse toward the broadly inclusive and generalized definitions of regionalism that professional historians have been trained to apply, these students see the obvious. And of course they are not alone. I have noticed with interest the testimonies of westerners recorded in the anthology *A Society to Match the Scenery*, assembled at the Center of the American West at the University of Colorado in 1991 as an exercise in envisioning the future of the West. Among the voices appearing in that anthology were those of Dan Kemmis and Camille Guerin-Gonzalez, both residents of the Rocky Mountains (western Montana and northern New Mexico, respectively) and both inhabitants of similar topographies where federal landownership and management are everyday facts of life, where resource extraction and tourism prevail economically, and where water problems and an influx of wealthy newcomers dominate local discussion. Yet after listening to Kemmis's remarks about life in the northern Rockies, Guerin-Gonzalez claimed that the northern New Mexico she knows bears no relationship whatsoever to what Kemmis described.[9]

The answer to the puzzlement and to the denials of uniformity just expressed is clearly that in the case of the American West, no set of generalized definitions, regardless of how inclusive, accurately explains the loose cluster of subregions comprising the huge swath of continental topography and ecology that is the western United States. Neither aridity and its effects, federal landownership, economic integration into the global market at a time of mature industrial development, the presence of Indian reservations, proximity to Mexico or to the Pacific Ocean, nor a legacy of conquest captures the particularism that is the historical reality of *place* in the western U.S.[10] I doubt that broadly generalized interpretations work very well at capturing historical senses of place elsewhere, either, but I leave that for others to decide. As it has done for a century now, the country west of the Mississippi River continues to work extraordinarily well as a laboratory in humanities and ecology.

Let me here define my own terms and explain my argument by following three lines of investigation. First, if environmental historians grant that specific human cultures and specific landscapes can and do, in fact, intertwine to create distinctive places, then why not just return to writing local county histories of the type that western history generated in Wal-Mart quantities earlier in this century? What would constitute the rationale and the basis for writing human "bioregional history"? Second, if Webb's theories are largely passé and those of James Malin—the early Kansas environmental historian—often too obscure and eccentric to serve as models for writing about place, what kinds of approaches and ideas ought we to employ in writing the environmental histories of human places, in the American West or anywhere else? And finally, are there existing works that qualify as bioregional histories or other studies whose appearance seems inevitable or at the very least particularly valuable to the literatures of place in the United States?

On one semantic score, a defense of sorts may be necessary. I am aware that my use of the terms *bioregion*, *bioregional*, and *bioregionalism* may strike some as either an unnecessary resort to jargon or a surrender to fadism. On the contrary, I assert that for all its association in the United States with countercultural environmentalism, "bioregion" should in fact be recognized as a precise and highly useful term of art for environmental historians. The word appears to have had its genesis in the early 1970s with the writings of the Canadian Allen Van Newkirk, and since then it has been most closely associated with the California counterculture and back-to-the-land prophets such as Peter Berg, Raymond Dasmann, Gary Snyder, and Stephanie Mills.[11] In their publications and their journals—*CoEvolution Quarterly* and *Raise the Stakes*—bioregions act as the essential natural human stages for bioregionalism, which has come to stand for what Berg calls "a kind of spiritual identification with a particular kind of country and its wild nature [that is] the basis for the kind of land care the world so definitely needs."[12] The geographer James Parsons, calling the bioregional movement to the attention of academics in 1985, noted that it "has attracted a remarkably sensitive, literate group of adherents" who, while they might seem like "misty-eyed visionaries caught up in a New Age semantics . . . may be the unwitting architects of a new popular geography, a grass roots geography with 'heart.' "[13]

Bioregionalism as a modern social movement ought to be interesting to environmental historians in its own right. But it is not merely its focus on ecology and geography but its emphasis on the close linkage between ecological locale and human culture, its implication that in a variety of ways humans not only alter environments but also adapt to them, that ties it to some central questions of environmental history inquiry. Whereas the history of politics and diplomacy and (sometimes) ideas may be extracted from the natural stage and studied profitably, the kinds of subjects that attract contemporary historical study—legal, social, gender, ethnic, science, technology, and environmental issues—literally cannot be examined without sophisticated reference to place.

But there is an irony here. Professional history, especially in the United States, has long regarded an interest in place as limiting, provincial, and probably antiquarian. A lobotomizingly familiar conversation among American historians is that unless their interest has been New England or California, those who have spent their careers writing about locales, states, or regions have routinely been dismissed as "cow-chip historians"—rather too often by those who see themselves as writing about more universally crucial locales such as Concord or Languedoc.

Environmental histories of place have already made strides in eroding this sniffing condescension, and if we are good at what we do, we are likely to wipe it out altogether. In truth, to an extent all history is the history of place. But environmental history can go beyond traditional history and justify its reputation for new insight if we follow the lead of ecologists, geographers, and ecological anthropologists—and bioregionalists—in drawing the boundaries of the places we study in ways that make real sense ecologically and topographically. It ought to be agreed that with rare exceptions, the politically derived boundaries of counties, states, and nations are mostly useless in understanding nature. Naturally, historians continue to rely heavily on the documentary trail generated by political life, but we may have to concede that there are significant limitations inherent in that dependency.

The founders of American environmental history, such as Webb and

Malin, realized this and pointed the way toward a more ecologically oriented kind of study more than half a century ago.[14] Clearly, the first step in writing environmental histories of place is recognition that natural geographic systems—ecoregions, biotic provinces, physiographic provinces, biomes, ecosystems: in short, larger and smaller representations of what we probably ought to call *bioregions*—are the appropriate settings for insightful environmental history.

Without county/province/state/national borders to provide clues for delimiting place history, to what sources might historians turn for ideas about drawing natural boundaries? One of the best is the earliest. In 1890, John Wesley Powell laid before a congressional committee a remarkable and beautiful colored map of the arid West mapping out 24 major natural provinces and further subdividing the region into some 140 candidates for "commonwealth" status based on drainage and topographical cohesion.[15] Powell's ideas for the West conform with most twentieth-century delineations, such as those in Wallace Atwood's *The Physiographic Provinces of North America*, and indicate in startling fashion the century-old genesis of the current rage among western resource bureaus for ecosystems management.[16]

More ecologically precise and recent than Powell's scheme is Robert Bailey's *Ecoregions of the United States* (1976) and its accompanying map. Assembled by the U.S. Department of Agriculture from a broad range of soil, climate, floral, faunal, and topographic sources, this work created a taxonomy of North American nature ranging from the macro (5 "domains" further differentiated into 12 "divisions") to the micro (the divisions further refined into 31 "provinces," which are themselves subdivided into 45 "sections"). The most refined category, the sections, range considerably in size. Bailey's southeastern mixed forest section, for example, extends across parts of nine states; his grama-buffalo grass section covers all of the southern and central Great Plains. On the other hand, the superficially homogeneous Rocky Mountain region is conceptualized by Bailey as being composed of eight ecologically unique sections. His system does not specifically locate and bound ecosystem corridors such as major rivers and their drainages, which demonstrably have played key roles in human history. But in his taxonomy, an area that historians might be tempted view in the round, such as the intermountain Great Basin, is subdivided into no fewer than five distinct sections.[17]

Typically, scholars closer to the local ground tend to subdivide even more narrowly. Today, in my home states of Montana and Texas, for example, bioregional particularism is rife. In Rocky Mountain Montana, Bailey's two "sections" are carved by state ecologists into double that number of categories: the Columbian Rockies, the Broad Valley Rockies, the Yellowstone Rockies, and the Rocky Mountain foreland. More recently, ecosystems ecologists in Montana and Wyoming have recognized the dynamism of bioregions, as well as the role played by cultural developments, with their designation of two modern natural systems they call the greater Yellowstone ecosystem and the northern Continental Divide ecosystem. As for Texas, the majority of ecologists see the state as an extraordinarily artificial creation that cobbles together no fewer than 10 individual bioregions (one more than Bailey recognizes).[18] Beyond providing scholars with a new axle upon which to spin environmental history, bioregional study at least partially explains why a "place" like Texas cannot decide whether it is properly southern, western, southwestern, or just Texan.

The particularism of bioregional places that is often so observable to travelers is explained by the geographer Yi-Fu Tuan's equation: space plus culture equals place.[19] Beyond ecological parameters, the second basis for bioregional histories is, of course, the diversity of human cultures across both time and space. Since I began this essay by invoking Webb, it is worth mentioning as a basic framework for understanding the dialogue between nature and culture that Webb was properly criticized in his day for resorting to environmental determinism in writing *The Great Plains*. As he saw it and wrote it, certain characteristics of nature on the Great Plains presented so many challenges to American settlement culture as it had evolved in the eastern woodlands that the Great Plains served up an "institutional fault line" that significantly modified the settlement strategies (by which Webb meant primarily materialist, economic culture) of the peoples who settled there. In something of a major misreading of history, Webb believed that Hispanic culture had failed to adapt to the plains to the same extent that horse Indians and Texans did, and that this adaptive failure explained why more technologically innovative Anglo-Americans seized the region.[20]

James Malin, among other historians, corrected Webb's naive assumptions about environmental determinism with his application of the interpretive framework of "possibilism."[21] Although Malin still accepted a dichotomy between nature and culture and had an inordinate faith in technological fixes, in possibilism human cultures bear a sturdier freight and responsibility in creating place. The possibilist idea is well understood now to imply that a given bioregion and its resources offer a range of possibilities from which a given human culture makes economic and lifeway choices based upon the culture's technological ability and its ideological vision of how the landscape ought to be used and shaped to meet its definition of a good life. Although the possibilist idea is scarcely new and in the social sciences has long been retired from the cutting edge of interpretation, I happen to believe that, carefully used, the idea has a continuing relevance for environmental historians, whose interest in the peculiar wake of events that follows ideologies and choices through time is particularly keen.[22]

Resting culture's role in bioregional history on a possibilist model was a reaction to environmental determinism. It continued the nature-culture dichotomy and carried with it the danger of playing to modernism's conceit that ever since the scientific and industrial revolutions, human culture had triumphed and nature hardly mattered anymore except as potential commodities. Too, possibilism alone can give the impression—and it is a frequent failing of the bioregional movement's own philosophy—that historical decisions about place are formed exclusively by local populations. Springing particularly from ecologist Eugene Odum's influential studies of ecosystems, since the early 1970s scholars of culture have had a set of mechanisms known collectively as systems theory to explain the diverse web of connections that tie local places to diverse economic and ideological systems.[23] Now at least a footbridge had been constructed to span the intellectual river between nature and culture: not only biological processes but cultural ones appeared to operate as functional and evolving systems.[24] Some of our best recent environmental histories—Worster's *Dust Bowl* (1979), Richard White's *Roots of Dependency* (1983), Bill Cronon's *Nature's Metropolis* (1991)—have made the concept of systems central to their analysis, primarily (but not exclusively, especially in White's case) in a materialist, market-integrative form.

One remaining element of cultural study that has relevance to bioregional history is the recent refinement of the cultural adaptation theory that Walter Webb brought to his work more than half a century ago. Inspired by the Odum-derived thought that human systems might, after all, be living organisms, in the late 1970s social scientists such as Karl Butzer and Roy Rappaport applied the precepts of organic evolution to cultural adaptation and tried to sort out its mechanisms in various simulations of real life—for example, in the feedback loops of a land management bureaucracy dealing with an environmental crisis.

While Butzer wondered whether entire human civilizations might not *be* organisms,[25] Rappaport went on to assert that ultimately adaptation's function is the same whether it occurs in organisms or societies. That function is to aid survival. "Since survival is nothing if not biological," he wrote, "evolutionary changes perpetuating economic or political institutions at the expense of the biological well-being of man, societies and ecosystems may be considered maladaptive."[26] In Rappaport's view, then, adaptation is critical to understanding the long-term successes (or the short-term failures) of human cultures in specific places, and positive adaptations are those that intertwine cultural choices with the dynamism of particular bioregions into a mix that "survives."

If contemporary environmental historians step up alongside geographers and anthropologists in their examinations of place and culture and begin to write what I am here calling bioregional history, how might we approach it? And since most of us are humanists, how will our work distinguish itself from what the social scientists do?

The process has to commence with the selection of a place based at least in part on the taxonomy I have already outlined, although given the natural human preference for ecotone edges, interesting settings for human history will not necessarily be bounded the way Bailey maps out his ecoregions. Pre-Columbian Indian cultures in North America hewed fairly closely to the larger bioregional divisions we now recognize, and the home ranges of individual bands often conformed roughly to slices of topography that Bailey identified in his "sections." On the other hand, cultural groups such as the New Mexican Hispanics and the Mormons occupied and adapted to several kinds of bioregions in the American West. Any number of intriguing possibilities exist when culturally distinctive groups such as Indian tribes, Hutterites, or Mennonites emerge, islandlike, in seas of cultural homogeneity such as the Great Plains, or when the political boundaries of different traditions cut across a bioregion with its own historical arc.[27]

In a recent article in *History News*, Hal Rothman expressed concern that historians of place might succumb to provincial biases if they write about places to which they have an emotional tie. That *could* be a danger for someone whose connections are to local culture, although one would have to look hard to find much native Kansan sympathy, say, in Worster's *Dust Bowl*. My own experience is the reverse: an emotional tie to a landscape made me considerably more critical of some of the human cultures occupying the Llano Estacado than a detached objectivity might have done. On the other hand, the best writing and most penetrating research consistently spring from passion, and places can summon that.[28]

Irrespective of the choice of bioregion, modern environmental historians are going to have to make their peace with a view of a natural world that is dynamic. Far from serving as some pristine baseline of climax harmony, the bioregions we study have to be accepted as endlessly evolving through time. Ecologists now speak of "internal change," "blurred successional patchworks," "moving mosaics." Disturbance is the natural state; hence adjustment is ongoing and fundamental. There has been some resistance to this among historians who have hoped, in the environmentalist tradition, that there *is* a harmonious, stable nature out there against which we might view human activity as arrayed in a destructive assault.[29] But whereas Daniel Botkin's view that "nature undisturbed [by human activity] is not constant in form, structure, or proportion, but changes at every scale of time and space" might be problematic for environmental romanticism, I don't see it so for history.[30] In fact, recognizing that the ground of the natural world is shifting and always has shifted can be another bridge tying the activity of human culture back into nature.

If they make any claim to apprehending reality, bioregional histories are going to have to capture that changeability. Indeed, in modern techniques such as repeat photography, fine-resolution remote sensing, and the manipulation of spatial information with computers (historic maps showing vegetation and fire patterns in the northern Rockies come to mind) we have the ability to track those changes at a denser grain than ever before. Personally, I do not in any case see how acknowledging the fact of ongoing natural disturbance in nature prevents us from critiquing human disturbances that were foolish in an anthropocentric sense or reprehensible from the perspective of the diversity of life.

Perhaps an element that ought to distinguish bioregional history from traditional histories, even of places, is a precise spatial application of Fernand Braudel's *longue durée*. We ought to aim for the "big view" not so much through wide geographic generalizations in shallow time but through analyzing deep time in a single place. As I attempted to show in a recent piece interpreting nineteenth-century Indian environmental history in the West, an accurate understanding of shallow time often isn't possible without the context of the *longue durée*.[31] Bioregional histories, then, should properly commence with geology and landform and then take up climate history, again using an array of modern approaches from ice cores and pollen analysis to packrat middens and dendrochronology. Climate has always been and remains one of the most visible forces interacting with human history. A climate record of place can then position us to understand the ebb and flow of floral and faunal species across space and time the way our eyes enable us to track cumulus clouds drifting across an open basin by the shadows they cast on the ground.

When we bring human culture into our stories of place, even if our search is, for instance, for clues to how earlier cultures in a place coped with a warmer climatic regime, I suspect we would do better to cease our quest for golden age utopias. Further, at every level of time we have to recognize that the supposed dichotomy between culture and nature is not, as Lévi-Strauss led us to believe, structurally basic to human consciousness but is a false dichotomy.[32] Preliminary studies in biophilia and biophobia indicate in striking ways that it is ludicrous to think that humans genetically ever stepped outside nature. Evolutionary psychology and sociobiology, for instance, remind us how rooted our social behavior is in the primate world. And studies of inherited biophobic responses (to snakes and spiders, for example), as well as genetically transmitted biophilic preferences (to savannas, parklands, certain tree shapes, and terrain scales), clearly center human fear of the natural world, as well as human settlement strategies and even aesthetics, in adaptations selected by evolution over deep time.[33]

As for ecological ideas in our cultures, they can be seen as adaptive packages of "captured knowledge" about living in place.[34] Our goal, then, should be to fashion a history that sees human cultural adaptation and knowledge transmission essentially as analogous to the natural selection of characteristics. As an example of how this might work in a historical case in the American West, geographer William Riebsame has recently melded some of the ideas of systems analysis and adaptation theory to distinguish between positive adaptation (a culture's willingness to change in the face of new circumstances) and cultural resiliency (a system's resistance to change and its tendency, if perturbed, to return rapidly to its former condition). In Riebsame's view, the southern Great Plains' response to the Dust Bowl of the 1930s is a classic instance of resiliency rather than adaptation. Although the tinkering with the system that went on in response to a major climatic swing and agricultural collapse could be seen as adaptive, Riebsame thinks that over the long term those actions will be seen as a resilient rebound to the status quo, and hence as maladaptive.[35]

The narrative line of bioregional history should be the story of different but sequential cultures occupying the same space and creating their own succession of places. Because it provides cause for so many observable effects, it is important to demonstrate in this kind of history that successive cultures inhabiting a space interact with a "nature" more or less altered by previous inhabitants.[36] Further, we ought to understand that the structure of the dialogue—and that is the proper way to describe it—between nature and human culture is the same kind of dialogue that exists between habitat and species in natural selection. Human cultures alter their places to shape them in accordance with their ideological visions, and in turn cultures are shaped by the power of their places.[37] As we now understand about organic evolution, significant change in bioregional history can be expected to occur as punctuations in equilibria, rapid ratchetings (or "ecological revolutions," to borrow Carolyn Merchant's term) to new conditions that have the spiraling effect of endlessly re-creating place.[38] Our investigation of the historical causes and results surrounding these ratchetings may be materialist, but the most penetrating sets of insights are more likely to spring from studies of ideologies, values, literature, and art.[39]

Having laid out something of a detailed theoretical structure for bioregional history, paradoxically I now have to express my preference for an approach to historical writing that would submerge this structure, if not render it all but invisible. What separates historical writing from the semantically challenged language of the social scientist is a greater burden to communicate. Unlike geographers, anthropologists, and sociologists, historians communicate generalities with stories of individuals whose experiences carry more of the scent of life for readers.[40] It is exactly the "fuzzy" propensity for anecdote, the discipline's inherent wish to tell stories, that drives history and makes it readable. Like all good writing, quality history "shows." It doesn't tell, and it doesn't seek obfuscation by resorting to jargon. As Yi-Fu Tuan wrote two decades ago in *Topophilia*, affection for history, as with place, tends to focus on smaller and more personal scales than the large political boundaries of the modern world, and human sense of place has everything to do with a shared sense of history.[41] I take this as confirmation that there is an eager audience for lovingly crafted bioregional history.

No one so far has written quite the kind of history I have in mind, but there are those who have come close. Despite its county focus and stuttering title, Richard White's first book, *Land Use, Environment, and Social Change: The Shaping of Island County, Washington* (1979), was a promising start to doing modern bioregional history. It showed well how a small place can encapsulate and exemplify many broader historical themes, and in the Pacific Northwest it is now regarded as basic to bioregional literature.[42] Worster's *Dust Bowl* of the same year is bioregionally centered and certainly explores adaptation, but because its topic is a specific event, it only superficially examines sequential cultures or deep time. The literature and art it explores are those of event rather than place. To the extent that it examines the Imperial Valley of California especially, the same can be said of Worster's *Rivers of Empire: Water, Aridity, and the Growth of the American West* (1985). These two books are almost indispensable to environmental history, but neither is quite a bioregional history.

Bill Cronon's *Changes in the Land: Indians, Colonists, and the Ecology of New England* (1983), Carolyn Merchant's *Ecological Revolutions: Nature, Gender, and Science in New England* (1989), Albert Cowdrey's *This Land, This South: An Environmental History* (1983), and Timothy Silver's *A New Face on the Countryside* (1989) have made the bioregions of the eastern United States perhaps the best studied on the continent. Cowdrey's book, although it mentions deep time, essentially is a broad geography–shallow time work. Cronon's, Merchant's, and Silver's are all more temporally focused and do explore processes and changes that create places across cultural lines, although without much reference to adaptation. Of the two New England books, Merchant's is the broader, but theory pokes through the fabric of the writing like rib bones through cowhide. Cronon's is the more readable by a wide margin. Finally, Philip Scarpino established a different and useful bioregional category with his *Great River: An Environmental History of the Upper Mississippi, 1890–1950* (1985), a book that in fact is somewhat narrow temporally as well as in its focus on industrial and bureaucratic developments.[43]

Among more recent, tightly focused books concentrating on bioregions in western America, geographer Robin Doughty's pair of works, *Wildlife and Man in Texas: Environmental Change and Conservation* (1983) and *At Home in Texas: Early Views of the Land* (1987), can be taken together as a shallow-time history of place, primarily of the bioregions of central Texas, although Doughty addresses only Anglo and German-American cultures there. Hal Rothman's *On Rims and Ridges: The Los Alamos Area Since 1880* (1992) and Peter Boag's *Environment and Experience: Settlement Culture in Oregon* (1993) are highly place-specific, describing the Parajito Plateau of New Mexico's Jemez Mountains in Rothman's case and the Calapooian Valley of Oregon in Boag's. Rothman's book, an exploration of growing competition for local resources into modern times, is effectively intercultural; Boag's is less so and is limited temporally to the nineteenth century, but it is an interpretively rich and imaginative work. My bioregional book on the southern High Plains, *Caprock Canyonlands: Journeys into the Heart of the Southern Plains* (1990), tried an experimental approach to history but did make an effort to incorporate most of the structure I have outlined here. But my choice for the best bioregional history anyone has written to date is William deBuys's *Enchantment and Exploitation: The Life and Hard Times of a New Mexico Mountain Range* (1985). It is place-specific (the

Sangre de Cristo range) and temporally deep, it examines environmental change across sequential cultures, and it deals with values and adaptation with an effortless style.

In its brief three decades, modern environmental history has made a name for itself primarily as a field that has offered up stimulating studies of environmentalism as a sociopolitical movement, of intellectual ideas about nature, and of specific environmental events of historical importance. For its theoretical framework it has mostly borrowed. Yet in the work of Turner, and particularly that of Webb and Malin, there existed from the beginning a focus on places and their history and at least the rudimentary foundations of how to approach that kind of study. As Malin put it 35 years ago, the "proper subjects of study" for a specific bioregion are "its geological history, its ecological history, and the history of human culture since the beginning of occupance by primitive men."[44]

Undoubtedly a serious mistake historians have made in writing about (let us say) a territory as vast and as topographically, ecologically, and culturally diverse as the American West has been to start with a single interpretive framework like Turner's or Limerick's and then to set about forcing the world around us into some facsimile of that model.[45] A more logical and enlightened approach, it ought to be obvious, is to go after the reality of the specific, to write sophisticated, deep-time, cross-cultural environmental histories of places—and after a sufficient number of such case studies have been done, *then* to look for patterns.

If we do that, we have a good deal of work to do, for much of the West, the continent, and the world awaits that kind of history.

NOTES

This essay previously appeared as "Place: An Argument for Bioregional History," *Environmental History Review* 18(4) (Winter 1994): 1–18. It is reprinted with the permission of the Environmental History Review.

1. Gregory Tobin, "Walter Prescott Webb," in *Historians of the American Frontier: A Bio-Bibliographical Sourcebook*, ed. John Wunder (New York: Greenwood Press, 1988), pp. 713–29; Walter Prescott Webb, *The Great Plains: A Study in Institutions and Environment* (Boston: Ginn and Co., 1931).

2. Fred Shannon, "An Appraisal of Walter P. Webb's *The Great Plains: A Study in Institutions and Environment*," *Critiques of Research in the Social Sciences* III, Bulletin

46 (1940). Webb's response and the comments of Arthur Schlesinger, Sr., Clark Wissler, C. F. Colby, and E. E. Dale are summarized on pp. 12–27.

3. See Tobin, "Walter Prescott Webb."

4. William Cronon, "Revisiting the Vanishing Frontier: The Legacy of Frederick Jackson Turner," *Western Historical Quarterly* 18 (April 1987): 157–76. Howard Lamar has observed more recently that within the past decade, a survey of historians of the American West indicated that it was Webb's ideas rather than Turner's that they found most stimulating. Howard Lamar, "Regionalism and the Broad Methodological Problem," in *Regional Studies: The Interplay of Land and People*, ed. Glen Lich (College Station: Texas A&M University Press, 1992), p. 25.

5. Donald Worster, "New West, True West," in Donald Worster, *Under Western Skies: Nature and History in the American West* (New York: Oxford University Press, 1992), pp. 23–24. Webb's assertion that the West is place rather than process is perhaps one point on which the new western historians and traditional western historians might agree.

6. Dan Flores, *Caprock Canyonlands: Journeys into the Heart of the Southern Plains* (Austin: University of Texas Press, 1990).

7. "Place" is a term attracting much attention of late, particularly in literature. Earlier assumptions about the demise of regionalism and place were predicated on their erosion at the hand of systems such as capitalist integration and communications. Bill Bevis, for example, has argued that "capitalist modernity seeks to create a kind of no-place center to which all 'places' . . . are marginal." William Bevis, "Region, Power, Place," in *Reading the West: New Essays on the Literature of the American West*, ed. Michael Kowalewski (New York: Cambridge University Press, 1996), p. 21. Similarly, in a chapter titled "Sense of Place" in Donald Worster's *Dust Bowl: The Southern Plains in the 1930s* (New York: Oxford University Press, 1979), Worster describes in graphic terms how Haskell, Kansas's, sense of place was subverted by American capitalism. On the continuing relevance of place, however, see works such Joel Garreau, *The Nine Nations of North America* (Boston: Houghton Mifflin, 1981); Samuel Hays, *Beauty, Health, and Permanence* (New York: Cambridge University Press, 1987): 36–39; John Wright, *Rocky Mountain Divide* (Austin: University of Texas Press, 1993); James Parsons, "On 'Bioregionalism' and 'Watershed Consciousness,' " *The Professional Geographer* 37 (February 1985): 1–5; Dan Flores, "The Rocky Mountain West: Fragile Space, Diverse Place," *Montana: The Magazine of Western History* (Winter 1995): 46–56.

8. Patricia Nelson Limerick, "The Unleashing of the Western Public Intellectual," in *Trails: Toward a New Western History*, eds. Patricia Nelson Limerick, Clyde A. Milner II, and Charles Rankin (Lawrence: University Press of Kansas, 1991), p. 72.

9. Dan Kemmis, "The Last Best Place: How Hardship and Limits Build Community," in *A Society to Match the Scenery: Personal Visions of the Future the American West,* eds. Gary Holthaus, Patricia Nelson Limerick, Charles F. Wilkinson, and Eve Stryker Munson (Boulder: University Press of Colorado, 1991), pp. 84–85; Camille Guerin-Gonzalez, "Freedom Comes from People, Not Place," *ibid.*, pp. 194–95.

10. To aridity, the causative factor in American western history for John Wesley Powell, Webb, and Wallace Stegner, contemporary historians have added the others I mention in the text. All these are singled out to explain western homogeneity in the articles in Limerick et al., *Trails:* see Limerick, "The Unleashing of the Western Public Intellectual," pp. 70–71; Elliott West, "A Longer, Grimmer, but More Interesting Story," pp. 103–111; Michael Malone, "Beyond the Last Frontier: Toward a New Approach to Western American History," pp. 139–60. See also David Emmons, "Constructed Province: History and the Making of the Last American West," *Western Historical Quarterly* 25 (Winter 1994): 437–59; Susan Neel, "A Place of Extremes: Nature, History, and the American West," in *A New Significance: Re-Envisioning the History of the American West,* ed. Clyde Milner II (New York: Oxford University Press, 1996).

11. Parsons, "On 'Bioregionalism' and 'Watershed Consciousness'," 1–5; Dave Foreman, *Confessions of an Eco-Warrior* (New York: Harmony Books, 1991), pp. 43–50; Stephanie Mills's lecture on bioregionalism, University of Montana—Missoula, November 16, 1993. For a good introduction to evolving bioregional thought, see the essays in *Home: A Bioregional Reader*, eds. Van Andruss et al. (Santa Cruz: New Society Publishers, 1990).

12. Peter Berg, "Strategies for Reinhabiting the Northern California Bioregion," *Seriatim: The Journal of Ecotopia* 3 (1977): 2.

13. Parsons, "On 'Bioregionalism' and 'Watershed Consciousness,' " pp. 2, 5.

14. Although Webb argued that his book was about the Great Plains, and he offered semiaridity, treelessness, and lack of topographical relief as the defining characteristics of the Plains, many readers have observed that his maps implied that virtually all of North America west of the ninety-eighth meridian belonged to the Plains province while most of his historical examples came from Texas. See Webb, *The Great Plains*, Map 1.

15. "Arid Region of the United States, Showing Drainage Districts," U.S. Geological Survey, *Eleventh Annual Report, 1889–90, Irrigation Survey, Pt. 2* (Washington, D.C.: Government Printing Office, 1891). See also Donald Worster, *An Unsettled Country: Changing Landscapes of the American West* (Albuquerque: University of New Mexico Press, 1994), pp. 15–16.

16. Wallace Atwood, *The Physiographic Provinces of North America* (Boston: Ginn & Co., 1940). See also Stephen Jones, "Boundary Concepts in the Setting of Place

and Time," *Annals of the Association of American Geographers* 49 (September 1959): 241–55.

17. Robert Bailey, *Ecoregions of the United States* (Washington, D.C.: Department of Agriculture, United States Forest Service, 1976).

18. Montana Environmental Quality Council, *Fourth Annual Report* (Helena: State of Montana, 1975). On the more recent ecosystem bioregions in the northern Rockies, see the chapter "The Yellowstone Ecosystem and an Ethic of Place" in Charles Wilkinson, *The Eagle Bird: Mapping a New West* (New York: Pantheon, 1992), pp. 162–86. On Texas bioregions, see *The Texas Almanac, 1994–95*, ed. Mike Kingston (Dallas: Dallas Morning News, 1994), p. 94. In Texas, the tenth bioregion that is not represented separately by Bailey is the blackland prairie.

19. Yi-Fu Tuan, *Space and Place: The Perspective of Experience* (Minneapolis: University of Minnesota Press, 1977), pp. 4–6.

20. Webb, *The Great Plains*, pp. 85–139.

21. See the various essays in James Malin's *History and Ecology: Studies of the Grasslands*, ed. Robert Swierenga (Lincoln: University of Nebraska Press, 1984), especially "Space and History: Reflections on the Closed-Space Doctrines of Turner and Mackinder" and "Webb and Regionalism"; Robert Berkhofer, "Space, Time, Culture and the New Frontier," *Agricultural History* 38 (January 1964): 21–30.

22. As Worster points out in "Doing Environmental History," in *The Ends of the Earth: Perspectives on Modern Environmental History*, ed. Donald Worster (New York: Cambridge University Press, 1988), pp. 289–307, this has been one of three major avenues of environmental history inquiry, and one that Roderick Nash and others have made especially influential in American environmental history.

23. See Eugene Odum, "The Strategy of Ecosystem Development," *Science* 164 (April 1969): 262–70; Eugene Odum, *Fundamentals of Ecology* (Philadelphia: W. B. Saunders, 1971). A good overview and introduction to the various systems models devised for human societies—modernization theory, dependency theory, world systems theory—can be found Thomas Hall's *Social Change in the Southwest, 1350–1880* (Lawrence: University Press of Kansas, 1985), pp. 11–32.

24. See Bruce Winterhaider, "Concepts in Historical Ecology: The View from Evolutionary Ecology," in *Historical Ecology: Cultural Knowledge and Changing Landscapes*, ed. Carole Crumley (Santa Fe: School of American Research Press, 1994), pp. 27–30.

25. Karl Butzer, "Civilizations: Organisms or Systems?" *American Scientist* 68 (September–October 1980): 517–24. In light of James Lovelock's Gaia hypothesis, Butzer's conception has some special interest.

26. Roy Rappaport, "Maladaptation in Social Systems," in *The Evolution of Social Systems*, ed. J. Friedman and M. J. Rowlands (London: Duckworth, 1977), pp. 69–

71. See Anne Whyte, "Systems as Perceived: A Discussion of 'Maladaptation in Social Systems,' " *ibid.*, pp. 73–78; Rappaport, "Normative Modes of Adaptive Processes: A Response to Anne Whyte," *ibid.*, pp. 79–88; Donald Hardesty, "Rethinking Cultural Adaptation," *Professional Geographer* 38 (February 1986): 11–18.

27. For example, Howard Lamar points out the striking differences in the Canadian and American responses to the dust bowl. Lamar, "Regionalism and the Broad Methodological Problem," 25–44. On North American cultural regions, see Alfred Kroeber, *Cultural and Natural Areas of Native North America* (Berkeley: University of California Publications in American Archaeology and Ethnology, 1939); Raymond Gastil, *Culture Regions of the United States* (Seattle: University of Washington Press, 1975); Garreau, *The Nine Nations of North America*.

28. Hal Rothman, "Environmental History and Local History," *History News* 48 (November–December 1993): 8–9.

29. See especially Donald Worster's three essays, "The Shaky Ground of Sustainable Development," "The Ecology of Order and Chaos," and "Restoring a Natural Order," in his *The Wealth of Nature: Environmental History and the Ecological Imagination* (New York: Oxford University Press, 1993), pp. 142–83, wherein Worster argues hopefully (p. 181) that ecology "will eventually come back with renewed confidence" to the older models!

30. Daniel Botkin, *Discordant Harmonies: A New Ecology for the Twenty-First Century* (New York: Cambridge University Press, 1990), p. 62. See also Winterhalder, "Concepts in Historical Ecology," 29–30.

31. Dan Flores, "Bison Ecology and Bison Diplomacy: The Southern Plains from 1800 to 1850," *Journal of American History* 78 (September 1991): 465–85. What *longue durée* history implies, of course, is that bioregional historians have a sound grasp of paleontology and archaeology as well as ecology and climate study.

32. See Alice Ingerson, "Tracking and Testing the Nature-Culture Dichotomy," in Crumley, *Historical Ecology*, pp. 43–66.

33. On human evolutionary psychology and sociobiology, see Jared Diamond, *The Third Chimpanzee: The Evolution and Future of the Human Animal* (New York: HarperCollins, 1992); Richard Dawkins, *The Selfish Gene* (New York: Oxford University Press, 1976); Diane Ackerman, *A Natural History of the Senses* (New York: Random House, 1990). The best recent introduction to biophilia and biophobia is contained in the essays collected in *The Biophilia Hypothesis*, eds. Stephen Kellert and Edward Wilson (Washington, D.C.: Island Press, 1993). For my points in the text, see the following articles in that volume: Kellert, "The Biological Basis for Human Values of Nature," pp. 42–69; Robert Ulrich, "Biophilia, Biophobia, and Natural Landscape," pp. 73–137; Judith Heerwagen and Gordon Orians, "Humans, Habitats, and Aesthetics," pp. 139–72. Ulrich (p. 125) concludes that genetic

biophilias and biophobias may be 20–40 percent determinative but probably have to be triggered by learning.

34. I derive the term "captured knowledge" from Joel Gunn, "Global Climate and Regional Bio-Cultural Diversity," in *Historical Ecology*, pp. 86–90. A useful new work in this area is an anthology, *Humans as Components of Ecosystems*, eds. Mark McDonnell and Stewart Pickett (New York: Springer-Verlag, 1993), with foreword by William Cronon.

35. William Riebsame, "Sustainability of the Great Plains in an Uncertain Climate," *Great Plains Research* 1 (1991): 133–51.

36. William Cronon has appropriated the term "second nature" to describe these culturally altered settings, but I would have to insist that for the last 11,000 years, very few human societies have interacted with anything else. See William Cronon, *Nature's Metropolis: Chicago and the Great West* (Oxford: Cambridge University Press, 1991), pp. 266–67. On the human shaping of North America before the arrival of Europeans, the best general discussion I have seen is William Denevan's "The Pristine Myth: The Landscapes of the Americas in 1492," *Annals of the Association of American Geographers* 82 (September 1992): 369–85.

37. See Winnifred Gallagher, *The Power of Place: How Our Surroundings Shape Our Thoughts, Emotions, and Actions* (New York: Poseidon, 1993); Dan Flores, "Spirit of Place and the Value of Nature in the American West," *Yellowstone Science* 3 (Spring 1993): 6–10.

38. Carolyn Merchant, *Ecological Revolutions: Nature, Gender, and Science in New England* (Chapel Hill: University of North Carolina Press, 1989), pp. 2–3. Clive Ponting prefers the term *ratcheting*. See Clive Ponting, *A Green History of the World: The Environment and the Collapse of Great Civilizations* (New York: St. Martin's Press, 1991), p. 38.

39. A summary of a classic and useful comparative study of cultural values in the context of environmental choice in a bioregion (in this case Pueblos, Navajos, Mormons, Texans, and Hispanics in the Southwest) is Evon Vogt and John Roberts, "A Study of Values," *Scientific American* 195 (July 1956): 25–30. On bioregional art, see the various essays in *The Desert Is No Lady: Southwestern Landscapes in Women's Writing and Art*, eds. Vera Norwod and Lois Rudnick (New Haven: Yale University Press, 1986). On literature, Bill Howarth, "Literature of Place, Environmental Writers," *ISLE: Interdisciplinary Studies in Literature and Environment* 1 (Spring 1993): 167–73. A particularly fine collection of place literature—using state borders as its parameters, however—is *The Last Best Place: A Montana Anthology*, eds. William Kittredge and Annick Smith (Seattle: University of Washington Press, 1988).

40. Bioregional historians who properly seek to bring their work to life by interweaving the stories of individuals should be aware, to quote Amos Hawley, that the

"basic assumption of human ecology . . . is that adaptation is a collective rather than an individual process. And that in turn commits the point of view to a macrolevel approach." Amos Hawley, *Human Ecology: A Theoretical Essay* (Chicago: University of Chicago Press, 1986), p. 126.

41. Yi-Fu Tuan, *Topophilia: A Study of Environmental Perception, Attitudes, and Values* (Englewood Cliffs: Prentice-Hall, 1974), pp. 93–112.

42. Conversation with Richard White, April 9, Missoula, Montana.

43. An effective model for river valley studies in the West, although written in social science language, is the article-length study by William Wyckoff and Katherine Hansen, "Settlement, Livestock Grazing and Environmental Change in Southwest Montana, 1860–1990," *Environmental History Review* 15 (Winter 1991): 45–72 (reprinted in a revised version in this collection).

44. The quote is in Swierenga's introduction (p. 129) to Malin's chapter titled "On the Nature of the History of Geographical Area," *History and Ecology*, 129–43. Malin goes on in this essay (p. 130) to assert that "the study of the history of the western United States as geographical area is *not* the study of 17, 20, or 22 separate *states* that lie within that area" (emphasis added). I would be remiss if I did not mention that Donald Worster described something like the history I am calling for in a 1984 essay titled "History as Natural History." It is reprinted in Worster, *The Wealth of Nature*, pp. 30–44.

45. Patricia Nelson Limerick, *The Legacy of Conquest: The Unbroken Past of the American West* (New York: W. W. Norton, 1987). I single out Limerick's book because of its influence, not because she believes that any one interpretive framework can explain all places in the West across time.

3 / A Natural History of the Puget Sound Basin

ARTHUR R. KRUCKEBERG

My title provokes two questions: What is "natural history," and what is the Puget Sound basin? An answer to the second question is easy and will be developed in depth later. But just what is natural history? I suspect we all have an intuitive grasp of its content and focus. My dictionary gives the term an all-embracing definition: "Natural history encompasses all the sciences, like botany, geology, zoology, that deal with the study of objects in nature." The objects can be living or nonliving things, the activities of things, and their histories. But such a factual definition obscures a major difference between any of the natural sciences and natural history. Botany, geology, and zoology are the domains of the specialist, usually professional and academic, whereas the arena of natural history knows no bounds and encourages generalists, often amateurs—in the true meaning of the word: lovers of nature. There is a passionate streak in one who practices natural history—a naturalist, no less.[1]

Natural history can embrace a spectrum of studies of natural entities, all the way from a single organism (the Olympic marmot or the calypso orchid) or a single geological phenomenon (the volcano Mount Rainier) to groups of natural objects (all terrestrial orchids or all the Cascade volcanoes). It can encompass a region ranging in size from Walden Pond or Lake Washington to the entire Pacific Northwest.

It is the regional kind of arena that is the showcase for this chapter—the Puget Sound basin.[2] Let it be said that any place on our beleaguered planet merits its own natural history—an island, a mountain, a bioregion, or even a continent. I take the Puget Sound basin for a model of a regional natural history: its living and nonliving features that join to make a unique mix of land, water, and life. To know one's place on the planet through its natural history is to be blessed with the knowledge necessary to cherish, nurture, and preserve it. So we begin with Puget Sound's spectacular physical setting: its inland sea, its terrain and geological makeup, as well as its geological history and the basin's unique climate.

LANDFORMS AND GEOLOGY OF THE PUGET BASIN

The Puget basin is grandly defined by the complex of streams and rivers draining into the marine waters of the Sound. The limits of this vast drainage system are set east and west by the Cascade Range and the Olympic Mountains; their summit crests are the lofty rims of the basin. Low hills serve as the southern perimeter of the Puget trough. To the north, only the confluence of the Strait of Juan de Fuca with the Sound sets an arbitrary northern boundary. Apart from the montane rim, landforms are gentle north-to-south trending low hills, liberally pockmarked by depressions, mostly watery: lakes and streams. The deepest is the trough of the Sound, the fjordlike inland sea. Maps of the Puget basin's drainage network tell the topographic story better than words (figs. 3.1, 3.2).

How this natural landform—the Puget basin—came into being is the fascinating domain of historical and structural geology. Three primary forces have conspired to make the basin: the collisions of continental and oceanic plates (plate tectonics), volcanism, and glaciation. These forces entice us to look at the history of the region in two time frames: "deep time," with spans of millions of years, and "shallow time," reaching back less than a million years. Whichever time frame we are in, the sense of geological time fosters the essential idea that geological process is ceaseless—ongoing in the past and present. Mountains are rising and eroding, lowlands are accreting with sediment, water courses are altering, and crustal plates are eternally on the move.

The prime geological processes over deep time are cast in the new geological model—plate tectonics. The region has been "assembled" from diverse slabs of the earth's crust—plates—that have "docked" onto western North America from far-off sources. The San Juan Islands and the northern Cas-

FIG. 3.1. (opposite) Topography of the Puget Sound basin (source: Kruckeberg, *Natural History of Puget Sound Country*).

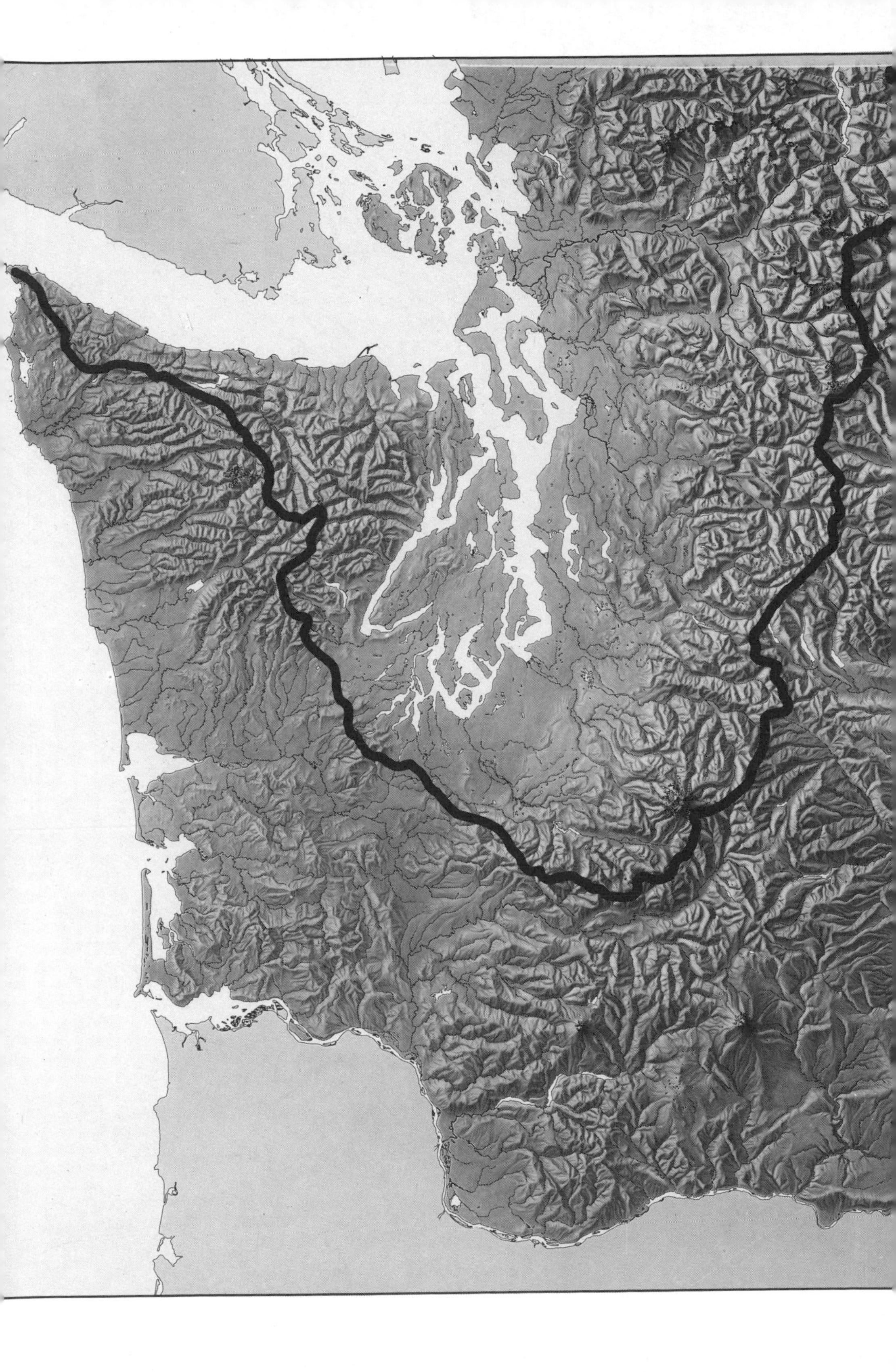

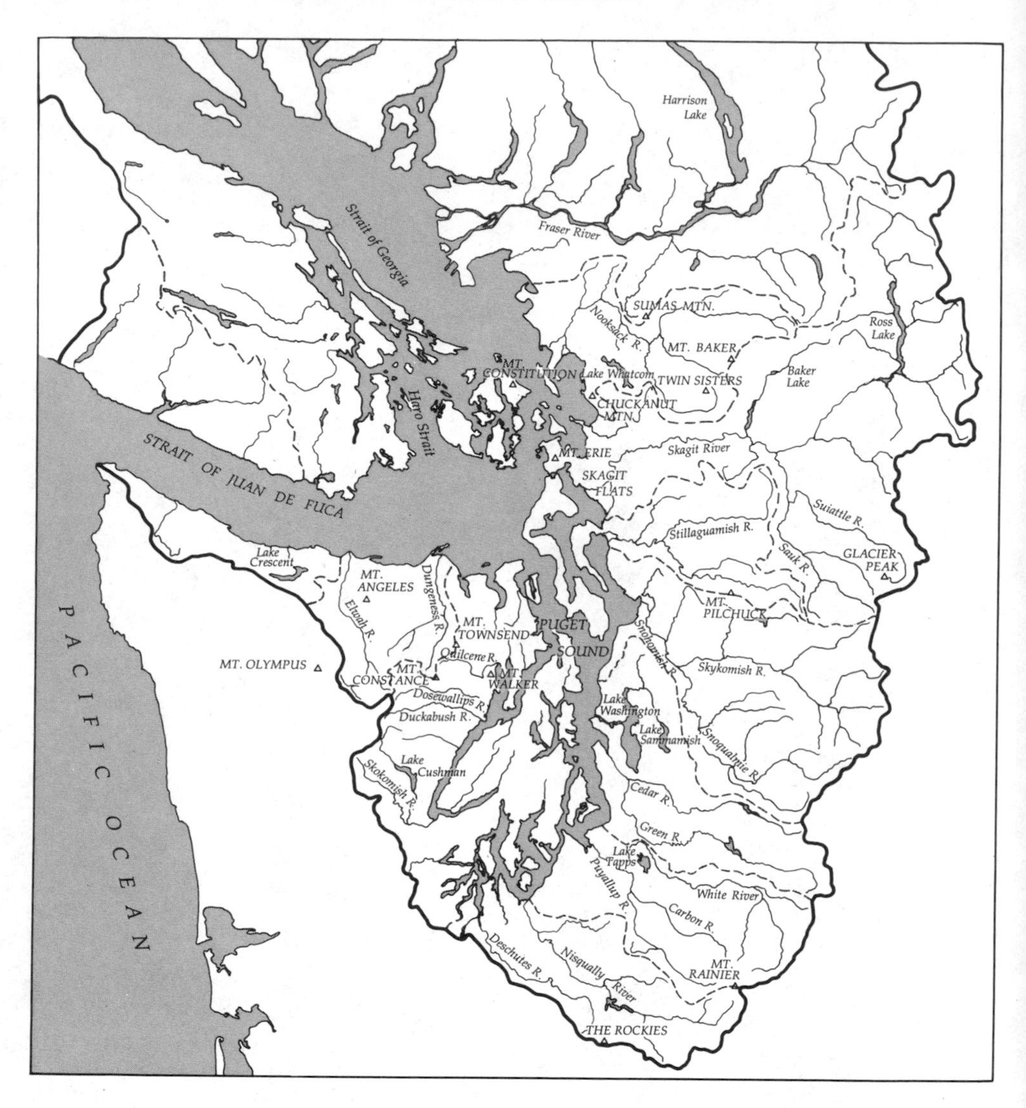

FIG. 3.2. The watershed of the Puget Sound basin (source: Kruckeberg, *Natural History of Puget Sound Country*).

cades have been knitted together from a variety of microcontinents (terranes) from as distant as eastern Asia. When plates or terranes "dock," they create complex patterns of rock, faults, and landforms and often yield explosive results—lava flows and volcanoes.[3]

The latest mountain-building event, the emergence of the Cascadian volcanoes, is superimposed over deep-time geologic events. Only in the last few millions of years have the spectacular volcanic peaks bordering Puget Sound come into full display: Mount Rainier, Glacier Peak, Mount Baker, and Mount St. Helens all rise well above the largely concordant heights of the older Cascades.[4] No other geologic event better exemplifies the incessant change to landform and geologic structure than does volcanism. Steam vents on Mount Baker, as well as Mount St. Helens' spectacular pyrotechnics of May 1980, remind us of this uneasy instability.

The western rim of the Puget basin—the Olympic Mountains—boasts no volcanoes. But it has had its share of turbulent plate movements. The eastern flank of the Olympics came out of the ocean originating as a submarine lava flow; then it became upended and plastered against the older Olympic core rocks. These marine lavas are easily recognized as basalt formed into "pillows" under water.

After the major plate movements and volcano-making had etched the contours of the basin, there came the most recent and decisive land-forming event—Pleistocene glaciation. All, yes all, of the basin's contemporary topography was fashioned—massively or delicately—by successive continental glaciers. The shallow time of the Pleistocene, the last million years, witnessed the sculpturing of the terrain by ice sheets and montane glaciers, giving us today's versions of mountains, hills, and valleys and the deep trench of Puget Sound. Every square mile of the earth's crust below 5,000 feet elevation, both above and below sea level, came under the scalpel and bulldozer of continental ice.

The last of the several glaciations, called the Vashon stade of the Fraser glaciation, came out of Canada, a massive wall of moving ice that inched its way south to below Olympia and west along the Strait of Juan de Fuca (fig. 3.3). This last Ice Age event occurred around 15,000 years ago, when the ice sheet had reached its maximum extent. The glacier was 3,000 feet thick at Seattle but 5,000 feet deep at the United States–Canada border. Evidence for this amazingly high wall of moving ice comes from a variety of traces on the land: glacial striations (parallel scratches on bedrock), massive amounts of rock, gravel, and silt forming deposits of glacial till, and displaced Canadian rocks, called "erratics." It is the appearance of the ice-rafted erratics

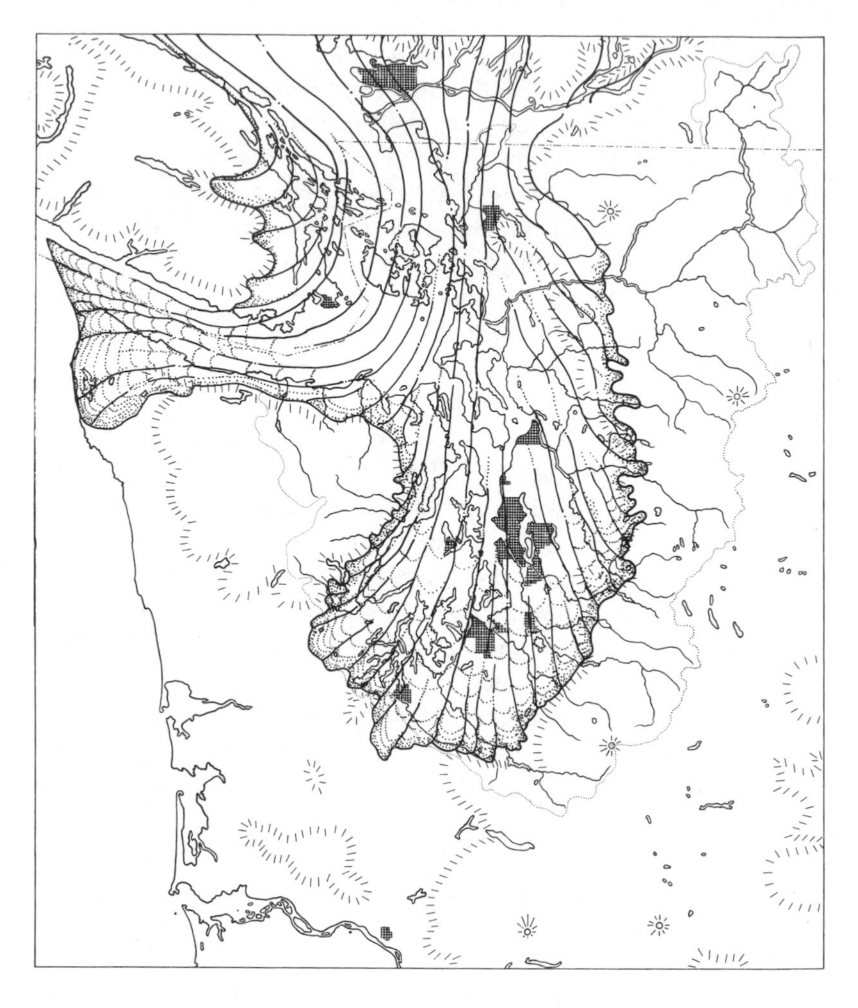

FIG. 3.3. The Fraser glaciation of the Puget Sound basin (source: Kruckeberg, *Natural History of Puget Sound Country*).

from Canada at 5,000 feet in the northern Cascades that is decisive evidence for the maximum height of the continental ice sheet. All landforms in the basin trace their latest shapes, sizes, and orientations to the last glaciation—and deglaciation. Low-lying hills trending north to south, U-shaped mountain valleys, and the many small depressions appearing now as lakes or ponds are all glacial in origin.

Then, in a brief space of shallow geologic time, the continental ice vanished. Recession of the glacier began around 14,000 B.P. (before present), and by 11,000 B.P. the ice sheet had "retreated" to the Canadian border. The vast melting of the ice had colossal effects on the land. Glacial lakes formed in lowland basins, only to break their ice and till dams and then flood south along the channel of the Chehalis River. The chasm that was to be the vessel for the marine waters of the inland sea, Puget Sound, once again became ice-free.

The geologic history of Puget Sound itself spans the 65 million years of the Tertiary era.[5] Radiocarbon dating of organic fragments in the interglacial sediments, as well as marine invertebrate fossils, give the genesis of the trough its time scale. Several geologic activities conspired to produce the submarine canyon. First, subsidence of rocks of the earth's crust defined the Puget lowlands along a north-south axis. During the Eocene, 56 million years B.P., the Pacific Ocean entirely covered parts of northwestern Washington and Oregon—there was no inland sea, simply ocean continuous to just east of Seattle. By the Pliocene (5 million years B.P.), the marine waters had receded to the Strait of Juan de Fuca, the mountainous rim of the Sound was in place, and the trough itself was filling in with sediments. Later, the successive glaciations of the Pleistocene further gouged out the trough; the last, or Fraser, glaciation applied the finishing flourishes, rasping out the trench to its present depths. It took at least four successive surges of glacial advance to fashion the trough and allow seawater to flow into the Sound.

The Puget basin, then, was a vast drainage, devoid of life, waiting to be recolonized after the Ice Age. The land and water ecosystems, teeming with life, viewed by early explorers such as Captains Vancouver and Malaspina in the late eighteenth century had reached their maturity and lavish diversity in less than 10,000 years. Recolonizing of the newly deglaciated terrain was slow but inexorable. Land plants near the border of the Vashon ice sheet were poised to reenter once-inaccessible land. Mature forest eventually replaced pioneer vegetation, a process of plant succession well known to plant ecologists worldwide.

Our knowledge of post-Pleistocene revegetation comes from a unique branch of biology called palynology. The palynologist identifies tiny fossils recovered from bogs and lake sediments. From bottom to top layers of sediments, kinds of plants and their community structure can be recognized microscopically. The microfossils used in such reconstructions are mostly pollen grains, though other plant fragments can also be used. The surfaces of pollen grains take on patterns that are distinctive for a plant group (a species or more often a genus). Thus pine pollen can be distinguished from oak or grass pollen. Kinds and frequencies of occurrence of pollen grains usually differ with depth; the tedious counting of pollen grains at each level of a sedimentary core is translated into a pollen diagram.

For most sampled sites in the Puget basin, the sequence of past floras is much the same. First, there are grasses, other herbaceous plants, and pine (usually thought to be shore pine, *Pinus contorta*). Next in sequence comes spruce, then true firs, Douglas fir, and finally western hemlock, western red cedar, and red alder as the most recent forest types. Two breaks in this stately progression of vegetation have been recognized. About 8000 B.P., a significant warmer epoch intruded upon the prevailing cool climate; this so-called hypsithermal period witnessed a sharp increase in grassland vegetation. Then came a return to the cool maritime environment. The other "blip" in the sequence is the ubiquitous appearance of a layer of volcanic ash at about 6000 B.P.; vulcanologists recognize it as from the eruption of Mount Mazama (now Oregon's Crater Lake). Eruptions of Glacier Peak and Mount St. Helens also have left their ashy fingerprints in bog sediments.

The postglacial sequence of vegetation types permits rough reading of past climates over approximately 11,000 years. The earliest grass-pine stage is interpreted as a cold period. The later grass–herbaceous perennial interlude (the hypsithermal) was a warm period, followed by a cooler maritime climate much like today's.

Can the pollen record as read by palynologists tell us anything about environmental change with the entry of humans into the Puget basin? Prior to the arrival of Europeans (beginning in the late 1700s), the coastal Indian occupancy made hardly a detectable sign in the pollen record. Possibly some bogs record an increase in fire, which Indians were known to use in perpetuating certain habitats, mostly the coastal prairies. But by the mid- to late 1800s, Euro-Americans' impact on the land became clearly discernible in the pollen record. Alder pollen leaps out in the later pollen profiles. Conifer forests denuded by logging, fire, and other agents were replaced by the pioneer red alder. Another kind of evidence of European presence is that pollen of plants of European origin appears in the bog sequences, and lead—from the tetraethyl lead in gasoline that entered roadside habitats—is found in the uppermost levels.

PUGET BASIN CLIMATE: PAST AND PRESENT

The cool, maritime climate of today has existed in the Puget basin for only a scant few thousand years. The contemporary climate has been called "modi-

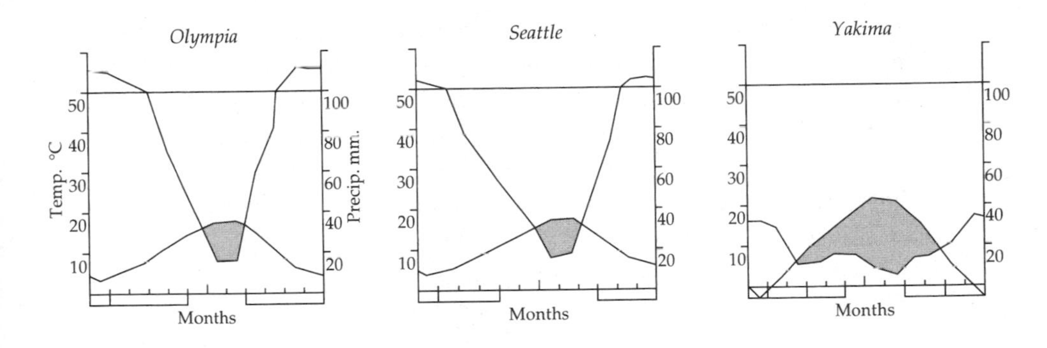

FIG. 3.4. Annual temperature and rainfall patterns in the Puget Sound basin (source: Kruckeberg, *Natural History of Puget Sound Country*).

fied Mediterranean." Forest ecologists recognize the critical Mediterranean segment of the basin's annual weather—the summer drought—as the key influence on its natural vegetation; its forests have evolved to cope with drought. Given the substantial rainfall of late fall to late spring (fig. 3.4), the term "modified Mediterranean" is apt; the true Mediterranean climate worldwide is warmer and drier than that of the Puget basin. But summer drought is its link to other Mediterranean climates. And ecologists have inferred that summer drought is the main reason the basin's forests are coniferous evergreen, not the deciduous types of the summer-wet East Coast.

What were past climates like in the basin? Like a region's topography, climate is subject to change. The Puget Sound region has seen climates in the early Tertiary (around 50 million years B.P.) that fostered a near tropical vegetation; fossil palms are even found in the old sedimentary rocks of the Chuckanut Formation. In those early Tertiary times there were no major mountain ranges to produce the wet-dry and east-west contrasts in climate that now characterize the Pacific Northwest. All was subtropical, and the Pacific Ocean's shore lay just east of Seattle. The climate and the swampy vegetation of that early time left a telltale imprint—occasional coal deposits. As the mountains rose to create the Puget trough, the climate became more variable, creating rain-shadow conditions (drier on the lee side of the mountains) here and there. Rain shadows still create local dry areas in the sound (e.g., at Port Townsend and Sequim).

Just before the Ice Ages of the early Pleistocene (approximately 2 million years B.P.), a cool-temperate climate arrived to usher in conifer forest vegetation nearly like that of the present. Then at least four surges of continental ice inundated the Puget basin, each followed by an interglacial. Less severe climate prevailed during the times between ice sheets; these interglacial periods, each a few thousand years in duration, had climates that allowed the return of temperate flora and fauna. Then another cold interlude would intrude to wipe out all, or nearly all, life.

Our present "interglacial" has lasted approximately 11,000 years, and in that span of shallow time, climates have fluctuated. At the perimeter of retreating ice, tundralike cold conditions existed. Then, early on, even when all the ice of the Vashon glaciation was gone, there were surges of ice; these minor advances finally gave way to land awaiting colonization by flora and fauna. The pollen record reveals climate change right up to the present. A major warm interval (the hypsithermal, 8500–3000 B.P.) was followed by the now-prevalent cool maritime climate, with predictable summer droughts. Minor deviations from this normal climate could last fewer than 100 years.

One such deviation is recorded by changes in montane vegetation. Jerry Franklin, a forest ecologist once dubbed the "arch-druid of old-growth forests", interpreted the following change in forest vegetation as due to a minor climate change.[6] The grand panoramic parkland habitat with its "timbered atolls" (clumps of subalpine trees) and intervening sweeps of mountain meadowland underwent a brief transformation owing to a short-term shift in climate earlier in the twentieth century. Tree saplings, mostly subalpine fir and mountain hemlock, invaded the meadows during a warmer and drier climatic episode less than 50 years in duration. One can still see vestiges of this tree invasion in parkland meadows, as at Mount Rainier; it is now arrested and probably yielding again to treeless meadows.

The basin's climate will continue to change, either as minor blips on the chart of normalcy or as more notable causes for alarm. A change such as the return of continental ice or global warming is sure to come, whether it is caused by human activities (especially emission of gasses from burning fossil fuels), by sunspots, or even by the capricious behavior of the earth's atmospheric envelope. Puget lowlands could then be invaded by ice, drought, or volcanic "winter" or inundated by waters from melting polar ice. To paraphrase Will Durant, "civilization exists by climatic consent, subject to change without notice."

PUGET SOUND: THE INLAND SEA

The grand perimeter of the Puget basin—created by its spectacular rim, the northern Cascades and Olympic Mountains—would be enough to call it a place of world-class wonders, with its glaciers, forests, lakes, rivers, and alpine meadows. But the capstone of the place is its inland sea, Puget Sound itself. The Environmental Protection Agency (EPA) declared it an "Estuary of National Significance."[7]

What is an estuary? Wherever freshwater meets the sea, an estuary results. Because so many rivers and creeks empty into the sound, it is really a multiple estuarine system. The sound's intricate shape and variable depths give it a host of topographic features. Its shoreline—nearly 2,000 miles of it—varies from steep bluffs to gently sloping sandy or rocky beaches and numerous tide flats and salt marshes. Not only does the sinuousness of the mainland coast along the Olympic Peninsula and the lowland base of the Cascades amplify the length of the shoreline, but so do the many islands, from Vashon, Whidbey, and the San Juans to barely exposed islets such as Peapod Rocks. Below the intertidal zone, a vast submarine canyon plunges from shallow sills like that at Admiralty Inlet, at 218 feet, to a depth of 930 feet off Point Jefferson. Two other submarine dams or ridges lie at the entrance to Hood Canal (South Point sill, at 175 feet) and at the Tacoma Narrows (145 feet). These sills impede the flushing of seawater that the daily tidal flux attempts to bring about.[8]

The Puget Sound system of estuaries is captured in this colorful definition: "An estuary is where the rivers' currents meet the seas' tides." And tidal displays are everywhere in the sound. The daily ebb and flow of seawater can be dramatic compared with other Pacific coast sites: the difference between the highest and lowest tides at Seattle is 13 feet, and down at Olympia, 15 feet, in contrast to Astoria, Oregon (8.2 feet), and Los Angeles, California (only 5.4 feet). Tidal changes combined with wind fetches can augment or dampen tidal effects on shore. It is the intertidal zone that is at the mercy of changes in sea depth; shoreline life is daily exposed—and safely submerged—by tidal contrasts.

How long the Sound's massive reservoir of seawater stays put is a complicated matter. Location, depth, the effects of the barrierlike sills all affect "retention time." On average, the Sound flushes twice a year; the cul-de-sac terminus of Hood Canal flushes only once a year. As a result, some waters at certain depths may stagnate and decrease biological productivity.

I have already touched upon the geologic history of the Sound and the adjoining basin. The deep, U-shaped submarine trough of the Sound is mostly a product of successive Pleistocene gougings by waves of continental ice—"God's great plow" as it was aptly termed by Harvard paleontologist Louis Agassiz. Some depression of the Sound's borders was due to the sheer weight of the ice—3,000 feet thick at Seattle 15,000 years ago. Then, as the last continental glacier retreated and melted, seawater began to replace the ice, no doubt mixed with freshwater from the extensive meltwater lakes bordering the Sound. By the time Captain Vancouver (1792) had charged Lieutenant Puget with exploring and charting the Sound, its contours, volume, and rich marine life were fully in place.

The early native peoples had of course already found the bountiful basin; it had been a rich source of food, transport, and cultural-spiritual rapport for more than 5,000 years. The aboriginal inland sea had become a "fertile fjord" after the ice retreated. The resilience of this inland marine ecosystem in the face of disturbance is best exemplified by its fate since the coming of Europeans. In less than 200 years of occupancy, Euro-Americans have dramatically altered the Sound. They have caused upland erosional runoff, dredged and filled in tidal estuaries, dumped solid and fluid (often toxic) wastes into the Sound, and altered shorelines with harbors and docks. The manifold impact on the Sound has been massive and destructive. Yet it is still resilient and can be saved, even restored, by a watchful and resourceful citizenry.

MARINE LIFE IN PUGET SOUND

"Fertile fjord" is an apt a name for Puget Sound. It teems with plant and animal life, from microscopic plankton to the awesome orca, or killer whale. The marine life of the Sound admirably illustrates many basic ecological concepts. First is *habitat* versus *niche*. Habitats—places where organisms live—run the gamut from the uppermost intertidal or "splash" zone down to the bottoms of bays and inlets and the main "canyon" of the Sound. Some habitats are narrow zonal bands or locally confined places such as intertidal rocks with barnacles, mussels, sea lettuce, and rockweed (*Fucus*). Others are expanses of open water throughout the water column for free-swimming creatures (salmon, harbor seals, and killer whales). Planktonic animals and plants (the "helplessly" floating microscopic algae and animal plankton) occupy open water at various depths.

To habitat add the notion of *niche*—the way organisms carry on their lives. In a nutshell, habitat is an organism's address, and niche is its occupation. The natural history of the starry flounder nicely exemplifies the habitat-niche linkage. This bottom fish mostly inhabits shallow sandy places inshore, where its remarkable body form adapts it to its home. It and most other flounders evolved a flat shape, yet shifted the two eyes to the top of the head. It is sluggish of movement and hence is camouflaged to avoid predators. It makes its diet out of a wide variety of invertebrates close at hand. Thus its occupation (its niche) admirably matches it address (its habitat).

Another ecological truism for the Sound's biota is the complexity of nutritional transfers—food webs or food chains. Food webs are well understood only for commercially important fish such as salmon. Yet for all animals, including zooplankton, the food chain begins with photosynthetic algae—mostly phytoplankton (microscopic algae), though many invertebrate and vertebrate animals live off larger algae (kelp and its kin). The phytoplankton flora of the Sound is rich in microscopic algae: green, blue-green, diatoms, and desmids, to name a few kinds.[9] Autotrophism (self-feeding) sets photosynthetic plants off from their heterotrophic (other-feeder) dependents. The essence of the nutritional network is captured in the pithy saying, "Biology is the study of plants and their parasites."

Nongreen plankton (zooplankton) are usually the second link in the food chain. These tiny animalcules eat phytoplankton and in turn are eaten by larger animals. The Chinese proverb says it well: "Big fish eat little fish, little fish eat bugs, and bugs eat mud [read, algae]."

The cool-temperate seawater and the constant addition of nutrients from estuaries and upslope runoff foster a great diversity of plant and animal life. No one has a firm count for the number of species of plankton that occur in the Sound; new ones continue to be discovered. For fish diversity we do have a count: 212 species in many different families, from the tiny rockpool sculpin to jumbo fish such as salmon and sharks.

Two marine plants, bull kelp and eel grass, illustrate the photosynthetic way of life in the Sound and portray both habitats and niches. Everyone who visits the seashore encounters bull kelp (*Nereocystis luetkeana*); the long, tubular corpses of this brown alga are washed ashore after a storm or at the end of their annual existence. The natural habitat of bull kelp is just offshore in submarine depths below the intertidal. There their cordlike stipes, as much as 80 feet long, are tipped by flattened blades and bulbous "floats." It is the bladder and blade that one sees at the surface in masses just offshore. Their ropelike stipes (the holdfast) attach to the seafloor and are massed to form a veritable forest. These plants in their submarine thickets come into being as tiny planktonic "larvae" (zoospores) and grow to their enormous biomass in one year. The bull kelp "forest" is home to a rich array of other plants and many animals. Some, known as epiphytes, live on the stipe; they can range in kind from those dependent on the sea bottom's holdfast to those living at the water surface. Other animals swim about in the forest, feeding, hiding from predators, and reproducing.

Eel grass (*Zostera marina*) is another marine plant that plays host to a community of dependent organisms. Eel grass has a special evolutionary history: unlike the usual algal flora of marine waters, it is a flowering plant. Its ancestors were land plants, first adapted to freshwater wetlands and ultimately evolving tolerance to the marine environment. Beds of eel grass commonly occur from the lower intertidal to the upper subtidal zones. These marine "meadows" are food and shelter to a host of organisms, including the shore bird known as the black brant and many sedentary plants and animals attached to the grass blades. Padilla Bay at Anacortes has a fine display of eel grass beds; the interpretive center on the east shore of the bay is well worth a visit.

So many and diverse are the marine animals of the Sound that I must single out just a few as examples.[10] Of the many kinds of fish (212 species), most intriguing are two family groups, the rock fish (Scorpionidae) and sculpins (Cottidae). They come in a wide range of sizes and shapes, from the tiny tidepool sculpin (4 inches long) to the giant cabezon (up to 2.5 feet long and weighing up to 25 pounds). Some of these curious fish, including the cabezon, are well-known for their edibility; others simply astound the fisherman when they are caught. Often grotesque and elaborately ornamented, these fishes occupy a variety of habitats, mostly sea bottoms of silt, sand, gravel, or rock outcrops. We can only assume that the many sculpins (36 species) and rock fish (24 species) tend to occupy distinct niches; little is known of their niche preferences. It is a working hypothesis in animal ecology that closely related species living in the same area must occupy distinct niches. If not, one would outcompete the other. This is known as the competitive exclusion principle.

Much more is known about the biology of the salmon clan (family Salmonidae).[11] Ten members of the family can occur in the Sound: trout, steelhead, and several kinds of salmon. That their life histories are well documented stems from their importance in commercial and sports fisheries. All are migratory, some making vast voyages from the North Pacific to breeding grounds in freshwater streams. I return to the present sad state of salmon populations a bit later.

The wildlife catalog for Puget Sound would be incomplete without mention of the marine birds and mammals that are linked to the inland sea.[12] Marine birds can either be denizens of the shoreline or birds that spend much of their lives on and over open water. More than 100 marine birds have been recorded for the Sound. Some, such as the western grebe and Bonaparte's gull, are common visitors; others, such as the eared grebe and the Caspian tern, are uncommon.

The fertile fjord is home to 10 or more kinds of marine mammals—the top of the food chain. The most familiar are the harbor seal and the killer whale. Their natural histories, including behavior and habitat-niche peculiarities, have been extensively observed.[13] Both of these marine mammals create emotional responses in human observers: they are seen either as superb seagoing animals or as rapacious predators of salmon.

From the very first, Europeans viewed Puget Sound as an economic resource, and its marine productivity was exploited early.[14] Salmon and oysters, for example, were harvested and shipped to San Francisco. One result has been a loss of biotic diversity; another has been the introduction of alien species into the Sound, often in competition with indigenous flora and fauna. The indigenous oyster was harvested to extinction and replaced by the Japanese oyster. Along with the introduction of oyster spat came several other alien shellfish: the Manila clam and the softshell clam.[15] Although only a few introduced fish persist in the Sound (e.g., American shad), many more have infested freshwater lakes.[16] The alien fish often outcompete the indigenous species. The most threatening of plant introductions is the East Coast cord grass (*Spartina alterniflora*), which has invaded salt marshes in the Willapa Harbor area. A close relative, the hybrid *S. townsendii*, is now known from northern Puget Sound. These introduced salt-marsh plants can drastically alter tide-flat habitats, crowding out native vegetation upon which native fauna depend.

In recent years, fish and other marine animals have been found to be afflicted by cancerlike growths attributed to toxic chemicals in the water and sediments. Loss of entire species is now imminent; probably some less-well-known invertebrate and fish species are already extinct.

PLANT COMMUNITIES OF THE PUGET BASIN

The terrestrial vegetation of the lowland Puget basin is far from homogeneous, even more so in the late twentieth century than when Europeans arrived in the late eighteenth century. In the absence of major human intervention, the lowland landscapes offered sufficient variation in topography, soils, and local climate to have elicited a diversity of vegetation types.

These include several kinds of forest communities as well as oak woodlands, prairies, streamside vegetation, and the plant life of wetlands (bogs, swamps, tideflats, etc.).

Ecologists have hypothesized for years that under optimal conditions of soil, climate, and landform, a forest dominated by western hemlock and western red cedar would reign over the land. Indeed, the entire lowland landscape has been called the western hemlock zone,[17] although it is best known as the Douglas fir zone for its most commercially valuable and dominant species. In the real world, however, realization of the dominant forest type has been tempered by local conditions. Thus some forest types are dominated by Douglas fir or even western white pine. In the absence of natural or human disturbance, all the variant forest types grow to reach a state of apparent equilibrium, the climax forest, in which the dominant trees are replaced by their own species after minor disturbances such as wind or light fires create gaps in the canopy. Forests can reach this unsteady equilibrium as old growth in 200 to 500 years. Their massive tree trunks become widely spaced as suppressed trees die out, allowing the development of a rich understory of shrubs and herbaceous perennials. Displays of old-growth forest are still accessible here and there: Longmire at Mount Rainier National Park, Federation Forest along U.S. Highway 410 east of Enumclaw, and the Asahel Curtis Natural Area, just off Interstate 90 west of Snoqualmie Pass.[18]

By the late twentieth century, most old growth had been harvested, to be replaced by even-aged stands of Douglas fir, either deliberately planted or from naturally reseeded sites with competing vegetation suppressed. Most such young stands are densely stocked with stems of a single species; their canopy cover is so dense that understory vegetation is sparse or absent—the so-called green shade effect.

In southern Puget Sound country, a special plant cover appears. The gravelly outwash plains derived from the receding Vashon continental ice sheet created ideal conditions for grassy prairies and open Garry oak woodlands, both liberally interspersed with conifer forest. These landscapes are known as "Tacoma prairies," dominant mostly in the Fort Lewis area but elsewhere in Pierce and Thurston Counties as well.[19] The prairies are frequently embellished with evenly spaced hillocks, the world-renowned Mima mounds landform. The prairies and oak woodlands were probably destined to be replaced by Douglas fir and other conifers in time, but plant succession was arrested in pre-European times as Indians used fire to keep the prairies open for getting game and for nurturing patches of camas, a favorite foodstuff. In recent times, the intentional suppression of fire has allowed succession to proceed. As a result, Douglas fir is encroaching on the prairies.

Forest most often dominates landscapes right up to the crests of the Cascades and Olympics. The composition of the forests, however, changes with altitude. At around 3,000 to 4,000 feet, western hemlock, western red cedar, and Douglas fir are largely supplanted by Pacific silver fir. Upslope this narrow zone gives way to the highest continuous forest zone, the mountain hemlock–Alaska cedar zone. But before timberline is reached, the most spectacular scenic landscape of all appears, the parkland subzone of the mountain hemlock–Alaska cedar zone. There, "islands" of trees, picturesquely called "timbered atolls," dot mountain meadows; the tree species are much the same as in the continuous forest just below: mountain hemlock, Alaska cedar, and subalpine fir. These parklands are richly diverse, especially in the lush wildflower meadows. They are best seen on the three volcanoes bordering the Sound; Mount Rainier National Park even names them "parks" (Grand Park, Spray Park, etc.). All throughout forested landscapes, stream and lake borders interrupt continuous forest with riparian trees and shrubs: alder, willow, and cottonwood.

Only at the highest reaches of the mountainous rim of Puget Sound does forest give way to alpine vegetation. It is mainly on the Cascade volcanoes that the treeless alpine zone manifests itself. Intervening between parkland and the alpine zone is that dramatic tension zone, timberline. Trees become dwarfed, gnarled shrubs subject to the severe climate. Timberline's elfin forest is universally known as "Krummholz" (crooked-wood), a vegetation type found on all continents. The same tree species of the parkland inhabit timberline, though in some drier areas whitebark pine gets the Krummholz treatment.

Two miracles strike us when we contemplate the world's richest temperate rain forest in the Pacific Northwest. First, this awesome forest ecosystem reached its late-eighteenth-century grandeur in the short span of post–Ice Age time, less than 10,000 years. Second, some of that magnificence is still left here and there in the Puget Basin after 200 years of occupancy by Europeans.

The early explorers—Spaniards with Malaspina and Britons with Vancouver—were confronted with a nearly continuous sweep of massive old-growth forest from the montane slopes to the saltwater's edge. Yet their appreciation of it was hardly aesthetic or ecological. Rather, they saw a resource to be exploited. Naturalists such as Archibald Menzies, traveling with Captain Vancouver, and José Moziño, voyaging with Malaspina, saw the forest in terms of how it might benefit settlers yet to come.[20]

And when settlers came they were eager to exploit what seemed to be an inexhaustible resource. Logging was the kingpin of their enterprises; as early as 1853, Henry Yesler started the first steam-powered sawmill on Puget Sound,[21] quickly followed by Pope and Talbot's mill at Port Gamble and other mills at Seattle, Alki, Port Ludlow, and Appletree Cove.[22] By the 1870s San Francisco had become a major market for the Sound's timber. Extraction of logs was an effort of manual labor during the nineteenth and early twentieth centuries, using ax and hand saw, skid roads and animal power. Logging railroads entered the forest scene in the 1880s, and by the turn of the century, Frederick Weyerhaeuser had come from Minnesota to acquire forest land in the Puget basin. In the 1920s began the age of the pulp mill. During World War II, extraction of old-growth timber accelerated because of both wartime needs and new technologies—the chain saw and massive logging equipment, including the most ubiquitous sign of log removal, the logging truck. Through the late nineteenth and the twentieth centuries, great forest fires were commonplace. The St. Helens fire of 1868 burned 300,000 acres, and in 1902 the more disastrous Yacolt burn covered 600,000 acres and turned Seattle from daylight to dark.

Yet as the great temperate rain forests slip away from us, some brakes on forest harvesting have been applied. Major tracts of old-growth forest were set aside: Mount Rainier National Park in 1899 and Olympic National Park in 1938, as well as wilderness areas within the national forests and old-growth remnants in state parks. But by the late twentieth century most of the old growth in the Puget basin was gone, some replaced by tree farms but most giving way to agriculture and urbanization. It has been remarked that the largest clear-cut of all is metropolitan Seattle! The capstone of this massive extraction of timber has been confrontations between the timber harvest industry and environmentalists who want to preserve what remains of old growth. Just emerging now is a rethinking of how to manage forests—not only as a commodity but also as a renewable biotic community. The "new forestry" challenges us to look on forests as ecosystems that tie trees to their physical environment and all other life in the forest. Will it work? Perhaps. The prospect for new forestry remains to be judged.

Historian Richard White provides a fascinating vignette of Euro-Americans' shaping of a piece of Puget country in his book on the environmental history of Whidbey and Camano Islands.[23] White records the changes to the land: logging, clearing for farming, urbanization, military installations, and tourism, and their consequent impacts on natural systems. What has happened on Whidbey Island has also been the story of change for the Puget mainland for nearly two centuries.

ANIMAL LIFE IN THE PUGET BASIN

Seeing animals in natural landscapes takes patience and special alertness, even sensitivity. Unlike animals on coral reefs or the Serengeti Plain of Africa, animals in the Puget Sound basin seem to be inconspicuous and secretive. No great herds of ungulates or flocks of birds reward the naturalist. But the animals are there: bear, cougar, elk, and deer command places in the forest ecosystem, as do a host of rodents, birds, and insects.[24] Without many of them, forests would be impenetrable. Tunneling, burrowing, and browsing animals, especially herbivores, modify and shape the green landscapes. Further, animals are an integral part of the essential and universal food chain. Green plant producers feed animal consumers that in turn feed decomposers. This pathway is the crux of sustained life on the planet. And for animals there is a hierarchy of consumption (heterotrophy): herbivores (deer, elk, rodents, many birds and insects) serve as food for carnivores (cougars, bobcats, raccoons, raptorial birds, etc.). Envision this hierarchy as a food pyramid, with plants forming the broad base, herbivores in the middle, and carnivores at the narrow apex.

Like forests and other types of vegetation, animal populations have changed through time, both in kinds (species) and in numbers. The fossil record for terrestrial animals in the Puget trough prior to the Pleistocene is thin. Some primitive mammals of the Tertiary have been unearthed. Then, during the Ice Age interglacials, horses, bison, caribou, woolly mammoths, and mastodons roamed the temperate lowlands, although finds of their remains have been scanty. And surely some of our present-day vertebrates were present, yet no fossils of them have been recovered to date.

Animal diversity and population densities, too, have changed dramatically since the coming of Europeans. In earliest settlement times, game animals became locally scarce; deer were hunted to extermination around Fort Vancouver for food and to protect crops. Yet deer have rebounded dramatically, becoming plentiful again in the twentieth century. Several environmental changes conspired to foster black-tailed deer increases. First, pressure of predators (except human hunters) decreased as wolves, mountain lions, and coyotes were hunted as big game or bounty "vermin." Then dense old-growth forest gave way to brushy open fields and early second-growth forest, ideal for foraging deer. The number of black-tailed deer in Puget Sound was estimated to be 97,450 in 1983, following their drastic decline in earlier years.

The most publicized loss of diversity has been that of salmon. Owing to a host of alterations to fish habitats (dams, siltation of streams, etc.), some salmon migrations no longer exist in certain streams. And because of habitat degradation, some salmon species are classed as threatened or endangered for the Puget Sound drainage system.

Changes in diversity of bird species illustrate well the impact of humans on lowland habitats.[25] Some native bird species have become less common (e.g., marbled murrelet, spotted owl, peregrine falcon); others have become more abundant (e.g., American crow, American robin, blackbirds). Several bird species, both alien and American (e.g., starling, pheasant, Canada goose), have been naturalized in the basin as resident local populations.

The same can be said for plant species diversity. There have been increases, decreases, and massive invasions of aliens (e.g., Scot's broom, gorse, English ivy, tansy ragwort, and many European grasses); the alien plant list is substantial. Like some animals, some plant species, such as the prairie herbs yellow Indian paintbrush and white-top aster, have become rare and even endangered.

AMERICAN INDIANS OF PUGET SOUND

Coastal groups of American Indians occupied many different locales in the Puget basin.[26] What effects did they have on the natural environment? The full answer may never be known, since evidence for their early, precontact occupancy is scanty and largely biodegradable—objects of wood, fiber, and animal products. Coastal Indians probably arrived in the Sound area some 8,000 years ago, either via the North Pacific coast or from the dry interior (Plains Indians). For the most part, these Salishan-speaking peoples lived near the water, on the borders of lakes, rivers, and the Sound itself. Their communities were small and consisted mostly of permanent dwellings—the longhouse. The bulk of their resources for food, clothing, and shelter came from nearby forests and bodies of water. The ubiquitous, massive cedar trees and the seemingly endless supply of salmon were the essentials of their domestic economy.

With this simple yet bountiful resource base, pre-Columbian native cultures presumably were self-sustaining. There is only scant evidence that they ever overexploited a resource: a known shellfishery was abandoned in the

San Juan Islands when it became depleted.[27] Given stable, low population numbers and a spartan livelihood gained by a simple but efficient manual technology, there is little reason to doubt that native impact on the environment was modest. What is harder to judge is the sentimental notion that Indians (and other nonindustrial peoples) practiced sound resource management, either by intuitive grasp of the workings of natural systems or by virtue of a sacred and spiritual consciousness about accommodating to the functioning of the natural world. Recent archaeological and anthropological findings suggest that some pre-European cultures substantially altered their natural surroundings (e.g., loss of bird species and forest habitats in Polynesian, especially Maori, areas) well before European contact. And for North American pre-Columbians, fire-setting, resource harvesting, and other practices altered nature in many ways.

I take up this enigma—whether or not the Indians evolved a land ethic—at some length elsewhere.[28] The answer in brief is conjectural and problematic. There is no clear evidence of a practiced environmental ethic. These peoples had no legacy of written history to inform us, and the use of native elders as informants is limited to post-Columbian survivors and can be unreliable. The best that can be said is that the very nature and magnitude of their cultures enabled them to avoid gross disturbance and loss of natural resources. Small communities of more-or-less stable populations were matched by simple, minimally destructive ways of gathering resources from nature. The result: coastal Indians lived in and with the natural environment, causing only minimal and repairable disturbance. This seeming equilibrium came to an end in the late eighteenth century. Diseases imported by Europeans decimated Indian populations; those who survived began to appropriate new goods and technologies (metal tools, guns, alcohol, etc.) from whites. Their lives were irrevocably changed, and in turn Indians assumed much of the white man's attitude toward exploiting nature.

THE EUROPEAN INVASION

It took several thousand years for the Puget trough to recover from the inundation by the last continental ice sheet. By the late eighteenth century, rich temperate forests and aquatic ecosystems held sway. Since that time, Europeans have attempted to return the bioregion to its early, postglacial barrenness. As I wrote in an earlier work, "the ice had radically altered life and land. Today a second catastrophic epoch is in full swing, one hardly less destructive than the great erasures wrought by ice. A human invasion now rolls over the land—a self-centered enterprise that leaves little untouched. Europeans are now altering the living historic monuments of towering forest and vital inland sea that existed only for 13,000 years, between the last Ice Age and their arrival."[29]

Changes in habitats in post-Columbian times are all too familiar—a litany of drastic alterations or destructions of natural systems. This litany, we all know, includes forest changes and losses due to logging, fire, and the clearing of land for other human uses. Urbanization has spread over the landscape like a cancer, the "built" environment replacing the natural. Attendant on these deteriorations and losses of habitat are air, land, and water pollution and the alteration or obliteration of natural landforms. And wherever Euro-Americans have disturbed natural ecosystems, their "camp followers"—alien organisms—have found the disturbed sites compatible with their broadly tolerant "lifestyles." As it has been said, weeds are plants that thrive under human disturbance.[30] These drastic changes to natural systems in the basin are amply documented elsewhere.[31] It should suffice here simply to highlight some of the environmental changes and their effects on the health of the bioregion and on us—the makers of change.

Statistics for change and loss of forests tell part of the story. In 1840, the pre-European timber stands for Washington State were estimated at 578 billion board feet. In 1973, inventory figures for western Washington old growth stood at 50 billion board feet, and for second growth, 100 billion board feet. These recent figures do not distinguish old growth under protection in national parks, wildernesses, and other reserves from that still available for harvest. Further, note the large volume of second growth, a type of forest vastly different ecologically from old growth in reduced biodiversity, stability, and biomass. Loss of forest acreage is another approach to measuring change in the timber resource. Forest loss in western Washington from 1933 to 1992 was about 1.5 million acres, or 11.2 percent.[32] Beyond volumes or acreages of timber in recent times, there is the matter of fragmentation of forest stands. The survival of many species, especially animals, depends on continuous expanses of forest. Fragmentation, resulting in "islands" of forest surrounded by other landscapes, can result in rarity or even extinction of certain biota.

After logging, the clearing of forest land for human settlement—farmland, villages, towns, cities, roads, industries—has caused the greatest loss of native climax forest. In many places we humans have built right over the land most productive for forestry and agriculture, the alluvial flats. Witness the loss of the Kent Valley, south of Seattle: once a prime agricultural region, it is now covered with industrial and suburban sprawl. In his *Design with Nature,* Ian McHarg urges keeping the rich alluvial plains for open space and agriculture; he insists on putting the industrial parks and housing developments on the hillsides.

Pollution is everywhere in the Puget basin—in the air, in the water, in and on the soils. Pollution should be put in its proper ecological context: it is deflected production, a resource out of place, the by-products of activities—human and natural—that resist recycling. The many sources of pollutants include industrial wastes (by-products of pulp mills and chemical manufacturers), treated sewage effluent, and non-point-source runoff (storm drain waters from farms and homes). Just a few of the many pollutants are the inorganic (metallic) compounds containing lead, arsenic, and mercury, all toxic. Organic chemicals are much greater in variety; they range from plastics and synthetic rubber by-products to herbicides, pesticides, and the halogenated (chlorine-containing) hydrocarbons such as DDT and PCBs.

On a clear day you once could see forever. Now summer smog obscures the view and troubles the eyes and lungs. Air pollution can be especially bad when a temperature inversion puts a "lid" on the air in the basin. Dumping wastes on soils inevitably contaminates them, often with noxious leachates that may get into the groundwater. But the worst is water pollution. Freshwater everywhere has been the recipient of wastes and the reservoir of toxins. Once-pristine mountain streams are now contaminated with *Giardia*, a pathogenic protozoan. And the inland sea fares little better; just read the annual reports of the Puget Sound Water Quality Authority. Inorganics such as lead and mercury, along with many toxins of organic origins, pose health risks and reduce the biological productivity of the waters.[33] Some sites, including Tacoma's Commencement Bay, are badly contaminated with chemical by-products from private residential and industrial sources. Because the chemical wastes become embedded in the bottom sediments, cleanup is difficult, to say the least.

For many watery basins, especially lakes, the most obvious change caused by continuous inputs of human wastes is eutrophication, or nutrient enrichment. The lakes' biological productivity dramatically increases when they are "fertilized" with nutrients from upland sources, mostly human. A well-fed (eutrophic) lake may foster blooms of noxious, unwanted plankton as well as replacement of native plants and animals with undesirable ones. The case history of Lake Washington is instructive. The lake was well on its way to eutrophication when the public realized that the trend could be reversed. The persistent teamwork of biologist W. T. Edmondson and public official

James Ellis (he who formed METRO, Seattle's regional planning agency) has now aided the return of the lake from its near-cesspool state to a cleaner, more habitable body of freshwater.[34]

Efforts to clean up the Sound, to prevent further pollution, and to protect as yet fairly natural habitats have grown apace in recent years. Government agencies at the federal level (EPA, U.S. Fish and Wildlife Service), state level (Department of Ecology, Fish and Wildlife), and local level are serving both regulatory and prohibiting functions and are undertaking major cleanup tasks as well. Private, nonprofit groups such as People for Puget Sound, the Seattle Audubon Society, and the Washington Environmental Council are active in promoting the environmental integrity of the Sound. These private organizations—the independent sector of any society—serve the "quis custodiet" function in our culture: they are "watching the watchers."

On top of pollution, there is the problem of invasion by alien plant species. Nearly every square yard of disturbed land in the basin is infested with plants from other lands, mostly introductions from Europe. Some were intentional—foxglove, Scot's broom, English holly, English ivy; these probably escaped from gardens. Others, including June grass, hawkweed, sow thistle, cat's ear, and dandelion, were introduced accidentally, often as unintended passengers in ships' ballast or in agricultural products. Though usually unwanted, from the ecologist's view weeds are biological success stories; unlike native flora, they are adapted to succeed under the most stressful conditions. But their very ability to succeed puts native species at competitive risk. Just witness the degradation of some wetlands when purple loosestrife takes over or when clear water becomes a murky tangle of introduced milfoil.

Introduced animals are not far behind their alien plant counterparts. Starlings, English sparrows, Norway rats, and a host of insects all usurp habitats and resources. Even the native Douglas squirrel and western gray squirrel are no match for the "weedy" eastern gray squirrel.

And so it goes. Environmental quality is degraded as the human enterprise accelerates, right down to the end of our century. At least two attributes set the impacts of precontact coastal Native Americans apart from those of Europeans. For the Indians, population sizes were small and technologies for resource extraction were simple and fairly benign. In contrast, Euro-Americans outnumber Indians in population and technological potential a thousandfold. As we enter the twenty-first century, ever-swelling populations and highly advanced technologies continue to accelerate environmental change (read, degradation) in the Puget drainage basin. The oft-touted "most livable place" in the country may slip into the sad state of other metropolitan areas in time. Is there any hope or solution?

The answer is a qualified yes. First, the theme of this collection of essays is *environmental history*. Are we not supposed to learn from history? The environmental history of the Puget Sound basin does hold many lessons for us—what to do and what not to do to preserve and enhance a livable environment. "What to do" has precedents: land-use planning, but with teeth for implementation, not just with platitudes; protection of remaining natural landscapes and seascapes from intrusive development and overuse; protection of existing natural areas (parks, preserves, open spaces, wetlands, etc.); and strong pollution controls. Recent past attempts to save environments have met with some modest successes. Public agencies such as the Environmental Protection Agency and the Fish and Wildlife Service, with sustained urgings from independent groups such as People for Puget Sound, the Washington Environmental Council, and the Nature Conservancy, have kept some areas from degradation.

In terms of "what not to do," the first admonition is of course not to repeat the follies of the past. Previous insults to the environment embrace all that humans do in the absence of a land ethic: encourage unbridled growth in population and urbanization, alter or destroy wildlands, and contaminate air, soil, and water. The litany of humanity's assaults on natural environments is nearly endless. Learning from these past mistakes is relatively easy; putting into practice what is learned is the crucial challenge, now and in the future.

Let me close by paraphrasing the epilogue of my earlier book on the Puget basin's natural history.

AN ECOLOGICAL IMPERATIVE FOR PUGET SOUND

The enjoyment of nature has been in and out of fashion over the years. Those living in the late twentieth century, perhaps more than any other human generation, have had the leisure time, the material wherewithal, and the urge to seek enjoyment from natural environments. Witness the weekend exodus of people from the built environment of cities to camp, ski, hike, climb mountains, fish, hunt, and just relax in wild nature. Though most who partake of outdoor life might equate the experience with other leisure pursuits such as chess, wine-making, or music, the outdoor experience could have far more significance than simple diversion from day-to-day toil. The trip to the ocean, the desert, or the mountains can serve to affirm the inseparability of humans and the rest of the natural world.

The pursuit of natural history can be a source of great personal enjoyment. And from it, the lively mind can perceive, by simple observations in forests, on beaches, or in a tide pool, the vital interconnectedness of things. An appreciation of natural processes and phenomena in Puget Sound country can lead to wiser, more peaceful coexistence between people and the elements of their natural environments. From such appreciation comes a sense of belonging to the natural world and of deriving both material and spiritual sustenance from it. Only when those living in a region develop an awareness of their natural heritage will they be in a position to preserve it. Conservation on a national or global scale may be too intangible a goal for the individual. Rather, those of us who live in the Pacific Northwest can begin with our own spectacular Puget Sound setting: to understand it, to appreciate it, and to preserve it from further degradation.[35]

Environmental history can both deal with the past and point out the pathway for the future. This sentiment is eloquently stated by Larry Harris: "Natural resources are not given to us by our fathers, but loaned to us by our children."[36]

NOTES

1. John T. Nichols, "What is a Naturalist Anyway?" *Natural History* (November 1992): pp. 6–10.

2. Arthur R. Kruckeberg, *Natural History of Puget Sound Country* (Seattle: University of Washington Press, 1991). This book is a primary reference for this essay.

3. David Alt and Donald W. Hyndman, *Northwest Exposures: A Geological Story of the North West* (Missoula: Mountain Press Publishing Co., 1995). This is the most recent popular account of Northwest geology.

4. Stephen L. Harris, *Fire and Ice: The Cascade Volcanoes* (Seattle: Mountaineers and Pacific Search Press, rev. ed., 1980).

5. For the geological history of Puget Sound, see Alt and Hyndman, *Northwest Exposures;* Parke D. Snavely, Jr., and Holly C. Wagner, *Tertiary Geological History of Western Oregon and Washington* (Olympia: Division of Mines and Geology, Report of Investigations no. 22, 1963); Don J. Easterbrook and David A. Rahm,

Landforms of Washington: The Geologic Environment (Bellingham: Western Washington State College, 1970).

6. Jerry F. Franklin and C. T. Dyrness, *Natural Vegetation of Oregon and Washington* (Corvallis: Oregon State University Press, 1988), p. 172. Basic reference on vegetation types in the Puget Sound area.

7. Kruckeberg, *Natural History of Puget Sound Country*, p. 63.

8. On Puget Sound marine environments, see Kruckeberg, *Natural History of Puget Sound Country*, ch. 3; Robert Burns, *The Shape and Form of Puget Sound* (Seattle: Washington Sea Grant Publications and University of Washington Press, 1985); John Downing, *The Coast of Puget Sound: Its Processes and Development* (Seattle: Washington Sea Grant Publications and University of Washington Press, 1983); Richard M. Strickland, *The Fertile Fjord: Plankton in Puget Sound* (Seattle: Washington Sea Grant Publications and University of Washington Press, 1983).

9. Strickland, *The Fertile Fjord*.

10. For further reading on marine life, see Eugene N. Kozloff, *Seashore Life of the North Pacific Coast* (Seattle: University of Washington Press, 1983); and Steve Yates, *Marine Wildlife of Puget Sound, the San Juans, and the Straits of Georgia* (Chester, Conn.: Globe Pequot Press, 1988).

11. Kruckeberg, *Natural History of Puget Sound Country*, pp. 94–106.

12. Tony Angell and Kenneth C. Balcomb III, *Marine Birds and Mammals of Puget Sound* (Seattle: Washington Sea Grant Publications and University of Washington Press, 1982).

13. *Ibid.*; Michael A. Bigg, Greame M. Follis, John K. B. Ford, and Kenneth C. Balcomb, *Killer Whales* (Nanaimo, B.C.: Phantom Press, 1987).

14. For more on the history of Europeans in Puget Sound, see Daniel Jack Chasan, *The Water Link: A History of Puget Sound as a Resource* (Seattle: Washington Sea Grant Publications and University of Washington Press, 1981); J. G. Swan, *The Northwest Coast, or Three Years Residence in Washington Territory* (1857; reprint, Seattle: University of Washington Press, 1972); Murray Morgan, *Puget Sound: A Narrative of Early Tacoma and the Southern Sound* (Seattle: University of Washington Press, 1981); Helmut K. Buechner, "Some Biotic Changes in the State of Washington, Particularly during the Century 1859–1953," *Research Studies of the State College of Washington* 21 (2) (1953): 154–192.

15. Buechner, "Some Biotic Changes in the State of Washington"; Yates, *Marine Wildlife of Puget Sound*.

16. Yates, *Marine Wildlife of Puget Sound*, p. 21.

17. Franklin and Dyrness, *Vegetation of Oregon and Washington*.

18. See Kruckeberg, *Natural History of Puget Sound Country*, for accounts of old

growth forest. A useful guidebook to stands of old growth can be found in Dittmar Family, *Visitor's Guide to Ancient Forests of Western Washington.*

19. For prairie ecology, see Kruckeberg, *Natural History of Puget Sound Country,* ch. 7.

20. Archibald Menzies, *Menzies' Journal of Vancouver's Voyage* (Victoria, B.C.: Archives of British Columbia, Memoir no. 5, 1928); Jose Moziño, *Noticias de Nutka: An Account of Nootka Sound in 1792,* ed. Iris H. Wilson (Seattle: University of Washington Press, 1970).

21. See Kruckeberg, *Natural History of Puget Sound Country,* Appendix 3, for chronologies of events.

22. See chronological tables in Chasan, *The Water Link,* pp. 162–71; Kruckeberg, *Natural History of Puget Sound Country,* pp. 430–35.

23. Richard White, *Land Use, Environment, and Social Change: The Shaping of Island County, Washington* (Seattle: University of Washington Press, paperback ed., 1992).

24. For more on animal life, see Kruckeberg, *Natural History of Puget Sound Country,* ch. 6.

25. For changes in bird fauna, see *ibid.,* pp. 244–45.

26. For more on Puget Sound Indians, see *ibid.,* ch. 10.

27. *Ibid.,* p. 402.

28. *Ibid.,* p. 32.

29. *Ibid.,* pp. xii–xiii.

30. Here I paraphrase H. G. Baker, "Characteristics and Modes of Origin of Weeds," in *The Genetics of Colonizing Species,* eds. H. G. Baker and G. L. Stebbins (New York: Academic Press), pp. 147–72.

31. *Ibid.,* ch. 11 and Appendix 3; Chasan, *The Water Link;* Morgan, *Puget Sound.*

32. Recent statistics on Washington forests can be found in C. L. Bolsinger et al., *Washington's Public and Private Forests* (Portland: Forest Service, Pacific Northwest Research Station Bulletin PNW-RB-218, 1997).

33. *State of the Sound,* annual report of the Puget Sound Water Quality Authority, 1988.

34. Kruckeberg, *Natural History of Puget Sound Country,* pp. 254–63.

35. *Ibid.,* pp. 415–21.

36. Larry D. Harris, *The Fragmented Forest: Island Biogeography Theory and the Preservation of Biotic Diversity* (Chicago: University of Chicago Press, 1984), p. i.

4 / From Where We Are Standing

The Sense of Place and Environmental History

WILLIAM L. LANG

The idea of place is surprisingly complex. More than location or the relationship of one site to another, place is a puzzle that to be understood requires delving into cognition, belief, emotion, and imagination. The essential, most direct understanding of place is as a self-evident physical reality, but the word carries little meaning if it is empty of human experience. It is in the relationship of experience to locality that the idea of place is most powerful and persuasive. Novelist Eudora Welty put it succinctly when she wrote: "It is by knowing where you stand, that you grow able to judge where you are."[1]

Welty's construction is purposefully ironic, for she tells us that enlightenment about place comes only *after* you discover precisely what you believe and what you hold dear—that knowledge of place is as much about knowing *who* you are as *where* you are located on a cartographic grid. Welty underscores the complexity of place, a complexity in which definitions of home and personal space become elaborations of the idea and where the focus is on a cultural intimacy that makes connections between human experience and specific geographical areas. As the geographer E. V. Walter explains: "A place has no feelings apart from human experience. . . . [It] is a location of experience. It evokes and organizes memories, images, feelings, sentiments, meanings and the work of imagination."[2]

The layered depiction of place suggested by Welty and Walter is as slippery as it is complex. After all, there is much less certainty in the dimensions of human experience and perception than in descriptions of locations by range and section numbers or by navigational coordinates. It is this slipperiness that makes the idea of place potentially too undisciplined to be of much use in environmental history. It is too relativistic, unlike descriptions of place that rely on scientific, economic, and political measurements—the concrete descriptions that dominate environmental history.

In addition, the terms *place* and *sense of place* have become nearly idiomatic in American culture, appearing in writing about such disparate subjects as photographic art, urban planning, and spiritualism.[3] In natural history writing, literary criticism, and regional fiction, place has become an especially dominant theme. Powerful reflections on place are compelling and vivid elements in writings by Wendell Berry, Barry Lopez, Wallace Stegner, Frederick Turner, and Terry Tempest Williams. In the Pacific Northwest, evocative writing about place makes the essays, fiction, and poetry of Mary Clearman Blew, Ivan Doig, William Kittredge, Craig Lesley, William Stafford, and Elizabeth Woody powerful and ultimately a reflection of artistic investment in the region.[4]

The diverse ways these writers use the concept of place seem to make its use in environmental history even more problematic, because so much of the focus is on the internal as contrasted with the external. The domain of place is within the perceiver, the storyteller, the poet, not "out there" on the landscape. It resides, as Frederick Turner labeled it, "beyond geography." Despite these apparent discouragements, however, the subject of place has actually had a significant role in some of our best environmental studies, including William Cronon's *Nature's Metropolis*, Donald Worster's *Dust Bowl*, and Richard White's *Land Use, Environment, and Social Change*. These superb investigations are, in part, extended and pointed histories of place. Nonetheless, the idea of place and the sense of place are far less developed in those studies than are histories of place that emphasize ecological, economic, and social relationships in Cronon's Chicago, Worster's southern plains, or White's Island County, Washington. The measured, not the perceptual, landscape dominates.[5]

Where we hear some of Eudora Welty's advice, however, is in recent environmentalist literature. Her caveat to know where you stand on humanistic and social issues as an essential component in understanding place, for example, is at the heart of a recent call from environmentalist and legal scholar Charles F. Wilkinson to develop "an ethic of place" as a new social architecture for the American West, one that consciously learns from the past and utilizes environmental history as a guide. With a sense of urgency born of this generation's recognition of environmental distress, he argues for this new ethic because humans are too distanced from their environs and thereby need the "subtle, intangible, but soul-deep mix of landscape, smells, sounds, history, neighbors, and friends that constitute a place, a homeland." From environmental writer Barry Lopez comes similar advice. "As deeply ingrained in the American psyche as the desire to conquer and control the land," Lopez reminds us, "is the desire to sojourn in it. . . . [But that] requires not only time but a kind of local expertise, an intimacy with place few of us ever develop." Lopez and Wilkinson echo the plea Aldo Leopold made more than four decades ago that Americans adopt a "land ethic" as a guide for all human activity and, more importantly, as an interpretive orientation toward their own homelands.

Wilkinson's proposal is part of our *fin de siècle* wrestling with what has happened to the American West in this century, especially the massive alterations to the landscape that include what Robert Pyle calls a "ravaged land" in the humid Pacific Northwest forests and the construction of what Donald Worster calls a "hydraulic empire" in the arid regions. This viewpoint is as much about what we have done to the land as about what the postindustrial culture may bring in the future. It is a perspective that stimulates activism and a demand for new ways of seeing place. Wes Jackson, the founder of the Land Institute in Salinas, Kansas, for example, has recently challenged university educators to teach students about the area where they live out their lives, about home, about place. "It has never been our national goal to become native to this place," Jackson explains. "It has never seemed necessary even to begin such a journey. And now, almost too late, we perceive its necessity."[6]

In historical studies, Wilkinson's and Jackson's suggestions have not been taken up with much enthusiasm, but a couple of exceptions in the recent literature come to mind. Dan Flores, in an important article on place that appears in this volume, advocates adopting a bioregional approach to environmental history, arguing that historians ought to "write sophisticated, deep-time, cross-cultural environmental histories of place." Bioregional divisions, such as watersheds and intermontane basins, can be studied in more depth than political or economic divisions, Flores maintains, because of the long-term relationships between human activity and the landscape that can be discovered, documented, and analyzed. The results of these bioregional studies might well disclose larger patterns of historical change.[7]

A recent analysis of environmental policy conflicts over the disposition of old-growth forests in the Pacific Northwest offers another example of a focus on place in environmental history. James Proctor's investigations have convinced him that competing ethics of place help explain the divergent perspectives expressed by political combatants in the old-growth controversy that divided so many communities in the Pacific Northwest during the 1990s. Participants in the dispute founded their ethic on "their lived experience in the region," on the intimate and powerful connections in their working lives that wedded them to place. This practical, utilitarian sense of place made a difference. It is unconvincing and patently wrong, Proctor argues, to "come up with some ethic in the abstract and then apply it to Pacific Northwest forests. This would be an ethic from nowhere."

What we need, he advises, is historical analysis that includes a "sensitivity to place" without being limited to place. "An ethics of place is not a spatially limited ethics," Proctor intones, "it is a spatially based ethics." To create an ethic that responds to issues at the center of the environmental debate over the fate of Pacific Northwest forests, Proctor argues, requires attention to "details of [the residents'] personal geographies" because they make a difference. "So do all the little details of the physical geography of the region: its natural and cultural fire history, climate, and vegetation changes over the last century and since the last glacial maximum. All of this helps us get a sense of place—a very practical sense we can weave our moral frameworks around."[8] Environmental historians should take notice. What Flores and Proctor seem to be saying is that *where* something happens makes as much difference as *what* happens.

It is common sense that knowledge about the location of environmental change is essential to describing and understanding it. But "sensitivity to place" is about something quite different. The focus is on the participants' perceptions and understandings of their surroundings and the role those perceptions and understandings play in human choices. Robert Pyle's recounting of the logging history of the Willapa Hills on the Lower Columbia River in Washington State is a case in point. In what Pyle calls "the sack of the forest," timber managers cut with abandon, leaving a wasteland in their wake. He objects to it all on behalf of the forest species obliterated in the operations, loggers who lost their livelihoods, and local communities that dissipated. "I am as abashed and offended by the callous crunching of these communities and their families, businesses, schools, and life-style as I am by the over-zealous savaging of a wood." It would not have happened, Pyle argues, if the managers had paid any attention to the ecological, social, and economic values of place. It is in this kind of commentary and analysis that we find the broadest range of possibilities and the greatest benefit for environmental historians in pursuing the meaning and importance of place.[9]

Among all social science and humanities disciplines, perhaps, geography has paid the most attention to the concepts of place and sense of place. Most historians, especially environmental historians, are familiar with the work of Carl Sauer and Donald Meinig. The American school of cultural geography Sauer created at the University of California at Berkeley during the 1920s held sway until the 1960s, and his work has had an enormous impact on how we understand cultural geography.[10] Sauer and his students investigated premodern cultures and offered explanations of their histories that distinguished between the origins of cultural orientations toward the environment and the diffusion of those ideas and activities to other places. In part Sauer and his protégés were reacting to the fallacies inherent in geographical determinism, which emphasized the directorial role of environmental conditions in human behavior and suggested that place determined social structure. Sauer's ideas also reflected his belief that although the physical environment stimulated cultural invention, it was not a constant casual force in human culture. Sauer pioneered geographers' analyses of the effects of European migration to North America and the specific environmental effects of Euro-American economic development of natural resources, a theme that has been adopted and brilliantly extended by Alfred Crosby in *The Columbian Exchange* and in *Ecological Imperialism*.[11]

Donald Meinig's focus has been on regional distinctiveness and the creation of regions in North America. His explanation for the evolution of cultural landscapes emphasizes the cultural contact and conflict that resulted in the disruption or destruction of earlier cultures and significant revisions in the relationships between human and nonhuman environments. It is a process that Meinig forcefully labels imperialistic. In this way, Meinig agrees with Crosby's interpretation of the dynamics and results of landscape change in North America. Meinig's macrogeographical analysis of the "shaping" of North America—now in three of four planned volumes—suggests a new casting of the story of American development that defines place, in part, as fundamental and generative in spatial relationships.[12]

Since the mid-1960s, cultural geographers have greatly expanded our understanding of landscapes, in some cases adding to the ideas of Sauer and Meinig, for example, and in other cases incorporating the findings of anthropologists, philosophers, historians, and natural scientists. In most of these new studies, time and space are assumed to be critical elements in understanding the meaning of place. Most cultural geographers argue that deciphering the meaning of place requires investigating its physical dimensions (which includes scientific description), analyzing the correspondences of its parts or regions, interpreting its symbolic representations, and reflecting on its history. "To view the landscape historically," Michael Conzen writes, "is to acknowledge its cumulative character; to acknowledge that nature, symbolism and design are not static elements of the human record but change with historical experience; and to acknowledge too that the geographically distinct quality of places is a product of the selective addition and survival over time of each new set of forms peculiar to that region or locality."[13]

In the Pacific Northwest, for example, histories of places that have long had environmental significance, such as Celilo Falls and Kettle Falls on the Columbia River, should include discussion and analysis of their ecological characteristics, the native fisheries that thrived there, and their destruction as a result of damming the river. As Conzen suggests, the historical dynamic in these places is found partly in the changing relationships between human communities—native fishers to dam builders to irrigators who tapped impounded river water—and the falls. Once described in native oral stories as places where Coyote brought salmon for human use, Celilo and Kettle Falls were places of immense cultural and economic power for native people, where intertribal places connected people with a lived space, where their fishing and other activities fit into a larger explanation of life and the meanings behind physical locations. After non-Indian cultures invaded the land and oral stories became replaced with written accounts, the places themselves, as David Abram argues, "often come to seem wholly incidental to the tales, the wholly arbitrary backdrops for human events that might just as easily have happened elsewhere." The non-Indians had no time for oral stories; they focused on the falls as impediments to transportation and economic intercourse. They were places that diminished environmental benefits, and their elimination was seen as an improvement on nature. The falls, as geographer Rob Shields explains, had become a "place-myth" that had been "displaced by radical changes in the nature of [the] place . . . becoming 'dead metaphors,' while others [were] invented, disseminated, and became accepted in common parlance." This viewpoint argues that understanding Celilo Falls, for example, requires that we connect the physical changes at the falls with the cultural descriptions of place expressed by native and non-native, before and after the dramatic change in 1957, when the waters backed up by The Dalles dam inundated Celilo Falls.[14]

Cultural geographers share with historians a focus on time and change as critical elements of the story to be told. Donald Parkes and Nigel Thrift label their approach a "chronogeographic" view of landscapes that makes "movement and place" antithetical, where "place [is] time made visible." And we can best understand "place as a pause in movement." It is the effect of change on a specific landscape, the geographers argue, that explains attachment to place by identified populations and offers a key to the components of meaning attached to places. In geographer Yi-fu Tuan's explanation, the passage of time "allows a location to become a center of meaning."[15]

What can environmental historians use in these theoretical propositions posed in the new cultural geography? Three important questions cultural geographers offer can be synthesized as the following: How have we perceived place? What has place symbolized in our culture? And what have we accepted and discarded about earlier meanings of place? Asking these questions is important to our investigation of environmental change for several reasons. First, knowing the context and content of descriptions of place makes it less likely that we will misunderstand human actions and reactions and their relationship to environmental changes. Second, exploring the dimensions of symbolic cultural meaning vested in places reminds us that the physical environment is often understood best as a symbol that represents cultural values and perceptions invested in a place. Third, determining which meanings of place have been repeated, revised, or replaced over time can help explain the relationships between human desires for place and the changes that have occurred there.

In the case of Celilo Falls, knowing that specific fishing spots used by native fishers carried names that described physical characteristics of each location, referred to their mythical origins, and identified them as owned or controlled by specific families expands our understanding of the dynamics between the environmental, economic, and cultural realities of that place. The names inform us that these fishing spots inherently include more meaning than simply the economics of salmon fishing. Although the falls are under the impounded waters of Lake Celilo, the reservoir created by The Dalles Dam in 1957, the importance of the falls as a place in native culture remains strong, in part because it is symbolic of an ecological relationship with the Columbia River that is much diminished for native people and in part because the inundation of the falls is symbolic of modern engineering's power to reshape the physical environment. The meaning of Celilo Falls among native and non-native people in the region includes the changing relationships created by dam-building on the river and the privileging of electrical power generation and navigation over fishing.

For geographers and historians, as Nicholas Entrikin explains, digging out the meaning requires imagination and attention to ambiguities that are inherent in representing what is important about places. "We know that places differ," Entrikin writes, "and that these differences are not imaginary, but rather are actual features of the world. We also suggest that these differences matter, and we self-consciously employ this knowledge in our everyday lives.[16]

We know that places where important events took place, for example, are locations where human experience became directly connected to physical space—at a mountain pass, a river landing, a road junction—and when we ask questions about these happening-places we include much of what we have learned and thought about the events that transpired there. We also have questions about the relationships of the place to living people and to events that continue to take place there. Explaining significant threads of memory and meaning that are attached to that place requires identifying the characteristic and idiosyncratic relationships between human activity and place that have occurred over time and analyzing how they affected human decisions and potentially shaped subsequent events. Entrikin's caveat boils down to developing an intelligent awareness that places are receptors for many layers of human meaning and that all one-dimensional representations of place are incomplete and surely wrong.

Among cultural geographers who have thought and written about these interpretive challenges, perhaps no one has had more influence on how we perceive place than Yi-fu Tuan. In *Man and Nature* (1971), *Topophilia* (1974), *Space and Place* (1977), and *Landscapes of Fear* (1979), Tuan explores the idea of place understood as a dimension of human perception. We see place, Tuan emphasizes in *Man and Nature*, through a "screen of words" that transmogrifies its physical properties to fit human needs and convey meaning.[17] Foremost among cultural geographers, Tuan has emphasized that perspective makes the difference. He uses two broad categories to explain the centrality of viewpoint in understanding place, especially in comprehending how humans have experienced seeing places. For Tuan, place "is an ordering of reality from different angles," which he categorizes as the "vertical view" and the "side view."

> The vertical view sees landscape as domain, a work unity, or a natural system necessary to human livelihood in particular and to organic life in general; the side view sees landscape as space in which people act, or as scenery for people to contemplate. The vertical view is, as it were, objective and calculating. . . . The side view, in contrast, is personal, moral, and aesthetic. A person is in the landscape, working in the field, or looking [at it] . . . from a particular spot and not from an abstract point in space.[18]

The implications for historians are significant. The suggestion is that both external and internal viewpoints are required to fathom place. Measurements and descriptions of the ecological effects of change, for example, have to be merged with viewpoints that represent the internal human experience. This is the viewpoint Proctor suggests in his study of the Pacific Northwest old-growth controversy. The viewpoint is the "betweenness of place" that Entrikin explains as an understanding that is "between" a "centered" place—the subjective and experiential descriptions—and a "decentered" place—the objective and measured descriptions.

This merging or spanning of aesthetic and objective views, Tuan and others argue, is the essence of understanding place. To see place properly, functional and moral-aesthetic data "must be conjoined through an imaginary effort." It makes a profound difference, then, whether the perceiver is native to the place or a visitor, whether the describer sees the place as comparable or contrastable to her or his native place (a forest dweller seeing the plains, for example), or whether the image of the place is seen as bucolic, fearsome, home, or alien. E. V. Walter distinguishes among healthy places, sick places, profane places, and sacred places, arguing that "the nature *of* a place . . . depends on the qualities and distribution of nature *in* a place." The "sick places" in Walter's geography are localities that literally harbor disease, where sickness caused changes in human activity; the "sacred places" evoke profound spiritual reactions. The implication is that places are lived in and they affect people in myriad and powerful ways. "Extended residence in a place," folklorist Kent Ryoden writes, "tends to make us feel toward it almost as toward a living thing, for better or worse . . . to love it or hate it for its nurturing comforts or alienating discomforts." Because participants in environmental change have such perspectives on place, environmental historians need to understand their similarities and dissimilarities if they expect to penetrate the "screen of words."[19]

People might view Celilo Falls, to use our Columbia River example, as an economic, a sacred, or a home place. Their responses to proposed changes to the dam, to fishing restrictions, or to a drawdown of a reservoir can be understood as expressions of economic, social, or political interest, but they are also expressions about place and the sense of place the participants carry with them, day in and day out. If we fail to fully grasp their perspective of place, we may well misunderstand their actions or miss entirely an action taken that did not seem to express their political or economic interests.

More than an angle of vision on a locality, the meaning humans vest in place is deeply embedded in lived experience and the "feel" of locality. It is a meaning psychologist James Gibson calls the "haptic" sense or the power of the physical touch of the place. The intensity of expression about place, especially on contested environmental terrains, can be better understood by environmental historians if the haptic response is taken into account. Put another way, Abram describes the phenomenon that is "evident to our senses"—the "feel" of the river, plateau, or other landform—as "manifested." But sensations that are not quite evident, that are developing, he calls "manifesting." These are the intangible quotients, for example, that contribute to the intoned defense of "wild salmon" on the Columbia River, where dams and ecological disturbance have decimated native anadromous fish runs since 1900. Advocates of extreme measures to protect salmon often express as much about their "feel" of the river as place as they do about scientific measurements of the fate of salmon.[20]

Asking the question of meaning about place would be a much less complicated exercise for environmental historians if each individual, interest group, or community saw place with any uniformity. But it is profoundly otherwise, for it is one of the complexities and paradoxes of place that it integrates contending and disparate forces as perhaps no other factor in environmental studies. It resists reduction. As geographer Robert Sack explains, the forces inherent in nature, culture, and social relations converge on place, making "wilderness," "ghetto," and "farm" locations where natural forces mix dynamically with social and cultural forces to create distinctive and recognizable places, ones that carry with them a characteristic form and generate specific expectations about their contents. These forces are often in strident conflict, and the meanings of places that are described as "natural" or "industrial" are usually multiple and composed of mixtures of opposites and contradictions. "There is not a single inner force directing and coordinating [it] all," geographer Edward Relph cautions, "yet it seems as though there is an individuality which lies behind the forms and appearances and maintains a coherent identity. We know that [identity as] the spirit of place." The singular expression, in other words, includes a protean mixture that should make historians wary of carrying away simplistic versions of place or accepting descriptions that reveal a singular face.[21]

Nonetheless, "places and their contents," Nicholas Entrikin reminds us, "are seen as wholes" in a very special way. The "wholes" are essentially symbolic. Place symbolizes a range of reactions to location that include aesthetic, moral, or religious significances. A congeries of representations, from characterizations that isolate spiritual qualities in a place to descriptions that emphasize beauty, are combined to create a singular image of place. Celilo

Falls is one place, but its characterizing images include coyote tales, the power of falling water, and the fecundity represented by annually returning salmon. This symbolism, some geographers argue, emerges directly from human reactions to place. "We personify the environment or employ anthropomorphic descriptions," Douglas Pocock explains, and in that way "we are all animists to a degree."[22]

Place also becomes the reification of larger and more powerful forces and ideas, because we read "the natural world," Denis Cosgrove maintains, "simultaneously as [a] thing and as [a] sign of meaning beyond itself." Put another way, humans create mythic places to mark and signify the most important aspects of their lives. "Over and over," David Sopher writes, "we find that rather than being the generator of people, land at one scale or another serves as the chosen symbol of a people's being." The symbols range from cosmic to religious to oneiric to poetic. Singular items in place—built structures such as the Eiffel tower or Seattle's Space Needle or natural features such as an old-growth grove or a waterfall—often become symbolic and expand the power and presence of specific locations. Celilo Falls on the Columbia, even as it exists under the reservoir's waters, is symbolic of a complex Indian culture and history as it also is symbolic of the transformative power of modern engineering. Environmental historians miss the power and culturally generative importance of such sites, as geographer Alan Baker puts it, if they fail to understand that "all landscapes are symbolic in practice."[23]

These symbols are often embedded in a people's sense of identity. They become integrated images that represent place and can define a population and its intentions at the same time. As describers of the environment, these symbolic images can be very powerful. Parks, wild areas, and firescapes in the natural world and factories, dams, skyscrapers in the built world come to mind.

In the Northwest, the Columbia River gorge is a dynamic example. The creation of geological and hydraulic natural forces, the gorge incorporates a range of ecological and social areas—forested wilderness and scenic waterfalls cheek by jowl with towns and businesses—but the larger symbolic image is expressive of natural power, natural resource wealth, beauty and aesthetics, and human history. The changes that have taken place there, especially the channelization and damming of the river, have pitted utopian and dystopian perceptions of place against each other. The utopian symbol prevailed for dam builders and planners, as Richard White argues, because they dreamed of a decentralized industrial world that would create "living communities instead of the cancerous growth of the metropolis" in the

Columbia River basin. For the native fishers, the dystopian symbol dominated. As Umatilla tribal leader Antone Minthorn argues, a river without fish is an injury to Indian people, for "what was done to the water was done to the Indian." The destruction of the fishery was the destruction of cultural space. Different uses of the river preferred by engineers and fishers, in other words, involve distinct, symbolic images of place. To understand the division of interest in damming the Columbia in the gorge as primarily an economic issue is, in short, to miss the power and symbolism of place as a force in environmental choice.[24]

Part of the place-image power of the Columbia River gorge is its function as a landscape of home. Homelands and places of long-term residency are powerful because they integrate time, place, and history. Tuan observes, "Landscape is personal and tribal history made visible." Home is a symbolic understanding of place that incorporates personal and group identity. It is the product of narratives about experience and location that often emphasize a distinctive history. Stories that have been retained dominate in the social construction of home as place. "Social creation of a mythic home," David Sopher explains, "seeks consciously to play up the uniqueness of place by accenting small distinctions in the landscape, by modifying it idiosyncratically, or by instituting in it a code of local signature."[25] The community's consensus on home as place is the product of history understood as tradition. These traditions are a "community of memory," or the embodiment, as Entrikin puts it, of "a constant examination of what constitutes the good life." It is a tie to place that is meant to be shared by the community, and its acceptance is central to becoming part of the place. "Such a cultural core," Entrikin emphasizes, "provides the basis for the shared sense of sacred space."[26]

The role of memory in creating tradition and the constant revision of tradition link time and place. What is retained about the past as connected to place and what is rejected directly shapes what the "community of memory" includes and means. Places change with time and human experience. In the study of environmental change, there is often no more profound and ineluctable factor than disputes between "insiders" and "outsiders," between populations who invade or disturb and populations who resist and persevere, between those engaged with the landscape and those who claim the landscape.[27] Knowing what changes take place, how fast the landscape is altered, and what kind of "home" is retained after the transformations is central to understanding the history of environmental change.

What happens to place and what constitutes our sense of place is integral to our relationship with the environment. Place is a constructed complex of ideas, memory, and imagination that represents a physical landscape. Environmental historians can acknowledge the cultural dimensions of their subjects, include the viewpoints that Tuan and others suggest are part of our understanding of physical places, and investigate Entrikin's "betweenness of place" that links the objective descriptions to the subjective ones. The analysis of human choices on the land and about the environment can be greatly expanded and enhanced through a broader and more powerful appreciation of place. The way we understand it inclines our reading of yesterday's landscapes and the lives of participants. Eudora Welty's instruction to "know where you stand" before you "judge where you are" reminds us that knowing the place must come before interpreting our environmental past.

NOTES

1. Eudora Welty, *The Eye of the Storm: Selected Essays and Reviews* (New York: Random House, 1970), p. 128.

2. E. V. Walter, *Placeways: A Theory of the Human Environment* (Chapel Hill: University of North Carolina Press, 1988), p. 21.

3. For examples, see Fritz Steele, *The Sense of Place* (Boston: CBI Publishing, 1981); Winifred Gallagher, *The Power of Place: How Our Surroundings Shape Our Thoughts, Emotions, and Actions* (New York: Poseidon Press, 1993); Frederick Turner, *Spirit of Place: The Making of an American Literary Landscape* (Washington, D.C.: Island Press, 1989); John R. Stilgoe, *Common Landscape of America, 1580–1845* (New Haven: Yale University Press, 1982). For the idea of place in folklore studies, see *Sense of Place: American Regional Cultures*, eds. Barbara Allen and Thomas Schlereth (Lexington: University of Kentucky Press, 1990).

4. See, e.g., Wendell Berry, *Home Economics: Fourteen Essays* (San Francisco: North Point Press, 1987); Wallace Stegner, *Wolf Willow* (New York: Random House, 1962); Terry Tempest Williams, *Refuge: An Unnatural History of Family and Place* (New York: Pantheon, 1991); Barry Lopez, *Arctic Dreams: Imagination and Desire in a Northern Landscape* (New York: Scribners, 1986); William Kittredge, *Owning It All: Essays* (St. Paul: Graywolf Press, 1989); Mary Clearman Blew, *All but the Waltz: A Memoir of Five Generations in the Life of a Montana Family* (New York: Viking, 1990); Ivan Doig, *This House of Sky: Landscapes of a Western Mind* (New York: Harcourt, Brace, Jovanovich, 1979); Craig Lesley, *River Song* (Boston: Hough-

ton Mifflin, 1989); William Stafford, *Stories That Could Be True: New and Collected Poems* (New York: Harper and Row, 1977); Elizabeth Woody, *Luminaries of the Humble* (Tucson: University of Arizona Press, 1992).

5. Frederick Turner, *Beyond Geography: The Western Spirit against the Wilderness* (New Brunswick, N.J.: Rutgers University Press, 1983); William Cronon, *Nature's Metropolis: Chicago and the Great West* (New York: W. W. Norton, 1991); Donald Worster, *Dust Bowl: The Southern Plains in the 1930s* (New York: Oxford University Press, 1979); Richard White, *Land Use, Environment, and Social Change: The Shaping of Island County, Washington* (Seattle: University of Washington Press, 1980).

6. Charles F. Wilkinson, *The Eagle Bird: Mapping a New West* (New York: Pantheon, 1992), p. 137; Barry Lopez, "The American Geographies," in *Finding Home*, ed. Peter Sauer (Boston: Beacon Press, 1992), p. 118; Donald Worster, *Rivers of Empire: Water, Aridity, and the Growth of the American West* (New York: Oxford University Press, 1985); Robert Michael Pyle, *Wintergreen: Rambles in a Ravaged Land* (New York: Scribners, 1986); Wes Jackson, *Becoming Native to This Place* (Lexington: University of Kentucky Press, 1994), pp. 2–3.

7. Dan Flores, "Place: An Argument for Bioregional History," *Environmental History Review* 18 (Winter 1994): 1–18, p. 14 (reprinted in a revised version in this collection). For other discussions of bioregionalism, see Donald Alexander, "Bioregionalism: Science or Sensibility?" *Environmental Ethics* 12 (Summer 1990): 161–74; Kirkpatrick Sale, *Dwellers in the Land: The Bioregional Vision* (San Francisco: Sierra Club, 1985).

8. James D. Proctor, "Whose Nature? The Contested Moral Terrain of Ancient Forests," in *Uncommon Ground: Toward Reinventing Nature*, ed. William Cronon (New York: W. W. Norton, 1995), p. 294. See also Yi-fu Tuan, "Attention: Moral-Cognitive Geography," *Journal of Geography* 68 (1983): 11–13.

9. Pyle, *Wintergreen*, p. 167.

10. For works by and about Sauer, see Carl Sauer, *Agricultural Origins and Dispersals: The Domestication of Animals and Foodstuffs* (New York: Columbia University Press, 1952); Carl Sauer, "The Agency of Man on Earth," in *Man's Role in Changing the Face of the Earth*, ed. W. L. Thomas (Chicago: University of Chicago Press, 1956); and J. Nicholas Entrikin, "Carl Sauer: Philosopher in Spite of Himself," *Geographical Review* 74 (Fall 1984): 387–408; Martin Kenzer, ed., *Carl O. Sauer: A Tribute* (Corvallis: Oregon State University Press, 1987).

11. Alfred Crosby, *The Columbian Exchange: Biological and Cultural Consequences of 1492* (New York: Greenwood, 1972); Alfred Crosby, *Ecological Imperialism: The Biological Expansion of Europe, 900–1900* (Cambridge: Cambridge University Press, 1992). See also Alfred Crosby, *Germs, Seeds, and Animals: Studies in Ecological History* (Landsdowne, Md.: M. E. Sharpe, 1995).

12. For representative works by Meinig, see Donald W. Meinig, *The Shaping of America: A Geographical Perspective on 500 Years of History* (New Haven: Yale University Press, 1986, 1993, 1998), and *The Interpretation of Ordinary Landscapes: Geographical Essays*, ed. Donald W. Meinig (New York: Oxford University Press, 1979).

13. Michael P. Conzen, ed., *The Making of the American Landscape* (Boston: Unwin Hyman, 1990), p. 4, as quoted in Alan R. H. Baker, "Introduction: On Ideology and Landscape," in *Ideology and Landscape in Historical Perspective: Essays on the Meanings of Some Places in the Past*, eds. Alan R. H. Baker and Gideon Biger (New York: Cambridge University Press, 1992), p. 7.

14. David Abram, *The Spell of the Sensuous: Perception and Language in a More-than-Human World* (New York: Pantheon, 1996), p. 182; Rob Shields, *Places on the Margin: Alternative Geographies of Modernity* (London: Routledge, 1991), pp. 60–61.

15. Don Parkes and Nigel Thrift, *Times, Spaces, and Places: A Chronogeographic Perspective* (New York: John Wiley and Sons, 1980), p. 28; Yi-fu Tuan, *Topophilia: A Study of Environmental Perception, Attitudes, and Values* (Englewood Cliffs, N.J.: Prentice-Hall, 1974), p. 20. On the power of perspective, see also Robert Lloyd, *Spatial Cognition: Geographic Environments* (Boston: Kluwer Academic Publishers, 1997), and R. Golledge and R. Stimson, *Analytical Behavioural Geography* (London: Croom and Helm, 1987).

16. J. Nicholas Entrikin, *The Betweenness of Place: Towards a Geography of Modernity* (Baltimore: Johns Hopkins University Press, 1991), p. 13.

17. Yi-Fu Tuan, *Man and Nature* (Washington, D.C.: Geographical Review, 1971), p. 4; Tuan, *Topophilia*, pp. 60–63.

18. Yi-Fu Tuan, "Thought and Landscape: The Eye and the Mind's Eye," in Meinig, *The Interpretation of Ordinary Landscapes*, (pp. 89–90.

19. Walter, *Placeways*, p. 44; Yi-Fu Tuan, *Landscapes of Fear* (New York: Pantheon Books, 1979), p. 6; Entrikin, *Betweenness of Place*, p. 134; Kent Ryden, *Mapping the Invisible Landscape: Folklore, Writing, and the Sense of Place* (Iowa City: University of Iowa Press, 1993), pp. 65–66. For another viewpoint on demarcated places, see Robert David Sack, *Human Territoriality: Its Theory and History* (Cambridge: Cambridge University Press, 1986).

20. James Gibson, *The Senses Considered as Perceptual Systems* (Boston: Houghton Mifflin, 1966), pp. 97–100; Abram, *Spell of the Sensuous*, p. 192.

21. Robert David Sack, *Place, Modernity, and the Consumer's World: A Relational Framework for Geographical Analysis* (Baltimore: Johns Hopkins University Press, 1992), pp. 15–16; Edward Relph, *Rational Landscapes and Humanistic Geography* (London: Croom Helm, 1981), pp. 173–74. For a recent example of the complexities of place in an environmental history, see Robert Sullivan, *The Meadowlands: Wilderness Adventures at the Edge of a City* (New York: Scribner, 1998). For a theoretical

framework on place, see Robert David Sack, *Homo Geographicus: A Framework for Awareness and Moral Action* (Baltimore: Johns Hopkins University Press, 1997).

22. Entrikin, *Betweenness of Place*, p. 11; Douglas Pocock, "Humankind-Environment: Musings on the Role of the Hyphen," in *The Behavioural Environment: Essays in Reflection, Application, and Re-evaluation*, eds. Frederick W. Boal and David Livingstone (London: Routledge, 1989), pp. 84–85. See also Akira Y. Yamamoto, *Culture Spaces in Everyday Life: An Anthropology of Common Sense Knowledge* (Lawrence: University of Kansas Press, 1979), p. 10.

23. Denis Cosgrove, "Worlds of Meaning," in *Re-Reading Cultural Geography*, eds. Kenneth E. Foote, Peter J. Hugill, and Kent Mathewson (Austin: University of Texas Press, 1994), p. 389; David E. Sopher, "The Landscape of Home: Myth, Experience, Social Meaning," in Meinig, *The Interpretation of Ordinary Landscapes*, p. 133; Baker, "On Ideology and Landscape," p. 9.

24. Richard White, *The Organic Machine: The Remaking of the Columbia River* (New York: Hill and Wang, 1995), p. 65. For Antone Minthorn's comments, see Joseph Cone, *A Common Fate: Endangered Salmon and the People of the Pacific Northwest* (New York: Henry Holt, 1995), pp. 231–36. On the Columbia River gorge, see Carl Abbott, Sy Adler, and Margery Post Abbott, *Planning a New West: The Columbia River Gorge National Scenic Act* (Corvallis: Oregon State University Press, 1997).

25. Sopher, "The Landscape of Home," p. 138.

26. Entrikin, *Betweenness of Place*, p. 66.

27. For a discussion of this kind of change, see Richard White, "Are You an Environmentalist or Do You Work for a Living?: Work and Nature," in *Uncommon Ground*, pp. 171–85.

5 / Willamette Eden

The Ambiguous Legacy

WILLIAM G. ROBBINS

"More than we ever believe," the British writer Raymond Williams remarked, "we understand life from where we are."[1] We use material from our immediate social experience as the groundwork for our larger stories. In *Landscape and Memory*, cultural historian Simon Schama emphasized a point that scholars have argued since the works of Carl O. Sauer and Henry Nash Smith: "The landscapes that we propose to be most free of our culture may turn out to be its product."[2] I begin by citing Williams and Schama to argue that our perceptions of human cultures, of time and place, are intertwined in an intricate choreography involving memory, experience, and understanding.

That is certainly true of the narrative constructions that have centered on Oregon's Willamette Valley over the last 150 years or so. The intruders who came to dominate the valley by the mid-nineteenth century carried with them powerful cultural prescriptions about the proper ordering of their new surroundings. Those mostly white newcomers who entered the Willamette Valley in increasing numbers after 1830 imposed on their adopted homeland a new language, a discourse steeped in the politics of power, that defined one people's relations with others, vested landscapes with special meaning and purpose, and carried with it heavy suggestions about transforming the land itself. The collective force of that new cultural vision introduced new bounds of reckoning with the landscape, new perceptions about place, and a literary tradition that emphasized optimism and the prospects for human betterment.

We should be cautious, however, about accepting that story at face value as a straight-line narrative of success, because there are complexities and ambiguities in its telling. Indeed, those myth-bearing Interstate 5 signposts south of Portland that read "The Eden at the End of the Trail" carry powerful hidden meanings. They suggest a single-dimensional, selective memory and fail to acknowledge that the celebrated immigrant trek to the Oregon country held tragic consequences for native people. One people's Eden, in that sense, was another's tragedy, and some would say that the consequences of the newcomers' heroic odyssey was to despoil Eden itself.

At the westernmost end of their journey, Lewis and Clark and the Army Corps of Discovery wintered amid the towering fir trees, marshlands, and what they deemed an inhospitable climate on the lower Columbia River. Through fog, wind, and storm, the Corps suffered a season of mildewed clothing, spoiled meat, and mind-numbing boredom. But the captains' gloomy stories about the land west of the Cascade Mountains stand in sharp contrast to the literature left by the successive waves of American venturers who followed them. Later comers to the Oregon country, most of them heading for the green valley of the Willamette, were ecstatic about the mild climate, long growing season, and fertile soil, the Edenic qualities that by the 1840s made Oregon, in the words of historian James Ronda, "the fabled land of promise." For the ambitious and those who dreamed of Elysian fields, Ronda points out, Oregon was a mythical place with the potential "to fulfill the dreams and calculations of every prophet, promoter, and visionary."[3]

Those early euphoric descriptions of the valley, I should add, also fit nicely with the mythical representations that one finds in the writings of the distinguished American historian Frederick Jackson Turner. Turner, whose most productive years embraced the period from 1890 to 1920, is arguably America's most celebrated theorist of material, cultural, and political progress. In Turner's stories an escalating progression from primitive beginnings to great achievements characterized the westering movement of white Americans. The representations of progress in his writings were literally ubiquitous. Hunting for subsistence needs, trapping for fur-bearing animals, and life centered around the rough-hewn log cabin quickly gave way to frame buildings and, in the case of the Willamette Valley, to fenced and ditched prairie land and fields of wheat. The plowed field, the church, the schoolhouse, the stage-coach, and the railroad were further signs of forward movement.[4]

The tall, bespectacled writer-politician Richard Neuberger captured the mood of that mythical Turnerian world for the Pacific Northwest in his

1938 book, *Our Promised Land:* "The farms were lonely and far between, but they were farms and not trappers' bivouacs. The settlements were small and isolated, but they were actually settlements and no longer trading posts. The plow and hoe were superseding the trap and rifle. To the farms there was some permanence. They might last only a year or so. . . . Yet these shacks on their plots of crudely cultivated ground represented the genesis of a civilization."[5]

Neuberger, of course, was doing little more than echoing the voices of nineteenth-century narratives that emphasized the natural abundance of the western valleys and the notion that white settlers were in the process of building a civilization and laying the groundwork for progress. A mid-century Vancouver, Washington, newspaper saw the hand of Providence at work: "The western coast is to be peopled; the treasures of her forests, her rivers, her rich soils, are to be developed for the expansion of civilization." In an 1850 article the *Oregon Spectator*, the first newspaper published in the region, put progress and civilization in a racial cast: "The Indian retreats before the march of civilization and American enterprise; the howling wilderness is fast becoming fruitful fields." One year later the reverend Gustavus Hines viewed the dramatically declining Indian population through a similarly focused ethnocentric lens: "The hand of Providence is removing them to give place to a people more worthy of this beautiful and fertile country." And A. W. Campbell, speaking at the opening of Oregon State Normal School in Monmouth in 1883, contrasted the Indian and the Anglo Saxon and remarked that the wigwam always retreated before the American schoolhouse.[6]

A careful reading of early journals and diaries makes it clear that white settlers in the Willamette Valley firmly believed they represented the rising trajectory of those stories. At the onset of the twentieth century those whose adulthood bridged the previous fifty years recalled the Indian-shaped environment of the 1840s: the autumn burning of the prairies; the meandering, debris-filled rivers; and the seasonal floods that turned the landscape into a vast wetlands. And when they recounted those experiences, they expressed great pride in what they had accomplished. No matter how exaggerated their achievements, historian Richard White points out, "their story became the American story." To the writer Dayton Duncan that form of national narrative has embraced elements of determination, of movement, of prospects, of getting there, "the quest for the unreachable." In the name of progress, Duncan argues—to obtain free land, to gain access to grazing and mining areas, to extend the bounds of an expanding nation—the intruding whites displaced the First Americans and confined them to colonial reserves in their once vast homelands.[7]

Although the outlines of that story are familiar, the details become less clear, even ambiguous, the closer we move to the present. For the Willamette Valley, we now recognize that the human experience has been more complex than an unmitigated tale of achievement and success. Indeed, we recognize today that there were (and are) multiple stories to be told about the groups who have inhabited the valley landscape. For the greater West, historian Elliott West urges us to look beyond the "pioneer experience" that has dominated the telling of the region's history to a more complex view of the western story, "its severe limitations and continuing conflicts, its ambivalence, and its often bewildering diversity."[8]

If we move past the heroic, romantic blush of the conventional accounts about the Oregon country, we recognize that stories of human tragedy preceded the triumphal accounts of those who traveled the Oregon Trail in the 1840s and 1850s. I refer to the epidemic contagions that rearranged the human geography of the lower Columbia River country. Elliott West refers to the catastrophic native population decline across North America as "the greatest die-off in the human record."[9] By the time a series of seasonally recurring malaria outbreaks had run their course in the early 1830s, the infections had contributed to an Indian mortality rate in the neighborhood of 90 percent. Anthropologist Robert Boyd estimates a pre-epidemic native population for the lower Columbia River and the Willamette Valley of about 14,000 people. The world of the Great River of the West was changing. When he visited the region in 1841, Lieutenant Charles Wilkes counted 575 Chinook survivors on the Columbia and 600 Kalapuyans in the Willamette Valley.[10]

In retrospect, we can say with some certainty that epidemic diseases had thinned the Indian population at a critical period in the post-Columbian history of the Oregon country. Sometime during the mid-1830s the region passed a demographic, cultural, and ecological divide of sorts, one that was clearly apparent to contemporary observers. Writing to the governors of the Hudson's Bay Company in a lengthy memorandum on November 18, 1834, chief factor John McLoughlin reported the recent arrival at Fort Vancouver of the Boston merchant Nathaniel Wyeth. On the following day, two Methodist missionaries, Jason Lee and his nephew Daniel Lee, and two naturalists, Thomas Nuttall and John Kirk Townsend, also called at the fort.[11] But the growing number of American citizens on the move to the Oregon country during the 1830s was little more than the advance guard in what soon would become a floodtide of overland immigrants pouring into the Willamette Valley. And through that critical period and after, the mortality and fertility rates and reproductive capacity of the invading groups were moving in the opposite direction from that of native people.

There are several explanations for the large-scale movement of Americans to the Oregon country in the 1840s—cultural, political, and economic. Although fur trappers and others who passed through the Willamette Valley provided promising accounts about the region's commercial prospects, it was left to others with wider-ranging and more inventive minds to embellish the region's Edenic qualities and carry that message to receptive audiences. Of those Oregon publicists, the most persistent—and some would say unmoored from reality—was Harvard-educated Hall Jackson Kelly. One day in 1817, according to his own account, Kelley read the journals of Lewis and Clark and immediately became obsessed with the Far West. It was in that year, he said, that "the word came expressly to me to go and labor in the field of philanthropic enterprise and promote the propagation of Christianity in the dark and cruel places about the shores of the Pacific."[12]

Kelley subsequently organized the American Society for the Encouraging of the Settlement of the Oregon Territory in 1824; he petitioned Congress to promote colonization in 1828; and two years later he published his famous pamphlet *A Geographical Sketch of That Part of North America Called Oregon*. Plagiarizing from earlier journals and using his imagination "to fill in the gaps," Kelley crafted a glowing description of a "New Eden." The mountains in Oregon were "peculiarly sublime and conspicuous"; the climate was "salubrious." Oregon was "well watered, nourished by a rich soil, and warmed by a congenial heat," and it was "exactly accommodated to the interests of its future cultivators." Unhinged from reality as he was, Kelley—who once referred to the Hudson's Bay Company as a "monstrous evil"—eventually traveled to the Northwest and visited Fort Vancouver. For explanations that reasonable people would understand, Kelley was kept at a distance and departed without ever seeing the inside of the fort.[13]

But the rhetorical embellishments of missionaries, politicians, and publicists like Kelley set in motion a great demographic movement that ultimately contributed to the disruption of existing human *and* ecological communities in the Oregon country, and first and foremost in the Willamette Valley. To explain those connections, historian Alfred Crosby uses the metaphor of folklore—the delightful old song about Sweet Betsy from Pike County, Missouri, who crosses the western mountains "with her lover, Ike, with two yoke of oxen, a large yellow dog, a tall Shanghai rooster, and one spotted hog." The symbolism in Betsy's party, Crosby points out, is the host of colonizing species in her entourage (mixed in gender, of course!). "Betsy came not as an individual immigrant," he points out, "but as part of a grunting, lowing, neighing, crowing, chirping, snarling, buzzing, self-replicating and world-altering avalanche."[14]

In the conventional histories of the Oregon country, heroic men and women built a civilization from a primitive but abundant landscape. Those narrative accounts begin with sawmills, gristmills, and increased acreages of "improved land," and then the story moves to the region's first sizable white American settlement at Oregon City. Sawmills and gristmills, of course, are mechanisms for turning trees into lumber and wheat and other grains into flour, products to be sold at market. As such, they imply varying degrees of ecological disturbance to places far beyond the sites where logs were sawed into lumber and wheat was ground into flour.

To the recently arrived white settlers, the new valley towns and the changes taking place on the prairies and elsewhere were signs of advance and forward movement. In that sense, progress was the antithesis of the status quo, the enemy of the Indian world, environments that whites defined as primitive, pristine, and natural. "White settlement," Richard White observed more than a decade ago, "destroyed the Indian Northwest."[15] But to a people with acquisitive instincts and a talent for turning the bounty of new places to personal advantage, the natural world represented opportunity, a second chance, a place to accomplish great things.

The essence of the newly emerging demographic, cultural, and ecological makeup of the Willamette Valley at mid-century is reflected to a considerable degree in statistics: the increasing number of immigrants, the acreages of improved land, the productions of the soil, and the growing herds of livestock being grazed on the prairies and foothills. To protect their domestic animals and crops, settlers waged wholesale campaigns to exterminate wolves, cougar, bear, and elk. Killing wolves, of course, was a sign of progress, a symbol that served to extend the bounds of civilized space in the valley. An optimistic and confident lot, settlers saw everything to gain and nothing to fear from those activities.

In an essay published in the *Oregon Historical Quarterly* early in this century, James R. Robertson claimed that Oregon had advanced from fur trading, an industry adapted to the support of a small population, toward "forms of social life that were better and higher." That was a remarkable accomplishment, a "work of heroism . . . as great as anything ever done." The migration of 1843 initiated the agricultural stage of Oregon's industrial life, he contended, and then with the California gold rush and the opening up of commercial markets, it shifted to a more mature stage. Robertson's use of an evolutionary metaphor to describe agricultural progress was consistent with the popular view that measured human accomplishment in terms of plowed fields, cleared and fenced pastureland, growing trading and commercial centers, and improved overland and water transport systems.[16]

The appearance in 1869 of *Willamette Farmer*, a weekly newspaper devoted to agriculture, signaled to a growing audience that land was little more than a commodity, a vehicle of production for the valley's expanding market output. Farmers in the valley also organized agricultural societies to disseminate up-to-date information about wheat culture, animal husbandry, horticultural advances, and a variety of marketing news. Those organizations also suggest more fundamental changes that were taking place as an increasing volume of wheat moved toward the warehouses and shipping facilities at Portland. Historian Dorothy Johansen eloquently underscored the city's influence in the late nineteenth century: "Portland has been appropriately described as the 'city that gravity built.' Down the Willamette flowed the produce of the Valley; down the Columbia came the immigrants and the riches of Idaho mines and the interior's wheat fields. California ships brought goods to Portland for distribution to the Valley and for transshipment on sternwheelers up the Columbia."[17]

To speed the movement of wheat to Portland, upriver shippers on the Willamette system funded the construction of a steam-powered "snag puller" to improve travel for steamboat traffic. Its operation marked the onset of human efforts to channelize and streamline the river for commercial traffic. Diking and revetment work soon followed in an effort to confine the Willamette to a single course. The removal of obstructions in the river, the building of a canal and locks at Willamette Falls, and the construction of dikes set in motion major changes to the contours of the river. Although they initiated the progressive industrialization of the Willamette waterway, those nineteenth-century modifications were modest in comparison with changes that took place after the Second World War.[18]

The completion of the nation's first transcontinental railroad, the famous Union Pacific–Central Pacific line, in the spring of 1869 captured the fancy of the Portland *Oregonian*. Sixty years earlier, the newspaper observed, it took one month to travel from Philadelphia to Pittsburgh: "But now one week will suffice for the whole journey across the middle of the continent from the Atlantic to the Pacific. Time and space have been actually annihilated by the industry of man."[19] The building of a railroad up the Willamette

Valley in 1870 and the extension of a transcontinental line to Portland a decade later brought the awesome forces of the industrial world to the Northwest and further integrated the region with the world beyond.

Without acknowledging the apparent hyperbole, the *Oregonian* portrayed railroads as "an indispensable adjunct of civilization." The more earthbound *Willamette Farmer* argued that "the progress of new countries depends upon their railroads, and their connection with extensive railroad systems." But virtually every railroad promoter acknowledged what the Oregon booster David Newsom called its one great virtue: "developing our vast, dormant resources." At the annual meeting of pioneers in Aurora in 1874, *Willamette Farmer* mused that the valley's Edenic qualities were "made by the joint hands of Nature and of man." The pioneers, it claimed, were celebrating "the triumph of man over Nature in Oregon."[20]

As the nineteenth century drew to a close, the Portland *Oregonian* praised the industrial potential of western Oregon. Even in the still sparsely populated valleys across the Coast Range, "the smoke of the settler's hut and the sound of the woodsman's axe [indicate] that the reign of Civilization is at hand." It pointed to the importance of the Willamette River as an "exhaustless storehouse of manufacturing power." With its rich soils, endless forest wealth, and unmatched scenery, the newspaper concluded, "Oregon is America in miniature."[21] As the decades passed, true believers extended those early visions of an agrarian and industrial paradise to embrace more environmentally intrusive forms of activity. But what is most fascinating about this period is the language defining human relationships with the natural world—the powerful idea that the destinies of both people and natural systems were intertwined.

By the early years of the twentieth century—in the midst of a period of explosive industrial growth—influential Americans were developing an almost transcendental belief in the efficacy of the unlimited manipulation of the physical world. That practical and commercial view of nature embraced Progressive-era conservationism: a belief that orderly, systematic scientific and engineering approaches toward the natural world would bring an endless bounty of riches, and a conviction that human technical genius would combine with an abundant landscape to improve the quality of life for present and future generations. That scenario, played out in the Pacific Northwest until well after the Second World War, elicited few questions about unseen and unintended consequences. The nation and the region, in fact, were entering an era of supreme self-confidence about technical solutions to human problems.

A careful reading of that literature reveals another caveat. For the Willamette Valley to better serve human objectives, visionaries argued, a comprehensive and integrated infrastructure, including a wide array of "improvements," was required: energy-producing facilities, water- and land-based transportation systems, educational institutions, and an enhanced ability to shape the regional landscape. Here the world of developer and engineer, of promoter and scientist, the uninhibited language of the booster and the cool rhetoric of the bureaucrat became one. The federal government itself would eventually be a full partner to development, providing the capital and much of the technical expertise to those willing to "develop" the Willamette region for agricultural and industrial purposes.

For the most part, boosters and promoters did not look back. Indeed, we might call them epic poets of a kind, harbingers of change, visionaries whose imaginations knew few restraints other than those dictated by the most obvious limits of geography and technology. In their efforts to turn rivers into dredged and channelized streams, trees into lumber, and salmon into slabs of meat, the commercial community worked closely with like-minded professionals in state and federal agencies—with the state water bureau, state departments of agriculture and forestry, and federal offices such as the Bureau of Reclamation, the Army Corps of Engineers, the Forest Service, and the Fish and Wildlife Service. Collectively those groups produced a strikingly like-minded literature in terms of its rationalized and scientific approach to the natural world. While the focus of many ambitions centered on the Columbia River, other strategies fastened on the Willamette Valley: reclaiming wetlands, flood control, hydroelectrical development, and applying the techniques of scientific management to forest and agricultural land. The rationale for those efforts was civic "improvement"—fully developing the valley's resources to advance the public welfare.

The Willamette River itself was beginning to reflect the influences of the industrial forces that were in the process of reshaping the valley landscape. The Willamette had become a commercial trafficway and dumping ground: lumbermen used it to float logs to downstream mills, and the growing cities and towns that lined its banks emptied untreated sewage directly into the river. To make matters worse, an increasing number of sawmills, pulp and paper plants, and canneries cycled water through their manufacturing processes and returned it directly to the river. Those conditions prompted Oregon's newly created Board of Health to declare in 1907 that the Willamette River was an open sewer. As the pollutants dumped into the waterway continued to mount, the Willamette eventually became the primary source of contaminants in the lower Columbia River.

To mitigate the damage from seasonal floods along the Willamette, agricultural interests and major urban centers eventually pressured the Army Corps of Engineers to build an extensive system of dikes in an effort to keep the seasonal rise of the river within its banks. Rivers, however, have a way of misbehaving—of defying attempts to control them—and annual flooding continued to plague those who lived along the Willamette floodplain. Finally, with the onset of a national economic crisis—the Great Depression—powerful economic development groups in the Willamette Valley combined with the army engineers to propose an engineering design for the river and its tributaries that would turn the system into a mostly regulated waterway. Its promoters believed that building dams would end flooding, alleviate pollution, improve navigation, produce hydropower, increase recreational possibilities, and provide water for irrigation. The era of the great multipurpose dam projects had arrived. Although federal monies for megaprojects on the Columbia dwarfed expenditures for the Willamette River, the big dams on its tributaries, while less spectacular, wrought a dramatic transformation in the valley's landscape.

By most measures, Lewis Mumford is one of the more significant American thinkers of the twentieth century. An intellectual, a social progressive and visionary, and a man of immense humanistic learning, Mumford gained public attention during the Great Depression for promoting regionwide planning as a socially redeeming virtue. He based his argument on the premise that human modifications of nature were "a part of the natural order." The moment was ripe, in his view, for the remodeling of both nature and society. The new power dams and the highway projects being built across the country, in his view, represented "the thrust and sweep of the new creative imagination." Regional planning, he believed, had the capacity "to bring the earth as a whole up to the highest pitch of perfection and appropriate use."[22]

Mumford's ideas attracted the attention of the Pacific Northwest Regional Planning Commission, especially its director, Benjamin Kizer, who wrote Mumford that his book *The Culture of Cities* was "causing young men to see visions and old men to dream dreams." At Kizer's suggestion, Mumford agreed to tour the region and then advise the commission. In his journey from the McKenzie River to Puget Sound in 1938, he reported "breath-taking landscapes, the great simplicity of the towering Douglas firs, the genial farm-and-orchard landscape of the Willamette Valley, the subtle and manifold beauties of the Columbia Gorge." But there was something unsettling as well, "a sense of unoccupied space," as though one still might expect to see a covered wagon and "a weatherbeaten pioneer family" walking alongside. "The whole 'Oregon Country,'" Mumford concluded, "is a region that has been partly defaced but not yet, one feels, fully mastered."[23]

The Willamette Valley Project that took shape during the 1930s fit Lewis Mumford's prescription for regional planning as rational, scientific enterprise. Firmly grounded in earlier engineering surveys, the series of dams in the original blueprint was designed to promote the efficient development of *all* the resources in the abundantly endowed Willamette basin. As such, the project represented the combined efforts of farmers, developers, and industrialists who worked with the Army Corps of Engineers to gain federal funding for a series of multipurpose dams. R. H. Kipp, a Portland-based executive, phrased the issue in the most aggressive terms at the onset of the depression: "It is not a question of 'can' or 'will' we develop these Northwest rivers, we must 'must' do it in order to place ourselves on an equal with other similar river districts."[24]

Supporters for the original seven-dam scheme argued that the Willamette Valley Project would bring great benefits to citizens from Cottage Grove to Portland. The Willamette River was badly polluted; only 10 of 69 communities along the river provided any sort of sewage treatment, and the unprocessed wastes were killing thousands of fish. In a unique twist of language, an Army Corp of Engineers pamphlet in 1938 reportedly stated: "Primitive conditions have been destroyed by the advance of civilization." Building dams and regulating stream flow, therefore, would improve navigation during the low-flow months *and* "would have the effect of delaying the need for [sewage] treatment works at some of the cities in the middle and lower valley."

The *Eugene Register-Guard* weighed in with some heavy rhetoric of its own in support of the project, or "environment refinement" as it referred to the effect of the dams: "From virtually a raw area, wherein man has done fairly well with conditions as nature left them, the valley will assume the polish of a regulated, planned region, designed by the ingenuity of man to provide the best possible living conditions for human beings." Except for a few sport and commercial fishers, William L. Finley, a nationally known naturalist, was the only person who spoke out early and often in those prewar years against building high dams on the Willamette system. One of Finley's critics, who accused the naturalist of throwing a "monkey wrench" in the way of progress, wrote to Senator Charles McNary, "We place human life above that of fish." Another of Finley's detractors referred to him as a "worshipful devotee of fish."[25]

In the mid-1960s and toward the close of the dam-building frenzy in the Northwest, the aging Lewis Mumford was still writing with confidence that scientific ideology had provided "both the means and the justification for achieving external control over all manifestations of natural existence."[26] Although those sentiments may seem utopian and naive today, the Willamette Valley is a visible example of how Mumford's general ideas about planning, about attempting to control the natural world, have prevailed. The valley's changing landscape following the end of the Second World War reflects the supreme self-assurance of more than two generations of thinking—the widespread belief that technical expertise could be used to manipulate the natural world and to improve the material and social condition of humankind.

In its customary and effusive style, *Time* magazine reported in 1950 that human expertise had created a new frontier in the Pacific Northwest "made ready for man by spectacular engineering." The magazine praised the scientific work of the Forest Service and the engineering genius of the Bureau of Reclamation and the Army Corp of Engineers, whose combined achievements meant that the United States could "expand almost indefinitely within its present boundaries." *Time* gave its full endorsement to federal agencies who were "making rivers behave," whose dams were accomplishing great things through "geographical judo."[27]

Such thinking represented the technocratic optimism of the postwar era, the idea that human genius could overcome all physical limits, that it was the mission of scientists and bureaucrats to overcome limits, not to establish them. Through those years a general aura of optimism prevailed—the notion that the natural world could be continually manipulated without consequences. Warren Smith of the Oregon Academy of Sciences expressed that sense of confidence, disputing the "prophets of doom" such as Oswald Spengler and Jose Ortega y Gasset who called for restraint. "We have a responsibility," he said, "to use our brain" to promote the welfare of our citizenry. With "greater knowledge of the physical universe," Smith remarked, "we can guide our evolution toward something better."[28]

To be sure, the multipurpose dams on the Willamette River system brought dramatic change to the valley's landscape. At least for the short term, the dams reduced pollution in the river by storing seasonal precipitation, which is released during the low-water months to maintain a relatively even flow in the main river. This procedure flushes contaminants downstream and out of sight. The dams changed the valley in other ways: the water impounded behind the reservoirs keeps most of the seasonal floods from the valley's cities and valuable farmland, and the once extensive floodplain now supports the most productive cropland in the valley—made possible, of course, with irrigation water released from tributary reservoirs. The broad fields of grass, the extensive fruit orchards, and the abundant farmlands that expanded across the old floodplain after 1945 have all been incredibly productive enterprises.

And following the logic of those powerful new cultural prescriptions, writers continued to praise the abundance and natural beauty of Oregon's most productive western valley. Peter James, a British planner who visited the state in 1978, remarked that "Oregon would be one of the most likely spots on the globe" in which to situate an ecotopian society, because it had one of the most productive agricultural valleys anywhere, seemingly endless stretches of evergreen forests, and "salmon teeming rivers." Writing for the *New Scientist*, he emphasized that the people were an outdoor folk who loved nature and thought they were superior to their more material-oriented neighbors. Using the state of Oregon as a real-life metaphor for Ernest Callenbach's science-fiction best-seller *Ecotopia*, James informed his readers that this "almost frontier country" had good reason to call itself "the most ecologically conscious area in the world."[29]

At the moment Peter James made his brief visit, the Oregon landscape was something less than a harmonious and pristine paradise. Indeed, his romantic description of the state would appear to fit Yi-Fu Tuan's precept that myths tend to prosper where precise knowledge is lacking, that stories take on a life of their own when they lose touch with material reality.[30] But the British visitor's exaggerated rhetoric, however embellished and devoid of substance, followed a grand narrative tradition and suggested that Oregon's mythical qualities of abundance continued to enjoy wide currency.

More recently another planning expert, Kevin Kasowski, offered a more sober and informed assessment in the pages of the state's leading newspaper, the Portland *Oregonian*. By the late twentieth century, Kasowski observed, the celebrated wagon ruts left by the 350,000 migrants who traveled west on the Oregon Trail had become symbols of ambivalence toward the natural world. Although many citizens cherished the famous ruts as a valuable cultural treasure, to others the wheel marks left by the wagons represented a defacing of the landscape. Kasowski's principal worry was that "our own much larger footprints on the land . . . are scarring the land's beauty and productivity." Once scenic wonders—American Falls on the Snake River and Celilo and Kettle Falls on the Columbia—had been flooded behind giant dams, all in the name of human progress and material comfort. And although hydropower production had made possible technological advances that had eliminated much of the drudgery of daily life, Kasowski argued that those innovations and similar modifications to the landscape had exacted a heavy price in sharply dwindling salmon runs and a generally worsening biodiversity crisis.[31]

Lest these concluding remarks have the ring of penitential history, a simple morality tale, I would remind readers that the floodwaters that inundated much of the Willamette Valley in February and again in November of 1996 do speak to the shortcomings of an overweening confidence in the human capacity to find technical and engineering solutions to natural phenomena. The consequences of that techno-vision are also reflected in the recent findings of deformed fish in the Willamette River and high incidences in bottom sediments of residue from the herbicide atrazine and the insecticide Diazinon. Indeed, the waterway may be a miner's canary of sorts. If culture is now a factor in climate change, if instrumentalism is the ultimate mark of value, and if human ingenuity oftentimes leads so easily to tragedy, then I agree with Simon Schama: we may be caught up in an out-of-control mind-set. When the Willamette system overflowed its banks in 1996, the meandering waters not only exposed connections to an earlier, nineteenth-century landscape but also revealed the flaws of a culture that assumed it had bested the river.

NOTES

1. Raymond Williams, *Towards 2000* (London: Chatto and Windus, 1982), p. ix.

2. Simon Schama, *Landscape and Memory* (New York: Alfred A. Knopf, 1995), p. 9.

3. James P. Ronda, "Calculating Ouragon," *Oregon Historical Quarterly* 94 (Summer–Fall 1993): 121–40, p. 121.

4. These ideas are adapted from Richard White's essay "Frederick Jackson Turner and Buffalo Bill," in *The Frontier in American Culture*, ed. James R. Grossman (Berkeley: University of California Press, 1994), pp. 7–66.

5. Richard W. Neuberger, *Our Promised Land* (1938; reprint, Moscow: University of Idaho Press, 1989), p. 14.

6. Vancouver *Columbian*, 12 February 1853; Gustavus Hines, *Life on the Plains of the Pacific: Its History, Condition, and Prospects* (Buffalo: G. H. Derby and Co., 1851), p. 319; *Oregon Spectator*, 19 December 1850. Campbell is quoted in Ellis A. Stebbins

and Gary Huxford, *Since 1856: Historical Views of the College at Monmouth* (Monmouth: Western Oregon State College, 1996), p. 47.

7. White, "Frederick Jackson Turner and Buffalo Bill," p. 26; Dayton Duncan, *Out West: An American Journey* (New York: Viking Penguin, 1987), pp. 392–93.

8. Elliott West, *The Way to the West: Essays on the Central Plains* (Albuquerque: University of New Mexico Press, 1995), p. 83.

9. *Ibid.*, p. 86.

10. Robert Boyd, "Disease as a Factor in Native Population Decline," a talk presented at Oregon State University, February 1988; Robert Boyd, "Demographic History, 1774–1874," in *Handbook of North American Indians, vol. 7: Northwest Coast*, ed. Wayne Suttles (Washington, D.C.: Smithsonian Institution Press, 1990), p. 139.

11. *The Letters of John McLoughlin from Fort Vancouver to the Governor and Committee*, ed. E. E. Rich (London: The Champlain Society for the Hudson's Bay Company Record Society, 1941), pp. 125–26.

12. For Kelly's exaggerated rhetoric, see Carlos A. Schwantes, *The Pacific Northwest: An Interpretive History* (Lincoln: University of Nebraska Press, 1989), pp. 78–79.

13. These passages are quoted in Oscar Osburn Winther, *The Old Oregon Country: A History of Frontier Trade, Transportation, and Travel* (1950; reprint, Lincoln: University of Nebraska Press, 1969), p. 6.

14. Alfred W. Crosby, *Ecological Imperialism: The Biological Expansion of Europe, 900–1900* (New York: Cambridge University Press, 1986), pp. 193–94.

15. Richard White, "The Altered Landscape: Social Change and the Land in the Pacific Northwest," in *Regionalism and the Pacific Northwest*, eds. William G. Robbins, Robert J. Frank, and Richard E. Ross (Corvallis: Oregon State University Press, 1983), p. 111.

16. James R. Robertson, "The Social Evolution of Oregon," *Oregon Historical Quarterly* 3 (1902): 1–37, 11.

17. Dorothy O. Johansen and Charles M. Gates, *Empire of the Columbia: A History of the Pacific Northwest* (1957; reprint, New York: Harcourt, Brace, and World, 1967), p. 284.

18. *Oregonian*, 1 January 1868; *Willamette Farmer*, 28 June 1869.

19. *Oregonian*, 8 May 1869.

20. *Oregonian*, 7 October 1865; *Willamette Farmer*, 2 February 1872; 20 June 1874.

21. *Oregonian*, 1 January 1890.

22. Mumford is quoted in Richard White, *The Organic Machine: The Remaking of the Columbia River* (New York: Hill and Wang, 1995), pp. 64–67.

23. Lewis Mumford, *Regional Planning in the Pacific Northwest* (Portland: Pacific Northwest Regional Planning Commission, 1939), pp. 1–2.

24. R. H. Kipp to Senator Charles McNary, March 17, 1931, in Charles McNary Papers, box 6, Oregon Collection, Knight Library, University of Oregon.

25. *Eugene Register-Guard*, 31 July 1938; *Salem Capitol Press*, 5 May 1939. The Army Corp of Engineers pamphlet is quoted in the *Register-Guard*.

26. Lewis Mumford, *The Pentagon of Power* (New York: Harcourt, Brace and Co., 1964), p. 106.

27. *Time*, 30 July 1951.

28. Smith is quoted in the *Oregonian*, 1 September 1949.

29. Peter James, "Ecotopia in Oregon?" *New Scientist* 81 (January 4, 1987): 28–30.

30. Yi-Fu Tuan, *Space and Place: The Perspective of Experience* (Minneapolis: University of Minnesota Press, 1977), p. 85.

31. *Oregonian*, 25 April 1993.

6 / Footprints and Pathways

The Urban Imprint on the Pacific Northwest

CARL ABBOTT

My title is a shorthand for introducing two key approaches for thinking about the urban imprint on the western landscape. One stance is to look *within* cities for the ways in which town-making and urbanization have directly altered urban sites, replacing natural landscapes with manipulated cityscapes—the footprint part. The second stance looks *outside* cities to examine the interactions between urban systems and the regional economy—the ways in which city dwellers and their activities have helped to organize and reorganize the human uses and adaptation of large natural regions. This is the pathways part.

FOOTPRINTS

Architects and city planners use the term *footprint* to describe the land that a building actually occupies—the ground that lies beneath its walls and foundations. By analogy, the "footprint" of a city is the acreage that has been converted to urban uses or intimately incorporated into its fabric. It is the buildings, the roads, the parking lots, and even the carefully tended parks. To talk about the urban footprint is to talk about the weight of cities on their landscapes.

Cities alter their sites in many of the same ways that farming alters the rural landscape. City-makers (to paraphrase the Book of Isaiah) raise up the valleys, lay low the mountains, and make the rough places smooth. They trim down hills, drain marshes, and fill ravines for economic use.

They divert natural flows of water to serve the needs of consumption and production, and they alter the character of those flows by channelizing rivers, burying creeks in culverts, and impounding streams for power, irrigation, and flood regulation. What Portland and Tacoma and Everett have done to their pieces of the Northwest is not all that different from what nineteenth-century farmers did in draining and cultivating the margins of the Willamette River, or twentieth-century irrigators in bringing new land of the Columbia Plateau into cultivation.

The remainder of this essay suggests two points about the urban footprint. First, there was an important historical change from the first generations of settlement, when the landscape strongly determined patterns of urban growth, to later generations, when Americans mobilized their capacities to adapt cityscapes to their own ends. Second, northwestern cities actually sit relatively lightly on the land.

Reading the Landscape

In the first decades of urban settlement, the preexisting landscape substantially controlled city-making in the Pacific Northwest. English-speaking traders and settlers arrived with limited capital, limited time, limited technologies, and limited access to energy sources. Out of necessity, they "read" the rivers and hills and wetlands and allowed those landscape features to dictate the locations of their first settlements.

Members of the first generation thus sought natural saltwater harbors where deep water reached close to shore and ships would not mire on mud flats. On the midshore of Puget Sound, it took only three wintery months (November 1851 to February 1852) to decide that Alki Point had good shelter for neither boats nor people. In 1852 Arthur Denny, William Bell, and Carson Boren used a horseshoe to sound the east side of Elliot Bay for the deepest water—conveniently close to a point where high hills lowered to meet the Duwamish mud flats—and then optimistically platted streets across the muck.[1]

Town-makers also found the natural transfer points for inland transportation. Falls and rapids interrupted navigable streams and required portages or transshipment, along with merchants, warehouses, and workers. The same falls also opened opportunities for manufacturing by driving mill wheels and machinery. One result was transportation towns such as The Dalles, mill towns such as Oregon City, and nascent cities such as Great Falls. The areas where routes across the plains and mountains converged at natural trading points evolved into cities such as Walla Walla and Pendleton. The latter in its early years was an assembly point for stock drives across the Blue Mountains to Idaho and Montana.[2]

Speculator-settlers searched out riverside locations where well-drained land sloped directly to navigable water, without marshes and sandbars to get in the way. In my own part of the Northwest, one result was the location of Fort Vancouver on the north side of the Columbia and not on the slough-and-island-choked south side, which had confused even the intrepid Meriwether Lewis and William Clark. Another example is Portland's having been located in a dry clearing on the west bank of the Willamette rather than on an east bank that alternated deep marshes and bluffs. Even before Asa Lovejoy and Francis Pettygrove cast their ambitious eyes on the site, it was already in use as an early highway rest area—the easiest spot on the water route between Oregon City and Fort Vancouver to pull a canoe out of the river, cook a meal, or make repairs. We can revisit the site through the eyes of young Jesse A. Applegate, who passed through in 1843 just before the founding of the city: "We landed on the west shore, and went into camp on the high bank where there was little underbrush. . . . No one lived there and the place had no name; there was nothing to show that the place had ever been visited except a small log hut near the river, and a broken mast leaning against the high bank."[3]

Town-makers who read the landscape incorrectly paid with wasted investments and shattered dreams. For example, Portland beat out rival townsites because its location made best use of the details of the lower Willamette Valley. Oregon City and Milwaukee lost out because they were upriver from sandbars that blocked passage of profitably sized ocean-going ships. Merchants downstream at St. Helens were on the deep channel but had to contend with the steep rampart of the Tualatin Mountains, which blocked trade with the farmers of Washington County. Portland merchants had the best of both worlds—the effective head of deep-water navigation and a usable pass through the hills to the grain fields of the Tuality Plains (building a plank road that was the predecessor of U.S. 26).[4]

The Leveling Impulse

By the second generation, Northwest Americans began to take the time and allocate the resources to *reshape* their landscapes to suit the needs of cities. No longer did city builders always accede to the constraints of their townsites. Instead, they began to remake those sites to human need. Because the real-estate market placed a premium on proximity (especially in the days of muscle-powered transportation), there was a high demand for useable land close to established town centers. The result was what I call the *leveling impulse*, because it often involved massive movement of dirt off hills to create level, buildable land around original core settlements.

Perhaps the best-known example in the Northwest is Seattle's long battle with its unpromising site—gifted with deep water but hemmed in by hills to the north and east and tide flats to the south. As R. H. Thomson wrote scornfully of anti-levelers, "Some people seemed to think that because there were hills in Seattle originally, some of them ought to be left there, no difference how injurious a heavy grade over a hill may be to the property behind that hill."[5]

As we know, the levelers rather than the skeptics won the day. In the first decade of this century, Seattle sluiced Denny Hill into Elliott Bay. Pumps raised 20 million gallons of water a day from Lake Union to the top of the hill to blast dirt and rocks into flumes and eventually the bay. The enterprise leveled 62 blocks. Further south, city-makers cut a notch between First Hill and Beacon Hill (for Jackson and Dearborn Streets). The muck turned much of the tide flats into Harbor Island.

Less famous is Portland's similar transformation of its Northwest End between the West Hills and the Willamette River. In the mid-nineteenth century, the river's floodplain was dotted with broad, stagnant seasonal lakes that were refreshed with high water each spring and slowly dried during the summer. Couch's Lake, close to downtown Portland, was filled in the 1880s and 1890s to serve as the Northern Pacific freight yards and is now scheduled for 6,000 housing units as a new in-town neighborhood. Guild's Lake, another mile downriver, was picked as the site of the Lewis and Clark Centennial Exposition and Oriental Fair of 1905, the first world's fair on the Pacific coast. For a handful of months, fresh water pumped from the Willamette kept the lake a blue backdrop for the exhibition halls surrounding it. With the end of the fair and the demolition of its buildings, however, Portlanders began to fill the site with dredge spoils and dirt washed from the overlooking ridge, creating an industrial and warehouse district.[6]

Also important were landscape modifications that we do not see because they are underground or underwater. Every city is built over a complex network of underground channels where natural streams have been buried in tubes and tile to make new building sites or serve as concealed storm sewers. Only in the last decade of the century are some cities considering "daylighting" a few of the old streams to restore their ecological functions. The Riverwalk through downtown San Antonio, Texas, is an envied model.

In an urban system dependent on ocean-going trade, dredging and filling have been essential activities. I have already noted the way Seattlites carved and filled a marshy delta for a vast port and industrial district. Tacomans did the same with the swampy end of Commencement Bay, dredging and filling to turn a natural estuary into squared-off channels and fingers of land for shipping and bulk storage businesses.

Portlanders have engaged in continual efforts for more than a century to deepen the original 12-foot channel of the Columbia.[7] Dredges have repeatedly clawed up the bottom of the Columbia, most recently to remove the huge sandbars that washed off the erupting Mount St. Helens. Portland now depends on regular dredging to maintain the Columbia's 40-foot channel, adequate for huge post-Panama ships that now reach 132 feet in width (the Panama Canal is 106 feet wide) and 931 feet in length and carry more than 2,500 containers. Political pressure is mounting for another round of dredging to reach 43 feet.

Dredge spoil is a natural for filling squishy waterfronts to get more industrial land. In Portland, it created the central eastside and Mock's Bottom. Portland even rerouted the Willamette to make Swan Island into Swan Peninsula. In the 1980s and 1990s the Port of Portland continued to create new industrial land at the intersection of the Willamette and Columbia.

Sometimes landscape wins. In 1942, the Columbia River floodplain north of Portland blossomed with hundreds of new apartment buildings to house a torrent of wartime shipyard workers and their families. At its peak in 1944, the instant community of Vanport housed more than 42,000 people, making it the second-largest "city" in Oregon. Four years later, the entire community was gone, washed away by rising waters that breached a protective dike on Memorial Day 1948, swirled through the streets, and spun buildings off their foundations.[8]

Planning and Redevelopment

The early twentieth century brought self-conscious efforts to develop good city planning to go along with sluicing and dredging. Indeed, a major theme of urban growth in this century has been the desire to find ways to better utilize already urbanized land.

Systematic park plans from the turn of the century in cities such as Spokane, Portland, and Seattle have shaped the basic outlines of city park systems to the present day. Park plans recognized the ecological and recreational value of conserving the highest and lowest land rather than converting it. Park systems usually identified hilltops, ridges, and riverside lowlands—the reserve land that seemed to have limited development potential but strong natural virtues.[9]

"City beautiful" plans aimed to design compact cities that would make efficient use of the site without extensive sprawl. Both the Bogue plan for Seattle and the Bennett plan for Portland proposed tightly focused cities oriented around rail transit systems, but what attracted the most attention were proposals for elaborate civic centers (three in Portland) served by radiating boulevards.[10] One poetic critic put it this way before Seattlites rejected Bogue's plan:[11]

> Aye, tear the city inside out
> And turn her upside down
> We want a Civic Center
> In this good old Potlach town
> We've won the wonder of the world
> With Thomson for our mentor
> But what's a world metropolis
> Without a Civic Center
> We want arterial highways
> In our Civic Center Plan
> That radiate to everywhere
> Converging like a fan
> When Bogue regrade assessments
> Their batteries unlimber
> They will give our population
> Short cuts to tallest timber.
> Seattle hasn't streets enough,
> She's hampered by her highways;
> She needs new angling avenues
> New slant arterial byways.

We can see the impulses both for "leveling" and for comprehensive planning in the Northwest's premier planned city. The Long-Bell Lumber Company built Longview, Washington, from scratch along the Cowlitz River in 1921–22 to serve the needs of its Washington timberlands. The site required diking and drainage, allowing a slough to be converted into a boulevard-bracketed park. The town itself was defined by the park, by a civic center, and by the arterial from civic center to train station.[12]

In turn, the sequel to city beautiful planning has been the impulse to *reuse* urban land. Because of the continuing social and economic importance of centrality—nodality, to use geographers' jargon—city centers have an immense inertia. The power of downtown business interests in local politics supports the continual reuse of core city land.[13] The same economic imperatives that "leveled" urban landscapes in earlier decades of Anglo-American settlement now recycle them. Successive generations erase parts of the center to rebuild and reuse them, conserving land in the process. Since the late 1950s, local government has often been an active partner, mobilizing urban renewal programs and public investments to recycle downtowns. The scale of the undertaking varies, but the essential goals have been the same from Seattle to Boise to Helena.

Particularly impressive examples have been the Northwest's two *modern* world's fairs, designed in part to reinvigorate rundown or economically underused core areas. Seattle's Century 21 Exposition reshaped the city's downtown fringe for more intensive use.[14] Spokane's fair in 1974 remolded its industrially derelict riverfront, creating public facilities and a riverside park to reinforce the attractions of the central business district.

How Much Land Do Cities Use?

Statistics on actual patterns of land use confirm the point that is implicit in my comments on comprehensive planning and downtown recycling. The numbers reassure us that Northwest cities are still compact after a century or century and a half of growth. They use their land conservatively, avoiding the dead zones and abandoned districts that ring the centers of many eastern cities. In aggregate, less than 1 percent of the territory of the four northwestern states directly supports human settlement in the form of urban and built-up land—compared with 23 percent of New Jersey.[15]

Only one-half of 1 percent of the Northwest lies within the 19 urbanized areas that the U.S. Bureau of the Census defined for 1990. Unlike the better-known term *metropolitan area* (an economic concept that uses counties as units), the *urbanized area* is the land that is actually settled at urban densities—the aggregate footprint, if you will. Urbanized areas in 1990 ranged in size from 20 square miles for Great Falls and 21 square miles for Idaho Falls to 426 square miles for Portland-Vancouver and 604 square miles for Seattle. In total, these urbanized areas in 1990 held 5,122,000 people—

that's 54 percent of the regional population on 0.5 percent of the regional land.[16]

The amount of land within urbanized areas grew only 18 percent in the 1980s—identical to those areas' 18 percent increase in population.[17] This is substantially different from elsewhere in the United States, where the consumption of urban land has increased much faster than population. Indeed, public policies in parts of the Northwest support high density, low-impact cities—particularly Oregon's Senate Bill 100 (1974) and the Washington Growth Management Act (1990–91). Both measures require planning for urban containment rather than sprawl and provide a context for local planning.

At the local level, public policies of investment and growth management now favor centers and compactness—whether downtown Seattle or attenuated versions such as downtown Bellevue. The Puget Sound Council of Governments in 1990 planned for an anticipated 1.4 million new jobs for the Seattle-Tacoma region by adopting a regional growth and transportation strategy called VISION 2020. The plan calls for "a regional system of central places framed by open space," concentrating new jobs and housing in central Seattle, in the metropolitan cities of Bellevue, Bremerton, Everett, Renton, and Tacoma, and in subregional centers along transit corridors.[18] In 1990, Mayor Norman Rice introduced a complementary idea to Seattle, calling on Seattle's notoriously no-growth voters to consider capturing a share of new regional growth in tightly defined "urban villages." The proposal evolved into a pending Comprehensive Plan that defines five "urban centers" to absorb 30,000 households and 100,000 jobs, along with smaller "hub urban villages" to pick up further development.[19]

Metro, the regional planning and service agency for Portland, engaged in a similar "Region 2040" process in the mid-1990s. The most popular alternative combined intensified development in the downtown with similar development along light rail lines and in regional centers. The goal is to focus development in a hub-and-spoke pattern that minimizes the urban impact on farms, forests, and environmentally sensitive land.[20]

My conclusion to this discussion of the urban footprint is reassuring. Northwest cities at the end of the twentieth century sit relatively lightly on their landscape. They do not sprawl excessively. They retain strong centers. If recent plans are an indication, they are also learning to respect natural barriers. We can hope that the level-and-fill mentality of the late nineteenth and early twentieth centuries is an anomaly, and that the careful reading of the landscape that marked the first generation of city-building will be the norm in the twenty-first century as well.

PATHWAYS

If "footprints" are the impress that urban nodes make on the landscape, "pathways" are the networks of transportation and communication that link urban nodes with the rural communities and natural resource districts in between. They allow economic specialization within the Northwest, and they allow the Northwest to specialize within the global economy.

In both past and present, cities have initiated, created, and funded the pathways. To continue my paraphrase of Isaiah, cities not only have made rough places plain but also have made straight the highways through the wilderness. Cities have created the navigation systems and organized the shipping services, planned and controlled the roads, built the wharves and piers, and constructed the airports. Pathways radiate out from cities. In doing so, they organize economic and social patterns into flows of people, goods, and information into and out of cities.

In this second section of my essay I want to contrast two ways in which cities have organized the regional economy—a nineteenth-century "Columbian" pattern and a late-twentieth-century "Cascadian" pattern.

The "Columbian" World

In the nineteenth century and first half of the twentieth century, the tools of regional organization were water transportation, telegraph lines, and railroads (which hugged stream valleys wherever they could). They created a Northwest that we can think of as "Columbia." Literally, Columbia was the Columbia River basin, with its ranching, mining, farming, and logging enterprises tied to markets by the major railroads and shipping lines. Columbians ran sheep and cattle, grew fruit and grain, cut trees, and scooped fish from the cool waters.

In this nineteenth-century world, Portland was the first and prime organizer—the regional capital that dominated the interior of the Columbia basin. Portland was the point where riverboats shifted their cargoes to ocean-going steamers and lumber schooners. It was the channel and source for investment capital that ran rail and telegraph lines up the Willamette Valley and over the Siskiyous to California. It was the main source of supply for miners and farmers in the great interior.[21]

Seattle filled a gap in this early regional system as organizer of the Puget Sound resource economy. Its role in western Washington and then in serving the Alaska resource economy mirrors that of Portland, creating a second off-center "Columbia" that began to develop in the 1880s as it grabbed a railroad connection and then took off with proximity to the Klondike gold rush.[22]

Portland's "partner" was Spokane. This younger city developed in the 1880s and 1890s as an organizing center for the upper Columbia River region. For a few decades it seemed that all railroads led to Spokane—from Coulee City, Pasco, Moscow, Bonners Ferry, and Nelson, B.C. Rail connections made Spokane the aspiring capital of an "Inland Empire," but one that never quite threw off its dependence on Portland and Seattle.[23]

By the early twentieth century these large cities sat at the top of a tightly integrated hierarchy of trading centers. The regional economy began at the country crossroads store, funneled through farm market towns and county seats, and continued on to regional rail centers and riverboat towns with wide hinterlands and enough traveling salesmen to support substantial hotels and large office buildings in the early twentieth century. H. L. Davis's *Honey in the Horn*, while canvassing Oregon's early-twentieth-century agriculture (sheep herding, hop picking, haying), makes a stop at an unnamed Columbia River town that is probably The Dalles, where Davis spent some of his youth:[24]

> Times were livening up in the Columbia River towns that fall, because the upper country was getting not one railroad, but two, and old E. H. Harriman and James J. Hill were out letting contracts, buying rights of way, and banging out court injunctions back and forth. . . . Men were already piling into the middle river ports to be on hand when work opened, every side-hill freight station in the upper country was petitioning to be the county seat of a new county, and windows of real-estate offices were loaded with maps of Jonesville and Wilkinsburg and Petersonville, Cherry Vale and Apple Heights and Gooseberry Villas and Sweet Pea Home Sites, all right in the path of future development, and all requiring only the investment of a little small change to make a man a capitalist for life. There was a carnival on all over the streets, and deckhands and cowboys and shovel-stiffs and real-estate promoters elbowed their way around under the arc-lights with mobs of street-show pitchmen and girls on the prowl picking at their flanks. Steamboats snorted and boomed their whistles from the river, a merry-go-round tooted and wiggled its varnished ponies.

The "Cascadian" World

The last fifty years of the twentieth century layered a new Northwest on the old Northwest that Davis described so vividly. We can call this second

Northwest "Cascadia." It is a product of automobiles, airlines, and long-distance transmission of hydroelectric power. In contrast to the east-west orientation of "Columbia," "Cascadia" is oriented north-south, from Anchorage through Victoria and Vancouver to Bellingham, Seattle, Tacoma, Olympia, Portland, Salem, and Eugene. It is marked by the high-volume movement of people and information. It is networked internally along the Interstate 5 axis, and its cities act as lenses to focus intercourse between the Asian and North American economies.[25]

In other writing I have talked about the western impacts of the "second urban revolution" of the nineteenth century and the "third urban revolution" of the late twentieth century.[26] The second urban revolution created vast industrial cities in Europe and eastern North America. In the American West it called forth systems of commercial cities and towns to gather and forward raw materials to these industrial metropolises and to distribute manufactured goods to facilitate further production. This is the essence of the "Columbian" economy I have just described.

The third urban revolution is still upon us. It involves the rise of new technologies of communication, the replacement of manufacturing workers with service workers, and the increasing importance of information as a commodity in itself. The booming "Cascadian" economy and its metropolis-centered social system is the northwestern manifestation of this revolution.

Seattle is the prime organizer of this new economy, with other cities of the Interstate 5 corridor in supporting roles. I attribute this rise of Seattle to several key decisions in the 1950s and 1960s: staging a successful world's fair, turning the University of Washington into a top-ten research institution, and investing in facilities for containerized cargo (to get long-distance shipments rather than bulk resource commodities).[27]

In the new economy, smaller cities can also connect directly to global networks. For just one example, Boise is a direct node in national networks. Corporate headquarters in Boise control thousands of jobs in locations scattered around the continent. The city boomed in the 1980s by attracting smaller businesses seeking an alternative to California.[28]

Participants in the Cascadian economy use the rest of the Northwest for recreation and refreshment as much as for resources and markets. It was a "Columbian" impulse in the 1890s to create the Northwest's system of national forests, with their orientation toward resource production. It is a Cascadian impulse to create a North Cascades National Park or a Columbia River Gorge National Scenic Area.[29] It is the Cascadian impulse that brings ex-urbanites to Sandpoint, Idaho, and Joseph, Oregon, to interact tensely with ranchers and loggers. It is the Cascadian impulse that has made Missoula a center for environmental policy-making and a significant point on the national intellectual map and given a cosmopolitan overlay to places like Hood River and Anacortes.

Conclusion

The urban imprint on the northwestern landscape is lighter than many of us might think. Northwest cities *are* compact. They certainly alter their environment and eat up new tracts of land, but they also recycle their cityscapes more effectively than do most United States cities.

We must also remember that it was city networks and city services that organized the traditional resource economy and are reorganizing the Northwest around an emerging information economy. For better or worse, cities shape the ways in which northwesterners have used their region as a whole.

This point, in turn, suggests that we think about the "imprint" of the urban Northwest in institutional or political terms as well as physical terms. On one side of the environmental ledger, the urban Northwest has organized and benefited from the development of extractive industries, with all their side effects. On the other side, it has been the seedbed for new ideas about environmental regulation and management and a source of political force behind those ideas. Cities house the headquarters of clear-cutting timber companies and environmental action organizations, the offices of transportation moguls and environmental regulators. In the end, we are left with the unavoidable physical presence of large cities, with the organizing power of their economic networks, and with their role as generators of old and new ideas about how best to utilize and manage the northwestern landscape.

NOTES

1. Roger Sale, *Seattle: Past to Present* (Seattle: University of Washington Press, 1976), p. 8.

2. Donald W. Meinig, *The Great Columbia Plain: A Historical Geography, 1805–1910* (Seattle: University of Washington Press, 1968); John Reps, *Cities of the American West: A History of Frontier Urban Planning* (Princeton: Princeton University Press, 1979), pp. 344–89.

3. Jesse A. Applegate, *Recollections of My Boyhood*, quoted in Howard McKinley Corning, *Willamette Landings: Ghost Towns of the River* (Portland: Binfords and Mort, 1947), pp. 16–17.

4. Eugene Snyder, *Early Portland: Stumptown Triumphant—Rival Towns on the Willamette, 1831–1854* (Portland: Binfords and Mort, 1970).

5. Sale, *Seattle*, p. 74.

6. Carl Abbott, *The Great Extravaganza: Portland and the Lewis and Clark Exposition* (Portland: Oregon Historical Society Press, 1996).

7. E. Kimbark MacColl, *Merchants, Money, and Power: The Portland Establishment, 1843–1913* (Portland: Georgian Press, 1988); E. Kimbark MacColl, *The Growth of a City: Power and Politics in Portland, 1915–1950* (Portland: Georgian Press, 1979).

8. Manly Maben, *Vanport* (Portland: Oregon Historical Society Press, 1987).

9. John Fahey, "A. L. White: Champion of Urban Beauty," *Pacific Northwest Quarterly* 72 (October 1981): 170–79.

10. Carl Abbott, *Portland: Planning, Politics, and Growth in a Twentieth-Century City* (Lincoln: University of Nebraska Press, 1983); Mansel Blackford, *The Lost Dream: Businessmen and City Planning on the Pacific Coast, 1890–1920* (Columbus: Ohio State University Press, 1993); William H. Wilson, *The City Beautiful Movement* (Baltimore: Johns Hopkins University Press, 1989).

11. Sale, *Seattle*, p. 101.

12. John M. McClelland, Jr., *Longview: The Remarkable Beginnings of a Modern Western City* (Portland: Binfords and Mort, 1949); Carl Abbott, "Longview," *Columbia: The Magazine of Northwest History* 4 (Summer 1990): 14–20.

13. John Fahey, "The Million-Dollar Corner: The Development of Downtown Spokane," *Pacific Northwest Quarterly* 62 (1971): 77–85; Neil O. Hines, *Denny's Knoll: A History of the Metropolitan Tract of the University of Washington* (Seattle: University of Washington Press, 1980).

14. John M. Findlay, *Magic Lands: Western Cityscapes and American Culture after 1940* (Berkeley: University of California Press, 1992).

15. U.S. Department of Commerce, *Statistical Abstract of the United States: 1989* (Washington, D.C.: Government Printing Office, 1989), p. 195.

16. Information from 1990 U.S. census data tapes.

17. U.S. Bureau of the Census, *Census of Population: 1980, U.S. Summary: Number of Inhabitants*, PC80-1-A1.

18. Puget Sound Council of Governments, *Vision 2020: Growth and Transportation Strategy for the Central Puget Sound Region* (Seattle: Puget Sound Council of Governments, 1990); King County Growth Management Planning Council, *Countywide Planning Policies* (Seattle: 1992).

19. Seattle Planning Department, *A Citizen's Guide to the Draft Comprehensive Plan* (Seattle, 1993); Timothy Egan, "Seattle Has a Plan: Urban Renewal for Fun," *New York Times*, 4 April 1993.

20. Metro, *Region 2040: Concepts for Growth* (Portland: 1994); *Region 2040: Recommended Alternative Decision Kit* (Portland: 1994).

21. MacColl, *Merchants, Money, and Power*.

22. Norbert MacDonald, *Distant Neighbors: A Comparative History of Seattle and Vancouver* (Lincoln: University of Nebraska Press, 1987); Richard Berner, *Seattle, 1900–1920: From Boomtown, Urban Turbulence to Restoration* (Seattle: Charles Press, 1991).

23. W. Hudson Kensel, "Inland Empire Mining and the Growth of Spokane, 1883–1905," *Pacific Northwest Quarterly* 60 (April 1969): 84–97; Belle L. Dickson, "The 'Why' of Spokane," *Journal of Geography* 30 (1931): 151–60; John Fahey, *The Inland Empire: Unfolding Years, 1879–1929* (Seattle: University of Washington Press, 1986); D. W. Meinig, "Spokane and the Inland Empire: Historical Geographic System and a Sense of Place," in *Spokane and the Inland Empire*, ed. David Stratton (Pullman: Washington State University Press, 1991).

24. H. L. Davis, *Honey in the Horn* (New York: Morrow, 1935), p. 329.

25. Alan F. J. Artibise, "Achieving Sustainability in Cascadia: An Emerging Model of Urban Growth Management in the Vancouver-Seattle-Portland Corridor," in *North American Cities and the Global Economy: Challenges and Opportunities*, eds. Peter Karl Kresl and Gary Gappert (Thousand Oaks, Calif.: Sage Publications, 1995), pp. 221–50; Theodore H. Cohn and Patrick J. Smith, "Developing Global Cities in the Pacific Northwest: The Cases of Vancouver and Seattle," *ibid.*, 251–85; John Hamer and Bruce Chapman, *International Seattle: Creating a Globally Competitive Community* (Seattle: Discovery Institute, 1992).

26. Carl Abbott, "The American West and the Three Urban Revolutions," in *Old West/New West: Quo Vadis?*, ed. Gene Gressley (Worland, Wyo.: High Plains Publishing, 1994), pp. 73–99.

27. Carl Abbott, "Regional City and Network City: Portland and Seattle in the Twentieth Century," *Western Historical Quarterly* 23 (August 1992): 293–322.

28. Allan Pred, *City-Systems in Advanced Economies: Past Growth, Present Processes, and Future Development Options* (New York: John Wiley, 1977); David Heenan, *The New Corporate Frontier: The Big Move to Small Town, USA* (New York: McGraw-Hill, 1991).

29. Carl Abbott, Sy Adler, and Margery Post Abbott, *Planning a New West: The Columbia River Gorge National Scenic Area* (Corvallis: Oregon State University Press, 1997).

PART II / FIRST PEOPLES

No one knows with any certainty how long humans have called this place, the Pacific Northwest, home. The archaeological record indicates that people have lived in the region at least since the retreat of the glaciers at the end of the Pleistocene epoch, some 10,000 years ago. At that time, salmon began to populate the rivers of the coast and the Columbia basin, and the sea level rose to its current height. In the succeeding 10 millennia, the climate has by turns grown warmer and drier and colder and wetter. Over the past 10,000 years, people have lived and thrived in the Pacific Northwest. How they did so is one of the themes of the essays in part two. Although Euro-Americans commonly considered the land they initially encountered as untouched wilderness, it was in fact a place that had been manipulated by indigenous people to produce what they needed for the life they chose.

Douglas Deur begins the section by examining the evolving subsistence strategies of the sedentary people who lived along the Pacific coast. He demonstrates that they built elaborate cultures based on a detailed understanding of the resources available to them. People in the southern portions of the region, for example, modified their environment with fire to open clearings and produce crops such as camas that were not indigenous to the coastal zone. Farther north, people constructed elaborate raised beds in coastal estuaries in which they grew a variety of plants. In combination with the rich marine environments, agriculture provided coast residents with a surplus that influenced their evolving political and social organization. Deur argues against the traditional view that Northwest Coast natives lived in such an abundant environment that they did not develop plant cultivation. Although resources were abundant, they were often seasonal and dispersed, or "patchy." This patchiness contributed both to the need for cultivation and to the creation of strictly enforced household property rights in resource gathering sites.

Eugene Hunn examines a very different subsistence strategy that was employed by people who lived on the Columbia Plateau. While coastal people developed sedentary villages, plateau people adopted a migratory life in which they used a variety of widely dispersed food sources on an annual trek that took them into the mountains as spring turned to summer and then back to the rivers as autumn turned to winter. While coast people developed well-defined property rights, plateau people tended to treat resources as communal. Hunn compares and contrasts these alternatives and concludes that the life adopted by the plateau people allowed them to live in an environment in which key resources were dispersed even more widely than along the coast—and to do so without depleting those resources over the course of millennia.

Alan Marshall also examines the subsistence strategies of a group of plateau people, the Nez Perce. Marshall, like Deur, argues that our view of native people is faulty—that our understanding is befogged by the stories Euro-Americans have told themselves to justify the theft of the land from its occupants. Those stories have from the beginning focused on the lack of agriculture as a sign of the lowly status of American Indians: Indians, the colonists argued, were not entitled to the land because they did not farm it. But as Marshall demonstrates, the Nez Perce engaged in a form of agriculture—horticulture—that produced a wide variety of food plants. It was the inability of Euro-Americans to recognize anything other than European crops and straight furrows as "agriculture" that made the Nez Perce gardens "unusual."

Paul Martin and Christine Szuter examine a different problem: they seek to understand the impact the region's native people had on the distribution of large game animals. The evidence they marshal is drawn from the journals of Lewis and Clark and other Euro-Americans who traveled in the region during the early eighteenth century. Occasionally reduced to eating dogs purchased from the Indians, these early travelers uniformly reported a near-total absence of game west of the Rocky Mountains. The situation was reversed in the upper Missouri River valley, where game was astonishingly abundant. Martin and Szuter argue that the status of tribal relations explains this pattern of abundance and scarcity. The upper Missouri was a buffer zone between warring nations, where few hunters dared to go; as a result, game flourished. Conversely, in the Columbia basin the Indian nations were generally at peace, allowing hunters to travel freely in search of game; as a result, game was scarce. This phenomenon has been documented in other areas of North America, too. Martin and Szuter thus argue that humans—whether industrial or preindustrial—are the "ultimate keystone species."

Unlike the first four chapters, the final essay by Barbara Leibhardt Wester examines the environmental effects of the Dawes Act—a federal statute that opened up Indian reservations across the West to Euro-American settlers. The act was advertised as a means of assimilating Indians into American culture; by giving them an allotment of land, they would become self-sufficient farmers plowing straight furrows. Focusing on the Yakama Nation, Wester tells a tale of greed and loss: at the same time Euro-Americans were increasingly disrupting traditional Indian subsistence activities such as fishing, the federal government forced Indians to accept small parcels of land that were generally insufficient to provide a living—even if the federal government had provided the necessary infrastructure such as irrigation systems. The Dawes Act, Wester notes, was "part of the same legal fabric that consistently functioned to channel land, water, and other resources toward those who would most expeditiously develop them for market."

The essays in this part thus offer a variety of perspectives on the relationship between the region's indigenous peoples and the environment. They also demonstrate that the "wilderness" the Euro-Americans initially encountered was actually a well-settled land.

7 / Salmon, Sedentism, and Cultivation

Toward an Environmental Prehistory of the Northwest Coast

DOUGLAS DEUR

Conventionally, environmental history emphasizes the occupation and transformation of the North American landscape by people of European origin. Its narratives begin on fecund frontier landscapes and progress to the despoiled environments of the present. Its object, in part, has been to reveal the environmental causes and effects of the larger sweep of history, while its subjects have included the particular actors, motives, and mechanisms that defined the interplay between alien peoples and endemic landscapes.

The topic discussed here—the use and modification of particular environments by the native peoples of the Northwest Coast of North America—does not conform to this model. Instead of a narrative of abrupt encounters between immigrants and the land or of rapid and industrialized resource depletion, this is a story of millennia spent surviving with only rare inputs of people, materials, or technologies from outside the region. Most of the actors are anonymous; one can infer their motives only vaguely. This raises the question, can we recover an environmental history of nonliterate people? I believe we can. Its recovery is problematic, but its outlines can be inferred through the integration of ecological, archaeological, and ethnographic evidence. I employ this method here to illuminate the changing relationship between the peoples of the Northwest Coast and their environment.

RETRACING THE HISTORY OF AN IDEA

Since the nineteenth century, the relationship between the native peoples of the Northwest Coast and their environment has frequently received passing mention in the literature of anthropology and other disciplines (fig. 7.1). Although this relationship has been described consistently, it has been widely misrepresented. Indigenous cultures were of interest to early anthropologists and travel writers, but they did not address indigenous peoples' use of the environment as a central theme. Most of them—following the lead of the prominent Northwest Coast anthropologist Franz Boas—described human uses of the coastal environment on the basis of brief visits and post-hoc analyses of indigenous oral literatures. Impressed by the region's abundant marine and terrestrial life, these writers concluded that food was plentiful at all times and in all places.

Thus, in books such as anthropologist Ruth Benedict's *Patterns of Culture*, food gathering was depicted as a simple, even leisurely affair. Speaking of the "Kwakiutl" Indians of the central British Columbia coast,[1] Benedict noted: "They were a people of great possessions as primitive peoples go. Their civilization was built upon an ample supply of goods, inexhaustible, and obtained without excessive expenditure of labour."[2]

If, as Benedict wrote, the Northwest Coast was a place of inexhaustible abundance, how do we account for the elaborate system of land and sea tenure found along much of the coast, a system in which trespassers were commonly fought or killed? Why were apparently long and brutal wars fought over particular territories and resource sites?[3] And why did displacement from traditional fishing and gathering sites result in such material hardship for the native peoples of the region?[4]

The answer to all three questions—and the key to success in Northwest Coast aboriginal subsistence—was location. Although the natural resources of the Northwest Coast were in some manner "abundant," they were not ubiquitous. They were unevenly distributed in time and space, and native peoples' valuations of particular landscapes varied accordingly. Indeed, the prehistory of Northwest Coast subsistence appears to have involved peoples' efforts to maximize output from, and tribal control over, a finite range of conveniently located, resource-rich sites.

THE NORTHWEST COAST: ENVIRONMENTS AND RESOURCES

Although the environment of the northwestern coastal zone has often been depicted as relatively homogeneous, it actually is characterized by consider-

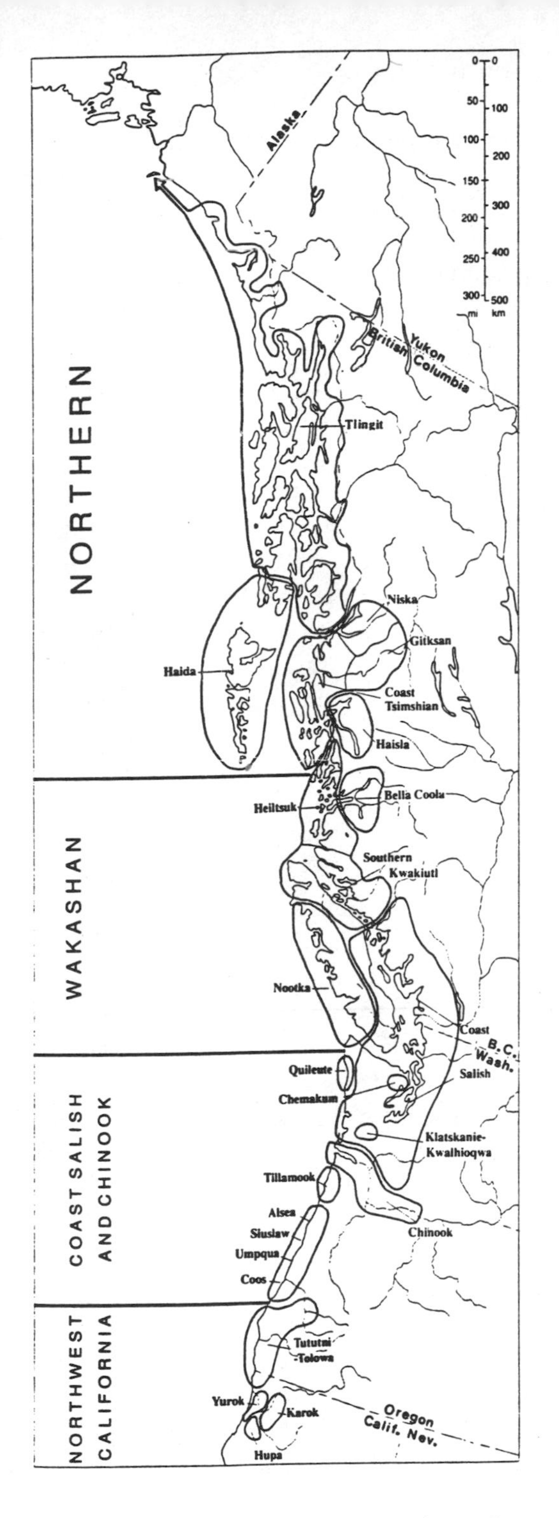

FIG. 7.1. Tribal distribution in the Northwest Coast cultural area.

able diversity. Along the coasts of Oregon, Washington, British Columbia, and southeastern Alaska, mountain peaks rise abruptly from sea level to over 10,000 feet. Shorelines, forests, and alpine areas house quite distinct environments. Marine air moving inland distributes precipitation unevenly, with temperate "rainforests" on windward slopes and relatively dry "rain shadows" on leeward slopes. Within a single tribe's territory, annual precipitation can vary more than 100 inches, with a corresponding variation in biota. Landslides, glaciers, wind, fires, and human clearing of the land have produced a diverse assortment of open spaces containing grasses and herbaceous plants. Marine environments vary from open ocean to ocean shoreline and protected inlets extending far inland. Sediment size, water temperature and salinity, and ocean currents vary from place to place. In combination, these factors produce a wide variety of biotic communities at different locations.[5]

As a result, the Northwest Coast has a variety of "patchy," small-scale environments—discontinuous concentrations of forest, rocky outcrops, grassy clearings, sandy beaches, mudflats, rocky intertidal areas, and so forth. Animals feed and breed, and plants grow, in a finite range of habitats, and their geographical distributions tend to be patchy as well. Without modern technologies for transporting food over long distances, native people who depended upon these plants and animals tended to locate near places where food plants and animals were naturally abundant, and they attempted to control access to these resource sites.

The availability of the staple plants and animals of the Northwest Coast diet varied both seasonally and geographically. Streams with large populations of salmon or eulachon (the oily and dietarily important "candle fish") could be tens of miles apart. Level, rocky areas on which seals or sea lions gathered might be even more widely spaced. Elk or deer were relatively scarce outside of the forest clearings where they grazed. Each type of mussel or clam was restricted to sites with specific rock or sediment sizes, levels of salinity, and wave action. Plant foods were also unevenly distributed: many berries grow poorly under the dense forest canopy that blankets much of the zone but flourish along streambanks or in small, moist clearings. Edible roots, bulbs, and rhizomes were limited to finite natural environments: camas, bracken ferns, and many lilies grew well only in clearings, whereas other root foods grew most densely along the margins of salt marshes, creating isolated food gathering sites so important that they can still be identified on the basis of indigenous place names.[6] Even the western red cedar (*Thuja plicata*)—from which the peoples of the coast made their homes, canoes, bark clothing, and the majority of their other durable goods—was irregularly distributed along the coast.[7] This discontinuous geography of natural resources presented a host of logistical challenges to the indigenous peoples of the Northwest Coast.

THE MOVE TOWARD SEDENTISM

During the earliest human settlement, the uneven distribution of resources placed limits upon patterns of human subsistence and settlement. Judging from evidence from a very small number of very old archaeological sites, most of the resources already mentioned—including salmon, several other freshwater and marine fishes, seals and sea lions, and large and small land mammals—were used by the native inhabitants not long after they first arrived on this coast more than 10,000 years ago.[8] Plants biodegrade rapidly in this region and thus are poorly preserved in the archaeological record; one can assume, however, that the earliest inhabitants extensively used the region's plant foods, which were more easily obtained without specialized technologies than were most animal resources. The earliest archaeological sites show little evidence of settled life or of intensive use of the resources of any single location. Food gathering therefore involved much movement between environments in which key resources were concentrated.

Resource use patterns changed significantly over time. In contrast to the earliest inhabitants, the peoples who occupied this coast at the time of European contact lived in large, sedentary villages and were highly dependent upon specific resource sites. In addition, at the time of contact, the region's most powerful tribes were those who had devised means to control, harvest, and trade the products found in these diverse and productive environments. The environmental prehistory of the Northwest Coast thus is characterized by a regionwide transition in which the migratory "resource generalists" of 10,000 years ago became the more sedentary "resource specialists" encountered by Euro-Americans in the eighteenth and nineteenth centuries.

The first archaeologically detectable movement toward greater sedentism and resource specialization occurred roughly halfway through this known 10,000-year history. Though the timing is unclear and probably varied from place to place, the hunter-fisher-gatherers of the Northwest Coast gradually increased their exploitation of the region's most abundant and accessible food resources, particularly salmon. This process apparently began along the lower stretches of particularly productive salmon rivers between 6,000 and 4,000 years ago.[9] There probably were several causes for this transformation:

improved technologies for fishing, hunting, and food storage, rising populations, and increasing social complexity that allowed organized, large-scale fishing and food storage. In addition, runs of salmon and other fish increased with the expansion of estuaries and their highly productive ecosystems as sea levels stabilized following the last Pleistocene "Ice Age."[10] Although salmon ostensibly were the primary basis for large settlements, some groups settled at other productive and predictable resource sites. For example, some major seasonal villages developed to hunt marine mammals, and others focused on the harvesting of shellfish, plant foods, or other varieties of fish.[11]

Forsaking some of the dietary diversity offered by more mobile hunting, fishing, and foraging, the inhabitants of the coast spent increasing amounts of time at productive estuarine sites. Increasingly sophisticated technologies for resource harvesting enabled them to exploit many of these resources with growing proficiency. Villages appeared with greater density around these estuarine environments, which provided the most abundant concentrations of salmon, other fishes, shellfish, waterfowl, and an assortment of other edible animals and plants. The geographical distribution of the human population, therefore, increasingly mirrored the uneven distribution of its primary food resources.

While all these factors were important to the emergence of sedentary villages, sedentary life would have been impossible on the Northwest Coast without effective methods of food storage. Salmon arrived in runs that lasted only days or weeks; without techniques to preserve and store the bounty, fewer fish could have been utilized. The development of methods for preserving and storing seasonally abundant foodstuffs made it possible to sustain a population throughout the year.[12] By the time large sedentary villages appeared, people had developed both methods of smoking and drying fish and the technology with which to manufacture watertight baskets and cedar boxes for food storage.

The social consequences of this transition to sedentary villages were profound.[13] People's gathering into "villages" created a potential for organized social, political, and subsistence activities. In the process, it is likely that political elites emerged to organize tasks such as house construction and food procurement and storage.[14]

Importantly, sedentism also contributed to the creation or augmentation of bounded tribal territories. The emergence of spatially fixed villages and hunting and gathering territories would have increased human dependence upon a finite constellation of resource sites. This placed a premium on productive sites of salmon, shellfish, game, berries, or roots. Seasonal migration patterns likely grew more restricted, and people became increasingly dependent upon the few plants they could control.[15] The need to have uninterrupted access to specific resource sites, to prevent others from using them, and to maintain their use by one's kin likely led to or enhanced a tradition of resource site ownership during the transition to sedentism. In turn, this limited the extent to which people could relocate to more productive resource sites without generating conflict with neighboring groups.

The successful exploitation of salmon and other marine resources by sedentary villagers created the potential for unprecedented food abundance but also raised the specter of potential scarcity. Some resources, such as natural plant food sites, would have been too widely distributed to be foraged effectively. Further, irregular or poor salmon harvests on a single river system could have become increasingly devastating to a settled population without access to food sources outside its territory.[16]

Still, there is little evidence of malnutrition in the skeletons found at archaeological sites throughout the region.[17] We can speculate on the reasons for this apparent good nutrition: estuaries provided uniquely low-risk environments with a wide assortment of concentrated food resources—shellfish, flatfish, marine mammals, and edible roots and berries—that could be relied upon when higher-priority foods such as salmon ran low. This contrasts with interior, upriver areas where salmon might be similarly plentiful but secondary resources relatively scarce; these areas had only a fraction of the population densities found in the coastal zone. Effective mechanisms for the intervillage exchange of food and other resources also co-evolved with increasing dependence upon finite territories. Indeed, such mechanisms for intervillage exchange, including the potlatch, were central cultural institutions among the peoples of the Northwest Coast at the time of their first encounters with Europeans.

CIRCUMSTANCES OF RESOURCE INTENSIFICATION AFTER 4000 B.P.

The successes of sedentary fishing strategies fueled the coast's first archaeologically detectable population boom, a rapid increase that took place between roughly 4,000 and 3,000 years ago.[18] In turn, sedentism and population growth contributed to rapid "cultural intensification," which involved increasing social complexity, ceremonialism, and economic stratification. These changes placed demands on local resources beyond those originally posed by the rise of sedentism, facilitating war, elaborating patterns of trade and wealth redistribution, and leading to increasingly sophisticated methods of food storage and resource specialization. Moreover, the display of wealth and status through surplus food production or the use of totemic crests probably grew in importance as sedentary villages increasingly functioned as independent polities and individuals' socioeconomic roles became more specialized.

Together these factors, both environmental and cultural, facilitated the intensified exploitation of many resources. Whereas the growth of sedentary villages had originally centered on productive fishing sites, the gradual increase in specialized food harvesting facilitated a growing secondary pattern of settlement at sites of "secondary resources," such as whale or seal hunting sites, eulachon or halibut fishing grounds, and clam, berry, or root gathering sites.[19] Thus, the settlement pattern that emerged during the last 3,000 years can be described as a network of large winter villages and essentially permanent, resource-specific sites, between which the people traveled seasonally.

Although I describe these changes as generalized across the region, they did vary somewhat along the coast. As one moves from the southern to the northern end of the region, opportunities for food gathering become more restricted. Marine productivity remains high, but terrestrial productivity declines. With cooler temperatures, increased seasonal variability in sunlight, and generally greater moisture in the north, certain edible plant and terrestrial animal resources become less abundant, and the ability to use fire to maintain forest clearings is diminished. Accordingly, although villages achieved roughly equal levels of resource wealth, their terrestrial hunting and gathering territories—and indeed, entire tribal territories—were larger at the northern end of the Northwest Coast culture region than at the southern end.[20]

Perhaps most importantly, salmon runs are shorter in duration in the northern portion of the coast. Prehistoric salmon runs in the southern Northwest Coast persisted discontinuously for months. On the rivers of southeastern Alaska, however, runs lasted only a few days. As a result, people of the northern coast were compelled to catch and store the bulk of their salmon in a matter of only a few days. This may partially explain why many north coast peoples, such as the Tlingit, appear to have developed sedentary salmon-fishing village sites comparatively recently. When they did, the organizational feat of conducting their major food-gathering activity in such a compressed time seems to have called for greater authority among elites than was found on the southern coast. This likely contributed to the northern coast's comparatively more hierarchical social structure and to more ostentatious displays of power and wealth by its elites at the time of European contact.[21]

PRECARIOUS POLITIES, PRESTIGE, AND RESOURCE PRODUCTION

What we refer to today as "tribes" had no formal political existence prehistorically. Ceremonial, economic, and kinship ties may have bonded villages together, but the "Kwakiutl" and the "Nootka," for example, were not "tribes" but ethnolinguistic groups. The village was the primary political unit. Each was largely autonomous in matters of politics, war, and trade, and each appears to have formed intervillage allegiances based upon mutual, intervillage interests.[22] Individual households—essentially extended kin groups and their slaves—were the fundamental social units in ceremony, food procurement, and resource ownership. Households were largely independent economic units that controlled a subset of resource sites within a larger, usually village-owned resource territory.[23]

This social structure was sustained by a number of distinctive cultural institutions. Perhaps most important among these at the time of European contact was the potlatch, a ritual exchange of goods that accompanied significant events such as marriage, death, personal shame, or puberty. Through these exchanges, elites repaid or generated debts, mediated and ameliorated interpersonal disputes, and enhanced their personal and household status. Goods were exchanged among elites from one or more villages, and these goods often were partially redistributed within the elites' households. Although the emergence of the potlatch has been difficult to document archaeologically, there is persuasive archaeological evidence that its rudiments had emerged by as early as 3,500 years ago.[24]

This emphasis upon elite legitimation greatly influenced indigenous patterns of resource use. Resource production was required not only to meet daily dietary needs but also to produce a surplus that could be displayed and distributed to maintain political and economic structures. To maintain or enhance their position, elites had strong incentives to acquire wealth and enhance the productivity of resources within their finite territories.[25] Scholars have long debated whether the goal of the potlatch was social status or resource wealth, but this debate has created a false dichotomy. Rather, resource wealth could not be obtained without high social status, and high social status could not be obtained without access to resource wealth.

Although most Northwest Coast elites derived their position from family lineage, their hold on power appears to have been precarious. If elites did not uphold the duties associated with their rank—notably the maintenance of group wealth and prestige through the potlatch and other forms of socio-economic validation—they could easily lose their ability to influence people or command the resources of their household or village. Without standing militaries or the capacity for forceful internal controls, elites needed popular support to maintain leadership. They were compelled to demonstrate and redistribute their wealth and prestige to their constituents. The potlatch, therefore, was not, as some have suggested, simply a means of "banking" resource wealth—of giving away resources so that one might receive them back during lean times. It was also a form of "profit sharing" for household resource production.[26] As beneficiaries of their household or village elites' fluctuating wealth and prestige, commoners also sought to enhance resource output. Cumulatively, resources became the commodities of both ceremonial and mundane trade relationships that likely facilitated localized resource specialization and strong ties and exchanges between ethnolinguistically diverse coastal peoples.[27]

THE INTENSIFICATION OF FINITE RESOURCE SITES

Between 3,000 and 1,500 years ago, several factors combined to change the ways in which the native peoples of the Northwest Coast made a living. Population pressures on finite territorial resources, trade and potlatch exchanges of goods, and the continuous need for elites' legitimation combined to facilitate attempts to increase resource output. This was achieved through the development of a number of "ingenious and complex technologies."[28]

Among the better-known examples are the salmon weirs and traps found along the entire coast. Aligned where fish ascended the streams in great numbers, these partially submerged weirs allowed fish to pass inside unharmed but barred their exit or ensnared them in labyrinthine baskets. Fish traps made of V-shaped or semicircular rock piles in the mid to low intertidal zones of estuaries served to trap fish and shellfish on the outgoing tide. Dip nets allowed the capture of smelt and eulachon as they entered rivers along the coast, and fresh-cut hemlock boughs were submerged in inlets to capture the edible eggs of herring.[29] Beds of clams and other bivalves were expanded by the removal of rocks and possibly the construction of silt-trapping rock alignments in the lower intertidal zone; such beds were sometimes subdivided into owned plots.

These resource enhancement strategies greatly improved localized food-gathering prospects and probably co-evolved with sedentary fishing villages. Furthermore, Northwest Coast oral history is peppered with stories of indigenous people managing animals such as clams and salmon smolts or "transplanting" them from places outside of easy or defensible access to places near their villages or within their own resource territories. As Darryl Forde noted of Nootka salmon management: "If the run on a particular stream began to fail, they actually restocked it, obtaining spawn from another river at the breeding season and carrying it back in moss-lined boxes to start a new generation in the depleted steam."[30]

Plant resources were also managed within discrete tribal territories in order to enhance their output. By replicating and enhancing natural conditions that fostered plant growth—such as the use of fire to promote the growth of meadow plants such as camas—people developed patterns of low-intensity plant cultivation. Among all indigenous strategies of resource intensification, these attempts to intensify the localized output of particular plants generally have been overlooked. The remainder of this chapter seeks to remedy this oversight.

THE QUESTION OF CULTIVATION

By most accounts, people of the Northwest Coast of North America did not cultivate plants prior to European contact. There is, however, a body of evidence suggesting that precontact people of this region did in fact engage in cultivation. In large part, these indigenous practices expanded the amount and productivity of land within a given resource territory on which coveted plant resources could grow. In the process, people altered the ecology of the coastal zone and, it seems, significantly augmented their diet.

Challenging the European biases of past studies, most scholars now assert that "emergent agriculturists" often merely *augmented* hunting and gathering with plant food production, thus increasing the abundance and predictability of their food supply.[31] Accordingly, "cultivation" has been redefined to include a continuum of practices that involve the repeated and purposeful manipulation of plants and their environments to enhance productivity. Cultivation is often indicated by the presence of "agro-ecosystems": human-constructed environments such as the small swiddens of indigenous slash-and-burn agriculture or the terraces, canals, and impoundments of wetland cultivation.[32]

The coastal forests of the Northwest have one of the highest levels of terrestrial biological productivity in the world, if productivity is measured as carbon stored in plant materials per unit of area. Nonetheless, there are serious obstacles to the growth of nutritionally valuable plants in this zone, most notably the low levels of sunlight under dense forest canopies and the region's leached soils.[33] Native people of the Northwest Coast overcame both solar and nutritional obstacles to food plant production by modifying existing plant communities—creating agro-ecosystems, in essence—to provide villages with nearby concentrations of staple food plants that might otherwise have been dispersed too widely to justify simple foraging.

Europeans were quick to note that people of the Northwest cultivated two familiar crop plants: tobacco and potatoes.[34] In 1787, 13 years after the first documented contact between natives and Europeans, both Archibald Menzies and George Dixon found the Haida and Tlingit cultivating a variety of tobacco (*Nicotiana quadrivalis*) in patches of ground near their primary settlements. The plant was probably native to the drier, interior Northwest rather than to the moist coastal zone, and it appears to have gone extinct in the coastal zone almost immediately after indigenous cultivation ceased. The seeds of the plant reportedly were collected and replanted with digging sticks; the leaves were chewed with lime made from burned shells to enhance the narcotic effect—a process without parallel elsewhere in the northern Pacific. Genetic evidence, archaeological remains, and the presence of tobacco in indigenous oral traditions suggest that the cultivation of introduced tobacco on this coast was a practice of considerable antiquity.

The potato also arrived in many portions of the coast before direct European contact. It spread through intertribal trade networks and was often grown in piles of rotting marine detritus. The diffusion of the potato appears to have happened so early that locally distinctive varieties may have emerged prior to the arrival of Europeans. Both tobacco and potatoes were grown in clearings presumably created by fire and manual clearing of native vegetation.

European explorers believed that these were the only cultivated plants, but coastal peoples also cultivated several native plants using the same methods—plants that were alien to the European eye and therefore were consistently overlooked in travel narratives and later ethnographic accounts. Although the potato was clearly introduced and tobacco was probably imported from the interior Northwest through trade networks, the planting practices used in their cultivation were very likely long-standing and quite possibly endemic.

CAMAS AND FIRE: AN EXAMPLE FROM THE SOUTHERN NORTHWEST COAST

People of the Northwest Coast tended a host of native plants that received little attention from early explorers, missionaries, and traders, who failed to see evidence of intentional "cultivation." These plant communities did not look like "agriculture" to the European eye: they usually lacked the rectilinear shape of European gardens and contained a polyculture of unfamiliar species.

Despite its poor representation in the literature of anthropology, there is clear evidence along much of the southern Northwest Coast for native cultivation of camas (*Camassia quamash* and *Camassia leichtlinii*) through transplanting and the use of fire. Burning was carried out on a grand scale in the interior plateaus and the Willamette Valley and Puget Sound lowlands, where it contributed to the creation of grasslands hundreds or thousands of square miles in extent.[35] Camas production appears to have been one of several intended outcomes of this process, although these cases may not warrant the designation "cultivation." A clearer case of cultivation comes from the more densely settled spruce-hemlock zone—the narrow strip of temperate rainforest that abuts the ocean coast. The inhabitants of this zone burned small plots, often of only a few thousand square feet, on south-facing slopes at the fringes of the dense, moist coastal forest. Although people of the northwestern interior burned patches of the landscape to facilitate the growth of wild edible plants, improve habitat for game, increase village defense and mobility over the land, and reduce insects, the small plots of the outer coast would have had negligible value for most of these purposes. Although burned prairies might have aided hunting in the coastal forest, fire served primarily to maintain small plots of volunteer and introduced plants, notably camas, by temporarily enhancing soil fertility and eliminating plants that competed with camas for sunlight.[36]

For example, the Tillamook—the Salish-speaking inhabitants of the northern Oregon coast—burned small clearings to create camas prairies; the earliest explorers' accounts of their territory describe a large number of such prairies.[37] According to the field notes of Elizabeth Derr Jacobs, an ethnographer who worked among Tillamook elders early in the twentieth century, camas from these village-owned clearings served as the "primary root food of the Tillamook" prior to contact and also were an important source of medicines.[38] Other plants that occupied these clearings, such as bracken fern, huckleberries, crab apples, and certain lilies, were important for both subsistence and trade purposes. As elsewhere on the coast, women were responsible for most camas digging. Because their status in the community depended in no small part on their ability to procure roots and berries, oral traditions suggest that they were motivated to intensify these resources through the use of fire and the tending of camas plots.

Camas is native to the drier interior and rocky montane areas of the Northwest. It is unlikely that it was native to the inhabited coastal fringes of Tillamook territory or to the spruce-hemlock zone, where it was common in the late eighteenth century. Oral traditions suggest that camas was a popular trade good imported from outside of the moist coastal zone by the Tillamook and other peoples. Its presence on the coast was due to the continuous use of fire, selective harvesting, the annual turning of the soil, and intentional or unintentional transplanting.

According to Tillamook creation tales, camas was originally brought by a transformer culture hero who dispersed it within Tillamook territory, saying of each place, "That will be a big camas field. They will always grow there."[39] Unfortunately, they no longer grow there, and the prairies on which they grew are disappearing rapidly.[40] Camas began to disappear from Tillamook territory once burning ceased, allowing successional vegetation to crowd them out of existence. The local presence of this staple food plant thus appears to have been entirely dependent upon human intervention—or "cultivation." Indeed, today camas is almost entirely absent from Tillamook territory, where once it served as a staple food to a population of thousands.

RHIZOME CULTIVATION: AN EXAMPLE FROM THE CENTRAL NORTHWEST COAST

Another, more intensive form of plant cultivation that predated European contact involved root-cropping, particularly of two endemic plants: springbank clover (*Trifolium wormskjoldii*) and Pacific silverweed (*Potentilla anserina* ssp. *pacifica*). People of the Oregon and Washington coasts used these two plants extensively; on the southern portion of the coast, patches were collectively owned and selectively weeded and harvested by people from individual villages. It appears, however, that among people from Vancouver Island northward (where terrestrial resources may have been under greater population pressure), plots of these plants were managed in a manner that can only be described as cultivation. Indeed, early writers—notably the anthropologist Franz Boas, who compiled voluminous notes on the practice among the Kwakiutl—termed these plots "gardens."[41] They occupied estuarine gravel beds, and the *Potentilla* and *Trifolium* were ordinarily grown together. They were harvested, usually by women, for their edible, starchy rhizomes or "roots," which served as a staple source of dietary starches among many Northwest Coast peoples. The rhizomes grow to almost one-half inch in diameter and accumulate into dense, sodlike clusters.[42]

These gardens were produced by a labor-intensive process. Beds commonly were constructed by removing rocks and small boulders to create a level surface. Rocks were piled along the boundaries of each plot (in a manner similar to the intertidal rock fish traps of the region), and these borders were often lined with split boards.[43] Undesirable plants were regularly weeded out. Significantly, testimony from several geographically distant peoples of this region recounts that the ends of the rhizomes of desirable specimens were consistently "placed back in the ground so they would grow the following year."[44] This pattern of management was observed and occasionally documented from the earliest European presence. For example, in September 1792, Archibald Menzies noted Nootka women tending abundant plots of *Trifolium*-containing soil that was "regularly turned over every year" and probably managed in other ways as well.[45]

Available data suggest that cultivated rhizome gardens occupied large areas, with several acres of gardens being found in association with single winter villages. Despite the abundance of these gardens, demand for the plants remained high. Among the "Kwakiutl," Boas noted that each garden was owned by an individual clan, and each plot within the garden was owned by a family subunit of the clan. Unauthorized people were prohibited from trespassing or utilizing the plants grown in the gardens. Nancy Turner describes recurring stories among the Nootka of an important clan chief who had six to ten slaves guard a particularly productive garden to ensure that no unauthorized person would dig there. This rigid system of land tenure reflects, in part, the fact that the demand for these resources exceeded the readily foraged supply on segments of the Northwest Coast.[46]

The demand for these rhizomes and resulting efforts to intensify their production were not strictly a function of their dietary significance. In addition, these practices may be partially attributable to the significant role the plants played in Northwest Coast ceremonial subsistence. The rhizomes were commonly depicted as the high-status food of supernatural beings in Kwakiutl oral traditions. Boas noted the use of these rhizomes as an important part of the Kwakiutl ceremonial barter economy. Several crates of rhizomes, for example, were a regular part of bride price. The Kwakiutl, Nootka, and others also regularly held ceremonialized feasts devoted to the consumption of either *Potentilla* or *Trifolium* rhizomes. The anthropologist Philip Drucker noted that his Nootka consultants expressed pride in great rhizome feasts of the past in which piles of rhizomes were reputedly so vast that their ancestors had to climb to the roofs of their longhouses to pour the water required for their steaming. The longest, thickest rhizomes of both plants were associated with high status, were actively sought in the gardens by cultivators, and were given to the "chiefs" of each Kwakiutl clan as a form of tribute. It thus appears that localized scarcity and ceremonialism were mutually reinforcing reasons for plant cultivation: local scarcity of food plants enhanced their value and ceremonial significance, which in turn increased demand for these plants for socioeconomic purposes, resulting in increased harvests, more scarcity, and attempts to increase production of the rhizomes.[47]

THE ECOLOGICAL BASIS FOR ESTUARINE RHIZOME CULTIVATION

Rhizome cultivation represented an adept ecological solution to environmental and social needs for intensified plant production. Today, collecting rhizomes from naturally occurring patches requires several days of intense labor to prepare for a traditional meal for a single family.[48] The spatial concentration of these rhizomes in gardens lessened the time required for foraging and placed dependable concentrations within each tribe's defensible territorial control. But this does not entirely explain the functions and placement of rhizome gardens.

Although rhizomes of sandy sites reportedly were preferred for their size and the ease with which they could be dug, gardens were most commonly found on rocky tidal flats where considerable labor was required to clear and maintain them.[49] While this placement may have improved access to the gardens from estuarine village sites, it also situated them in an ecologically unique position in the estuary: in a narrow band where, in Eugene Odum's terms, the "energy subsidy" from outside sources was particularly high but the cumulative stresses on the plants were relatively low.[50] Indeed, this strategy harnessed the tremendous biological productivity of the mid-latitude estuary—one of the world's most productive ecosystems—in a way that has had few parallels elsewhere in the world.

Like camas cultivation, this plant management strategy overcame the dual limitations of low soil fertility and the region's dense forest canopy. The narrow band of upper estuary where these gardens were located is high enough to avoid daily inundation by brackish estuarine water but low enough to be flooded periodically by peak annual tides and stream flows. As a result, the upper estuary is one of the few places in the spruce-hemlock zone where soil fertility remains high, because the periodic inundation deposits detritus in this slow-moving backwater. Furthermore, the detritus contains a wide assortment of organics such as plankton, bacteria, marine algae, estuarine eel grass, animal remains and wastes, and particulate and dissolved matter from upstream sources. Roughly one-quarter of the total organic material that enters the estuaries of the south and central Northwest Coast is redistributed to the upper marsh in the form of detritus; this occurs most intensively in the fall or early winter, when there is a die-off of estuarine plants and animals, and materials are uprooted and transported to the high tide line by floodwaters and heavier wave action. The region's high rainfall dilutes the salinity of this estuarine margin enough to allow the entrance of terrestrial decomposers such as earthworms, but salinity periodically rises high enough to eliminate most terrestrial plants—so the forest canopy is kept at bay.

While most of the region's terrestrial vegetation must contend with very low levels of phosphorous and nitrogen, the plants of the upper estuary benefit from the nitrogen-fixing bacteria on the decomposing rhizomes of marsh plants and the binding of phosphorous to the estuarine detritus. Plants and sediments in places with reduced stream velocities—especially the backwaters of the upper estuary—thus are nutrient sinks, capturing and storing the estuary's organic output.[51] The main obstacle to the productivity of *Potentilla* and *Trifolium* in this environment is the presence of rocks. Gardeners constructed short rock terraces that removed rocks from the growing bed and expanded the overall area of this high marsh zone. These terracelike features often created a high marsh zone several times the size of that found prior to cultivation.[52]

Not only did these rockworks slow and impound water, acting as efficient settling ponds for detritus and sediment, but small dams and interwoven hemlock boughs appear often to have been placed a short distance seaward from the gardens for the express purpose of trapping sediment.[53] According to twentieth-century Kwakiutl informants, their ancestors mounded this detritus soil on gardens to produce a planting surface, presumably retained by the short rock and wood enclosures. The rock gardens thus functioned in a manner reminiscent of other low-intensity agricultural practices indigenous to the Americas that relied on the accumulation of nutrient-rich shoreline deposits, such as the floodwater fields constructed along streambeds or lakes by Indians of the desert Southwest or the American tropics.[54] Churning of the garden soil with digging sticks also aerated and mixed detritus into the soil, facilitating the growth of the valuable larger rhizomes. The estuarine gardens consistently produced more nutrient-rich soil than would have been found elsewhere in the spruce-hemlock zone. Indeed, the Kwakiutl term for the gardens described here translates as "manufactured-soil," suggesting that their role in soil accumulation was understood and intentional; the term is semantically distinct from those used for naturally occurring patches of rhizomes.[55] The product of these practices was a crop of rapidly growing starchy rhizomes, larger than those of naturally occurring plants and with a nutritional value higher than that of the potato.[56]

Clearly, *Potentilla* and *Trifolium* were "cultivated." But were they also "domesticated"? In other words, did the long-term selective propagation of the plants produce genetically distinct populations of these rhizomes?[57] Ethnobotanical studies demonstrate that some coastal peoples recognized two types of estuarine *Trifolium*. Similarly, in places with a documented tradition of intensive rhizome gardening, there are distinct varieties of *Potentilla*.[58] Although these may represent responses to natural differences in growing conditions, it is an intriguing but as yet unexamined possibility that prolonged protection and propagation may have produced genetically distinct populations of *Potentilla* and *Trifolium*.

CONTACT AND CONTRACTIONS

The arrival of Europeans in the late eighteenth century caused dramatic changes in patterns of indigenous cultivation. Diseases spread throughout the Northwest Coast. As mortality rates soared, local population pressure on endemic resources diminished, as did incentives to enhance the productivity of local resource sites.[59] As populations plummeted along the coast, survivors regrouped with survivors from other villages in new, multitribal villages, a trend that continued well into the twentieth century. Slave labor diminished as a result of high mortality followed almost immediately by forceful colonial opposition to this practice.[60] Formerly land-rich tribal elites became social and economic exiles as relocation uprooted them from their traditional territories and they lost control of their resources. Although traditional foods were still eaten by Northwest Coast groups, labor-intensive food production strategies were largely abandoned in the course of relocation. For example, although salmon fishing continued, the use of fish traps and weirs rapidly diminished over the course of the nineteenth century. Endemic, intensive traditions of plant cultivation also began to disappear.

The abandonment of traditional cultivation practices also paralleled the simultaneous decline of the fur trade. As fur-bearing animals were depleted, native peoples looked for new items to trade with Europeans arriving by ship. Potatoes began to replace furs as the coast's premier trade good. Potatoes began to be cultivated more widely than they were before contact, rapidly replacing most endemic root crops in the native diet.[61] This transition was also fostered by missionaries who, not recognizing true "agriculture" in the unfamiliar plants and patterns of indigenous cultivation, promoted the Indians' adoption of European agriculture as a necessary early step in the progression from "savagery" to "civilization." Since few other introduced crops grew well on the perennially drizzly Northwest Coast, potatoes became the primary new crop. While it was palatable to native people, the potato and other European crops also became status symbols—a sort of "European merchandise"—which aided the plants' rapid dispersal.

As Indian settlements reconsolidated near missions, schools, canneries, and sawmills, traditional foods were increasingly replaced by foodstuffs purchased from European settlers.[62] With the rising dependence of native people upon the cash economy and the declining ceremonial significance of native foodstuffs, only a portion of traditional subsistence methods—such as salmon fishing, berry picking, and shellfish harvesting—persisted among the majority. Rhizome gardens fell into disuse and were slowly destroyed by diking, floods, waterborne logs, tsunamis, and artifact hunters. New settlers displaced native people throughout large sections of the spruce-hemlock zone. Only in the more remote native communities did the staple use of endemic food plants, as well as the use of fire in vegetation management, continue into the present.

Although traditional foodstuffs such as salmon, berries, and shellfish retained an important role in the native diet of the early twentieth century, the resurgence in indigenous identity during the late twentieth century has heightened interest in traditional foods and food preparation. Along with this trend has come a revived interest in traditional ethnobotany. Traditional plant foods are being experimentally readopted. Some of the native communities that abandoned traditional cultivation practices in the nineteenth century now have begun to contemplate reviving some of the cultivation practices of their ancestors.[63]

CONCLUSION

The prehistoric Northwest Coast commonly has been described as a land of resource abundance. Still, it may well have been the "patchy" distribution of resources and localized resource scarcity that served as catalysts not only for the emergence of complex political institutions and forms of landownership on the Northwest Coast but also for the emergence of low-intensity forms of plant cultivation. Paradoxically, perhaps, the abundance of marine resources appears to have both facilitated the emergence of plant cultivation and inhibited its emergence as the region's primary form of subsistence. The peoples of the Northwest Coast cultivated food plants in places they modified to provide adequate sunlight and fertile soil for higher plant productivity—places such as the upper intertidal zone and burned meadows. This intensified production served not only as an important foundation of the native diet but also as a means to many social and ceremonial ends.

Nineteenth-century writers who dismissed the Northwest Coast as a place lacking cultivation did so without appreciating the sophistication of Northwest Coast plant use.[64] This label has persisted despite the fact that less intensive practices elsewhere have been identified as "cultivation" or "agriculture" in contemporary ethnographic studies.[65] If the Northwest Coast was not "agricultural," it was, at the very least, a place of considerable plant tending and cultivation. And although the peoples of this coast generally ate well, talk of resource superabundance oversimplifies indigenous subsistence. Their long history of dietary success reflects the presence of effective cultural strategies for resource procurement as much as it reflects the fortuitous presence of these peoples in a resource-rich region of North America.

NOTES

1. The people whom Euro-Americans, including anthropologists, have known as the Kwakiutl and the Nootka call themselves the Kwakwaka'wakw and the Nuu-Chah-Nulth, respectively. Because most readers are familiar with the old usages, they have been retained in this chapter.

2. R. Benedict, *Patterns of Culture* (Boston: Houghton Mifflin, 1934), pp. 173–74.

3. The Kwakiutl, for example, appear to have taken neighboring tribes' land by force, both before and after the arrival of Europeans. See H. C. Taylor, Jr., and W. Duff, "A Post-Contact Southward Movement of the Kwakiutl," *Western Washington State College Research Studies* 24 (1956): 133–47; J. D. Vaughan, "Toward a New and Better Life: Two Hundred Years of Alaskan Haida Cultural Change" (Ph.D. diss., University of Washington, 1984).

4. For example, when the Spanish garrison at Nootka Sound occupied an important Nootka fishing site, Francisco de Eliza reported that Chief Maquinna and his people "do not cease in coming to question me daily about when we will leave," showing their emaciated bodies to the Spaniards as proof of their inability to find

adequate food elsewhere on their lands. F. de Eliza, "Costumbres de los naturales del Puerto de San Lorenzo de Nuca" (April 1791), translated and quoted in C. I. Archer, "Seduction before Sovereignty: Spanish Efforts to Manipulate the Natives in Their Claims to the Northwest Coast," in *From Maps to Metaphors: The Pacific World of George Vancouver*, eds. R. Fisher and H. Johnston (Vancouver: University of British Columbia Press, 1993), pp. 149, 153.

5. See S. T. Schultz, *The Northwest Coast: A Natural History* (Portland: Timber Press, 1990).

6. F. Boas, *Geographical Names of the Kwakiutl Indians*, Columbia University Contributions in Anthropology 20 (1934); J. P. Harrington, *Tillamook Ethnographic and Linguistic Notes* (Washington, D.C.: Smithsonian Institution Microfilms, John Peabody Harrington Collection, n.d.); D. Deur "Chinook Jargon Placenames as Points of Mutual Reference: Discourse, Environment, and Intersubjectivity in an Intercultural Toponymic Complex," *Names* 44 (1996): 291–321.

7. See R. J. Hebda and R. W. Mathewes, "Holocene History of Cedar and Native Indian Cultures of the North American Pacific Coast," *Science* 225 (1984): 711–13.

8. C. Carlson, "The Early Component at Bear Cove," *Canadian Journal of Archaeology* 4 (1979): 177–94; R. G. Matson, "The Glenrose Cannery Site," *National Museum of Man Mercury Series, Archaeological Survey Papers*, 52 (1976); C. E. Borden, "Origins and Development of Early Northwest Coast Culture, to About 3000 B.C.," *National Museum of Man Mercury Series, Archaeological Survey Papers* 45 (1975).

9. Evidence for sedentism at the earlier extreme of this chronology can be found in A. Cannon, *The Economic History of Namu* (Burnaby, B.C.: Simon Fraser University Department of Archaeology Press, 1991). Evidence for a more recent emergence of sedentism is provided in R. G. Matson, "The Evolution of Northwest Coast Subsistence," *Research in Economic Anthropology* (Supplement) 6 (1992): 367–430.

10. K. R. Fladmark, "A Paleoecological Model for Northwest Coast Prehistory," *National Museum of Man Mercury Series, Archaeological Survey Papers* 43 (Ottawa: National Museum of Man); Cannon, *The Economic History of Namu*.

11. See D. H. Mitchell, "Seasonal Settlements, Village Aggregations, and Political Autonomy on the Central Northwest Coast," in *The Development of Political Organization in Native North America*, ed. E. Tooker (Washington, D.C.: American Ethnological Society, 1983), pp. 97–107.

12. R. F. Schalk, "The Structure of an Anadromous Fish Resource," in *For Theory Building in Archaeology*, ed. L. R. Binford (New York: Academic Press, 1978), pp. 207–49; Matson, "The Evolution of Northwest Coast Subsistence."

13. See K. M. Ames, "The Northwest Coast: Complex Hunter-Gatherers, Ecology, and Social Organization," *Annual Review of Anthropology* 23 (1994): 209–29.

14. K. M. Ames, "The Evolution of Social Ranking on the Northwest Coast of North America," *American Antiquity* 46 (1981): 789–805.

15. See A. Richardson, "The Control of Productive Resources on the Northwest Coast of North America," in *Resource Managers: North American and Australian Hunter-Gatherers*, eds. N. M. Williams and E. S. Hunn (Boulder: Westview, 1982), pp. 93–112.

16. L. Donald and D. H. Mitchell, "Some Correlates of Local Group Rank among the Southern Kwakiutl," *Ethnology* 14 (1975): 325–46.

17. See J. S. Cybulski, "Human Biology," in *Handbook of North American Indians, vol. 7: Northwest Coast*, ed. W. Suttles (Washington D.C.: Smithsonian Institution Press, 1990), pp. 52–59.

18. D. R. Croes and S. Hackenberger, "Hoko River Archaeological Complex: Modeling Prehistoric Northwest Economic Evolution," *Research in Economic Anthropology* (Supplement) 3 (1988): 19–85. Space does not permit a thorough discussion of subregional trends, which often varied considerably; therefore, archaeological and cultural trends are discussed here in very general, regionwide terms.

19. See R. Minor, "Aboriginal Settlement and Subsistence at the Mouth of the Columbia River" (Ph.D. diss., University of Oregon, 1983); G. Thompson, *Prehistoric Settlement Changes in the Southern Northwest Coast: A Functional Approach* (Seattle: University of Washington Department of Anthropology, 1978).

20. R. F. Schalk, "Land Use and Organizational Complexity among Foragers of Northwestern North America," *Senri Ethnological Studies* 9 (1981): 53–75.

21. Schalk, "The Structure of an Anadromous Fish Resource."

22. P. Drucker, "Ecology and Political Organization on the Northwest Coast of America," in Tooker, *The Development of Political Organization in Native North America*, pp. 83–96. For an overview of oral accounts of inter- and intragroup warfare along the entire coast, see R. B. Ferguson, "A Reexamination of the Causes of Northwest Coast Warfare," in *Warfare, Culture, and Environment*, ed. R. B. Ferguson (New York: Academic Press, 1984), pp. 267–328.

23. For an overview of the antiquity and socioeconomic significance of large-scale longhouses, see K. M. Ames, "Life in the Big House: Household Labor and Dwelling Size on the Northwest Coast," in *People Who Lived in Big Houses: Archeological Perspectives on Large Domestic Structures*, eds. G. Coupland and E. B. Banning (Madison, Wis.: Prehistory Press, 1995), pp. 131–50; J. C. Chatters, "The Antiquity of Economic Differentiation within Households in the Puget Sound Region," in *Households and Communities*, eds. S. MacEacharn, D.J.W. Archer, and R. D. Garvin (Calgary: University of Calgary Anthropology Association, 1989), pp. 168–78; G. Coupland, "Household Variability and Status Differentiation in Kitselas Canyon," *Canadian Journal of Archaeology* 9 (1985): 39–56.

24. R. L. Carlson and P. M. Hobler, "The Pender Island Excavations and the Development of Coast Salish Culture," *BC Studies* 99 (1993): 25–50.

25. H.D.G. Maschner, "The Emergence of Cultural Complexity on the Northern Northwest Coast," *Antiquity* 65 (1991): 924–34.

26. D. R. Croes, "Exploring Prehistoric Subsistence Change on the Northwest Coast," *Research in Economic Anthropology* (Supplement) 6 (1992): 337–66; John W. Adams, *The Gitksan Potlatch: Population Flux, Resource Ownership, and Reciprocity* (Toronto: Holt, Rinehart, and Winston Canada, 1973); Ames, "Life in the Big House"; Drucker, "Ecology and Political Organization."

27. Long-standing intraregional exchange patterns are reflected in shared motifs in the region's material culture. See M. A. Holm, "Prehistoric Northwest Coast Art: A Stylistic Analysis of the Archaeological Record" (M.A. thesis, University of British Columbia, 1991). Specialization and trade in both foodstuffs and durable goods appears to have been common. See, e.g., D. R. Huelsbeck, "The Surplus Economy of the Northwest Coast," *Research in Economic Anthropology* (Supplement) 3 (1988): 149–77.

28. Drucker, "Ecology and Political Organization," p. 89.

29. See M. Eldridge and S. Acheson, "The Antiquity of Fish Weirs on the Southern Coast: A Response to Moss, Erlandson, and Stuckenrath," *Canadian Journal of Archaeology* 16 (1992): 112–15; M. L. Moss, J. M. Erlandson, and R. Stuckenrath, "Wood Stake Weirs and Salmon Fishing on the Northwest Coast: Evidence from Southeast Alaska," *Canadian Journal of Archaeology* 14 (1990): 143–58; J. A. Pomeroy, "Stone Fish Traps of the Bella Bella Region" in *Current Research Reports*, ed. R. L. Carlson (Burnaby, B.C.: Simon Fraser University, 1976), pp. 165–73.

30. C. D. Forde, *Habitat, Economy, and Society: A Geographical Introduction to Ethnology* (New York: Harcourt, Brace, 1934), p. 78.

31. Some of this reevaluation can be attributed to late-twentieth-century research that has demonstrated that hunter-gatherers were not uniformly plagued by either chronic scarcity or enervating workloads and thus were not always motivated to engage in a "revolutionary" adoption of cultivation, even if they were aware of the potential for cultivation. See M. Sahlins, *Stone Age Economics* (Chicago: Aldine-Atherton, 1972). See also D. M. Pearsall, "The Origins of Plant Cultivation in South America," in *The Origins of Agriculture: An International Perspective*, eds. C. W. Cowan and P. J. Watson (Washington, D.C.: Smithsonian Institution Press, 1992), pp. 173–205; P. B. Griffen, "Hunting, Farming, and Sedentism in a Rain Forest Foraging Society," in *Farmers as Hunters: The Implications of Sedentism*, ed. S. Kent (Cambridge: Cambridge University Press, 1989), pp. 60–70.

32. See J. J. Parsons and W. A. Bowen, "Ancient Ridged Fields of the San Jorge River Floodplain, Columbia," *Geographical Review* 56 (1966): 317–43; G. C. Wilken,

Good Farmers: Traditional Agricultural Resource Management in Mexico and Central America (Berkeley: University of California Press, 1987); W. Denevan, "Stone vs. Metal Axes: The Ambiguity of Shifting Cultivation in Prehistoric Amazonia," *Journal of the Steward Anthropological Society* 20 (1992): 153–65; W. Doolittle, "Agriculture in North America on the Eve of Contact: A Reassessment," *Annals of the Association of American Geographers* 82 (1992): 386–401.

33. See J. F. Franklin and R. H. Waring, "Distinctive Features of the Northwest Coniferous Forest: Development, Structure, and Function," in *Forests: Fresh Perspectives from Ecosystem Analysis*, ed. R. H. Waring (Corvallis: Oregon State University Press, 1980), pp. 59–85; T. Fujimori, *Primary Productivity of a Young* Tsuga heterophylla *Stand and Some Speculations about Biomass of Forest Communities on the Oregon Coast* (Portland: U.S. Forest Service, Pacific Northwest Research Report PNW-123, 1971).

34. On tobacco, see N. J. Turner and R. Taylor, "A Review of the Northwest Coast Tobacco Mystery," *Syesis* 5 (1972): 249–57. On potatoes, see W. Suttles, "The Early Diffusion of the Potato among the Coast Salish," in *Coast Salish Essays* (Seattle: University of Washington Press, 1987), pp. 137–51.

35. See R. Boyd, "Strategies of Indian Burning in the Willamette Valley," *Canadian Journal of Anthropology* 5 (1986): 65–86; R. White, *Land Use, Environment, and Social Change: The Shaping of Island County, Washington* (Seattle: University of Washington Press, 1980); H. H. Norton, "The Association between Anthropogenic Prairies and Important Food Plants in Western Washington," *Northwest Anthropological Research Notes* 13 (1979): 175–200; C. L. Johannesen, W. Davenport, A. Millet, and S. McWilliams, "The Vegetation of the Willamette Valley," *Annals of the Association of American Geographers* 61 (1971): 286–306.

36. There has been some suggestion that rotting marine detritus was used to additionally fertilize camas plots in a manner similar to that used for potato cultivation. On coastal camas plots, see E. Gunther, *Ethnobotany of Western Washington: The Knowledge and Use of Indigenous Plants by Native Americans* (Seattle: University of Washington Press, 1945, 6th ed. 1992), p. 24; A. B. Reagan, "Plants Used by the Hoh and Quileute Indians," *Transactions of the Kansas Academy of Science* 37 (1934): 55–70; Norton, "The Association between Anthropogenic Prairies and Important Food Plants in Western Washington"; N. J. Turner, "Burning Mountain Sides for Better Crops: Aboriginal Landscape Burning in British Columbia," *Archaeology in Montana* 32 (1991): 57–73; L. M. Gottesfeld-Johnson, "Aboriginal Burning for Vegetation Management in Northwest British Columbia," *Human Ecology* 22 (1994): 171–88.

37. Most valuable among these is that of Warren Vaughn, who noted the practices and effects of native people's maintenance of small prairies around Tillamook

Bay, where "not a bush or tree was to be seen . . . as the Indians kept them burned off every spring." W. Vaughan, "An Early History of Tillamook" (Tillamook, Oreg.: Tillamook Pioneer Museum Archives, n.d.), p. 119.

38. E. D. Jacobs, unpublished Tillamook field notes (University of Washington Special Collections, Jacobs Collection, n.d.), folder 106–8, p. 23.

39. E. D. Jacobs, *Nehalem Tillamook Tales* (Corvallis: Oregon State University Press, 1990), pp. 125–26; D. E. Deur and M. T. Thompson, "South Wind's Journeys: A Tillamook Epic," in *One People's Stories: A Collection of Salishan Myths and Legends* (Washington, D.C.: Smithsonian Institution Press, in press).

40. See J. Gritzner, "Native-American Camas Production and Trade in the Pacific Northwest and Northern Rocky Mountains," *Journal of Cultural Geography* 14 (1994): 33–50.

41. Boas described these gardens in some detail. See F. Boas, "The Kwakiutl of Vancouver Island," *Bulletin of the American Museum of Natural History* 8 (1909): 301–522; F. Boas, *Ethnology of the Kwakiutl*, Thirty-fifth Annual Report of the Bureau of American Ethnology, pts. 1, 2 (Washington, D.C.: U.S. Government Printing Office, 1921); F. Boas, *Kwakiutl Ethnography*, ed. H. Codere (Chicago: University of Chicago Press, 1966); Boas, *Geographical Names of the Kwakiutl.*

42. Boas, *Ethnology of the Kwakiutl*, pp. 527–43.

43. See Boas, *Geographical Names of the Kwakiutl Indians*, p. 37; Boas, *Ethnology of the Kwakiutl*, pp. 186–94.

44. N. J. Turner and B. S. Efrat, *Ethnobotany of the Hesquiat Indians of Vancouver Island*, British Columbia Provincial Museum Cultural Recovery Paper no. 2, Ethnobotanical Contribution no. 1 of the Hesquiat Cultural Committee (Victoria: British Columbia Provincial Museum, 1982), pp. 68, 73; see also N. J. Turner and H. V. Kuhnlein, "Two Important 'Root' Foods of the Northwest Coast Indians: Springbank Clover (*Trifolium wormskioldii*) and Pacific Silverweed (*Potentilla anserina spp. pacifica*)," *Economic Botany* 36 (1982): 411–32; G. Edwards, "Indian Spaghetti" *The Beaver* (Autumn 1979): 6; D. Deur, "Wetland Cultivation on the Northwest Coast of North America" (M.A. thesis, Louisiana State University, 1997).

45. C. F. Newcombe, *Menzies' Journal of Vancouver's Voyage: April to October 1792* (Victoria: W. H. Cullin, 1923), p. 117.

46. See Boas, *Ethnology of the Kwakiutl*, pp. 145–48; Turner and Efrat, *Ethnobotany of the Hesquiat*, p. 120.

47. Boas, *Ethnology of the Kwakiutl*, pp. 527–42, 1333–39; Boas, *Geographical Names of the Kwakiutl Indians;* Boas, "The Kwakiutl of Vancouver Island"; Edwards, "Indian Spaghetti"; P. Drucker, "The Northern and Central Nootkan Tribes," *Bureau of American Ethnology Bulletin* 144 (1951): 62.

48. I have found this to be true when collecting rhizomes for research purposes.

See also Turner and Kuhnlein, "Two Important 'Root' Foods of the Northwest Coast Indians," p. 423; Edwards, "Indian Spaghetti," p. 6; Boas, *Ethnology of the Kwakiutl*, pp. 186–94.

49. See Turner and Kuhnlein, "Two Important 'Root' Foods of the Northwest Coast Indians," p. 415.

50. See E. P. Odum, "Halophytes, Energetics, and Ecosystems," in *Ecology of Halophytes*, eds. R. J. Reimold and W. H. Queen (New York: Academic Press, 1974), pp. 599–602. For a widely accessible, classic work on these spatial and trophic flows of energy within estuarine environments, see also E. P. Odum, *Ecology* (New York: Holt, Rinehart, and Winston, 1963).

51. See Odom, "Halophytes, Energetics, and Ecosystems"; B. L Howes and J. M. Teal, "Nutrient Balance of a Massachusetts Cranberry Bog and Relationships to Coastal Eutrophication," *Environmental Science and Technology* 29 (1995): 960–74; R. M. Thom, *Primary Productivity and Carbon Input to Grays Harbor Estuary, Washington* (Seattle: U.S. Army Corps of Engineers, 1981); H. P. Eilers, "Plants, Plant Communities, Net Production, and Tide Levels: The Ecological Biogeography of the Nehalem Salt Marshes, Tillamook County, Oregon" (Ph.D. diss., Oregon State University, 1975).

52. This is well illustrated by the fact that the area covered by rhizomes shrinks considerably after these rock walls collapse and their soil erodes back to the precultivation marsh gradient.

53. Boas, "The Kwakiutl of Vancouver Island," fig. 139; Boas, *Geographical Names of the Kwakiutl Indians*, p. 37; Boas, *Kwakiutl Ethnology*, p. 35.

54. See, for example, W. Doolittle, "Agriculture in North America on the Eve of Contact: A Reassessment," *Annals of the Association of American Geographers* 82 (1992): 386–401; J. H. Steward, "Irrigation without Agriculture," *Papers of the Michigan Academy of Sciences, Arts, and Letters* 12 (1930): 149–56; K. Bryan, "Flood-Water Farming," *Geographical Review* 19 (1929): 444–56; Parsons and Bowen, "Ancient Ridged Fields of the San Jorge River Floodplain, Columbia."

55. Boas, *Geographical Names of the Kwakiutl Indians*, p. 37.

56. For nutritional analyses of these two plants, see H.V. Kuhnlein, N.J. Turner and P.D. Kluckner, "Nutritional Significance of Two Important 'Root' Foods (Springbank Clover and Pacific Silverweed) Used by Native People on the Coast of British Columbia," *Ecology of Food and Nutrition* 12 (1982): 89–95.

57. See for example, J. R. Harlan, *Crops and Man* (Madison, Wis.: American Society of Agronomy, 1975), pp. 63ff (discussion of the distinction between "cultivation" and "domestication").

58. See the discussion of both plants in N. J. Turner, J. Thomas, B. F. Carlson, and R. T. Ogilvie, *Ethnobotany of the Nitinaht Indians of Vancouver Island* (Victoria:

British Columbia Provincial Museum, Occasional Paper Series, no. 24, 1983). See also Turner and Kuhnlein "Two Important 'Root' Foods of the Northwest Coast Indians," pp. 415–16.

59. By the end of the smallpox epidemics of the 1860s and 1870s, the population of the entire region had dropped to approximately 10 percent of its precontact total. Boyd, "Demographic History, 1774–1874."

60. See B. M. Gough, "Send a Gunboat! Checking Slavery and Controlling Liquor Traffic among Coast Indians of British Columbia in the 1860s," *Pacific Northwest Quarterly* 69 (1978): 159–68.

61. See W. Suttles, "The Early Diffusion of the Potato among the Coast Salish." For general overviews of the tempo and character of cultural impacts of European settlement upon native peoples, with occasional mention of ecological consequences, see R. Fisher, *Contact and Conflict: Indian-European Relations in British Columbia, 1774–1890* (Vancouver: University of British Columbia Press, 2d ed., 1992); W. Duff, *The Indian History of British Columbia, vol. 1: The Impact of the White Man* (Victoria: British Columbia Provincial Museum, Memoir 5, 1965).

62. See M. Lee, R. G. Reyburn, and A. Carrow, "Nutritional Status of British Columbia Indian Populations, I: Dietary Studies at Ahousat and Anaham Reserves," *Canadian Journal of Public Health* 62 (1971): 285–96.

63. On cultural revivalism, see, for example, P. T. Amoss, "Strategies of Reorientation: The Contribution of Contemporary Winter Dancing to Coast Salish Identity and Solidarity," *Arctic Anthropology* 14 (1977): 77–83.

64. In a posthumous volume Boas hesitantly used the term "agriculture" to refer to what he described as the "somewhat careless clearing of ground" for *Potentilla* and *Trifolium*. Boas, *Kwakiutl Ethnography*, p. 17.

65. See, e.g., K. Anderson, "Native Californians as Ancient and Contemporary Cultivators," *Ballena Press Anthropological Papers* 40 (1993): 151–74; L. J. Bean, L. J. Lawton, and H. W. Lawton, "Some Explanations for the Rise of Cultural Complexity in Native California with Comments on Proto-Agriculture and Agriculture," *Ballena Press Anthropological Papers* 40 (1993): 27–54; Harlan, *Crops and Man*, pp. 64–68.

8 / Mobility as a Factor Limiting Resource Use on the Columbia Plateau

EUGENE S. HUNN

Native Americans have lived along the Columbia River in the great plateau east of the Cascade Mountains for at least 10,000 years, judging by archaeological evidence from sites such as Wakemap Mound near The Dalles and Windust Cave up the Snake River from its junction with the Columbia (fig. 8.1).[1] Whether the people who gathered roots and berries, fished, and hunted there 10,000 years ago spoke a language directly ancestral to the Sahaptin spoken today by Yakama, Warm Springs, and Umatilla tribal elders cannot be known for certain. The Sahaptin language today, however, has words for more than 500 local plant and animal species and more than 1,000 places scattered across these tribes' traditional territory.[2] This suggests a very long and continuous occupation.

The drastic environmental changes that followed Euro-American settlement of the plateau after 1850 force us to ask, how was it possible for these Indian people to live there for 10,000 years and not degrade their environment? They did not exterminate the salmon or the bitterroots or the black mountain huckleberries they harvested so abundantly each year in season. Did they simply lack the ability to do much harm, armed as they were with just digging sticks, baskets, hemp nets, and bows and arrows? Or were they simply too few ever to have had a significant environmental impact? If so, why did their numbers not increase after 400 generations? Perhaps it had more to do with motives than with means. Perhaps they saw the need to conserve their natural resources long before this was popular among Euro-American settlers.

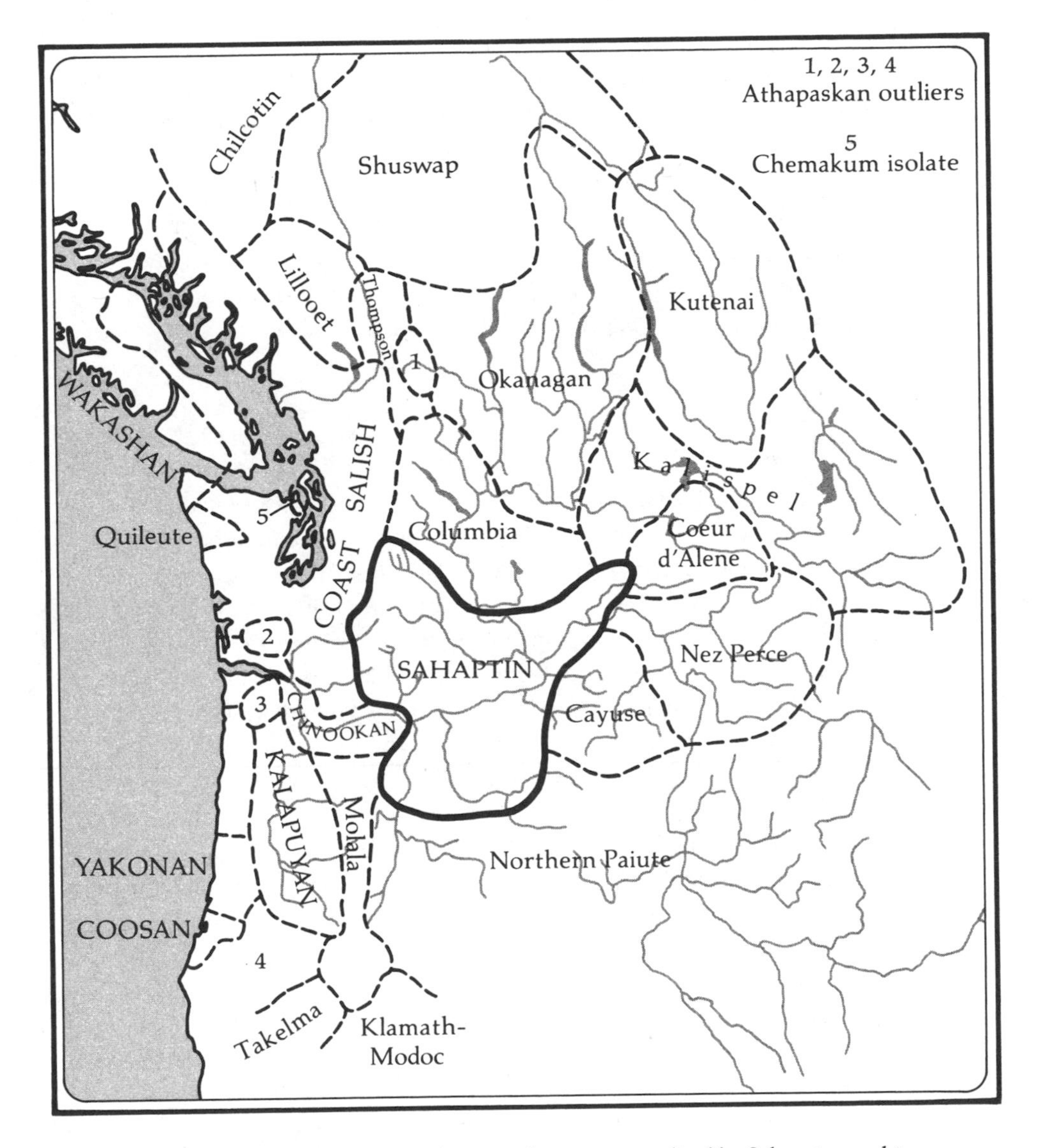

FIG. 8.1. The Pacific Northwest, showing the territory utilized by Sahaptin-speaking people (source: Hunn, *Nch'I-Wana*).

Perhaps they felt no need to harvest more than was required to sustain themselves and their families through the next winter, with a modest surplus to exchange with distant relatives and trading partners.

I address these issues by considering the evidence that Plateau Indians systematically managed their natural resources (such evidence is scant). The mid–Columbia River Indians appear never to have exceeded the capacity of the land to support them, even after 10,000 years. How were they able to maintain this enviable balance with nature for such a long time? Did they recognize the danger of excessive harvests and enforce rules to prevent "overkill"? Or was their life organized in such a way that conscious conservation measures were unneeded? In conclusion, I compare Plateau Indian societies with those on the nearby Pacific coast, where Indian populations were denser and conflict over resources was more intense.

SUBSISTENCE STRATEGIES IN THE COLUMBIA PLATEAU

The native subsistence economy was based on the harvest of anadromous fish and several species of roots, supplemented by resident fish, other vegetal products, and game, primarily mule deer (*Odocoileus hemionus*), in varying proportions.[3] Though there is evidence for some intensification of resource use and consequent population increase during this time,[4] the same basic resource types have been exploited throughout this long period. The local adoption of the horse after 1750 no doubt heightened mobility, but its major effects seem to have been more social than economic.[5] The basic plateau subsistence round was probably little affected by these changes.

At European contact the Columbia basin was home to Salishan and Sahaptin speakers. Ethnobiological research with contemporary Sahaptin informants conducted by David French and me has demonstrated the critical importance of plants as staple food sources, with root "crops" estimated to have provided on average more than 60 percent of local caloric needs.[6] The salmon harvest contributed on average some 30 percent of caloric needs and an abundance of protein.[7] At least 100 plant species were regularly used as food, suggesting the breadth of the human ecological niche in the plateau.

The following is a summary of patterns of resource use by Sahaptin-speaking people of the John Day and Umatilla areas; these patterns illustrate many features characteristic of subsistence patterns throughout the plateau region. The summary is organized by native categories of resources, roughly

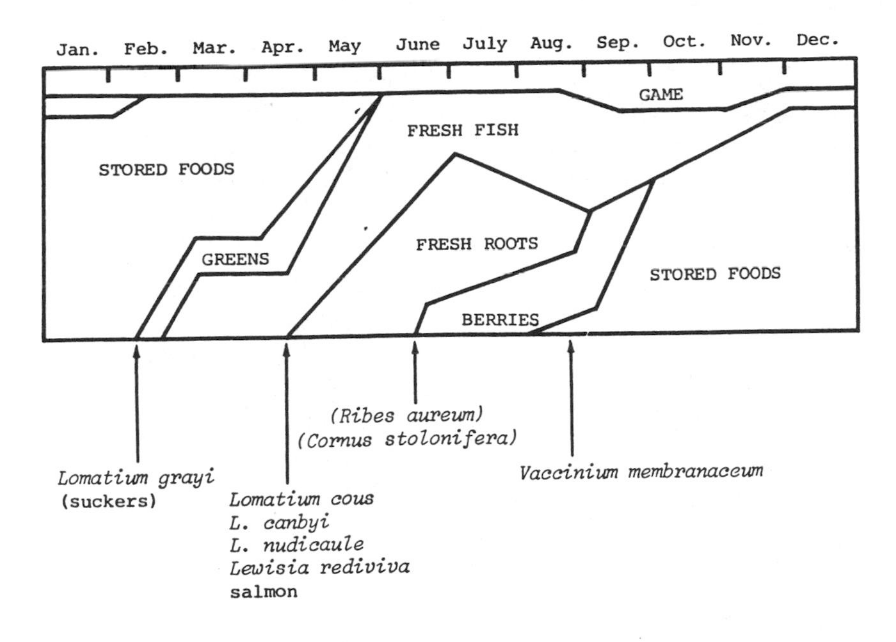

FIG. 8.2. Seasonal patterns of food utilization by Sahaptin-speaking people.

in the order of their harvest in the seasonal round. This information is illustrated schematically in figure 8.2.

Indian Celeries

"Indian celeries" is the English phrase used by contemporary Sahaptin speakers to refer to a set of plants that provide fresh edible sprouts, stems, and shoots. The first to appear is Gray's lomatium (*Lomatium grayi* C. and R.), with sprouts harvested as early as the first weeks of February. This perennial species is most abundant on talus slopes and along rocky stream courses at low elevations. The harvest season is limited to a few weeks at a given elevation, since the sprouts are considered worth digging only before the plants leaf out. The high cultural value assigned to this food is indicated by the fact that several longhouse congregations—the institutional focus of Native American religious expression in the plateau today—hold a thanksgiving feast to mark its first harvest. Other Indian celeries, in particular the bare-stemmed lomatium (*Lomatium nudicaule* [Pursh] C. and R.), are also important.[8]

Foods Which Are Dug

"Foods which are dug" constitute a named group of plants characterized by the action involved in their harvest. Most plants in this group are perennials with edible underground storage organs such as tuberous roots, corms, or bulbs. These plants provided the bulk of the carbohydrate sources of the aboriginal diet.[9] Such plants are most effective as food sources after the completion of vegetative growth and the annual reproductive cycle. Yet actual harvest periods represent a compromise with other considerations affecting harvestability. The three primary staples (considering the Sahaptin range as a whole) each represent a somewhat different compromise. Bitterroot (*Lewisia rediviva* Pursh) is harvested early in its cycle, shortly after the leaves first appear. By the time the buds open, the bitter skin of the root will no longer peel readily and the plants are deemed not worth harvesting. Canby's lomatium (*Lomatium canbyi* C. and R.), with its congener cous (*Lomatium cous* [S. wats.] C. and R.), is one of the earliest sprouting plants of the region. However, the tuberous roots are quite "soft" at this time and are not considered "mature" and harvestable until the petals have fallen. The plants then quickly dry and the "tops" blow away, dispensing seed but making the plants invisible until early the following spring.

A third staple that is "dug" is camas (*Camassia quamash* [Pursh] Greene), a lily with an edible bulb. Bitterroot and the lomatiums are spatially widespread but temporally restricted. By contrast, camas, having a less ephemeral "top," may be harvested at any time between early June and late September but is abundant at only a few favored wet meadow sites. These sites attracted large multi-"tribal" collocations from a wide surrounding area. If the early fur trader Alexander Ross is to be believed, upwards of 10,000 Indians were encamped in the Kittitas valley, digging camas, racing horses, and gambling in early June 1814.[10] Concentrations of more than 1,500 are reported for the Nez Perce–Sahaptin collocation at a camas prairie near present-day Moscow, Idaho.[11] These camas harvest gatherings typically lasted one to several weeks and provided a year's supply of this staple for the attending families.

Initial low-elevation root harvests were conducted from winter village or fishing camp as a home base.[12] The major root harvest season might be extended from early April through June if harvesters followed the upslope advance of spring. Root digging was suspended briefly at the peak of the spring Chinook salmon (*Oncorhynchus tschawytscha*) passage at Celilo Falls during late April and early May, since women's labor was essential then to clean, dry, and otherwise prepare the salmon harvested by the men. Following the spring salmon harvest, families moved in loose association through a series of camps at increasing elevation. For example, Umatilla and John Day Sahaptins continued to harvest bitterroot and cous until late June at elevations above 1,500 meters in the Blue Mountains of Oregon, some 80 kilometers from their winter villages. Large quantities of roots in excess of immediate needs were either cached, dried whole or as prepared cakes, or baked underground, depending on the species and variety involved. Transporting root stores exceeding 1,000 kilograms per family from root gathering sites to the winter villages must have been a major tactical problem of plateau subsistence prior to the adoption of the horse.

Fish

Although salmon provided the bulk of fish consumed, a variety of other species of fish were also important. Particularly notable are the suckers (or "mullet"), *Catostomus columbianus* and *C. macrocheilus*, which were harvested in large quantities from late February through April over much of the plateau.[13] These spawned conveniently in small streams tributary to the Columbia River adjacent to many winter village sites. The critical role of suckers as a fresh fish resource available two months prior to the spring Chinook salmon run was ritually and mythologically marked.[14]

The variety, predictability, abundance, and time of arrival of salmon varied dramatically within the region. The narrow, obstructed course of the Columbia River between The Dalles and Celilo Falls provided prime dipping and spearing sites. Runs there were probably relatively consistent year to year because all upstream populations of a river basin of 600,000 square kilometers, extending over 10 degrees of latitude, contributed to them.[15] Spring Chinook salmon passed quickly but in great abundance. After a hiatus of high water in late May and June came summer and fall Chinook runs, sockeye salmon (*O. nerka*) in late June, silver salmon (*o. kisutch*) in September, and some chum salmon (*o. keta*) in October, to provide salmon fishing along the middle Columbia River from July through October. Lewis and Clark, during their descent of the river on October 11–20, 1805, found Sahaptin people busily engaged in drying spawned-out salmon.[16] Not everyone was so engaged, however; whole villages were absent from the river, their residents occupied in hunting or other economic pursuits. Thus there were strategic choices available in the allocation of time and effort.

Above Celilo Falls there were numerous additional fishing sites at rapids on the Columbia or its major tributaries. Yet as one progressed upriver, the runs were delayed and the fat content of the migrating fish was progressively depleted.[17] Some runs were absent altogether from certain tributaries, and the increased dependence on a few localized breeding populations increased the likelihood of significant annual fluctuations in yield. The Dalles–Celilo Falls area therefore remained a center where surplus dried salmon could be obtained in trade for products of the hinterlands, such as dried roots and root products, Indian hemp (*Apocynum cannabinum* L.), and skins; later horses, bison robes, and slaves became important, since trade in these items was greatly enhanced by equestrian mobility. At Celilo and The Dalles, the resident population of a few hundred people might swell to 3,000 at the peak of the salmon harvest.[18] Lewis and Clark found the Indians at The Dalles still in possession of large quantities of dried salmon from the previous season in mid-April 1806,[19] indicating the value of this Columbia River entrepôt as a "hedge" against local shortages.

Other fish such as steelhead (*O. mykiss*) and whitefish (*Prosopium williamsoni*) were available nearly year-round, providing limited fresh winter rations. The abundance and reliability of the fishery is further indicated by the exercise of cultural preferences. The Sahaptins, for example, disdained sturgeon (*Acipenser transmontanus*),[20] an important food fish on the lower Columbia and Fraser Rivers.

Foods Which Are Picked

"Foods which are picked" is a Sahaptin food category that might be glossed as "fruits and berries" except that it also includes a lichen reminiscent of Spanish moss, which is "picked" from the branches of forest trees. The moss is rich in several mineral nutrients but requires elaborate processing.[21] Low-elevation species such as the early fruiting golden currant (*Ribes aureum* Pursh) and dogwood (*Cornus stolonifera* Michx.), available in June, and the chokecherry (*Prunus virginiana* L.), serviceberry (*Amelernchier alnifolia* Nutt.), and blue elderberry (*Sambucus cerulea* Raf.) were harvested in conjunction with summer salmon fishing. Sahaptins, however, generally placed greater emphasis on the various huckleberry and blueberry species (*Vaccinium* spp.), which are most abundant near timberline in the Cascade Mountains. Most families forsook salmon fishing in favor of berrying from mid-August through September. French reports that at least in the Mount Hood and Mount Jefferson region, systematic burning was practiced to open up montane areas and improve huckleberry yields.[22] Many fruits were dried for storage over winter.

Game

Hunting engaged the men while the women gathered huckleberries. The late summer upslope movement of human beings seeking fruits paralleled the migration of their major ungulate prey, the mule deer.[23] Hunting was practiced year-round, as fishing-and-root-and-berry-oriented population movements allowed. The timing and techniques of hunting were as varied as the prey, which ranged from diminutive Townsend's ground squirrels (*Citellus townsendii*), extracted from their burrows in spring by being twisted on a willow stick, and cottontail rabbits (*Sylvilagus nuttallii*), the object of cooperative net drives, to black bear (*Ursus americanus*), bearded in their winter dens.

In sum, the plateau provided abundant and varied subsistence resources to people willing and able to range seasonally over an extensive area of diverse habitat, from desert basins to the alpine zone. Stockpiling substantial surpluses of processed food for winter consumption was an essential component of this subsistence strategy.

RESOURCE MANAGEMENT: PLATEAU AND NORTHWEST COAST COMPARED

The Columbia Plateau contrasts sharply with the adjacent Northwest Coast culture area in terms of actual resource-use restrictions and conservation practices by Native Americans. On the coast, the rights of families and local village groups to control and harvest specific resources at specific sites were highly developed and often heritable.[24] A wide variety of resources was so controlled, including salmon fishing sites, halibut shoals, shellfish beds, camas, bracken, clover and berry patches, crabapple orchards, and mountain goat hunting ranges. If permission were formally requested, and if the owner had harvested a sufficient quantity, non-owners might share the resource.[25] However, warfare motivated by and/or resulting in territorial aggrandizement apparently was common.[26] Environmental manipulations designed to improve yields, such as burning and various types of cultivation, were also widely reported.[27]

In contrast, resources in the plateau were harvested in common. Individuals apparently were free to travel over great distances and to harvest resources freely in the course of their travels.[28] Typical are accounts of several thousand men, women, and children peaceably sharing the abundance of a camas field, a bitterroot-lomatium zone,[29] a favored berrying and hunting ground,[30] or a concentration point for migrating fish.[31] At these sites, families might camp in traditional locations[32] and harvest within traditionally limited sections of the camas meadow,[33] but there is no evidence of conflict over use rights or of efforts by owners to restrict access by others to the resources at these collocations. Nor are there reliable reports of violent confrontations between plateau groups over access to indigenous resources.[34] Accounts of controlled burning,[35] cultivation, or explicit conservation practices are few and scattered, though the general proscriptions against waste associated with animistic belief systems are acknowledged in myth.[36]

There is one notable exception to this contrast between coast and plateau approaches to resource management. At Celilo Falls and at The Dalles in Chinookan territory immediately downstream, a family controlled each fishing platform and well-situated point of rock;[37] these rights were heritable. A "salmon chief" was empowered to open and close the fishing daily to allow escapement and to regulate the onset of the spring Chinook salmon harvest.

This intensity of resource management is anomalous in the larger plateau context, and it is tempting to attribute it to the fact that the salmon resource at The Dalles and Celilo Falls was uniquely concentrated and predictable—and thus particularly defensible.[38] The typical plateau pattern of use-in-common held force, however, at the region's second largest fishery, at Kettle Falls in the Okanogan Salish life-range. At Kettle Falls, salmon were distributed equally to all comers, apparently without regard for group of origin or kinship connection. The salmon chief's role there seems to have been essentially ritualistic rather than managerial.[39] Similarly, the explanation of restricted access to a resource in terms of the predictability and concentration of that resource cannot account for the plateau pattern of camas utilization, since camas is an eminently concentrated and predictable resource. Nor does this explanation account for the fact that camas, berries, shellfish, and game were harvested by both coastal and plateau people but were explicitly owned and managed on the coast while being exploited in common and apparently not regularly managed in the plateau.

A more convincing explanation must account not simply for the intrinsic characteristics of particular resource species, such as their concentration or predictability, but also for the interaction of the subsistence strategy with those resource characteristics. Here again plateau and coastal people contrast. The coastal people tended to be more sedentary and to exhibit a greater tendency toward subsistence specialization than did upriver groups along the Skagit,[40] the Cowlitz,[41] the Nooksack,[42] and other coastal rivers. A similar difference can be seen between groups occupying the outer coasts and those of the sheltered inland waters.[43] This variation resulted not so much from the degree of concentration of any *particular* resource or from the abundance of all the resources within the region; rather, it resulted from the fact that the shorewise concentration of a *variety* of key resources favored exploiting closely spaced rather than widely separated habitat patches.[44] This strategy was also consistent with a seasonal round that was independent of variations in elevation.

In the plateau, a balanced and varied diet was obtained by means of extensive upslope movements in spring and fall. The major caloric resources, bitterroot and the lomatiums, apparently defined this basic seasonal rhythm by virtue of their annual growth characteristics. Their restricted temporal availability combined with their widespread but patchy spatial distribution to protect them from overexploitation by requiring their human predators to be highly mobile. Perhaps the plateau people were incapable of depleting bitterroot and lomatium populations during their brief annual availability. In such cases, direct resource management through territoriality or conservation does not pay because it is superfluous. Furthermore, to get the most from the key plateau energy sources, Sahaptin households had to adopt a strategy of movement that incidentally limited their impact on more concentrated and less temporally restricted staples such as camas and salmon. In short, *time* rather than any specific resource may have been the factor limiting the intensity of resource exploitation in the plateau—the time required to move from one widely separated resource concentration to the next.

As Lewis and Clark descended the Columbia River in 1805 they noticed a dramatic change from mat-covered to cedar plank lodges just below Celilo Falls.[45] This material culture boundary coincided with the linguistic distinction between Sahaptin and Chinookan; it also marked a transition from Columbia Plateau to Northwest Coast cultural orientations. One key aspect of this was a transition to a more specialized, river-oriented subsistence economy and a less mobile lifestyle toward the coast, symbolized by the degree of portability of material used in house construction. Perhaps the anomalous pattern of resource management at Celilo Falls is best interpreted as a reflection of this coastal cultural influence rather than as an ecological response.

CIRCUMSTANCES THAT LIMIT RESOURCE USE: AN EVOLUTIONARY PERSPECTIVE

There are at least three situations in which humans may exploit a resource species at a level below that which in theory could be sustained without the species experiencing a population decline. Each type of restricted resource use has evolutionary implications. The three types of restricted resource use may be divided between *direct* and *epiphenomenal*, and direct limitations on resource use can be further divided into *exclusive* and *inclusive*.

Direct and exclusive limitation on resource use involves *territoriality*, which may be defined as any restriction on access to a resource based on group membership. Typically, territoriality involves "ownership" of land or resources with enforceable rights to restrict access by non-owners. Direct and inclusive limitation involves *conservation*, which may be defined as a culturally sanctioned pattern of restricted resource use imposed upon the members of the resource-controlling group. Conservation may be manifested in a variety of ways: by an ideology that mandates supernatural sanctions against waste; by harvest practices designed to spare a resource, such as leaving a portion of a tuber for regeneration; by selective hunting practices designed to spare reproductive females; and so on. Epiphenomenal restrictions on resource use differ from direct restrictions because they are by-products of some other conduct. For example, the use of an animal species may be limited by the species' migratory behavior.

In both types of direct restricted resource use, a direct cost is incurred. In a territorial system, the resource owner must pay the cost of preventing outsiders from using the resource through force or litigation. In the case of conservation, there is the "conservation cost" of some added labor per unit of resource harvested—for example, the cost in efficiency of harvest consequent to digging up a tuber and then leaving a portion for regeneration.

The theory of natural selection requires that the costs incurred be justified by a compensating selective advantage to the individuals who must pay the cost.[46] For example, in the case of territoriality, those who pay the cost of defending the territory are the same people who benefit from the rights of ownership, which may assure them a more secure food supply. Those who avoid the costs of territorial defense by defending no territory may suffer in the case of shortages, since they will not be able to limit the harvests of others. Their population will decline relative to those who defend adequate resources. Thus territoriality may make evolutionary sense in an environment characterized by the expectation of shortages. In contrast, conservation is evolutionarily problematic because "wasteful" individuals may cheat the system, thus avoiding the costs of conserving while suffering no more from shortages than conservers do. If "conservation costs" are minimal, however, and cultural sanctions against cheaters are effective, a conservation strategy could evolve by natural selection.

Epiphenomenal restrictions on resource use differ from direct restrictions because no direct costs are incurred. The only "costs" are those of opportunities not taken. An example is the "buffer zone" phenomenon described by Harold Hickerson, in which resource conservation is an epiphenomenon of warfare: a no-man's-land between two warring tribes becomes a wildlife refuge because hunters do not visit the area for fear of becoming casualties in the war.[47] In this instance it is warfare, not conservation, that requires evolutionary explanation.

The familiar recourse to technological limitation to explain the sparing of resources by preindustrial people is a good example of an epiphenomenal restriction but a poor explanation: it is sometimes argued that Eskimo whale hunters limited to harpoons constructed of locally available materials, bison hunters of the Great Plains armed with bows and arrows or stone-tipped spears, or hunters of beaver and caribou in the Canadian forests lacking firearms and steel traps were simply technologically incapable of exterminating their prey. Calvin Martin argues, in contrast, that the Canadian forest hunters of beaver lacked not the technological capacity but rather the motivation required to pursue their prey to extinction—at least, that is, prior to the ravages of epidemic disease following European contact.[48] In any case, the sparing of resources is not an epiphenomenon of technological limitation, because the existing technology may always be employed by more people, thus circumventing the so-called limitation. The human use of technology is an integral part of a culturally defined strategy for making a living, which includes a more or less explicit "population policy." The sparing of resources may, however, be an epiphenomenon of the subsistence strategy.

Columbia Plateau resources were spared primarily as an epiphenomenon of the highly mobile subsistence strategy that characterized the region. Thus, neither territoriality nor a conservation ideology (as defined here) was the effective means of plateau resource management.

DISCUSSION AND CONCLUSION

I have argued that restrictions on resource use may either result from direct efforts at resource management or be incidental to strategic decisions that are motivated by factors unrelated to conservation per se. In the latter case, they are what I call here epiphenomenal restrictions. The striking contrast between the degree to which resources were directly managed in the Columbia Plateau as opposed to the Northwest Coast is a function of the contrasting subsistence strategies characteristic of the two regions. Coastal strategies focused on littoral resources within easy reach and control of a home base occupied for most of the year. Plateau strategies were characterized by extensive seasonal movements patterned after the temporal and spatial distribution of key carbohydrate resources.

As a general rule, if groups of equal size vary in mobility, the more mobile group will exploit its resources less intensively than the less mobile group. This is a simple consequence of not being able to be in more than one place at a time. The apparent lack of direct resource management practices in the aboriginal Columbia Plateau, in contrast to the Northwest Coast, may best be explained by reference to the greater mobility consequent to the plateau subsistence strategy. That strategy was tailored to the distributional facts of the plateau resources selected as staples, but it was not the only conceivable strategy for survival in this environment. Plateau people might well have adopted a more specialized riverine orientation comparable to that of their Chinookan neighbors downstream, which might have supported increases in overall population density. Alternatively, groups of plateau people might have developed increasing specialization coupled with regional commodity exchanges. They did not. The fact that direct resource management was not an obvious concern on the plateau suggests that the plateau resource base could have supported more intense exploitation and that this remained a viable evolutionary option at the time of European contact. In sum, the plateau case suggests that an evolutionarily stable subsistence strategy is possible in the absence of both territorial defense and altruistic conservation practices. The "tragedy of the commons" may prove to be a special case.[49]

NOTES

An earlier version of this essay was previously published as "Mobility as a Factor Limiting Resource Use in the Columbia Plateau of North America," in *Resource Managers: North American and Australian Hunter-Gatherers*, eds. Nancy M. Williams and Eugene S. Hunn (Boulder, Colo.: Westview Press, 1982). Reprinted with the permission of the American Association for the Advancement of Science.

1. Luther S. Cressman, *Prehistory of the Far West: Homes of Vanished Peoples* (Salt Lake City: University of Utah Press, 1977).

2. Eugene S. Hunn, *Nch'I-Wana "The Big River": Mid-Columbia Indians and Their Land* (Seattle: University of Washington Press, 1990).

3. *Ibid.*

4. C. M. Nelson, "Prehistoric Culture Change in the Intermontane Plateau of Western North America," in *The Explanation of Culture Change: Models in Prehistory*, ed. Colin Renfrew (London: Duckworth, 1973), pp. 371–90.

5. Verne F. Ray, *Cultural Relations in the Plateau of Northwestern America* (Los Angeles: Southwest Museum, Publications of the Frederick Webb Hodge Anniversary Publication Fund 3:13, 1939).

6. Eugene S. Hunn, "On the Relative Contribution of Men and Women to Subsistence among Hunter-Gatherers of the Columbia Plateau: A Comparison with Ethnographic Atlas Summaries," *Journal of Ethnobiology* 1 (1981): 124–34; Eugene S. Hunn and David H. French, "*Lomatium*: A Key Resource for Columbia Plateau Native Subsistence," *Northwest Science* 55 (1981): 87–95.

7. Gordon W. Hewes, "Indian Fisheries Productivity in Pre-Contact Times in the Pacific Salmon Area," *Northwest Anthropological Research Notes* 7 (1973): 133–55; Hunn, "On the Relative Contribution of Men and Women."

8. Its fresh growth is particularly high in vitamin C. E. M. Benson, J. M. Peters, M. A. Edwards, and L. A. Hagen, "Wild Edible Plants of the Pacific Northwest," *Journal of the American Dietetic Association* 62 (1973): 143–47; P. B. Keeley, *Nutrient Composition of Selected Important Plant Foods of the Pre-Contact Diet of the Northwest Native American Peoples* (M.S. thesis, University of Washington, 1980).

9. Keeley, *Nutrient Composition of Selected Important Plant Foods*.

10. Alexander Ross, *The Fur Hunters of the Far West*, ed. Kenneth A. Spaulding (Norman: University of Oklahoma Press, 1956), pp. 22–28. Ross's estimate of 10,000 seems excessive by an order of magnitude.

11. Alan G. Marshall, "Nez Perce Social Groups: An Ecological Interpretation" (Ph.D. diss., Washington State University, 1977), pp. 55–56.

12. *Original Journals of the Lewis and Clark Expedition, 1804–1806*, ed. Reuben G. Thwaites (New York: Dodd, Mead, and Co., 1905), 4: 294, 3: 173–74.

13. Eugene S. Hunn, "Sahaptin Fish Classification," *Northwest Anthropological Research Notes* 14 (1980): 1–19; *Original Journals of the Lewis and Clark Expedition*, 4: 290, 328.

14. Hunn, "Sahaptin Fish Classification."

15. Joseph Craig and Robert L. Hacker, *The History and Development of the Fisheries of the Columbia River*, National Marine Fisheries Service Fishery Bulletin 32 (Washington, D.C.: Government Printing Office, 1940), pp. 135–37.

16. *Original Journals of the Lewis and Clark Expedition*, 3:107–40.

17. D. R. Idler and W. A. Clemens, *The Energy Expenditures of Fraser River*

Sockeye Salmon during the Spawning Migration of Chilko and Stuart Lakes (New Westminster, B.C.: International Pacific Salmon Fisheries Commission, Progress Report no. 6, 1959); Hunn, "On the Relative Contribution of Men and Women."

18. Alexander Ross, "Adventures of the First Settlers on the Oregon or Columbia River," in *Early Western Travels, 1748–1846*, ed. Reuben G. Thwaites (1849; reprint, Cleveland: Arthur H. Clark, 1904), 7: 123.

19. *Original Journals of the Lewis and Clark Expedition*, 3: 288.

20. Cf. *Original Journals of the Lewis and Clark Expedition*, 3: 290.

21. Nancy Turner, "Economic Importance of Black Tree Lichen (*Bryoria fremontii*) to the Indians of Western North America," *Economic Botany* 31 (1977): 461–70.

22. David H. French, "Aboriginal Control of Huckleberry Yield in the Northwest," paper read at the annual meeting of the American Anthropological Association, Chicago, 1957.

23. Marshall, "Aboriginal Nez Perce Subsistence."

24. Allan Richardson, "The Control of Productive Resources on the Northwest Coast of North America," in *Resource Managers: North American and Australian Hunter-Gatherers*, eds. Nancy M. Williams and Eugene S. Hunn (Boulder: Westview Press, 1982), pp. 93–112.

25. Nancy Turner, *Food Plants of British Columbia Indians, part 1: Coastal Peoples* (Victoria: British Columbia Provincial Museum, Handbook no. 34, 1975), pp. 129, 152, 187, 191, 221.

26. E.g., Wayne P. Suttles, "Economic Life of the Coast Salish of Haro and Rosario Straits" (Ph.D. diss., University of Washington, 1951).

27. June M. Collins, *Valley of the Spirits: The Upper Skagit Indians of Western Washington* (Seattle: University of Washington Press, 1974), p. 55; A. B. Reagan, "Plants Used by the Hoh and Quileute Indians," *Proceedings of the Kansas Academy of Science* 37 (1934): 55–70, 56–57; Turner, *Food Plants of British Columbia Indians*, pp. 68, 81, 150, 164, 203; Richard White, "Indian Land Use and Environmental Change," *Arizona and the West* 17 (1975): 327–38, pp. 331, 333; H. G. Zenk, "Contributions to Tualatin Ethnography: Subsistence and Ethnobiology" (M.A. thesis, Portland State University, 1976), pp. 22–25.

28. Deward E. Walker, *Mutual Cross-Utilization of Economic Resources in the Plateau: An Example from Aboriginal Nez Perce Fishing Practices* (Pullman: Washington State University, Laboratory of Anthropology Report of Investigations no. 41, 1967).

29. N. Washington, "Tsukalotsa (*Lomatium canbyi*): Key to Understanding Central Washington Nonriverine Archaeology," paper presented to the twenty-ninth annual Northwest Anthropological Conference, April 10, 1976, Ellensburg, Washington.

30. George Gibbs, "Report of Mr. George Gibbs to Captain McClellan, on the Indian tribes of the Territory of Washington, Pacific Railroad Report," in *Report of the Secretary of War* (Washington, D.C.: Government Printing Office, 1854), vol. 1, p. 404; Helen H. Schuster, "Yakima Indian Traditionalism: A Study in Continuity and Change" (Ph.D. diss., University of Washington, 1975), p. 85.

31. Verne F. Ray, "Native Villages and Groupings of the Columbia Basin," *Pacific Northwest Quarterly* 27 (1936): 99–152, p. 142.

32. Haruo Aoki, "Nez Perce Texts," *University of California Publications in Linguistics* 90 (1979): 81–85.

33. Marshall, "Nez Perce Social Groups."

34. For a rare exception, see Click Relander, *Drummers and Dreamers: The Story of Smowhala the Prophet and His Nephew Puck Hyah Toot, the Last Prophet of the Nearly Extinct River People, the Last Wanapums* (Caldwell, Idaho: Caxton Printers, 1956), p. 312.

35. French, "Aboriginal Control of Huckleberry Yield."

36. M. Jacobs, "Northwest Sahaptin Texts, 1," *University of Washington Publications in Anthropology* 2 (1929): 175–244, 196–200.

37. Leslie Spier and Edward Sapir, "Wishram Ethnography," *University of Washington Publications in Anthropology* 3 (1930): 151–300, 175. A family's ability to restrict use was tempered, however, since individuals were entitled to exercise fishing rights on the basis of extended kin links. Because the peoples of the plateau frequently intermarried across dialect and language boundaries, fishing rights were not restricted to local residents. Similarly, the elderly were entitled to fish for their immediate needs on request.

38. R. Dyson-Hudson and E. A. Smith, "Human Territoriality: An Ecological Reassessment," *American Anthropologist* 80 (1978): 21–41.

39. R. Bouchard and D.I.D. Kennedy, "Utilization of Fish by the Colville Okanagan Indian People," *British Columbia Indian Language Projects* (Victoria, B.C.: Ms., 1975); Verne F. Ray, "The Sanpoil and Nespelem: Salishan Peoples of Northeastern Washington," *University of Washington Publications in Anthropology* 5 (1933): 69–75.

40. Collins, *Valley of the Spirits*.

41. Verne F. Ray, *Handbook of Cowlitz Indians* (Seattle: Northwest Copy Co., 1966).

42. Pamela Amoss, *Coast Salish Spirit Dancing: The Survival of an Ancestral Religion* (Seattle: University of Washington Press, 1978).

43. Philip Drucker, *The Northern and Central Nootkan Tribes*, Bureau of American Ethnology Bulletin no. 144 (Washington, D.C.: Government Printing Office, 1951).

44. Cf. Richard A. Gould, "To Have and Have Not: The Ecology of Sharing among Hunter-Gatherers," in Williams and Hunn, *Resource Managers*, pp. 69–91.

45. *Original Journals of the Lewis and Clark Expedition*, 3: 154.

46. Ronald Dawkins, *The Selfish Gene* (New York: Oxford University Press, 1976).

47. Harold Hickerson, "The Virginia Deer and Intertribal Buffer Zones in the Upper Mississippi Valley," in *Man, Culture, and Animals: The Role of Animals in Human Ecological Adjustments*, eds. Anthony Leeds and Andrew P. Vayda (Washington, D.C.: American Association for the Advancement of Science, 1965), pp. 43–66.

48. Calvin Martin, *Keepers of the Game: Indian-Animal Relationships and the Fur Trade* (Berkeley: University of California Press, 1978).

49. Garrett Hardin, "The Tragedy of the Commons," *Science* 162 (1968): 1243–48.

9 / Unusual Gardens

The Nez Perce and Wild Horticulture on the Eastern Columbia Plateau

ALAN G. MARSHALL

I use the term *unusual gardens* in my title because the Nez Perce and their relatives on the Columbia Plateau practiced a mode of agriculture that was so unusual as to be largely invisible to Euro-Americans. Rather than creating geometric, fenced, plowed fields of a single plant species, Nez Perce people developed irregular plots by hand labor and fire. This distinct and generally unfamiliar form of agriculture—known as "horticulture"—is commonly thought to be a feature of societies displaying greater social complexity than is usually ascribed to the Nez Perce.

In the past few years, the ethnography and history of Native Americans in many regions have undergone revision to correct biases in the description of native peoples. This is true of the literature concerning the native residents of the northwestern United States, although progress has perhaps been somewhat slower here. Ethnographers, for example, no longer speak of native culture "in its present debased form" or seek to present it "as nearly as possible in its form at the time of first contact of these Indians [Nez Perce] with the whites."[1] Historians now recognize that Euro-American eyewitness records emphasize non-Indian interests.[2] But most current work is built on historical sources, and those sources are so fundamentally biased that skepticism toward even seemingly well-founded interpretations is appropriate—and revision of them must continue.

This chapter briefly reviews the dominant ethnographic image of the Nez Perce, or *nimípu*,[3] showing how it reflects this bias. Then the evidence for Nez Perce horticulture is reviewed. Finally, the implications of this plant use for other aspects of Nez Perce social organization are discussed. My view is that the Nez Perce subsistence economy was more complicated and sophisticated than is generally recognized. This conclusion has implications for other Plateau groups, because the Nez Perce were intimately related to the other peoples of the plateau.[4]

IMAGES OF PLATEAU INDIANS

Robert Berkhofer's *The White Man's Indian* outlined the stages of development of outsiders' images of Native Americans, images that swing with increasing sophistication between the poles of debased, even hellish, chaotic savages and the noble savage, the natural man who lives in concert with the earthly paradise that was the New World.[5] These images are significant to non-Indians because they suit their ideological purposes: the first image affirms white American values and justifies the changes being wrought upon the land; the second discomfits the establishment and opposes the "manifest destiny" of industrial America. The development of ethnographic images of Plateau native peoples, including the Nez Perce, reflects the pattern described by Berkhofer.

Ethnographic, historical, and archaeological images of the *nimípu*, though mildly discordant, all assume an economy based upon subsistence hunting and gathering. This economic base supported a social infrastructure of politically independent villages, which were the primary social units of the Nez Perce.

The earliest ethnographic work established this image. In his 1908 ethnographic sketch, Herbert Spinden concluded that the "natural" environment was of overwhelming importance in shaping and limiting Nez Perce social organization. Spinden characterized the Nez Perce as lacking in any indigenous ideas or material culture.[6] Instead, he viewed Plateau cultures as transitional between the more "vigorous" cultures of the Great Plains and Pacific coast, which were the source of ideas and technology that diffused across the plateau.

This image was challenged by Verne F. Ray in the 1930s.[7] Ray thought that four outstanding "social culture" elements distinguished Plateau peoples from those of the Northwest Coast, Great Plains, and Great Basin culture areas: village autonomy, individualism, pacifism, and the guardian spirit complex. Ray's work presented a picture of isolated villages of people who lived from season to season by fishing, hunting, and gathering.

Deward Walker's ethnography of the Nez Perce was unlike those of his predecessors because Walker treated the Nez Perce as contemporary people with a vital past instead of in the static ethnographic present.[8] He recognized that the Nez Perce cultural tradition continued into the present—for him, the 1960s. Instead of an ethnographic reconstruction, Walker was interested in contemporary political and religious processes. But as a result, he relied on the points of view of his predecessors for a general ethnographic description: "It is clear that Nez Perce culture was characterized by what Steward has called a low level of socio-cultural integration. It was thus like most other hunter-gatherer societies, limited in organizational scope, strongly influenced by its biophysical environment, and only weakly elaborated in terms of its social structure."[9]

Historical works have begun with similar assumptions about the Nez Perce. Although the two best-known works focus on the Nez Perce–United States conflict,[10] both outline Nez Perce subsistence in terms of hunting, fishing, and gathering primarily for domestic use. In setting the stage for his story of the conflict, Alvin Josephy wrote: "The Nez Perces, like all the Northwest peoples, practiced no agriculture and suffered for it. The gathering of food supplies was an almost constant preoccupation, and although the villagers stored some of their surplus food, they knew lean periods. They were shrewd hunters and artful fishermen, but the wild game came and went, and fish were plentiful only during the seasonal runs."[11]

More recent analyses of the Columbia Plateau have found a somewhat more complex social life. Angelo Anastasio analyzed the network of social relationships among Plateau groups on the basis of detailed ethnohistorical data. He described a highly fluid social situation in which there was "no necessary relationship among kinship organization, political organization, language affiliation, ethnic characteristics, and group self-identity," and he argued that the plateau was a single social entity.[12] Yet when discussing the Nez Perce, he found little social organization above "regional" alliances of neighboring villages in times of "grave emergency."[13]

My own earlier interpretation of Nez Perce social life was also hampered by the prevailing ethnographic image of Nez Perces—although my examination of intervillage relationships did reveal that villages were not the largest groupings of people.[14] I found instead that villages were primarily consumption groups. That is, people dispersed into relatively small groups on the rivers of deep canyons during winter (when resources were generally limited), where they consumed the surpluses they had accumulated in summer and fall. Much larger "production groups" formed during summer. These were multiethnic in social character and were often established on the most productive vegetable food gathering grounds.

Walker, Anastasio, and I began to revise images of the Nez Perce and the other native residents of the Columbia Plateau. Anastasio showed that the Nez Perce were part of a single complex area. Walker added a historical dimension to Nez Perce society. I drew attention to a critical resource, plants, and to differences in the character of settlement types. Yet these views, too, are muddied by Euro-American biases.

REVISIONISM

Francis Jennings called attention to the "cant of conquest," the often-intentional bias in historical records concerning Native America.[15] Alice Kehoe suggested that the problem is far deeper: it lies, she concluded, in a cognitive tradition that constructs the data used to understand Native America.[16] Kehoe was especially concerned with ethnographic data because European viewpoints concerning the "primitive" still taint anthropological descriptions of Native Americans: "With few exceptions, American anthropological works continue uncritically to embed the Europeanist oppositional dualism of primitive and civilized. . . . Anthropologists tend, further, to accept the traditional European collocation of "primitive" with kin-based, nonagricultural, nonstratified, nonliterate, nonscientific, having shamans rather than priests, and lacking professions, commerce, or industry."[17] Her point is sharp, since most of these terms continue to be used for describing the Nez Perce.

In recasting the image of aboriginal North America, Kehoe argued that there were two major ecological regions: in the first, food production was universally present; in the second, food production was climatically impossible. Within the southerly, food-producing area she sketched another two "zones": one in which Mexican domesticates were grown, the other in which native plants were favored. She carefully avoided the term *agriculture* because it is slippery at best. Instead, she used the term *food production* to mean "techniques that increase the harvest of preferred resources beyond their natural abundance. These techniques include regular firings of grassland or camas meadow, transplantation of roots, tubers, or bulbs, irrigation, weeding, and hoe cultivation of undomesticated plants."[18]

The remainder of this chapter examines these questions in the context of

Nez Perce data. What is at stake is the image of the Nez Perce drawn in historical, ethnohistorical, and ethnographic studies.

NEZ PERCE PLANT MANAGEMENT

Nez Perce people used more than 32 plants for foodstuffs.[19] Although many additional plants were used for medicinal and industrial purposes, food plants are my focus.

To determine whether the Nez Perce employed food production technologies—as contrasted with simply harvesting naturally occurring plant foods—requires consideration of some basic questions: How much "manipulation" of food plants occurred? What technology was involved? Did such manipulation result in increased plant net productivity? (That is, did their numbers increase? Did their size increase?) Did the people know their efforts had such effects? And did they produce enough for their needs?

My analysis of these questions focuses on only three of the food plants used by the Nez Perce, because these three were central to the Nez Perce economy and thus were the most intensely managed. Other plants were less important to Nez Perces in terms of their intensity of use and "manipulation," and they are now ignored and unknown to many people. Some, such as spring beauty (*Claytonia lanceolata*), were gathered, but they formed only a minor, seasonal part of the diet. These "adventitious" plants (if that term may be used) were gathered by men in their travels when they were hungry.

In English the three most significant plants are glossed as "camas" (*Camassia quamash*), "biscuitroot" (*Lomatium kous*), and "snowdrops" (*Lomatium canbyi*).[20] The Nez Perce words are *qem'es*, *qaamsit*, and *q'eq'iit*, respectively. These and other plant foods were, and remain, the product of work by women and girls, except in some unusual instances.

Qem'es was the premier food plant for the Nez Perce. Lewis and Clark remarked upon it in their diaries. Almost every observer afterward commented upon its abundance and usage. Although he did not use Nez Perce terms, Spinden's description of a camas field has never been surpassed:

> The most important of all vegetal foods was camas. . . . This liliaceous bulb was gathered in enormous quantities in the wet upland meadows of Weippe prairie, Camas prairie, in the open country near the town of Moscow [Idaho], and in the Grande Ronde valley. It grows in the glades of piny woods, and in partly inundated prairie land. The bulbs lie very closely together, so that when the camas is in bloom, the flats from a short distance resemble lakes of fine, clear water.[21]

Spinden also described the essentials of the harvest and preparation of camas bulbs. Minor amounts were gathered in spring, but the new growth of leaves and stems of this perennial exhausted the stored energy of the bulb, so that the quality of the "root" was poor at this time.[22] The great harvest occurred after the seeds had matured sometime between June and September. The timing of the harvest depended upon elevation and microclimatic factors.[23]

Qem'es was dug using a "digging stick," or *tuuk'es*. Spinden described it well: "It was about two and a half feet long, with a fire-hardened point bent slightly forward. The handle consisted of a piece of bone, horn, or stone, from five to eight inches in length. This handle was usually perforated in the middle and lashed firmly at right angles to the stick. The stick was operated with both hands, one being placed on each end of the handle. When the weight was applied to the handle, the point penetrated the ground to a depth of about six inches, and then the curved end was directed forward, thus breaking the sod."

My information, based on photographs, interviews, and participate observation, varies from Spinden's. One difference lies in the technique of digging *qem'es*. The "roots" of this plant are so densely packed in preferred digging areas that taking them out one by one is impossible. Instead, large areas were loosened using the *tuuk'es* as a pry bar, then the chunks of the relatively soft, moist earth were broken up by hand and the "roots" sorted.

Sorting the "roots" remained an interesting process even in the 1970s. In my experience of digging "roots" in a *qem'es* field on Musselshell Creek, near Weippe, Idaho, the profusion of these plants became apparent.[24] Bulbs ranged from 3 millimeters to 3.5 centimeters in diameter. Only those 2 centimeters or more were taken. Thus many were replanted.[25] Furthermore, at least some families took only one "sex" of "roots," so that about one-half of the mature bulbs were returned to the soil.[26]

Qaamsit and *q'eq'iit* also remained important in the 1970s. Many authors refer to these and closely related plants of the genus *Lomatium* as "kouse" or "kaus," which is derived from the Nez Perce word *qaws*.[27] *Qaws* refers to the corms of these plants (and some other lomatiums) once they have been preserved by drying, the first and sometimes only step in preparation for eating.[28]

Qaamsit and *q'eq'iit* grow in a different habitat from *qem'es*. "These plants . . . flourish in dry rocky soil and were commonly gathered by the Indian women along the brows of steep hills. The harvest season was very early, most of the digging occurring during April and May."[29] They were dug with a *tuuk'es* as well. In some areas they grew so thickly that the technique used for gathering *qem'es* could be used.

A less well-known aspect of the "management" of these three plants was the use of fire. The importance of fire in maintaining prairies and improving food-plant production in the Northwest is known from western Washington State but has not been reported in the eastern portion of Washington or in Idaho.[30] In a study of the Bitterroot Forest Reserve (which was within the Nez Perce region), John Leiberg interviewed several Nez Perces who reported the intentional firing of forests in order to improve hunting and berrying.[31] My consultants asserted that the firing of prairies—including camas fields—was also intentional and common because it improved the quality and quantity of these three desirable foods.

In fact, these three plants, *qem'es, qaamsit,* and *q'eq'iit,* increase their net productivity when disturbed. Rexford Daubenmire, who conducted a detailed study of the region's plant ecology, noted that "all three plants grow in rather restricted habitats, and . . . their numbers, at least, do not decrease when disturbed."[32]

Of *qem'es* Daubenmire stated: "Great quantities were steam-roasted, with some then dried for later use. There is no indication of over-exploitation by the Indians. Possibly, the churning of the soil facilitated the establishment of new plants from seeds of those missed by the diggers."[33] He had no direct information about the process of digging *qem'es*, so he did not know that the preferred digging season was after the seeds set or that immature bulbs and half of the mature ones were returned to the soil. Indeed, my teachers, Cyrus Red Elk, Mrs. Cyrus Red Elk, and Samuel Watters, explicitly stated that these actions ensured a continued supply of *qem'es*. Daubenmire also thought that people were digging primarily for their own immediate use and was unaware of the surplus production involved in the work by women. Finally, Daubenmire only implied that burning and aerating the soils by digging sped up nutrient cycling and improved growing conditions for the plants.

Disturbance also increases the abundance of the lomatiums. These plants are most abundant on lithosols—shallow, rocky soils on bedrock.[34] These soils are subject to frost heaving, and these locations are often characterized as having "patterned ground."[35] Even disturbance by cattle seems to increase the abundance of these plants. It thus seems reasonable to conclude that the churning of the soil by "root diggers" increased the productivity of *qaamsit* and *q'eq'iit*.

My consultants independently agreed that the lomatiums were best collected after the seed was mature. They gave the same reason as that for *qem'es:* the "roots" were of higher quality. Furthermore, all reported that the "diggers" replaced undesirable roots in the ground and scattered seeds on the broken ground. The reason given to me was that it ensured the following year's production.

Such efforts resulted in a substantial surplus. Although a "good digger" could lay in and process a year's supply in less than a week, many worked for a week or more. The surplus was used for feasts, marriage "trades," and exchanges with *yeleptin*, "trading partners." Observers in the 1800s commented on the vast supply of *qem'es*. Lewis and Clark saw great amounts of camas collected in September 1805.[36] Isaac I. Stevens reported that four days' work was sufficient to lay in a year's supply of *qem'es*.[37]

DISCUSSION AND REVISION

Several biases are now apparent in conventional interpretations of Nez Perce "gathering" activities. Most dramatically, in being defined as "gathering," Nez Perce horticulture is set off against "agriculture"—a term that carries ideas about human control of plant genetics ("domestication"), about technological complexity, and about aspects of social organization such as landownership and control of production. Although European agriculture does indeed involve these elements—fields of wheat harvested with increasingly complicated machinery at the behest of landowners—this is only one particular social formation, one example from a range of possibilities. "Agriculture" may also be a form of food production that "simply" increases the productivity of food-plant resources. Both forms of food-plant production may produce a surplus that will support a larger population than is possible in a "pure" subsistence economy—and both will lead to different social formations.

The differences between European agriculture and other forms of surplus food-plant production are not natural but symbolic and social. Thus, biases in understanding Nez Perce food-plant production arose because Nez Perce gardens were so unusual to Europeans that they failed to recognize them as gardens. They were unusual because the plants grown in them were "wild" and undesirable to Euro-Americans. This symbolic bias has several elements. First is the idea that cereals are the most desirable agricultural products. Second is the idea that agriculture requires domesticated plants, that is, plants that require human intervention to reproduce successfully.[38] Third is the fact that Nez Perce food plants are native to the area rather than domesticates originating in a different ecosystem. Instead, agriculture ought to be defined as "human efforts to modify the environments of plants and animals to increase their productivity and usefulness."[39]

Nez Perce gardens were also unusual because they were established and maintained with a different, "simpler" technology than Euro-American gardens. "Simple" technology may be very powerful and have substantial effects. In the case of Nez Perces, perhaps the most powerful "simple" item of technology was fire. This has been a fundamental tool of native agriculture throughout most of the Americas and the world. Fire certainly was used to maintain at least some "prairies" and "meadows"—among them the "unusual gardens"—by discouraging the growth of an overstory. Indeed, fire created many of them.

This observation points to another significant bias in the perception of Nez Perce food-plant production: Euro-Americans regard fires as destructive, a priori. They actively suppress them and aggressively prosecute those who start them. Only recently has this attitude been challenged, and even the new attitudes reveal their ideological roots: many opponents of blanket fire suppression distinguish between "natural" and "man-made" fires. They favor extinguishing the latter in order to preserve the "natural" environment—which is in part man-made. The significance of fire to Nez Perce food-plant production has gone unrecognized because the idea that something inherently destructive might be used to produce food was unthinkable.

The gardens were unusual because Euro-Americans did not recognize the "digging stick" as an agricultural tool, even though it was familiar to them in other contexts. The *tuuk'es* is clearly a dibble like those used by many other "slash-and-burn" or "milpas" agriculturalists. The European term for such agriculture is *swidden*.

Nez Perce gardens also were unusual because they were not fenced. This did not mean, however, that anyone could go to any spot and start digging. The women who worked the gardens knew their boundaries very well. The use of sticks to mark locations during disputes makes this clear. Women returned yearly to the same locations, maintaining their rights through the effort of working in them.[40]

Finally, Nez Perce gardens were unusual because women, rather than men, created and owned them. Furthermore, under Nez Perce concepts of tenure and ownership, women controlled the distribution of garden products in both intra- and extrafamilial contexts. In other words, the surplus produced by women was distributed not only through kinship networks but in commerce as well. This may partially account for the considerable power exercised by Nez Perce women.[41] The androcentric bias of Euro-Americans is well known, and it is probable that most observers of native daily life were neither privy to nor interested in such "women's work"—which most Euro-Americans regarded as mere drudgery.

CONCLUSION

The Nez Perce of the nineteenth century have long been characterized as hunters, gatherers, and fishers. Consonant with this view of their subsistence, Nez Perce society is viewed as having been organized into bands led by individuals (mostly men but occasionally women) who had no authority to control the behavior of band members. This ethnographic image of the Nez Perce as a band society is riddled with exceptions, however, which are attributed to long interaction with the Northwest Coast, to recent influence from the Great Plains as a result of "horse culture," or to a reaction to Euro-American pressure and acculturation.

An alternative explanation, following Kehoe,[42] is that ethnographers and historians have mistakenly assessed Nez Perce subsistence and society. The Nez Perce used horticultural practices as well as gathering, hunting, and fishing. Horticultural technologies included the use of dibble sticks to prepare beds for favored plants. They involved the use of fire to create and maintain meadows in which favored plants were gathered, to create fields of huckleberries and grouseberries in the mountains, and to improve hunting conditions.

Understanding the Nez Perce as horticulturalists explains many sociocultural practices that differed from those of more traditional foragers. These differences included (1) communal housing and food storage, (2) villages with "men's houses" and perhaps "women's houses," (3) named, ranked leadership positions associated with recognized redistribution of resources, (4) massive raiding parties, and (5) encampments of more than 1,000 people from a variety of "ethnic" groups.

The last feature—the alliance of ethnic groups—suggests that the apparent cultural and social complexity of Nez Perce–speaking people was widespread. Indeed, all five of the features I have listed were found throughout the Plateau culture area. What is also important is that different ethnic groups stressed the control of different aspects of the environment. For example, people in the Yakama region depended more upon the production of several species of lomatium,[43] perhaps with similar social effects,[44] whereas those in the Colville region stressed a particular lomatium ("white camas," *Lomatium canbyi*).[45] These different groups were economically and symbolically linked in a variety of ways into a single society.[46]

This raises a complex theoretical and categorical problem. Stated most simply, it seems that the great divide between "civilized" and "primitive" has been reduced in contemporary subsistence studies to the presence or absence of domesticated plants within people's subsistence economies. Thus, the social and cultural world is divided between those who are dependent upon domesticated species (horticulturalists and agriculturalists) and those dependent upon undomesticated species (foragers).

The consequences of domestication are thought to be a kind of "settling down" of nomadic people. The domestication of plants, in other words, domesticated people by requiring them to give up their nomadic ways in order to farm. More productive food sources and a reliable food supply meant an end to hand-to-mouth existence and greater energy and nutrient flows into human populations, which also increased in response to these changes. More people, in turn, required more farming, greater central control to coordinate it, and another round in the cycle. Hence, "civilization" is rooted in the domestication of plants and the consequent appearance of agriculture. The great divide opens between us and our "contemporary ancestors," the foragers. This is the European "myth of ecology" as told in anthropology.[47]

This myth illustrates the penchant of members of the European tradition for sui generis causes. Once these causes have been determined, it is believed, then their effects can be calculated.[48] It also illustrates the comfortable assumption that domestication and agriculture provide a more stable food supply than do nondomesticated species and that our bargain with the plants has been a good one.

Of course, if we apply a fundamental ecological insight regarding the direct relationship between ecosystem diversity and stability, then we see that what domestication does is quite different from producing a more stable food supply. Indeed, simple ecosystems are often more unstable. Although explosive population growth of some domesticated species in such ecosystems clearly occurs, so does the devastating population collapse of others. These are signs of systemic instability, not stability. Furthermore, the species that "benefit" most from simplifying the system—that is, those whose growth is greatest—are also susceptible to the most devastating population collapses.

Reliance on domesticated organisms is associated with three processes of simplification. Most fundamental is the control of genetic diversity through control of reproduction: plants are bred in order to increase desirable features and decrease undesirable ones. Associated with this is the reduction of habitat diversity by creating environmental conditions favorable for the domesticated organisms' survival, at the expense of competing organisms. The third is the simplification of the food chain by reducing the number of food resources through reliance primarily on domesticated organisms.

The first of these processes, control of plant reproduction, is the sine qua non of domestication but not of agriculture per se, since agriculture need not involve the genetic changes associated with domestication.[49] The social relationship involved in domesticated agriculture is different from that of "complex hunters-gatherers," who are distinguished from "simple hunters-gatherers" because they store foods produced from nondomesticates. This allows complex hunter-gatherers to survive periods of low resource availability. Nez Perces not only stored foods but also intensely altered portions of their environment in order to raise the net productivity of favored plant resources, which were then transformed into storable food products.

This view clearly conforms with Kehoe's hypothesis that groups living in what is now the conterminous United States were agricultural. In this view, social complexity is an adaptation not to problems resulting from increased population but to increasingly unstable food supplies. It is a human attempt to reintroduce ecosystem complexity by expanding the boundaries of the system, which may be done through alliance and warfare. Those who control such exchanges control human relationships and the environmental dimensions of group life.

The absence of domesticated plants from a population's subsistence base does not necessarily make them foragers. Such a view is too simple. It inaccurately portrays the relationships between many societies and their constructed environments. Examining the social and material technologies of subsistence reveals that, despite the absence of domesticated plants, the *nimípu* were horticulturalists.

NOTES

An earlier version of this chapter was read in 1991 at the annual meeting of the American Anthropological Association in Chicago, Illinois. I am deeply indebted to Samuel M. Watters and his family for sharing their knowledge with me and encouraging my research and publication on these matters.

1. Herbert Joseph Spinden, "The Nez Perce Indians," *Memoirs of the American Anthropological Association* 2 (3) (1908).

2. Christopher L. Miller, *Prophetic Worlds: Indians and Whites on the Columbia Plateau* (New Brunswick, N.J.: Rutgers University Press, 1985).

3. "Nez Perce" refers to the people who spoke *nimiputimt*—"Nez Perce tongue" or language—and their descendants. The "Nez Perce Tribe" is a construct of United States–*nimipu* conflict. Prior to this conflict several politically independent groups of *nimipu* lived on the eastern edge of the Columbia basin, in present-day southeastern Washington, northeastern Oregon, and central Idaho.

4. Angelo Anastasio, *The Southern Plateau: An Ecological Analysis of Intergroup Relations* (Moscow: University of Idaho Laboratory of Anthropology, 1975), pp. 109, 199–200.

5. Robert F. Berkhofer, Jr., *The White Man's Indian: Images of the American Indian from Columbus to the Present* (New York: Vintage Books, 1978).

6. Spinden, "The Nez Perce Indians," p. 270.

7. Verne F. Ray, *Cultural Relations in the Plateau of Northwestern America* (Los Angeles: Southwest Museum, Publications of the Frederick Webb Hodge Anniversary Publication Fund, no. 3: 13, 1939). See also Verne F. Ray, "The Sanpoil and Nespelem: Salishan Peoples of Northeastern Washington," *University of Washington Publications in Anthropology* 5 (1933): 69–75.

8. Deward E. Walker, Jr., *Conflict and Schism in Nez Perce Acculturation: A Study in Religion and Politics* (Pullman: Washington State University Press, 1968).

9. *Ibid.*, pp. 15–17.

10. Francis Haines, *The Nez Perces: Tribesmen of the Columbia Plateau* (Norman: University of Oklahoma Press, 1955), pp. 8–22; Alvin M. Josephy, Jr., *The Nez Perce Indians and the Opening of the Northwest* (New Haven: Yale University Press, abridged ed. 1971), pp. 14–20.

11. Josephy, *The Nez Perce Indians*, p. 17.

12. Anastasio, *The Southern Plateau*, pp. 184–88, 191.

13. *Ibid.*, p. 112.

14. Alan G. Marshall, "Nez Perce Social Groups: An Ecological Interpretation" (Ph.D. diss., Washington State University, 1977).

15. Francis Jennings, *The Invasion of America: Indians, Colonialism, and the Cant of Conquest* (Chapel Hill: University of North Carolina Press, 1975).

16. Alice B. Kehoe, "Revisionist Anthropology: Aboriginal North America," *Current Anthropology* 22 (1981): 503–9.

17. *Ibid.*, p. 505.

18. *Ibid.*

19. Marshall, "Nez Perce Social Groups."

20. The reason these plants are significant has changed. In the past—at least until the 1940s for some families—native plant foods constituted a major portion of the diet. Over the last 50 or more years, however, their everyday use has declined and their importance as symbols of ethnic difference between Nez Perces and "whites" has increased. Lucy J. Harbinger, "The Importance of Food Plants in the Maintenance of Nez Perce Cultural Identity" (M.A. thesis, Washington State University, 1964).

21. Spinden, "The Nez Perce Indians," p. 201.

22. Nez Perces, when using English, refer to all vegetal foods from underground as "roots." All Nez Perce English words in this chapter are indicated by quotation marks.

23. Marshall, "Nez Perce Social Groups."

24. We avoided areas that were the "property" of other "root diggers," even though they were not present.

25. Unfortunately, I did not count the total number of bulbs discovered or the number returned to the soil. My estimate is that at least one-third were rejected and replanted.

26. The basis for the distinction by sex is unclear to me, and I heard conflicting statements about which "roots," males or females, were returned to the earth.

27. The popular name for these plants is "biscuitroot," a term rarely used by Nez Perces.

28. A corm is "a short, vertical, underground stem that is thickened as a perennating storage organ." In all, about nine species of *Lomatium* were consumed as food, but the "roots" of six species were considered edible, even in times of food shortage. Two species were exploited primarily for their shoots and leaves.

29. Spinden, "The Nez Perce Indians," p. 203.

30. Helen H. Norton, "The Association between Anthropogenic Prairies and Important Food Plants in Western Washington," *Northwest Anthropological Research Notes* 13 (2) (1979): 434–39.

31. John B. Leiberg, *The Bitterroot Forest Reserve,* Twentieth Annual Report of the United States Geological Survey, vol. 5 (Washington, D.C.: Government Printing Office, 1900).

32. Rexford Daubenmire, *Steppe Vegetation of Washington* (Pullman: Washington Agricultural Experiment Station, Technical Bulletin 62, 1970). Interestingly, Daubenmire was convinced that the region was *not* burned by native peoples because of the lack of written reports regarding the use of fire.

33. *Ibid.*, p. 78.

34. *Ibid.*, pp. 37–40.

35. Henry Smith, personal communication; Peter J. Mehringer, personal commu-

nication. Patterned ground is caused by frequent freezing of wet soils. In consequence, the soil particles are sorted by size. The larger particles form a netlike pattern while the finer particles form small mounds within the net of coarse particles; these mounds are sometimes called "mima mounds."

36. *Original Journals of the Lewis and Clark Expedition, 1804–1806*, ed. Reuben G. Thwaites (New York: Dodd, Mead, and Co., 1905), 3: 79, 12: 199.

37. Isaac I. Stevens, *Report of Explorations for a Route for the Pacific Railroad, Near the Forty-Seventh and Forty-Ninth Latitude from St. Paul to Puget Sound*, 33d Congress, 2d Session, House Executive Document 91.

38. Robert J. Wenke, *Patterns in Prehistory: Humankind's First Three Million Years* (New York: Oxford University Press, 1990), pp. 226–28.

39. *Ibid.*, p. 229.

40. Alan G. Marshall, "Nez Perce Tenure," ms. in possession of author.

41. Lillian A. Ackerman, "The Effect of Missionary Ideals on Family Structure and Women's Roles in Plateau Indian Culture," *Idaho Yesterdays* 31 (1988): 64–73.

42. Kehoe, "Revisionist Anthropology."

43. Eugene S. Hunn and David H. French, "*Lomatium*: A Key Resource for Columbia Plateau Native Subsistence," *Northwest Science* 55 (1981): 87–95.

44. Eugene S. Hunn, "On the Relative Contribution of Men and Women to Subsistence among Hunter-Gatherers of the Columbia Plateau: A Comparison with Ethnographic Atlas Summaries," *Journal of Ethnobiology* 1 (1981): 124–34.

45. Ray, "The Sanpoil and Nespelem."

46. Anastasio, *The Southern Plateau*.

47. Alan G. Marshall, "Prairie Chickens Dancing . . .: Ecology's Myth," in *Idaho Folklife: Homesteads to Headstones*, ed. Louie W. Attebery (Salt Lake City: University of Utah Press, 1985), pp. 101–7.

48. Eric R. Wolf, *Europe and the People without History* (Berkeley: University of California Press, 1982), pp. 6–7, 17.

49. Barbara Bender, "Gatherer-Hunter to Farmer: A Social Perspective," *World Archaeology* 10 (1978): 204–22.

10 / Megafauna of the Columbia Basin, 1800–1840

Lewis and Clark in a Game Sink

PAUL S. MARTIN AND CHRISTINE R. SZUTER

The diversity of wildlife in the Americas was forged both by extinction and by the hunting practices of native people. Historical ecologists probing radiocarbon time—the last 40,000 years—recognize two catastrophes that shaped the biogeography of large mammals in the American continents. The first was a megafaunal extinction around 13,000 years ago, when the Americas lost two-thirds of their large mammals, including all species of elephants, horses, glyptodonts, and ground sloths.[1] The second was a post-Columbian catastrophe that repeatedly decimated native human populations, owing to epidemics of Eurasian pathogens in tandem with cultural changes brought about by the arrival of new cultigens, metals, domestic animals, and religions.

Despite dynamic changes throughout the Holocene,[2] there were no additional losses of American megafauna (animals over 44 kilograms of adult body weight) beyond taxa of bison in the mid-Holocene. By 12,000 years ago, continental extinction had run its course. The survivors in western North America included bison (*Bison bison*), moose (*Alces alces*), elk, wapiti or red deer (*Cervus elaphus*), mule and white-tail deer (*Odocoileus hemionus* and *O. virginianus*), bighorn or mountain sheep (*Ovis canadensis*), mountain goat (*Oreamnos americanus*), pronghorn antelope (*Antilocapra americana*), and woodland caribou (*Rangifer tarandus*). Large predators, scavengers, and omnivores included the wolf (*Canis lupus*), grizzly bear (*Ursus arctos*), black bear (*Ursus americanus*), mountain lion (*Felis concolor*), and jaguar (*Panthera onca*).

Biogeographers have long regarded the ranges and numbers of all animals—both large and small—as essentially natural, with scant recognition that ranges and numbers of certain preferred prey might be shaped by native people. Recently historians have begun to reexamine the role of Native Americans as hunters.[3] Wildlife ecologist Charles E. Kay has replaced the concept of an "ecologically noble savage" with that of an "ultimate keystone species."[4] According to Kay's hypothesis, in regions of sizable and stable human populations that were supported primarily by other resources, hunters would be sufficiently numerous to drive populations of desirable "target species," or "preferred prey," such as buffalo, elk, and deer, to low levels or even to local extinction. The result would be "game sinks."

The historical record, however, also demonstrates that there were game-rich regions, or "game sources." Although this seems paradoxical if native hunters were highly effective, Harold Hickerson provides an answer.[5] He found that the forest-prairie border in Wisconsin and Minnesota, a favorable habitat for deer and other game, constituted a buffer zone between warring Chippewas and Sioux. In this no-man's-land, hunters themselves might become the hunted. As a result, game thrived until a peace treaty was enforced by the United States government, after which hunting of game intensified.

This essay provides a fresh look at historic documents of the contact period, seeking neglected indications of Native American influence on wildlife. Focusing on the Columbia River drainage, we test Kay's and Hickerson's concept that native people had a significant effect on populations of large animals.

As Kay has shown,[6] historic documents offer untapped resources for understanding wildlife dynamics—and the Columbia River basin has a rich load of such documents. The natural history data in Lewis and Clark's journals and those of other members of their party are truly outstanding.[7] For the Columbia River those records are matched by David Thompson's *Narrative* and *Columbia Journals*.[8] Less trustworthy are some parts of the narrative of Ross Cox, who was 19 when he got lost on a cross-country trek from Palouse Falls to Spokane House, a misadventure that lost no dramatic impact in the telling.[9]

Journals or narratives from various employees of fur trading companies, such as Ross Cox,[10] Daniel Harmon,[11] Alexander Henry,[12] Gabrielle Franchere,[13] Peter Skene Ogden,[14] Alexander Ross,[15] Robert Stuart,[16] and David Thompson,[17] yield insights not found in the journals of Lewis and Clark.

The fur traders spent decades with various tribes, learned to speak some of the native languages, and probed regions previously unexplored by Europeans. They married or lived with Indian women, raised families, joined hunts, lamented the destructive impact of epidemic diseases and of alcohol (even as they provided it), and witnessed the collapse of traditional beliefs. Exposed in small posts that were at best marginally defensible against any sustained attack, the fur traders out of necessity developed considerable diplomatic skill. They managed to provision themselves and to conduct business while maintaining a peace of sorts with a variety of Indian nations often hostile to each other and unimpressed by the claims of the alien traders that they came from lands of vast wealth and power.

All of these sources, of course, suffer from an unavoidable shortcoming: they postdate the onset of significant biocultural change triggered by Eurasian diseases, guns, horses, and so forth, the legacy of Columbian contact. Nonetheless, the information they contain offers evidence supporting the conclusion that native people had a substantial impact on populations of game species—and thus on the environment. We begin with an example of a game source or park and follow with examples of game sinks.

LEWIS AND CLARK IN A GAME PARK

On the Great Plains east of the Rockies, Lewis and Clark found large animals in abundance (fig. 10.1). Near Great Falls, Montana, Clark thought he saw 10,000 buffalo in one view,[18] and he estimated 20,000 animals on the Missouri near the White River.[19] According to an accounting by R. D. Burroughs, the Corps of Discovery shot 1,000 deer, 375 elk, 227 bison, and hundreds of individuals of other species,[20] not an excessive number when one considers the size of the party—30 men along with Sacagawea, a Shoshone woman, and her infant—and its reliance on wild game for subsistence.

In four months during the spring and summer of 1805, between the time they left the vicinity of the Hidatsa-Mandan villages north of modern Bismarck, North Dakota, and their rendezvous with Shoshone hunters at Lemhi Pass on the Continental Divide on the Idaho-Montana border, the Corps of Discovery encountered no other Indians. For the first hundred miles upstream, in Hidatsa and Mandan hunting grounds, game was scarce. Approaching the mouth of the Yellowstone River near the modern town of Williston in western North Dakota, conditions changed. Lewis reported that "the whole face of the Country was covered with herds of Buffalo, Elk & Antelopes . . . so gentle that we pass near them while feeding. . . . When we attract their

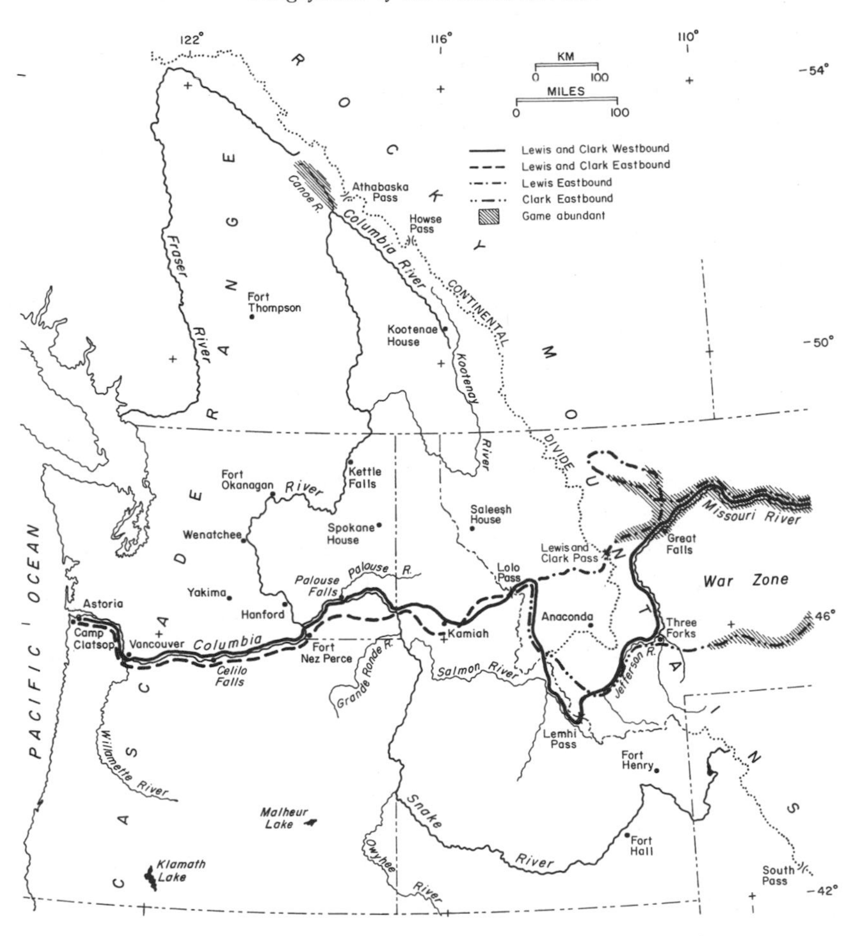

FIG. 10.1. Lewis and Clark's routes, showing areas of game abundance. Elsewhere game was scarce except for elk in the forest outside Fort Clatsop.

attention, they frequently approach us more nearly to discover what we are and in some instances pursue us a considerable distance."[21]

On May 8 Lewis claimed: "We can send out at any time and obtain whatever species of meat the country affords in as large quantities as we wish." On May 9 he added: "We saw a great quantity of game today particularly of Elk and Buffaloe, the latter are now so gentle that the men frequently throw sticks and stones at them in order to drive them out of the way." Wolves scavenging dead buffalo were so tame that Clark speared one.

The following year, during their return in the summer of 1806, Clark found game equally abundant on the Yellowstone River. They canoed past "emenc number of Deer Elk and buffaloes on the banks." Clark felt that "for me to mention or give an estimate of the different species of wild animals on this river particularly Buffalo, Elk Antelopes & Wolves would be increditable. I shall therefore be silent in the subject further."[22] His promise of silence on the subject proved impossible to keep. The next night the grunting noises around their camp—sounds of males in the buffalo herds in rut—ruined sleep. On July 27, buffalo and elk were "astonishingly numerous," the elk so gentle that the party passed within 20 or 30 paces without causing alarm.

Although they saw no Indians, Lewis and Clark found signs of them. On May 4 they discovered two abandoned Blackfeet war lodges. Twice the exploring party found stray Indian dogs. Southwest of Great Falls Lewis and Clark began to find brush huts used by the Shoshone. On their return a year later, they lost horses to unseen Indian horse thieves, probably Crow. Nevertheless, Indian activity was limited and hunting was easy.

Between April 25 and June 13, 1805, in 50 days of travel along the upper Missouri between Williston, North Dakota, and the mouth of the Marias River, Montana, Lewis and Clark's journals report that their hunters killed 79 deer, 50 elk, 44 bison plus 7 calves, 8 antelope, and 12 grizzly bears (table 10.1). In addition, they killed 9 mountain sheep and 3 wolves, took many beaver, and caught or killed a variety of small game. "We eat an immensity of meat," Lewis wrote.[23] "It requires 4 deer, an elk and a deer, or one buffaloe, to supply us plentifully 24 hours. Meat now forms our food principally as we reserve our flour, parched meal and corn as much as possible for the rocky mountains which we are shortly to enter, and where from the Indian account game is not very abundant." They also killed elk and deer for hides to trade, to cover an ill-fated iron-frame boat, and for moccasins and clothing.

LEWIS AND CLARK IN A GAME SINK

The transition from the upper Missouri to the Columbia River drainage was dramatic. As their Hidatsa Indian informants at Fort Mandan had predicted, they killed the last buffalo southwest of Great Falls on July 16. Farther upstream along the Jefferson River, near present-day Whitehall, Montana, they found only buffalo bones and dung. In the same area on August 2, 1805, the party killed its last elk in the Missouri drainage. From there onward the hunters found mainly deer and antelope, in diminishing numbers. From there to Tongue Point, near Astoria, on the mouth of the

TABLE 10.1
Lewis and Clark's Game Bag, 1805–1806

	Upper Missouri River, 25 April–13 July 1805	Columbia River, 18 Sept.–6 Nov. 1805	Camp Clatsop, 1 Jan.–19 Feb. 1806	Columbia River, 23 Mar.–11 May 1806	Upper Missouri River, 30 June–18 Aug. 1806
Deer	79	28	8	38	191
Elk	50	0	51	22	51
Bison	44	0	0	0	55
Antelope	8	0	0	0	9
Bear	12	0	0	1	12
Dog	0	101+	5	83+	0
Ration Units	111	7	40	26	150

SOURCE: Meriwether Lewis and William Clark, *The Journals of the Lewis and Clark Expedition*, ed. Gary E. Moulton (Lincoln: University of Nebraska Press, 10 vols. to date, 1983–97.

NOTE: Each sample spans 50 days. To compute ration units (excluding dogs), bison = 1, elk or bear = 3/4, deer = 1/4, and antelope = 1/8.

Columbia, elk were scarce. They managed to kill one in the Bitterroot Valley south of Missoula. From this experience and that of other explorers to follow, one might not anticipate the large numbers of elk now to be found throughout the Rockies. One would certainly not expect the recent irruption of elk such as that inside the Hanford reserve near Yakima, Washington, a treeless region of Great Basin shrub-steppe, a habitat not usually thought of as suitable for elk.[24]

In contrast, the vegetation along the Continental Divide—spacious grassy valleys and sagebrush steppe laced by streams draining conifer forests on north slopes—would seem ideal for most game species, including elk and bison. Despite the attractiveness of the habitat, the Corps of Discovery gradually passed from a land of abundant game (around Great Falls, Montana) to one of scarcity in Idaho along the Salmon River west of Lemhi Pass. The lack of game did not involve badlands or any major change in soil type, and in less than 20 years buffalo would overrun this country.

West of Lemhi Pass the Corps of Discovery could find little to hunt. While Clark scouted the Salmon, the wild "River of No Return," the rest of the party exhausted their food reserves. With Shoshone guides, the party proceeded north over 100 miles before turning to the west again through Lolo Pass into the Bitterroots. Hunting game with no success, they killed and ate colts and a horse. Cameahwait, the Shoshone chief, Sacagawea's brother, had warned them to expect scarcity. But Lewis reasoned that if the Indians, including women and children, could pass the mountains, the explorers could as well. Furthermore, thought Lewis, if there were large numbers of Indians living on the river below the mountains, "they must have some means of subsistence which would be in our power to procure in the same country."[25]

On September 20, after a week of unusually rugged travel and near starvation, an advance party led by Clark descended out of the Bitterroots. They found a beautiful, level pine country whose friendly inhabitants, the Nez Perce (known to Lewis and Clark as Chopunish), shared their traditional foods. These included dried fish (from the salmon season), berries, roots (bulbs) of camas or "quamash" (*Camassia*), and even a small piece of buffalo meat.

In his journal entry for September 22, Lewis allowed himself the rare luxury of self-congratulation: "The pleasure I now felt in having tryumphed over the rocky Mountains and descending once more to a level and fertile country where there was every rational hope of finding a comfortable subsistence for myself and party can be more readily conceived then expressed, nor was the flattering prospect of the final success of the expedition less pleasing." Despite his rational hopes, Lewis found no "comfortable subsistence." He had misjudged their situation. Many members of the party became seriously ill on the new diet, Lewis included. The trouble was that the rations of dry fish and roots that the explorers could readily obtain from the Nez Perce they could not stomach. The problems continued. At their "Canoe Camp" in October, Clark killed a horse to make soup for sick men. Nevertheless, trouble persisted: "Nothing to eate but dried roots dried fish, . . . which filled us so full of wind, that we were scarcely able to Breathe all night."[26]

Although it was hard to hunt in the dry hills above the Clearwater,[27] their hunters bagged some deer. Nevertheless, during the 50 days the Corps of Discovery canoed and camped on the Clearwater, the Snake, and the Columbia, they killed a scant 28 deer (table 10.1), half of the animals on the upper Clearwater and the rest over 200 miles to the west as they reached forested parts of the lower Columbia near the Cascades. Besides ducks, geese, and swans, they shot and ate a coyote. But game birds were not enough. Their total kill of big game averaged 0.15 ration units a day (table 10.1), less than a fifth of their desirable daily ration and an order of magnitude less than their 50-day bag on the upper Missouri. Despite the best efforts of their best hunters, for 26 days beginning September 30 they found no big game (no elk, antelope, or deer) at all. The main alternatives were dried fish, roots, and dogs.

Unable to stomach the native diet of dry fish and roots and unable to find enough game in this vast game sink, the Corps of Discovery, not surprisingly, turned to the domestic animals of the local people to meet their needs. In 30 days beginning in early October, their journals report that the explorers bought at least 100 dogs. On October 17 they obtained all the dogs they could—the exact number unspecified. On October 18, in exchange for "beads, wire and other trinkets of little value," Clark acquired 40 dogs.

On January 3, 1806, Lewis wrote that "our party from necessity having been obliged to subsist some length of time on dogs have now become extremely fond of their flesh; it is worthy of remark that while we lived principally on the flesh of this animal we were much more healthy, strong and more fleshy than we had been since we left the Buffalo County." In his own contribution to the expedition's journal that day, Clark admitted that he had not become "reconciled to the taste of this animal yet."

At Camp Clatsop at the mouth of the Columbia, Lewis and Clark wintered in the dense, wet forest of the Oregon coast. They sought a site with plenty of game. Unlike their experience upstream along the Columbia and the Clearwater, they found elk numerous, at least until spring when the animals retreated inland. Sergeant Gass calculated that between December 1, 1805, and March 20, 1806, the party's hunters bagged 131 elk and 20 deer.[28] The group also dined on wapato roots, candlefish, sturgeon, whale blubber, and a few dogs.

On the homeward journey, a few days spent near the mouth of Sandy River allowed hunters to forage far enough toward Mount Hood to shoot 14 elk and a bear—a surplus the corps held in reserve. Still, their kill of big game was about half their desired daily ration. Again dogs helped fill out the menu. On their trip down and up the Clearwater-Snake-Columbia River, the Corps of Discovery bought and ate more than 180 dogs and a few horses.

Members of the Corps of Discovery were not alone in their caniphagy. In 1811 and 1812 both Hunt (traveling west) and Stuart (returning east) purchased and ate dogs and horses on their perilous overland journeys to and from the Pacific Fur Company's Fort Astoria, pioneering a route that became the Oregon Trail.[29] In two weeks on the lower Columbia and the Willamette River in January 1814, Alexander Henry purchased 52 dogs for his party.[30] In December 1815, the Spokane brigade left Okanogan with 40 freshly killed dogs in the loading.[31] At villages between Celilo Falls and the land of the Walla Wallas, the North West Company traders purchased and ate 150 dogs,[32] which the Canadians preferred to horse flesh. Seeking provisions for his ravenous fur brigade in early December 1827 in the Klamath Lakes country of southwestern Oregon, Peter Skene Ogden discovered that the "Clammette Nation" had plenty of dogs to sell. In four days he bought 137.[33]

Although game was scarce, people were not—the reverse of what the Corps of Discovery had found in Montana. Lewis and Clark contacted Indians daily and camped with or near them every night. From the bank, large numbers of Indians watched Lewis and Clark float down the river. Some of the groups they mention or learned of in the Columbia drainage were the Palouse, Yakama, Wanapam, Cayuse, Umatilla, Okanogan, Yakama, Clatsop, and other Chinookans, besides those Indians they encountered first, the Shoshone, Flat Heads, and Nez Perce, in aggregate a population estimated at 80,000.[34] In 1824, Governor George Simpson of the Hudson's Bay Company found that "the population on the banks of the Columbia River is much greater than in any other part of North America that I have visited."[35]

Such was not the case in the dense forest at the mouth of the Columbia where the corps wintered. The Clatsop and other coastal tribes were in eclipse. More exposed to diseases than the interior nations, the Clatsop would soon disappear. Dense coastal populations of indigenous people, particularly those along the mouth of the Columbia, were highly exposed to viral and venereal diseases brought by trading vessels.[36] More than 100 ships had landed in the area before Lewis and Clark arrived overland.[37] By then, smallpox had already struck the Clatsop at least twice, reducing the coastal populations by 40 percent.[38] According to Lewis, diseases "would account for the number of remains of villages which we find deserted on the river and sea coast in this quarter."[39]

The scarcity of wild game that the corps encountered in the Columbia basin above Celilo Falls might have resulted from either seasonal movements of game, poor habitat, or heavy hunting by the natives. Although seasonal change in game might account for scarcity in the Bitterroots when Lewis and Clark traveled west, it would not account for scarcity on their return, when the explorers, the Nez Perce, and wild game alike were confined to the lowlands until the snowpack melted in late June. And the fact that horses thrived in the area provides indirect support for the conclusion that the problem was not poor habitat.

HORSE HEAVEN

Only one large herbivore was vastly more numerous along the Columbia than along the upper Missouri River. This was the horse. Horses reached the Columbia Plateau region roughly a century before Lewis and Clark.[40] After tens of millions of years of horse evolution in North America, the extinction of all native American species of horses by 13,000 years ago is indeed strange. The unwitting restoration of horses by Spaniards in the sixteenth century was hardly a biogeographical anomaly. Not surprisingly, the animals thrived.

According to North West Company clerk Ross Cox, "among the Flatheads, Cootonais, Spokans, etc., whose lands are rather thickly wooded, there are not more [horses] than sufficient for their actual use, and every colt, on arriving at the proper age, is broken in for the saddle. But in the countries inhabited by the Wallah Wallahs, Nez Perce, and Shoshones, which consist chiefly of open plains, well watered, and thinly wooded, they are far more numerous, and thousands are allowed to go wild."[41] According to Sergeant Patrick Gass of the Corps of Discovery, "between the Great Falls [Celilo Falls] of the Columbia and this place [the Canoe Camp on the Clearwater] we saw more horses, than I ever saw in the same space of country."[42] "These people have immense numbers of [horses]; one individual might own 50 to 100 head," wrote Meriwether Lewis.[43] Large horse herds in places such as the Horse Heaven Hills along the Columbia suggest a substantial carrying capacity for megaherbivores. At a carrying capacity of three to five animals per square kilometer, the potential population of the region may have approached half a million.

The productivity of the habitat was also noted by Lewis and Clark, each of whom had a woodsman's eye for evaluating such things. In the sagebrush–rabbit bush shrub-steppe, wrote Clark, "great numbers of the natives pass us on horse back."[44] Lewis remarked the next day that "the soil is not as fertile as about the falls [Great Falls of the Columbia], tho' it produces a low grass on which the horses feed very conveniently; it astonished me to see the order [good condition] of their horses at this season of the year when I knew that they had wintered on the dry grass of the plains and at the same time [were ridden] with greater severity than is common among ourselves. I did not see a single horse which could be deemed poor and many of them were as fat as seals."[45]

In 1836, the missionary Samuel Parker was equally impressed with how well his worn-down horses had wintered outside Fort Nez Perce. In addition, cattle kept outside the fort with nothing more to feed upon than what they found on the prairies thrived "in as good condition for market as oxen driven from the stalls of New England."[46]

FUR TRADERS, THEIR TRADING PARTNERS, AND WILDLIFE

The journals of fur traders and explorers of the early 1800s reiterate many of the observations of the Corps of Discovery. In the summer of 1807, a year after Lewis and Clark's return trip, David Thompson, his wife, Charlotte, their three small children, and 10 traders, along with other women and children, crossed the Canadian Rockies at Howse Pass to open a trading post in Kootenai territory. West of the Rockies on the upper Columbia, Thompson soon found he faced the same difficulty faced by Lewis and Clark—the lack of game. Even deer were scarce that summer: "We had very hard times and were obliged to eat several horses."[47] Once Thompson's party boiled and ate part of a wild horse killed the day before and left undisembowelled, "but it made us very sick, being half rotten."[48] When salmon arrived, they were scant and spent from the vast distance they had swum. The construction of Kootanae House, Thompson's new post, was delayed: "As we have no provisions and are too weak to build without food, the men went [fishing]."[49]

In the summer of 1811, Thompson and a party of 10 descended the Columbia from the forty-ninth parallel at Kettle Falls to its mouth at Astoria. Above the mouth of the Snake, Thompson mapped several hundred miles of previously unexplored river. Like Lewis and Clark, he found virtually no game.

Below Wenatchee, the natives "describe their country to the southward to being high dry and barren, without animals; to the northward the lands are good with Antelopes [deer], Mountain Sheep and Goats, of which their clothing is made, and of the fine long wool of the latter they make good rude blankets." They had a few bison robes, obtained by trade, and were better clothed than any other tribe Thompson had yet seen.[50]

Thompson also explored the uplands. Returning from Astoria, he left the Snake River at Palouse Falls, where he purchased horses, and proceeded overland 90 miles north to Spokane House. He must have crossed the Palouse Prairie, with its deep, fertile loess soils. Although the country became increasingly attractive, with groves of aspens and tender green grass on which their horses fed avidly, he found no game. "Provisions having fallen short and our Guide assuring us we should see no Deer, nor Indians to supply us, we had to shoot a Horse for a supply."[51]

Thompson was by no means the only one to resort to horse meat in the early days at Spokane. According to a footnote in Franchere's account, the traders wintering at Spokane House in 1812–13 had to live on horse flesh; they ate 90 horses.[52] Prior to his 1825 tour of inspection, Hudson's Bay Company Governor George Simpson discovered that to provision Fort Nez Perce near the mouth of the Snake, 130 miles southwest of Spokane, the traders had slaughtered some 700 horses in three years.[53]

Impressed with the sizable Indian population on the banks of the Columbia, which appeared to be lined with lodges, Simpson imagined that the natives were content to live on fish and roots alone and did not turn their attention to hunting.[54] In contrast, brigade leader Alexander Ross, who had lived with the Salishan tribes, wrote: "Hunting is a favorite exercise with all Indians, and the Oakinackens are very fond of displaying their dexterity in riding and decoying animals of the chase. I have seen a fellow get into a deer-skin, stripped for the purpose, with the skin of the head and horns complete, walk off on all fours, and get actually among a herd of deer without their taking notice of the deception."[55] Indians of the Willamette Valley,[56] the Nez Perce,[57] and Shuswap[58] made effective use of a deer's head and horns to imitate and decoy wild animals. The Shuswap used a number of different calls for game. In season they lured does by imitating the bleat of a fawn,[59] a technique familiar to Lewis and Clark's hunters.[60]

CONCLUSION: CONTROLLERS OF THE GAME

We can only imagine what the West might have looked like when its native mammoths, camels, and horses roamed the land. From early accounts of certain uninhabited and underhunted regions, we learn of abundant and tame bison, elk, deer, and wolves. Nevertheless, the "wild America" reported by Lewis and Clark is an epiphenomenon in the history of contact. The game-rich upper Missouri was no more natural than the game-poor Columbia Plateau.

The journals of early explorers and the late prehistoric archaeological record suggest considerable suppression of large game by human activity. This was especially evident along floodplains featuring biologically productive, if limited, habitats highly attractive to farmers, foragers, and fishermen. A population subsidized by the underground storage organs of wild plants and by shoals of andromadous fish will maintain heavy hunting pressure on preferred prey (large mammals), thereby generating a classic predator pit.[61]

On occasion, the journals of early explorers describe regions of abundance—game sources or game parks—in uninhabited land known or suspected to be war zones.[62] The introduction of horses created a new opportunity for nomadic raiders to generate or expand intertribal war zones. With chronic intertribal warfare fueled by guns from traders, sizable buffer zones developed, and in them game thrived. In these areas, the abundance of game simulated what one might expect in the absence of any human presence.

In the time of the mammoths, until 13,000 years ago, bison ranged coast to coast.[63] Historically, "it is probable that had the buffalo remained unmolested by man and uninfluenced by him, [buffalo] would have crossed the Sierra Nevada and the Coast Range and taken up an abode in the fertile valleys of the Pacific Slope."[64] "Unmolested by man," it is probable that throughout the West, from the Columbia Plateau and the intermontane region south into the Mexican Plateau, bison, elk, deer, antelope, mountain goats, bighorn, and javelinas, and their associated wolves, bears, and jaguars, would have ranged much more widely and in far greater numbers than they did historically. On a scale unimaginable in terms of historical observations, an unhunted fauna of bison, deer, elk, and the like would have proliferated, absorbing niches once held by the much more diverse megafauna of the late Quaternary.

If "wild" is to mean pristine or natural or essentially bereft of human influence, it vanished more than 12,000 years ago. The last "wild" West supported an unhunted megafauna of at least 39 species, including mammoths, mastodons, native camels, native horses, and ground sloths, three times the diversity of megafauna found since.[65] Whether or not the Clovis Paleoindian colonization initiated the extinctions of megafauna more than 12,000 years ago, numerous historical accounts suggest that the range and numbers of surviving large animals were profoundly influenced by the activities of Native Americans in the centuries before the intrusion of Europeans.

Although European settlement is commonly blamed for *all* important historic losses of wild game, the vital role of Native Americans in influencing numbers of large mammals can be detected in early historic accounts.[66] The implications for managers seeking to restore "wild America" to some imaginary "natural" condition are profound: not only did the full complement of native megafauna vanish around 12,000 years ago, but subsequent human activity controlled the population sizes and severely limited the ranges of the surviving buffalo, elk, moose, and other megafauna.

Historical experience demonstrates the effects of reduced hunting pressure in intertribal war zones such as the one reaching from the forest-prairie edge in Wisconsin west to the Red River of the North, and a second on the plains of the upper Missouri, including the Yellowstone. On August 29,

1806, near the end of Lewis and Clark's expedition, William Clark observed that "in the country between the nations which are at war with each other the greatest number of animals are to be found." Lewis, Clark, and others of the period found very few large animals except for domestic horses in the neighborhood of populated, relatively peaceful settlements of the Columbia basin. The view that prior to European expansion, "Native Americans were the ultimate keystone species that structured entire ecosystems"[67] is abundantly supported by historic records of big game in the Northwest.

NOTES

1. Paul S. Martin, "Who or What Destroyed Our Mammoths?," in *Megafauna and Man: Discovery of America's Heartland*, eds. Larry D. Agenbroad, Jim I. Mead, and Lisa W. Nelson (Hot Springs, S.D.: The Mammoth Site of Hot Springs, South Dakota, Inc., Scientific Papers vol. 1, 1990), pp. 109–17; Paul S. Martin and Richard G. Klein, eds., *Quaternary Extinctions: A Prehistoric Revolution* (Tucson: University of Arizona Press, 1984).

2. W. R. Dickinson, "The Times Are Always Changing: The Holocene Saga," *Geological Society of America Bulletin* 107 (1995): 1–7.

3. Dan Flores, "Bison Ecology and Bison Diplomacy: The Southern Plains from 1800 to 1850," *Journal of American History* 78 (1991): 465–85; Elliott West, *The Way to the West: Essays on the Central Plains* (Albuquerque: University of New Mexico Press, 1995).

4. A "keystone species" determines the composition of the plant and animal community in which it lives. If the species is removed, the community will change, often dramatically. Charles E. Kay, "Aboriginal Overkill and the Biogeography of Moose in Western North America," *Alces* 33 (1997): 141–64; Charles E. Kay, "Ecosystems Then and Now: A Historical-Ecological Approach to Ecosystem Management," in *Proceedings of the Fourth Prairie Conservation and Endangered Species Workshop*, eds. W. D. Williams and J. F. Dormaar (Edmonton: Provincial Museum of Alberta, Natural History Occasional Paper no. 23, 1996), pp. 79–86; Charles E. Kay, "Aboriginal Overkill: The Role of Native Americans in Structuring Western Ecosystems," *Human Nature* 5 (1994): 359–98; Charles E. Kay, "Yellowstone's Northern Elk Herd: A Critical Evaluation of the 'Natural Regulation' Paradigm" (Ph.D. diss., Utah State University, 1990); C. E. Kay, B. Patton, and C. A. White, *Assessment of Long-Term Terrestrial Ecosystem States and Processes in Banff National Park and the Central Canadian Rockies* (Banff National Park, 1994).

5. Harold Hickerson, "The Virginia Deer and Intertribal Buffer Zones in the Upper Mississippi Valley," in *Man, Culture, and Animals: The Role of Animals in Human Ecological Adjustments*, eds. Anthony Leeds and Andrew P. Vayda (Washington, D.C.: American Association for the Advancement of Science, 1965), pp. 43–66.

6. Kay, "Yellowstone's Northern Elk Herd."

7. Daniel Botkin, *Our Natural History: The Lessons of Lewis and Clark* (New York: G. P. Putnam's Sons, 1995).

8. *David Thompson's Narrative of His Explorations in Western North America, 1784–1812*, ed. J. B. Tyrrell (Toronto: The Champlain Society, 1916); *Columbia Journals: David Thompson*, ed. Barbara Belyea (Montreal: McGill-Queen's University Press, 1994).

9. Ross Cox, *The Columbia River*, eds. Edgar I. Stewart and Jane R. Stewart (1831; reprint, Norman: University of Oklahoma Press, 1957).

10. *Ibid.*

11. W. Kaye Lamb, ed., *Sixteen Years in the Indian Country: The Journal of Daniel Williams Harmon, 1800–1816* (Toronto: Macmillan, 1957).

12. Elliott Coues, ed., *The Manuscript Journals of Alexander Henry and David Thompson: Exploration and Adventure among the Indians on the Red, Saskatchewan, Missouri, and Columbia Rivers* (New York: Francis P. Harper, 1897).

13. Gabrielle Franchere, *Adventure at Astoria, 1810–1814*, trans. and ed. Hoyt C. Franchere (Norman: University of Oklahoma Press, 1967).

14. K. Davies and A. Johnson, eds., *Peter Skene Ogden's Snake Country Journal 1826–27* (London: Hudson's Bay Record Society, 1961); Glyndwr Williams, David E. Miller, and David H. Miller, eds., *Peter Skene Ogden's Snake Country Journals, 1827–1828 and 1828–1829* (London: Hudson's Bay Record Society, 1971).

15. Alexander Ross, *Adventures of the First Settlers on the Oregon or Columbia River, 1810–1813*, ed. Reuben G. Thwaites (Cleveland: Arthur H. Clark, 1904); Alexander Ross, *The Fur Hunters of the Far West*, ed. Kenneth A. Spaulding (Norman: University of Oklahoma Press, 1956).

16. *The Discovery of the Oregon Trail: Robert Stuart's Narratives of His Overland Trip Eastward from Astoria in 1812–1813*, ed. Philip A. Rollins (Lincoln: University of Nebraska Press, 1995).

17. *David Thompson's Narrative; Columbia Journals.*

18. Meriwether Lewis and William Clark, *The Journals of the Lewis and Clark Expedition*, ed. Gary E. Moulton (Lincoln: University of Nebraska Press, 10 vols. to date, 1983–97), 30 June 1805. We generally cite passages by date of entry rather than page number so that a variety of editions of the journals may be consulted. Quotations from the journals are not edited for spelling or grammar.

19. *Journals of the Lewis and Clark Expedition*, 29 August 1806.

20. R. D. Burroughs, *The Natural History of the Lewis and Clark Expedition* (East Lansing: Michigan State University Press, 1961).

21. *Journals of the Lewis and Clark Expedition*, 25 April 1805.

22. *Ibid.*, 24 July 1806 (between present-day Laurel and Billings).

23. *Ibid.*, 13 July 1805.

24. L. E. Eberhardt, L. L. Eberhardt, B. L. Tiller, and L. L. Cadwell, "Elk Population Increase," *Journal of Wildlife Management* 60 (1996): 373–80.

25. *Journals of the Lewis and Clark Expedition*, 14 August 1805.

26. *Ibid.*, 4 October 1805 (Clark).

27. *Ibid.*, 5 October 1805 (Clark).

28. *Journals of the Lewis and Clark Expedition*, vol. 10, p. 199.

29. *Discovery of the Oregon Trail.*

30. Coues, *Manuscript Journals of Alexander Henry and David Thompson.*

31. Cox, *The Columbia River*, p. 206.

32. *Ibid.*, p. 127.

33. Davies and Johnson, *Peter Skene Ogden's Snake Country Journal 1826–27.*

34. Lewis and Clark, *The Journals of the Lewis and Clark Expedition*, ed. Gary E. Moulton, 7: 488.

35. Frederick Merk, ed., *Fur Trade and Empire: George Simpson's Journal, 1824–1825* (Cambridge: Harvard University Press, 1968), p. 94.

36. R. T. Boyd, "Demographic History, 1774–1784," in *Handbook of North American Indians, vol. 7: Northwest Coast*, ed. Wayne Suttles (Washington, D.C.: Smithsonian Institution Press, 1990), pp. 135–48.

37. James P. Ronda, *Lewis and Clark among the Indians* (Lincoln: University of Nebraska Press, 1984).

38. Boyd, "Demographic History," p. 147.

39. *Journals of the Lewis and Clark Expedition*, 7 February 1806.

40. Lewis and Clark, *The Journals of the Lewis and Clark Expedition*, ed. Gary E. Moulton, 7: 260.

41. Cox, *The Columbia River*, p. 244.

42. *Journals of the Lewis and Clark Expedition*, vol. 10, p. 254.

43. *Journals of the Lewis and Clark Expedition*, 13 May 1806.

44. *Ibid.*, 24 April 1806.

45. *Ibid.*, 25 April 1806.

46. Samuel Parker, *Journey of an Exploring Tour beyond the Rocky Mountains* (1842; reprint, Moscow: University of Idaho Press, 1990), p. 279.

47. Jack Nisbet, *Sources of the River: Tracking David Thompson across Western North America* (Seattle: Sasquatch Books, 1994), p. 376.

48. *Columbia Journals*, p. 54.

49. *Ibid.*, p. 58.

50. *David Thompson's Narrative*, p. 484.

51. *Ibid.*, p. 530.

52. Cox, *The Columbia River*, p. 209; Franchere, *Adventure at Astoria*, p. 76.

53. Merk, *Fur Trade and Empire*, p. 128.

54. *Ibid.*, p. 94.

55. Ross, *Adventures of the First Settlers*, pp. 282–83.

56. *The Manuscript Journals of Alexander Henry and David Thompson*, 24 January 1814.

57. *Journals of the Lewis and Clark Expedition*, 15 May 1806 (Lewis).

58. J. A. Teit, *The Shuswap* (New York: American Museum of Natural History, Memoir 4[7], 1909), p. 524.

59. *Ibid.*, p. 520.

60. *Journals of the Lewis and Clark Expedition*, 23 June 1806 (Lewis).

61. R. Fritz, R. Suffling, and T. A. Younger, "Influence of Fur Trade, Famine, and Forest Fires on Moose and Woodland Caribou Populations in Northwestern Ontario from 1786 to 1911," *Environmental Management* 17 (1993): 477–89; Kay, *Yellowstone's Northern Elk Herd;* Kay, "Aboriginal Overkill."

62. Flores, "Bison Ecology and Bison Diplomacy"; Kay, "Aboriginal Overkill"; Paul S. Martin and Christine Szuter, "War Zones and Game Sinks in Lewis and Clark's West," *Conservation Biology* 13 (1999): 36–45.

63. Jerry N. McDonald, *North American Bison: Their Classification and Evolution* (Berkeley: University of California Press, 1981).

64. William T. Hornaday, *The Extermination of the American Bison, with a Sketch on Its Discovery and Life History* (Washington, D.C.: Smithsonian Institution, 1889), pp. 367–548.

65. Martin and Szuter, "War Zones and Game Sinks."

66. Fritz, Suffling, and Younger, "Influence of Fur Trade, Famine, and Forest Fires on Moose"; Kay, "Aboriginal Overkill"; Martin and Szuter, "War Zones and Game Sinks"; J. Truett, "Bison and Elk in the American Southwest: In Search of the Pristine," *Environmental Management* 20 (1996): 195–206; West, *The Way to the West.*

67. Kay, Patton, and White, *Assessment of Long-Term Terrestrial Ecosystem States*, p. xvi.

11 / Land Divided

Yakama Tribal Land Use in the Federal Allotment Era

BARBARA LEIBHARDT WESTER

The Dawes General Allotment Act of 1887 was the cornerstone of federal Indian policy well into the twentieth century.[1] Through the act, Congress dramatically altered its previous policy of isolating tribes on remote reservations. As the ready supply of fresh land and unexploited resources elsewhere diminished, white settlers clamored to open Indian land to farming, mining, and other uses. In the postfrontier age, federal Indian policy aimed for nothing less than the total destruction of Indian cultures and the incorporation of their people, land, and resources into the national economy. How best to absorb Indians into nineteenth-century United States society was commonly known as the "Indian problem." The Dawes Act represented one solution to it. By endowing each individual member of a tribe with his or her own land, federal policymakers believed Indian people would quickly learn to become self-sufficient, market-oriented farmer-citizens.

But the "Indian problem" had no single easy answer, and the Dawes Act also contained seeds of a quite different policy. The act, and subsequent legislation that provided for leasing, taxing, and alienating individual allotments, not only permitted but even encouraged Indians to sell their land to whites. Moreover, like the Homestead Act before it, the Dawes Act existed in a land system that promoted speculation in land rather than settlement. As the historian Paul Gates noted some 60 years ago, the goal of federal land policies was to convert public land into cash as quickly as possible.[2] This goal reflected the larger economic system, in which land was a commodity whose only value inhered in the rapidity with which people could turn it into personal wealth. This social and economic framework guaranteed that the liquidation of federal land would result in wild speculation first; settlement was a distant second. Although federal officials continued to espouse the rhetoric of fostering independent Indian farmers, the practical effect of the Dawes Act was to transfer Indian land rapidly to white control.

This chapter focuses on the implementation of the Dawes Act and the environmental and social effects it had on the Yakama Reservation in south-central Washington State from the 1880s to the Indian Reorganization Act of 1934.[3] On reservations such as the Yakama, few tribal members were fooled by the promise of the Dawes Act. Indians recognized the act as a "grafters' policy," designed by whites to deprive them of their legal entitlements. The Yakama correctly perceived that the Dawes Act would further dislodge their tenuous control over their reservation's land and resources, and they tirelessly lobbied officials at the Bureau of Indian Affairs to provide protection from the act's harsh consequences. In this effort, the Yakama were active participants in determining how the act would be interpreted on their reservation. Although they seldom prevailed, their struggles laid a foundation for the tribe's active role in natural resource management and advocacy for treaty-protected resources, a role that continues unabated to the present.

TRIBAL BACKGROUND

The confederated tribes and bands of the Yakama Indian Nation are an amalgamation of at least 14 distinct bands that were resettled on a reservation in south-central Washington under a treaty between the United States and the tribe that was signed in 1855 (fig. 11.1). The Yakama were hunter-gatherer-fishers with limited farming experience prior to the 1850s. Although some people tended small riparian gardens along streambeds as a supplement to their gathering activities, the people's constant movement on their annual subsistence trek from plains to river valleys and back precluded any long-term, labor-intensive cultivation. The Yakima basin, however, with its fine, volcanic ash–enriched soils, soon attracted scores of white settler-farmers who were eager to capitalize on the promise of large-scale irrigation works to reap agricultural riches from the Yakama's land.

Pursuant to the terms of the 1855 treaty, the Yakama ceded approximately 10 million acres to the United States in exchange for $200,000 and a 1.25-million-acre reservation.[4] The treaty provided that the Yakama could continue to practice their subsistence lifestyle, and to this end the treaty

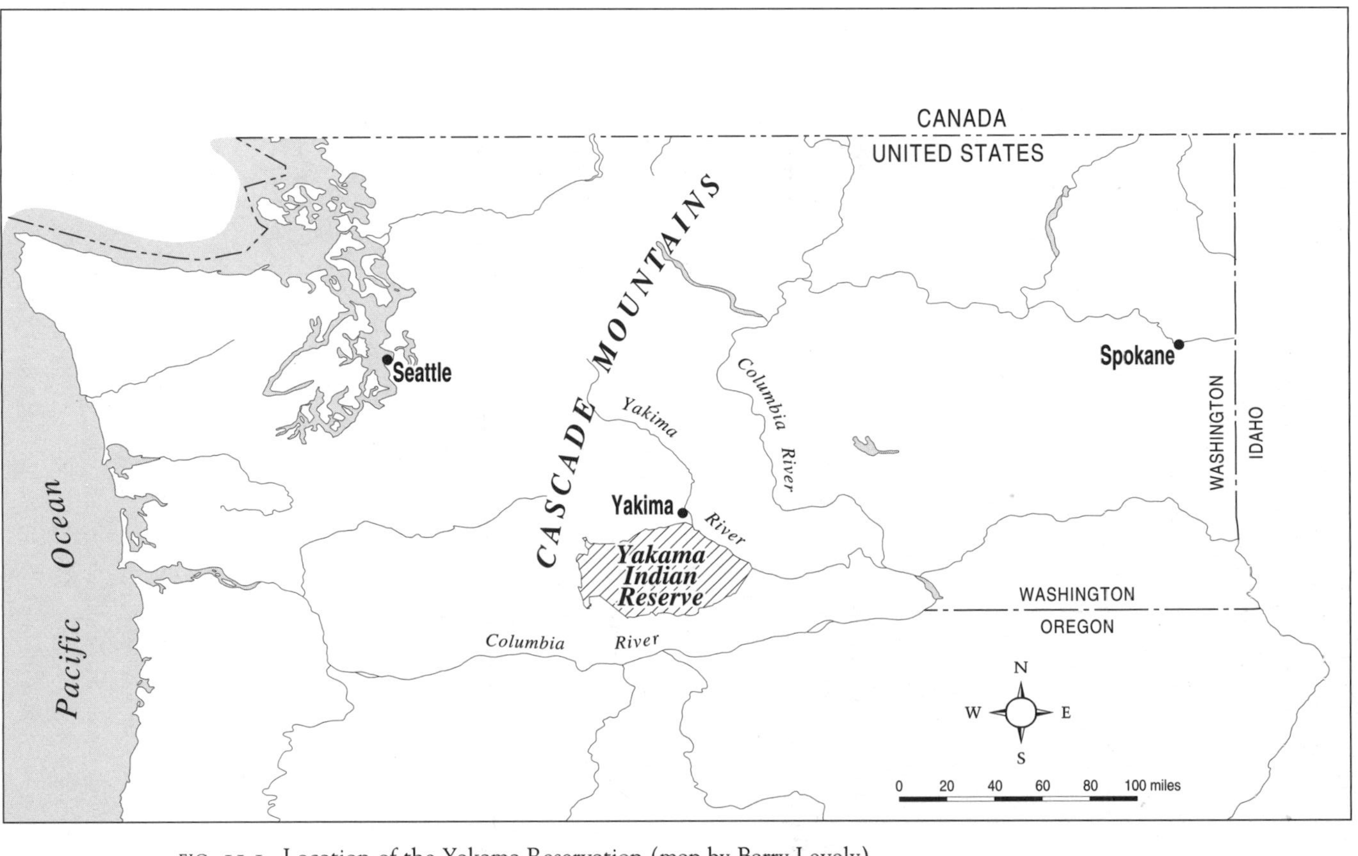

FIG. 11.1. Location of the Yakama Reservation (map by Barry Levely).

included provisions that protected the tribe's access to fishing, hunting, and gathering grounds both on the reservation and on ceded land. Significantly, the treaty also included provisions allowing the federal government to teach the basics of Euro-American culture to tribal members, and it permitted white farmers to settle on the reservation as an inducement to Indians to learn farming. The treaty also laid the foundation for a farming society: the federal government would survey reservation land and could make parcels available for assignment to individual tribal members. Additionally, the treaty provided that the tribe's first compensation payment of $60,000 would go entirely to "breaking up and fencing farms, building houses for [the Indians], supplying them with provisions and a suitable outfit."[5]

SEEDS OF FARMING

The first Yakama agent, R. H. Lansdale, wrote in his annual report to the commissioner of Indian affairs in 1858 that "with a fixed home, and with an individual right in the soil from which [the Yakama] will be instructed to derive their subsistence, they will be stimulated to . . . create an adaptation to civilized pursuits never to be acquired while the nomadic character is retained."[6] Lansdale and others believed that in order to become civilized, the Yakama must be "fix[ed] to the soil, as such domestication must always underlie any permanent progress in civilization."[7] By "fixing Indians to the soil," Lansdale really meant to wed them to the Euro-American legal notion of property and property ownership. Once Indians became self-sufficient, independent landowners, they would have no further need of their collective tribal culture, or so officials believed.[8]

In 1861, Agent W. B. Gosnell, Lansdale's successor, wrote that he had sent two teams—two men apiece, one white and one Indian—each equipped with a plow and oxen, to reservation settlements to plow all the tillable land. Sixteen years later, his replacement, J. H. Wilbur, enthusiastically reported the results. Significantly, his measurement was in terms of the tribe's social, rather than economic, progress:

> The Indians of the Yakama Agency were as low at our beginning with them as humanity gets without getting into the pit that is bottomless. They were taken from the war-path, gathered upon the reserve, . . . clothed with annuity blankets and goods, living in idleness, using the goods furnished as a gambling-fund, drinking whisky, running horses on the Sabbath, stealing each other's wives, and carrying out the practices of the low degraded white men to great perfec-

tion. The Bible and the plow (which must never be divorced) have brought them up from the horrible pit, and put a new song into their mouths, and new hopes into their hearts. They are washed and clothed and in their right minds. Between five and six hundred are accepted members of the Methodist Episcopal Church.[9]

Such reports were exactly what officials in Washington, D.C., wanted to hear, and they confirmed their belief that agriculture was the surest step to civilization.

THE DAWES ACT

Under the terms of the 1887 Dawes Act each tribal member was required to take an allotment on the reservation.[10] Each head of an Indian family received a quarter section (160 acres), and each single person over 18 was allotted a one-eighth section; other classes received smaller allotments.[11] The act imposed limits of a maximum of 80 acres of farmland or 160 acres of grazing land. The federal government held the land in trust for Indian allottees for a specified period of time, generally 25 years, during which the property was exempt from state taxation. At the end of the trust period, the federal government issued title to Indians deemed "competent." Patent in hand, the new owner received all rights and burdens of United States citizenship, including the responsibility to pay taxes. The new owner was also to relinquish all tribal ties.[12]

The Dawes Act and subsequent federal laws implementing it classified reservation land into agricultural, irrigable, grazing, timber, mineral, and "surplus" land. The government used this classification system to dislodge tribal control over land and resources. Pursuant to the act, land classification determined whether the federal government would hold the land in trust for the tribe, lease it to raise revenue for the tribe, or sell it to white settlers on the presumption that they could develop the resources more quickly than could the tribe. Land classification also determined whether an individual—tribal member or white settler—could acquire acreage in a particular class of land, and if so, how much. At Yakama, for example, the Office of Indian Affairs reserved timbered areas from entry by Indians and sold the timber to raise revenue for the tribe beginning in the 1880s.[13]

In practice, the act did not function as smoothly as planned. Time-pressed federal bureaucrats, traveling from one reservation to another, frequently performed their survey functions cursorily. Often individual tribal members were not consulted in the assignment of an allotment. Indian agents routinely assigned parcels of land to anyone who refused to select an allotment. Federal officials, unfamiliar with Indian names, could not always correctly identify allottees, correctly record which person had obtained which allotment, or even identify where the allotments were located.

THE DAWES ACT COMES TO YAKAMA

The Yakama had no choice regarding the allotment of their land. According to the minutes of an 1887 Yakama tribal council meeting, "Agent [Thomas] Priestly explained to Indians the law in regard to 'Indians taking land in severalty.' " When asked to comment, "many came forward for a talk with [the] agent, & all expressed themselves as decidedly against a 'survey' of their lands."[14] A group protesting the work of the special allotting agent in 1893 stated that it had "never received any notice of the allotting by the government, except prior to the time the allotting was begun, [when there] was a meeting of a few Indians with the agent and only four signed their names to an agreement to have their lands allotted."[15]

Early allotment apparently progressed at the agent's discretion, without enforcement of the acreage limitations as set forth in the Dawes Act. Thus, by 1892, according to Commissioner of Indian Affairs Thomas J. Morgan, some "crafty and selfish" Yakama controlled "large bodies of rich pasturage under fence."[16] It is not possible to determine whether these people truly were land grabbers or were instead families attempting to retain control over their territory.

There was no single "Yakama" viewpoint on property holding, either before or after the Dawes Act. Some took land allotments on the reservation under the provisions of the act. Another group, dubbed by Indian agents the "Wild Yakamas," refused to acknowledge the legitimacy of private property ownership altogether.[17] Still others staked out homesteads on public land under the Indian Homestead Act of 1875. By 1891, for example, there were more than 100 Indian homesteads in Klickitat County.[18] But white settlers viewed these off-reservation Indian homesteaders with prejudicial eyes and believed, as one put it, that such Indians were "in many ways detrimental to the interests of the settler, and [they] would be better off were they to live on the reservations."[19] Officials periodically attempted to round up off-reservation Yakama but generally were unsuccessful.[20]

In 1892, the Office of Indian Affairs (OIA) sent special agent John Rankin to take over the Yakama allotment process. Rankin's instructions were to promote allotment, allowing heads of families to select land for themselves and their children, and to make assignments "with reference to the best interests of the Indians, the choice portions of the reservation being given them."[21] Those who already had built on or cultivated land could select that land for their allotment if the choice did not conflict with another allotment. The OIA had instructed Rankin to ensure that all allotments were clearly marked and that each allottee "of sufficient age should be personally shown the boundaries of the allotment selected by him, so that he will understand exactly where the land selected by him lies, and every possible means should be taken to familiarize him with his boundary lines."[22]

When Rankin began his work in 1892, "quite a number of Indians . . . refused to have lands allotted to them."[23] Why? Before the late 1890s there was little incentive to take reservation allotments. The so-called Wild Yakama refused to acknowledge that people could own land.[24] Agent Erwin reported in 1895 that one-tenth of the Yakama population was "wild" and had refused to take up farming because the earth was their mother and "her bosom should [not] be scarred with section lines and subdivisions."[25]

Apart from religious reasons, there were plenty of other good reasons to stay off the reservation and avoid taking an allotment. Living on an allotment meant being under the control of the agent. Additionally, the long start-up time and cost involved in bringing land into production meant Yakama could not immediately obtain income from farming and would still have to hunt, fish, and gather food as the mainstay of their diet.[26] But as long as Yakama could live on the proceeds of fishing-hunting-gathering, they had no need to farm.

By 1898, however, Agent Lynch reported that "many [Yakama] . . . [had] changed their minds and . . . [now] expressed their willingness to come to this reservation and take land in severalty."[27] The change in attitude is traceable to relentless Euro-American pressure on Indians living on public land, which drove many back to the reservation. White Swan, the Yakama chief from 1868 to 1910, told two Indian commissioners in 1897, "You see at Yakima and Goldendale there is a city over there at both sides and the whites is pushing us on each side. . . . They are hurting us on both sides." Chief Louis Simpson told the same commissioners, "We are afraid of them [whites]."[28]

Yakama were returning to reservation land to gain a refuge from whites as well as a source of subsistence food and to obtain the small monthly annuity payments available to those residing on the reservation. With salmon fisheries decimated by huge commercial fishing operations that fed the canneries lining both sides of the Columbia, the enclosure by white commercial and private property owners of vast areas of the tribe's ceded territory, and the growing scarcity of game, hunting-gathering-fishing was an increasingly risky way of life. Because many federal officials believed that protecting treaty rights to subsistence resources would unnecessarily retard the Yakama's progress toward becoming a civilized, agricultural people, the OIA, and later Bureau of Indian Affairs, did little or nothing to assist the Yakama in the legal protection of their treaty resources. Consequently, many Yakama turned to farming because they had no other choice.[29]

Allotment affected individuals differently depending upon their family ties, familiarity with the system, and preallotment status. By arranging their allotments carefully, some families controlled large blocks of land.[30] Some used different names to capitalize on family connections and obtained multiple allotments at Yakama or other reservations.[31] Individuals without families, however, did not fare well, since their holdings were capped at 80 or 160 acres.[32] Those deemed "incompetent" by the allotting agent—that is, unable to make a choice by virtue of age, infirmity, or recalcitrance—were assigned allotments irrespective of their wishes. Those who protested the allotment of their land reported in 1893 that "the only heed paid to their protest was to be arrested by the agent, Mr. Lynch, and carried from one jail to another."[33] Lynch's superiors demanded an immediate report from him and reminded him that "no coercive measures . . . are permissible."[34] In any case, by 1914, when the allotment rolls closed at Yakama, 440,000 acres had been granted to just over 4,500 individuals.[35]

FROM FARMERS TO WAGE LABORERS

Once Indians obtained allotments, federal officials hoped that they would quickly and naturally adopt the ways of farmers. But it was not that easy. Many unanticipated problems cropped up: some people could not remember or did not know the boundaries of their assigned allotments; others did not remember the English names under which their trust patents were assigned; still others were unable to locate their land for other reasons.[36]

Even if an individual overcame these hurdles, successful farming still depended upon more than ability and inclination. Land quality, access to water, cash for tools, seeds, and water charges—if the land was within a reclamation project—all added to the cost of making a parcel into a farm. Agent Lynch estimated in 1902 that the cost of clearing, leveling, fencing, and irrigating sagebrush land amounted to an investment of $10.00 per acre before the farmer received any returns, a price steep enough to drive many from the reservation to find other income.[37]

Some Yakama chose favorable allotments and adopted Euro-American practices. Those who were fortunate—either through luck or by design—took allotments in the rich bottomlands along the Yakima River and other reservation streams. There they employed their available water supply to grow fine crops of corn, wheat, oats, barley, potatoes, and onions. Agent Jay Lynch reported in 1895 that "as a rule . . . [those people were] self-supporting, and some of their crops are equal to whites."[38]

Yakama also used springs and artesian wells for water,[39] but white farmers enclosed as many of these as they could, leaving hardly a drop even to drink. One Yakama, White Salmon Charley, recalled in 1904: "The children used to crawl outside of the [Winans brothers'] fence for [spring] water and tear their clothes. That is the way the Winans are stopping the Indians from getting in."[40] William Speedies, another Yakama, described the situation this way: "Mr. Winans told me that I must not go down to this spring to get any water, where we had been in the habit of getting water, and we had to go up the fence about half a mile, but some times we stole through the fence, crawled under the wire to get our drinking water. . . . That spring has been there ever since I can remember. . . . Now I have to go there and steal water."[41]

Aside from enclosure of springs, massive reclamation works under construction around the reservation dried up almost all available water by diverting it into large-scale irrigation projects on white farmers' land. Many Yakama wrote letters and petitioned the Indian commissioner to have access to the projects. But their demands literally and legally had little priority in the grand scheme of the basin's waters.

The Yakama's treaty of 1855 contained references to farming and fishing, both practices that required water, but nowhere did the document recognize an explicit water right. White farmers pointed out this omission in early lawsuits they filed to stop the Yakama from using water in streams bordering the reservation. Even after the United States Supreme Court affirmed Indians' right to appropriate water in the famous *United States v. Winters* decision in 1908, neither the courts nor federal lawmakers explained how Indians were to protect their water rights against encroachment by white farmers and non-Indian irrigation projects.[42]

Funding for federal reclamation works for Indian users never matched policymakers' rhetoric about the importance of irrigation to farmers. The

Office of Indian Affairs' irrigation budget was chronically underfunded, and a combination of state statutes and ambiguous state and federal case law left Indians without any special tribal status in their competition against non-Indian water users on streams bordering the reservation. White settlers and developers obtained their water rights as state citizens under a state statutory and common law scheme that granted priority to a claim according to the order in which the claimant made the appropriation from a given body of water. This scheme was referred to as "prior appropriation."

While federal lawmakers debated whether and how to define Indian water rights, non-Indian water users were busily and wildly competing in a virtually uncontrolled free-for-all of water speculation. Many streams literally ran dry because so many people were diverting water onto their holdings. Professor O. L. Waller, who studied basin water use for a 1904 report for Washington State Agricultural College in Pullman, reported with disgust that appropriative filings in the Yakima River basin were so far in excess of the available water supply that what actually remained in the streams was "not even enough to water a thirsty dog."[43] Waller estimated that "if all the water had materialized for which there were at one time valid filings, it would have been enough to supply about 2.2 feet of precipitation daily for the entire state, or if it could have been held on the ground for one year evenly distributed, the water would have stood about 800 feet deep over the state of Washington, or about 17 feet deep over the entire United States."[44]

State law protected the actual diversion and application of the water. Under such a system, the Yakama had no way to protect instream uses such as their fisheries. The emphasis on current productive use also assigned no priority (and therefore no right) to the as-yet-unquantified needs of the tribe for future farming or other uses.

Even those tribal members who had built irrigation works and were making productive use of water found themselves in conflict with white farmers. William Colwash, a Yakama who lived with his wife along Simcoe Creek, recounted in 1922 how a white man dammed his irrigation ditch to divert the water. When Colwash tore the dam out, a deputy sheriff arrived at his house and fired three shots at him before Colwash's wife chased the deputy away with an ax.[45]

After years of political wrangling, the Office of Indian Affairs finally secured several ad hoc agreements that restored water to the rivers bordering the reservation. Under the 1905 Agreement for the Yakima River, for example, the Yakama Tribe was limited to 147 cubic feet of water per second (cfs), whereas non-Indian water users off the reservation received 2,065 cfs. Under these agreements Yakama farmers also remained a minority of farmers, by acreage, in on-reservation reclamation projects.[46]

Not only did Indian farmers receive proportionately less water than their off-reservation, white counterparts, but also the pattern of water distribution to Indian farmers differed. Indian irrigators received most of their water from natural flow early in the growing season, before the low-flow period set in. Off-reservation users received more water, and their supplies extended through the low-flow period because they received water from storage later in the growing season. This disparate water distribution pattern ensured that Indian farmers would be limited to low-value crops that required little water while white farmers could grow the more valuable, water-intensive crops, such as orchard fruits. This pattern of disparate water use and its impact on Indian and white farmers' land and crop values remained constant through the 1930s.

Because the water agreements that the Office of Indian Affairs negotiated for the Yakama largely ratified the existing, skewed distribution of water, there was little incentive for those who had water to reduce waste.[47] Repeated overirrigation of off-reservation land north of the reservation created widespread swamping and alkalinization of vast tracts on the reservation.[48] Additionally, on-reservation non-Indian lessees were notorious for having neither the time nor the economic incentive to prepare or irrigate their land properly.[49] O. L. Waller wrote angrily in his 1904 report, "Water has been used in lieu of cultivation."[50]

In 1909, the Reclamation Service began draining excess water from flooded reservation land, but by that time, according to Special Indian Agent Thomas Downs, "nearly as much land is being rendered unfit for farming from seepage as is being benefited by irrigation." He estimated that 12,000 acres, mostly allotments, could no longer be farmed, and "the damage is still in progress."[51] By 1915, a report on the Wapato project stated that "approximately 40,000 acres below the irrigated area has become alkalied and partly submerged."[52] Drainage systems proved a double-edged sword. Efforts to install disposal canals and excavate alkaline land did bring areas back into crop production. But the same systems often injured Indian irrigators who had long depended upon some seepage to support their hay and pasture land. Sometimes Yakama irrigators received compensation for the loss of their water, but often they could not pay for water under the new replacement systems.[53]

The long-term effects of this allocation pattern were clear to all. Lucillus V. McWhorter, an advocate for Yakama water rights, noted in 1920, "It takes no careful observer to ride through the . . . Reservation . . . and pick out the Indian tilled lands from those of white owners and lessors. The former invariably present a withered appearance, while those of the whites show fine crops, resultant from sufficient water."[54]

Most Indian farmers grew wheat; those who produced hay or alfalfa could take at most two cuttings per season. "In other words," L. M. Holt observed in 1927, "the land produces about one half what it would with an adequate water supply."[55] The Bureau of Indian Affairs itself acknowledged in 1949 that "under present conditions of water use, the development of Indian lands is restricted primarily to alfalfa, grains and hay meadows. Little or no sugar beets, hops or fruit can be raised successfully."[56]

That the Yakama had little water for farming directly affected the ecology of their land. Water that was diverted into reclamation projects more often than not also diverted anadromous fish runs with it. As a consequence, it was not unusual to see the irrigation canals across vast areas of farmland clogged with dead and dying fish. Although the Reclamation Service made periodic attempts to require irrigators to screen irrigation canals, screening was expensive and difficult to enforce. Moreover, the numerous federal and state dams surrounding the Yakama's reservation diverted the natural courses of the Yakima River, Ahtanum Creek, and their tributaries, whose annual flooding had provided rich bottomlands for early Indian and white farmers. With water flow increasingly controlled by the needs of noncontiguous water users whose competition for water precluded any overall management of the water resource as a whole, the environmental consequence of irrigation was to create chronic problems with swamping, flooding, and salinization on thousands of acres of the reservation, while formerly riparian land dried up.

The Office of Indian Affairs was unwilling to advocate on behalf of Indian farmers in the basin because such advocacy would necessitate political conflict with powerful state interests. This unwillingness directly undercut the agency's self-professed policy of helping Indians to become self-sufficient farmers. Indeed, it made the agency an ally of white farmers in robbing the tribe of both its water and its land. While Yakama actively lobbied Indian agents at the Yakama Agency and in Washington, D.C., to protect their water claims, they did not have the authority to advocate on their own behalf until the mid-1930s. Consequently, their voices carried little weight in the distribution of the basin's waters.

SEPARATED FROM THE LAND: WAGE LABOR

There were many reasons why the Dawes Act failed to transform the hunter-gatherer-fisher folk of the Yakama reservation into farmers. Those Yakama with poor land and without money to develop it had few options. They could try to survive on federal annuity payments in the form of a meager monthly stipend; they could lease their allotments; or they could sell their land. None of these alternatives provided sufficient support, and most people cobbled together an existence by gathering what traditional subsistence foods they could and by selling land or goods. Many Yakama worked as hired hands, bringing in crops on the farms of their white neighbors.

By the end of the 1880s, even before allotment had gotten under way on the Yakama reservation, federal policymakers knew that simply allotting land had not turned Indians into farmers on any other reservations. Much allotted land remained unfarmed, and white farmers and ranchers continued to pressure Congress to open tribal land for non-Indian use. Federal officials began to espouse the benefits of leasing, which would provide both rental income and the inspiring example of white lessees working the land, at the same time diffusing the political pressure to open reservation land for development.[57]

In 1891 Congress authorized the leasing of allotments.[58] Leases were for three-year terms on farming and grazing land and for ten-year terms on land used for mining.[59] Because Congress could never decide whether it was more important for Indians to learn farming or for Indian land to be farmed—irrespective of who wielded the plow—federal laws alternately raised and lowered maximum term lengths for allotment leases.[60]

Leases went from "a few" in 1898 to a high in 1919 of 1,250, covering 86,000 acres. In 1928 leases numbered more than 950 and covered 57,000 acres.[61] Revenue from farming and grazing leases averaged between 50 cents and two dollars per acre per year, with improved land leasing for $2.50 in the 1890s and about $5.40 by the 1920s.[62] The majority of the lessees were whites—although Japanese farmers seeking to escape alien land laws also rented some land.[63] Not surprisingly, leasing failed to bring any long-term benefit to the reservation because lessees were interested in making large profits before their leases expired rather than in improving the land as federal officials had hoped. Agent Lynch reported in 1905 that few lessees had "made anything from their leases and some have lost money." Although the lessees said they needed more land and longer leases, Lynch believed they did not know how to farm or irrigate properly.[64] Overirrigation and overgrazing on leaseholds remained chronic problems through the 1930s.[65]

Many Yakama eventually sold their allotments. The Dawes Act prohibited allotment sales during the trust period, to protect new landholders from tax collectors and land sharks.[66] Without trust protection, the land invariably was sold to pay previously accrued debts, taxes, or living expenses.[67] An Indian Office official reported in 1910 that sales of Yakama allotments "are being constantly made to the highest bidder under sealed bids."[68] The seller received the proceeds from the sale through the Indian agent, who doled out a monthly payment: $10.00 in 1909—raised to $15.00 in 1914. No one could live on that kind of money. Yakama Louis Mann wrote angrily to Indian Commissioner Robert Valentine in 1909: "There are many poor Indians who have went to work and sold their . . . land through Sup[erin-]t[endant] or ag[en]t hand and this day many are suffering hard ship or starvation[.] Why[?] Because one ten dollars monthly payment is . . . very small[.] I know no one on this earth would go to work sell his lands and get his payment that way and make his business go right[.] . . . You law makers there you need not want Indians to starve."[69]

But federal officials justified the sales in rhetoric resounding with tones of Manifest Destiny. The commissioner of Indian affairs' *Annual Report* of 1904 proclaimed that "sturdy American citizens" were buying Indian land and settling among Indians, although the previous year's report had noted the rampant speculation and land monopolies generated by allotment sales.[70] Moreover, the 1904 report noted that other (perhaps less "sturdy") citizens "fleece the Indian of the last penny within a few hours after the agent has turned over to him the proceeds of a sale."[71] In any case, federal officials continued to promote the sales on the grounds that penniless and landless Indians would be forced by what they dubbed the "law of necessity" to survive in a capitalist world.[72]

CONCLUSION

Louis Mann noted with dismay in 1916 that land sharks, with the approval of the Department of Interior, were watching the reservation, ready to snatch up land.[73] Mann and other Yakama saw that federal law protected white interests, not those of Indians. Mann wrote angrily to Senator Miles Poindexter that year, "How can we be brought into Civilization by looking at our superiors let them rob us or do as they Damn please is this [what] you white man call a Civilization or a law[?]"[74] To Yakama such as Mann, Indian land laws were "grafting" policies that repudiated the promise in the 1855 treaty to protect Yakama economy, society, and land.

In 1934, federal policies began to recognize and even restore a measure of self-determination to Indian peoples.[75] By that time, everyone could see that federal policies had not turned Indians into yeoman farmers. Rather, officials implementing the Dawes Act had succeeded in alienating through lease and sale a large portion of the Yakama reservation. At the same time, federal and state policies that encouraged the rapid and excessive use of water, fish, and land resources by white settlers and commercial enterprises effectively undercut the Yakama's ability to continue their subsistence lifeways. The Yakama, like other Native Americans, were left to seek new methods of support in a market economy in which they were ill-prepared to participate.

It was as much the ecological transformation of the Yakama's homeland as the course set by federal Indian policy that forced the Yakama to become players in the larger state and national economies. Taxes on land, along with purchases of farming tools, materials, and food, required money. To obtain cash, Indians sold whatever they had, including labor and land.[76] Yakama field hands, for example, who picked hops during the September through October season, could earn from $1.50 to $2.00 per day in 1889—an amount equal to a year's lease on an acre of land.[77] What incentive had they to stay on the reservation?

The world created by federal allotment laws, beginning with the Dawes Act, was not one peopled by agrarian or economically self-sufficient Indians. When farming failed, the Yakama could not resort to subsistence ways because by the 1930s their reservation environment simply could not support a subsistence economy. The Yakama had been forced to cede the majority of their former territory, non–tribal members were privatizing through lease and sale increasing amounts of land both on and adjacent to the reservation, and regional water, fish, and game resources were declining. The Dawes Act and its progeny were part of the same legal fabric that consistently functioned to channel land, water, and other resources toward those who would most expeditiously develop them for market. In this era of privatization, political power and economic capital were what mattered, and Indian peoples collectively had little of either. To Indians, as the Yakama example shows, the system was doubly unfair because it robbed them of their land and resources despite the guarantees they believed had been written into their treaties with the United States. "No person should be prohibited from his food by law," Louis Mann wrote in 1916. To Yakama, the Dawes Act in particular and United States laws in general were white people's laws—not the laws of a civilized people.

NOTES

Parts of this essay were previously published in "Allotment Policy in an Incongruous Legal System: The Yakima Indian Nation as a Case Study, 1887–1934," in *Agricultural History* 65 (Fall 1991): 78–103. Those parts are reprinted here by permission.

1. Act of Feb. 8, 1887, ch. 119, 24 Stat. 389 (1887).

2. Paul Gates, "The Homestead Law in an Incongruous Land System," *American Historical Review* 41 (1936): 652–81.

3. The Yakama Nation has recently adopted the spelling of its name as it appears in the 1855 treaty. Thus, it is Yakama Nation but Yakima River and Yakima, Washington.

4. Treaty with the Yakama, June 9, 1855, 12 Stat. 951 (1863).

5. *Ibid.*, art. 4.

6. U.S. Department of the Interior, Bureau of Indian Affairs, *Annual Report of the Commissioner of Indian Affairs to the Secretary of the Interior* (Washington, D.C.: Government Printing Office, 1858), p. 275 (hereafter cited as *CIA-AR*); *CIA-AR* (1859), p. 384.

7. *CIA-AR* (1858), p. 275.

8. For a discussion of federal assimilation policies and their connection to Indian farm policies, see Frederick E. Hoxie, *A Final Promise: The Campaign to Assimilate the Indians, 1880–1920* (Lincoln: University of Nebraska Press, 1984); R. Douglas Hurt, *Indian Agriculture in America: Prehistory to the Present* (Lawrence: University Press of Kansas, 1987), pp. 136–37; Thomas R. Wessel, "Agriculture, Indians, and American History," *Agricultural History* 50 (January 1976): 9–20; Wilcomb E. Washburn, *The Assault on Indian Tribalism: The General Allotment Law (Dawes Act) of 1887* (Philadelphia: J. B. Lippincott Co., 1975); William T. Hagan, "Private Property: The Indian's Door to Civilization," *Ethnohistory* 3 (Spring 1956): 126–37.

9. *CIA-AR* (1878), p. 141.

10. Act of Feb. 8, 1887, Ch. 119, 24 Stat. 389 (1887) (codified as amended at 25 U.S.C. secs. 331–334, 339, 341, 342, 348, 349, 354, 381). The following discussion is based on the act and its amendments.

11. Act of Feb. 28, 1891, ch. 383, secs. 1–2, 26 Stat. 794 (codified as amended at 25 U.S.C. sec. 331). See also Felix S. Cohen, *Handbook of Federal Indian Law* (Charlottesville, Va.: Michie, 1982), pp. 132–36.

12. Act of Feb. 8, 1887, Ch. 119, 24 Stat. 389 (1887); Act of May 8, 1906, Ch. 2348, 34 Stat. 182.

13. Click Relander, ed., *Treaty Centennial, 1855–1955, The Yakimas* (Yakima: Republic Press, 1955), p. 53.

14. Minutes of Yakama tribal council meeting (hereafter YTC-WS), White Swan, Washington, Feb. 15, 1887, Charles (Click) Relander Papers, box 115, folder 18, Yakama Valley Regional Library, Yakima, Washington (hereafter cited as box:folder, CRP).

15. Charles Ike to U.S. President, enclosing petition, quoted in Acting Commissioner of Indian Affairs to Jay Lynch, May 27, 1893, file "Ahtanum Letters, 1881–1899," letters received regarding irrigation matters, 1903–8, YA, RG 75, BIA, NA-Seattle.

16. Morgan to John K. Rankin, Office of Indian Affairs (hereafter OIA) special allotting agent, April 14, 1892, 115:24, CRP. For a general discussion of pre-Dawes allotment attempts, see Paul Gates, "Indian Allotments Preceding the Dawes Act," in *The Frontier Challenge: Responses to the Trans-Mississippi West,* ed. John G. Clark (Lawrence: University Press of Kansas, 1971), pp. 141–70; Cohen, *Handbook of Federal Indian Law*, pp. 129–30.

17. *CIA-AR* (1881), p. 174. According to the Yakama agent's report in 1894, a group of about 100 people who refused to take allotments were in the mountains, but since the report was written in August, we do not know whether they were living in the mountains or were simply there to pick berries. See *CIA-AR* (1894), p. 327.

18. 18 Stat. 420, sec. 15, Act of March 3, 1875; Helen H. Schuster, "Yakima Indian Traditionalism: A Study in Continuity and Change" (Ph.D. diss., University of Washington, 1975), p. 246.

19. Petition, March 1886, 115:17, CRP.

20. See, e.g., Clay Wood, U.S. special agent, colonel U.S. Army, to James Wilbur, May 4, 1880, 114:23, CRP; U.S. Department of the Interior, Census Office, *Report on Indians Taxed and Not Taxed: 1890* (Washington, D.C.: Government Printing Office, 1894), p. 606.

21. Morgan to Rankin, April 14, 1892, 115:24, CRP.

22. Morgan to Rankin, April 14, 1892, 115:24, CRP. See E. Jane Gay, *With the Nez Perces: Alice Fletcher in the Field, 1889–92*, ed. Frederick E. Hoxie and Joan T. Mark (Lincoln: University of Nebraska Press, 1981), for a detailed account of how difficult such allotting instructions were to implement on the Nez Perce reservation.

23. *CIA-AR* (1898), p. 304.

24. Schuster, "Yakima Indian Traditionalism," pp. 253–54.

25. *CIA-AR* (1895), p. 321.

26. *CIA-AR* (1884), p. 172.

27. Lynch report in *CIA-AR* (1898), p. 304.

28. Statements of White Swan and Louis Simpson in "Report of Council with Indians at Yakima Reservation by Commissioners Barge and Goodwin, Fort Simcoe, Feb. 20, 1897," folder "Yakima Council 1898," Tribal Records, 1897–1952, YA,

BIA, RG 75, NA-Seattle. Many Indians expressed fear of whites throughout the period. I have indicated White Swan's tenure as chief of the Yakama Nation; Simpson was a headman, or subchief, and unfortunately I do not have the dates of his tenure.

29. *CIA-AR* (1884), p. 172; *CIA-AR* (1894), p. 326.

30. "Report on the Conditions of Yakima, Washington," file no. 120334-12-341, YA, Correspondence File 1907–39, Central Classified Files (hereafter CCF), BIA, RG 75, NA, 25.

31. John A. Bower, "The Hydrogeography of Yakima Indian Nation Resource Use" (Ph.D. diss., University of Washington, 1990), pp. 272–73.

32. Thomas J. Morgan to Rankin, April 14, 1892, 115:24, CRP.

33. Ike to U.S. President, May 27, 1893.

34. *Ibid.*

35. H. G. Barnett, *The Yakima Indians in 1942* (Eugene: University of Oregon, n.d.), pp. 6–7.

36. See, e.g., *CIA-AR* (1898), p. 305 (Instructions to agents in regard to family names, March 19, 1890); *CIA-AR* (1899), p. 363; case of Calvin Hale, whose son died with only 20 acres having been allotted him, and referring to a man who lived in his household who had received no allotment by 1898, "Report of Council of Commissioners and Yakima Tribe of Indians," n.d., folder "Yakima Council, 1898," Tribal Records, 1897–1952, YA, RG 75, BIA, NA-Seattle. See also Gay, *With the Nez Perces*.

37. *CIA-AR* (1902), pp. 368–69.

38. *CIA-AR* (1895), pp. 320–21.

39. Case of Yesmowit Yah-ho-tow-wit, who irrigated from springs on allotment no. 1587. "Report of Commission to Investigate the Claims of Indians along Simcoe Creek," part 1-A (hereafter cited as "Report of Simcoe Creek Commission"), transmitted by letter, June 9, 1924, file no. 77253-22-341, CCF, General Records 1907–39, BIA, RG 75, NA-Seattle, 5; Testimony of Hahise Sawyahill, who also used a spring on her allotment, in Report of Simcoe Creek Commission, p. 12.

40. Testimony in *United States v. Winans* transcript of record, Testimony and Final Report of Examiner, enclosed in Attorney General (AG) to Clerk of the Supreme Court, Jan. 21, 1904, Appellate Case File no. 19213, Records of the Supreme Court, RG 267, NA (hereafter cited as *United States v. Winans*, transcript of record), p. 135 (speaking through interpreter).

41. Testimony in *United States v. Winans*, transcript of record, pp. 176, 180 (speaking through interpreter).

42. For a discussion of the practical difficulties of implementing the *Winters* decision at Yakama, see Barbara Leibhardt, "Law, Environment, and Social Change in the Columbia River Basin: The Yakima Indian Nation as a Case Study, 1840–

1933" (Ph.D. diss., University of California, 1990), pp. 276–79. For a discussion of the difficulties in making the *Winters* decision workable, see Norris Hundley, Jr., "The Dark and Bloody Ground of Indian Water Rights: Confusion Elevated to Principle," *Western Historical Quarterly* 9 (1978): 454–82; Norris Hundley, Jr., "The 'Winters' Decision and Indian Water Rights: A Mystery Reexamined," *Western Historical Quarterly* 13 (1982): 17–42.

43. Osmar L. Waller, A *Report on Irrigation Conditions in the Yakima Valley, Washington* (Pullman: Washington State Agricultural College, Experiment Station Bulletin 61, 1904), p. 22.

44. *Ibid.*, p. 19.

45. Report of Simcoe Creek Commission.

46. For a detailed discussion of the impact of disparate water use by white and Indian farmers on the Yakama reservation, see Leibhardt, "Law, Environment, and Social Change," pp. 269–316. See also Bower, "Hydrogeography of Yakima Indian Nation Resource Use."

47. See, e.g., *United States v. Ahtanum Irrigation District,* 236 Federal Reporter 2d Series 321, 341 (U.S. Circuit Court of Appeals, Ninth Circuit, 1956).

48. George Otis Smith, *Geology and Water Resources of a Portion of Yakima County, Washington* (Washington, D.C.: U.S. Geological Survey, Water Supply and Irrigation Papers no. 55, 1901), pp. 13–14.

49. Office of Indian Affairs–Indian Irrigation Service (hereafter OIA-IIS), District no. 1, *Annual Report*, file no. 78600-15-341, YA, CCF, BIA, RG 75, NA; see also CIA-AR (1905), p. 370.

50. Waller, *Report on Irrigation Conditions*, p. 16.

51. Thomas Downs, OIA special agent, "Report of Inspection of Yakima Indian School," July 23, 1909, file no. 85193-07-341, YA, CCF, BIA, RG 75, NA.

52. OIA-IIS, Indian Irrigation District no. 1, *Annual Report*, 1915, file no. 78600-15-341, YA, CCF, BIA, RG 75, NA. Waller, in a 1904 report, estimated that the damage to lands along the lower Ahtanum totaled "several thousands of acres" alone; Waller, *Report*, 18. See also Snively to John McPhaul, OIA, Dec. 22, 1914, case file "Reopening of Yakima Indian Reservation, 1914–15," 1:8, Henry J. Snively Papers, University of Washington, Library Manuscripts Division, Seattle. Snively noted that "a number of allotments that three years ago were very valuable for agricultural purposes, at least worth $100.00 per acre, are now worth no more than probably $5.00 to $10.00 per acre for pasture, the salt grass having taken them." He advocated state-based drainage districts.

53. Leibhardt, "Law, Environment, and Social Change," pp. 308–9.

54. Lucillus V. McWhorter, *The Discards* (n.p., 1920), p. 5; see also Lucillus V. McWhorter, *The Crime against the Yakima* (North Yakima: Republic Print, 1913).

55. L. M. Holt, testimony for Inspector Trowbridge, June 4, 1927, quoted in American Indian Defense Association, Inc., April 5, 1932, "The Yakima Bill, Interior Department's Mock Moral Fervor Suddenly Evaporates," file no. 9225-36-341, Part O-3, YA, CCF, BIA, RG 75, NA.

56. A. L. Wathen to CIA, Memo, Jan. 1949 (no file no.), box 24, class 341, YA, CCF, BIA, RG 75, NA.

57. *CIA-AR* (1893), p. 339 (Lynch recommends leasing); *CIA-AR* (1898), 304–5; *CIA-AR* (1901), p. 394 (classification of leased lands on reservation); *CIA-AR* (1902), p. 369 (noting that lessees would make permanent improvements); *CIA-AR* (1909), p. 7 (stating that lessees would model a "progressive spirit" for Indians). See also Department of the Interior, General Land Office, "Indian Reservations to Be Opened" (Washington, D.C.: January 21, 1908), describing lands "to be opened as fast as they are ready" to entry by whites. file ([missing]—08-308.1), box 08-308.1/10198-18-308.3, DSF, YA, BIA, RG 75, NA; Nealy N. Olney, secretary Yakima Indian Commercial Club, to Franklin K. Lane, secretary of the interior, April 16, 1917, file 40795-17-308.1, DSF, YA, BIA, RG 75, NA, noting pressure to open Yakama reservation land. Hurt, *Indian Agriculture in America*, pp. 141–42; Cohen, *Handbook of Federal Indian Law*, pp. 134–36.

58. 26 Stat. L. 795, sec. 3 (1891).

59. Under the same act, lands purchased by individuals that were not needed for farming or grazing and were not used as individual allotments could be leased with the permission of the tribal council and Indian agent—subject to approval by the secretary of the interior—for periods of five years for grazing and ten years for farming; 26 Stat. 795, sec. 3.

60. Barbara Leibhardt, "Allotment Policy in an Incongruous Legal System," pp. 95–96.

61. The records of Yakama reservation leasing are scattered and have not yet been compiled in a systematic way. I have based my totals here on *CIA-AR* records and Estep to CIA, Nov. 4, 1928, file no. 52539-28-341, YA, CCF, RG 75, BIA, NA.

62. Petition signed by 70 individuals, Feb. 1913, 76:13, MCPP, requesting that lease money be distributed in larger installments. Leasing revenues remained about $1.75 per acre per year through at least 1911.

63. On Japanese lessees, see Miles C. Poindexter to Cato Sells, March 13, 1916, 79:1, MCPP. Poindexter quotes from a letter from F. Benz and Sons, Toppenish, Washington, "indicating the difficulties of the white citizens in that community who are leasing Indian lands, arising from competition of Japanese. The results of that competition would make a long story, but you are, I think familiar with the fundamental principles involved. It is really a great and serious problem." Poindexter recommended that OIA give preference to American citizens in leasing Indian lands.

64. *CIA-AR* (1905), p. 370.

65. Gretta Gossett, "Stock Grazing in Washington's Nile Valley, Receding Ranges in the Cascades," *Pacific Northwest Quarterly* 55 (July 1964): 119–27.

66. 24 Stat 389, sec. 5.

67. Sales of "noncompetent" Indian lands began in 1908. In 1909, one such 80-acre tract was sold for $6,420; in 1911, 18 tracts, totaling 1,367.37 acres, were sold for $37,866.

68. John T. Reeves, Acting Chief Land Division, to A. T. Thorson, Aug. 23, 1910, file no. 65118-10-308.1, DSF, YA, BIA, RG 75, NA.

69. Mann to Valentine, Nov. 8, 1909, file no. 7003-1909-307.4, DSF, YA, BIA, RG 75, NA. See also Cohen, *Handbook of Federal Indian Law*, p. 136.

70. *CIA-AR* (1903), p. 48; *CIA-AR* (1904), p. 29.

71. *CIA-AR* (1904), p. 63.

72. Hoxie, *Final Promise*, p. 183.

73. Mann to Poindexter, 2 February 1916, 76:12, MCPP; see also Mann to Poindexter, 7 February 1916, *ibid.*

74. Mann to Poindexter, 7 February 1916, 76:12, MCPP.

75. Indian Reorganization Act of 1934, ch. 576, sec. 16, 48 Stat. 987 (codified at 25 U.S.C. sec. 476).

76. See C. K. Jacobson, "Internal Colonization and Native Americans: Indian Labor in the United States from 1871 to World War II," *Social Science Quarterly* 65 (March 1984): 158–71.

77. *CIA-AR* (1889), p. 291; *CIA-AR* (1890), p. 234, reporting that agent discouraged hop-picking because Indians purportedly got drunk and left their farms.

PART III / RIVERS

Rivers are central to any understanding of the Pacific Northwest. The first peoples in the region drew much of their sustenance from the riverine environment. As both Ewert and Abbott noted earlier, Euro-Americans also located their towns and cities along the rivers. And as the essays in this section demonstrate, the region's rivers continue to define the Pacific Northwest's environmental history.

Dale Goble offers a legal history of the region's salmon and steelhead runs. He examines American legal traditions regarding use and ownership of natural resources as expressed in three different stories. The first story—the lawsuit of *Pierson v. Post* over a fox—forms the legal basis for the various resource rushes that have characterized much of American history: the first person to grab a resource becomes its owner; winner takes all. This tradition led to the rapid depletion of salmon during the peak of the salmon canning industry a century ago. The second story—of progressive conservation—sought to articulate an alternative to the chaos and waste produced by the first story. Rather than rushing to grab resources, progressives sought to plan resource development rationally. But their story was also cornucopian: resource scarcity could be avoided because nature could be engineered to produce more of what we desired. The progressive story thus also contributed to the demise of the region's salmon runs when salmon biology proved unamenable to engineering. The third story—embodied in the Endangered Species Act—rejects the cornucopian perspective and holds that nature is complex, chaotic, and interconnected in ways we cannot reengineer. Although the ESA was a dramatic break with the past, it has not prevented the extinction of the region's salmon runs.

Michael Blumm examines the creation of the Columbia River's hydroelectric system. Since the end of the nineteenth century, more than 100 dams have been constructed on the region's rivers. Initially the system was seen as a means of "turning our darkness to dawn," in the words of Woody Guthrie's ballad "Roll On Columbia." As Blumm notes, the Northwest was rapidly electrified. But the New Deal construction projects that put the country back to work created dams that had no customers for their electricity. The surplus led to the influx of war industries—including Boeing, Kaiser, and Hanford—lured to the region by cheap power. After the war, the dam-building continued, as did the effort to lure power-dependent industries to absorb the surpluses. When all of the promising dam sites had been developed, the region's utilities turned to nuclear power as the next cheap source of electricity. The result was the infamous WPPSS (pronounced "whoops") fiasco and regional bankruptcy. At the same time, it became apparent that another cost of the hydroelectric system was the destruction of the region's anadromous fish runs. Blumm's detailed history of the regional power system offers a persuasive analysis of how a system takes on a life of its own so that means become ends: a system intended to provide cheap electricity to improve people's lives has become a system operated largely to benefit the operators.

The final essay in part three examines the ethical issues implicit in both Goble's and Blumm's essays. Carolyn Merchant examines the demise of the region's fish runs as an example of failed ethical systems. The laissez-faire economic system that Goble examined as a corollary of the tale of Pierson, Post, and the fox, Merchant sees as an example of an "egocentric" ethical system. The progressive-era conservation that both Blumm and Goble discuss, Merchant categorizes as a "homocentric" ethic. She also finds the ethical perspective embodied in the Endangered Species Act—which she labels "ecocentric"—faulty for giving precedence to fish over humans. To replace these three failed ethics, Merchant proposes a "partnership ethic" based, in her words, on the idea that "people and nature are equally important."

12 / Salmon in the Columbia Basin

From Abundance to Extinction

DALE D. GOBLE

> The salmon begin to leap shortly after sunrise. At this time the Indians swim to the centre of the falls, where some station themselves on rocks, and others stand to their waists in the water, all armed with spears, with which they assail the salmon as they attempt to leap, or fall back exhausted. It is an incessant slaughter, so great is the throng of fish.
>
> —Washington Irving, *Astoria, or Anecdotes of an Enterprise beyond the Rocky Mountains*

In his 1836 book *Astoria*, Washington Irving described an Indian fishing station at Salmon Falls on the Snake River in what is now southern Idaho.[1] His description was only one of several. Paul Kane, a Canadian artist, described a similar scene at Kettle Falls on the upper Columbia River in 1846.[2] And as Lewis and Clark, with their Corps of Discovery, rode and floated down the Clearwater to the Snake and the Snake to the Columbia and the Columbia to the ocean, they noted that the rivers were "crouded with salmon." The shores housed an almost continuous string of lodges, and in front of each lodge were large scaffolds of drying fish.[3] The scenes were ancient. Indians had fished for salmon in the Columbia basin for at least 10,000 years;[4] their annual catch measured in the millions of tons.[5] Now, less than two centuries later, many of those salmon runs are extinct and others are listed for protection under the Endangered Species Act.

Salmon are not, of course, unique—either in the stories told about them or the consequences of human action they have endured. America is a story of mythic abundance transformed: of flights of passenger pigeons so numerous they darkened the sky, of herds of bison that shook the earth as they passed, of forests so dense that a squirrel could walk from the Atlantic Ocean to the Mississippi River without touching the ground.[6] The newcomers to this continent brought with them a cultural imagination that saw pigeons and bison and trees and salmon as commodities to be taken for personal gain. The stories they told themselves were of an industrious people taming a wilderness, of a godly society that existed to promote the rapid accumulation of personal wealth.[7]

One of these stories—an archetypal tale written by the New York Supreme Court—was a legal case involving Pierson, Post, and a fox.[8] Post was out riding one day with his hounds "upon a certain wild and uninhabited, unpossessed and waste land" when he came upon "one of those noxious beasts called a fox." After a strenuous chase, and just as Post was on the point of seizing the animal, Pierson mysteriously appeared and, knowing that Post was in hot pursuit, killed the fox and carried it off. Post was off to his attorney, who filed suit to recover the fox. The trial judge in Queens—for that was where this wasteland was located—awarded the fox to Post, and Pierson appealed to the New York Supreme Court. The court's opinion is a learned discussion of Puffendorf's *Law of Nature and of Nations*, of Blackstone and *Fleta* on the general proposition that "property in [wild animals] is acquired by occupancy only." Thus, "mere pursuit gave Post no legal right to the fox"; it became the property of Pierson, who had killed it—and who by killing it had taken possession.

Much of the story of the United States can be seen as an extended footnote to the saga of Pierson, Post, and the fox. One United States Supreme Court Justice, for example, generalized the tale thus:

> The wild bird in the air belongs to no one, but when the fowler brings it to earth and takes it into his possession it is his property. He has reduced it to control by his own labor, and the law of nature and the law of society recognize his exclusive right to it. . . . So the trapper of the plains and the hunter of the north have a property in the furs they have gathered, though the animals from which they were taken roamed at large and belonged to no one. . . . So the miners, on the public lands throughout the Pacific States and Territories, by their customs, usages, and regulations everywhere recognized the justice in this principle.[9]

The moral of the tale was not lost on gold miners or pigeon hunters or salmon fishers: the law rewarded individual initiative; the prize went to he who seized it. And the tale soon acquired the coda present in Justice Field's retelling: the government's role was to make resources available but not to restrict or regulate their use. Regulation, if there was to be any, was left to those exploiting the resource: users know best. But the tale produced unfortunate results. The passenger pigeon is extinct; salmon are endangered.

Pierson v. Post is the legal basis for the so-called tragedy of the commons, the destructive overuse of common-property resources such as clean air or fisheries.[10] Labeling the problem the "tragedy of the commons," however, resembles the magician's sleight of hand: it distracts and thus conceals. It is not the presence of common property that is problematic but rather the drive to market a commodity taken from the commons. A more apt phrase would be the "tragedy of the market." Common ownership of property itself can be an ecologically valid response to seasonal resource variation. For example, when resources vary significantly over the seasons, the population of a species dependent on those resources is limited by the minimum rather than by the total amount of resources available annually. Scarcity in winter, in other words, restricts the resource-dependent population and helps to maintain the abundance of resources in spring by preventing overharvesting. Thus, for a human community subject to seasonal scarcity, common property in an abundant resource would not lead to overharvesting. As Eugene Hunn points out in his essay on the Indians of the Columbia plateau, the first inhabitants of the Columbia basin harvested the region's abundant salmon runs as common property for some 10,000 years before the arrival of Euro-Americans.[11]

The tale of Pierson, Post, and the fox is not the only kind of story that has guided resource exploitation in America. A second influential narrative offers an account of the nature of nature in which nature is a great machine governed by natural "laws." This story—which Donald Worster has aptly characterized as "imperial science"[12]—can be traced to the mechanical view of Newtonian physics: nature is a collection of bodies moving in abstract, Euclidean space; it is quantity and mechanical regularity. Heaven and earth are united: the apple and the planets are subject to—or at least described by—the same mathematical relations. Francis Bacon, lord chancellor of England, gave voice to the optimistic vision inherent in this perspective: we humans could discover and manipulate nature's laws so that we—as mechanics standing outside the system—could reengineer it to produce what we desire. In Bacon's dramatic image, Newton's apple would reverse the effects of Adam's apple; science was the means of reacquiring the dominance that man lost in the Fall. "Let the human race recover that right over nature which belongs to it by divine bequest."[13]

The optimistic and mechanistic view remains influential in part because it was one of the core beliefs of progressive conservation as preached by Theodore Roosevelt and Gifford Pinchot.[14] The progressives believed that experts could manage resources for the public benefit. Centralized planning by impartial experts, they urged, would ensure that natural resources were used to provide the greatest good for the greatest number. When applied to "renewable resources" such as forests and fisheries, this translated into "multiple use" and "sustained yield." River basins, for example, were to be viewed as a whole and managed to provide an unending supply of trees and wheat and fish and electricity.

The technological optimism of this story of managerial expertise was particularly attractive to lawmakers and administrators because it allowed them to say "yes" to nearly everyone. There was no need to deny shippers or electric power interests or fishers because it was possible to have it all—to provide slack water for shipping, turbines for power production, and fish for fishers—through careful management of the rearranged parts of the machine called "nature."

Finally, there is a third story: a narrative about humans' propensity to alter nature in fundamental though often unintended ways.[15] It is a cautionary tale in which the mechanic is revealed to be Doctor Frankenstein. This story stresses complexity and interrelatedness; it views humans as one species among many, all of which have a claim to continued existence; it points out that we have exterminated other species without intending to do so and for reasons that even now are not apparent. It is a tale of the once impossibly abundant passenger pigeons that just melted away even after the slaughter ceased. It is a story of salmon that seem to be following the same trajectory.

Just as the judicial decision in *Pierson v. Post* embodied the laissez-faire philosophy that government ought not to intrude into economic activities, and just as the progressive conservation story was reflected in various statutes providing for "multiple use and sustained yield," so, too, the new story is captured in law—the Endangered Species Act.[16]

These three stories and the laws that embody them continue to play roles in the evolving fate of the region's salmon and steelhead runs. There is, however, an additional story: the biology of the fish themselves.

BIOLOGY OF COLUMBIA RIVER ANADROMOUS FISH

Anadromous fish are born in freshwater but spend most of their adult lives in the ocean before returning to freshwater to spawn.[17] Anadromy is an evolutionary response to seasonal variations in ecosystems similar to other forms of animal migration. It allows species to utilize the seasonal benefits of an ecosystem while avoiding its limitations during other seasons. It enables fish to use rivers and streams despite the limitations on food in such habitats during fall and winter in temperate and northerly latitudes. The seasonally limited food supply restricts the size of year-round fish populations. Nevertheless, the riverine environment offers relatively secure spawning and incubation conditions—because low population levels reduce the population of potential predators. Migration into rivers to reproduce makes use of these benefits, whereas migration into the ocean avoids the limitations and allows the species to mature in the far richer marine environment.

Six species of anadromous salmonids spawn within the Columbia River basin.[18] Although there are significant differences among them, they all share a general life history that begins as a fertilized egg deposited in a gravel nest, or "redd," in a stream or lake. After two to three months, the "alevin" hatches from the egg. It remains in the gravel, however, until its yolk sac has been absorbed and it has completed the physiological transformation into a "fry." With the emergence of the fry from the protective gravel, the life cycles of the species diverge. A-Type salmon (chum, pink, and sockeye) migrate from the natal stream soon after emerging from the gravel. Sockeye migrate into nearby lakes, where they remain for up to three years; most chum and pink salmon quickly migrate to the ocean. B-Type salmon (coho, chinook, and steelhead) have a stream-resident stage during which the fry disperse along the natal stream, establishing a feeding territory. The fry will grow into fingerlings and may remain in the stream for up to three years.

Eventually, the juveniles of all species begin the smolting process, drifting downstream to the estuary at the river's mouth. In this fertile mixture of freshwater and saltwater, they complete the physiological and metabolic adaptation to saltwater. With smolting completed, the salmon leave the estuary for the ocean and a variety of migratory paths. Some Columbia River salmon, primarily coho, turn south, feeding off the Oregon and California coasts. Others, primarily chinook, turn north and, following the narrow continental shelf, move in a northwesterly circle as far as the Aleutian Islands and the Gulf of Alaska. The adolescents will remain in the ocean, feeding and growing, for up to five years.[19] Then, in some as-yet-not-understood manner, the adult leaves this ecosystem of sea, plankton, and fish. Most return to the estuary at the mouth of the river which they left years before and begin the arduous upstream migration to the spawning grounds where they were hatched, to court, mate, and die.

Anadromy thus evolved as a way to overcome the limitations of nutrient-poor, seasonally limited freshwater ecosystems. But that evolutionary strategy has now become a liability. As one conservation biologist noted, "Migratory species are exposed to double jeopardy because they are subject to the pressures of change at both ends of their routes, and may have to run a gauntlet of polluted waterways and altered landscapes on the way."[20] Such is the case with salmon: their migratory life histories place them at risk from human actions on the high seas and on their spawning beds as well as in the rivers that link the two.

But it is not only salmon that are at risk. They are themselves crucial sources of food and nutrients in the aquatic and riparian ecosystems of the Columbia basin—ecosystems that are often nutrient-poor. Salmon are important sources of food for a wide variety of carnivores and scavengers. One study found 20 species of mammals and 23 species of birds that relied on salmon and their spawned-out carcasses.[21] Other studies have found salmon carcasses to be a significant source of primary nutrients such as nitrogen and carbon. Consider one small stream: Kennedy Creek on Puget Sound. The lower three miles of the creek are used by salmon for spawning. In 1991, an estimated 16,390 chum salmon spawned there. Since chum average 10 pounds each, this run provided the creek with nearly 82 tons of biomass.[22] Given the millions of salmon that once ascended the Columbia basin's rivers, the loss of nutrients is substantial—in the millions of tons. A final study demonstrates the ecosystem-wide impact of this loss: up to 90 percent of a grizzly bear's sustenance came from salmon even in the upper watersheds of the Columbia basin—and the salmon-eating bears each spread some 400 pounds of nitrogen and phosphorus in the forest. As a result, salmon accounted for approximately 20 percent of the metabolism of a tree. The loss of salmon runs thus will have impacts throughout the region.[23]

Descriptions of the abundance of salmon are commonplace in early travelers' accounts of the region. Harry Fisher, who spent three months in the later summer of 1890 exploring the Olympic Mountains in Washington, returned with such a tale—but with a twist that suggests the role of salmon in the region's ecosystems: "Great salmon threshed in the water all night long, in their efforts to ascend the stream. Wild animals which I could not see snapped the bushes in all directions, traveling up and down in search of fish. At every few yards was to be seen the remains of a fish where cougar, coon, otter, or eagle had made a meal."[24]

PIERSON, POST, AND THE FOX ON THE COLUMBIA RIVER

For those steeped in the ethics of *Pierson v. Post,* the Pacific Northwest was an abundance of natural capital waiting to be transformed into private wealth. Trees and grass and wildlife existed in cornucopian quantities. In the laissez-faire ethos of the time, the government stood by as individuals rushed to claim some part of the region's wealth. Fur trappers were followed by cattlemen and farmers and loggers and fishers in a free-for-all.

Columbia River Fishers

To Euro-Americans, the Columbia basin's fisheries offered possibilities of wealth—if only the problem of getting the fish to market could be solved. The Hudson's Bay Company considered salting salmon for trade to Spanish California in the 1820s.[25] In 1831, the brig *Owyhee* left the Columbia River carrying 53 hogsheads of salted salmon for the East Coast. The following year, a party led by Captain Nathaniel Wyeth crossed overland to the Columbia basin intending to begin a salmon fishery. But salted salmon did not ship well through the heat of the tropics. As the *San Francisco Bulletin* delicately put the matter in the 1850s: "The efforts to export salmon in barrels have not, so far as we can learn, been very successful. They, being very fat, do not keep well, and when the packages after the long voyage around the Horn are opened, they are not in good merchantable order."[26] Although a small salted-salmon industry existed into the 1880s, it was only marginally successful.

Canning changed this. It was the technology that enabled salmon to be mass-marketed because it solved the problems of palatability and transportability. Canning was also the technology that revealed the core sensibility of *Pierson v. Post:* it triggered a salmon "rush." As one government report dryly noted: "The prices received for the finished product were good and the business offered a quick profit from a moderate investment."[27] The *Oregonian* was blunter: before 1881, it concluded, "a fortune was expected and made each season."[28] Quick profits, of course, generate competition. The expansion of the industry can be graphed in production figures. In 1866, Andrew Hapgood and William, John, and George Hume opened the first salmon cannery on the Columbia River.[29] They packed 4,000 cases of 48 one-pound cans each. In 1882, more than 1 million cases were produced by Columbia River canneries. By 1895 the pack exceeded 2 million cases, and in 1901 the number grew to more than 5 million cases.[30]

The canneries' demand for fish fostered technological innovation.[31] Traps and fish wheels replaced the dip nets, harpoons, and weirs that had served the Indians for millennia. They, in turn, were replaced by gill-netters, purse seiners, and, most recently, ocean trollers. The canneries' demand also fostered overfishing: the chinook catch peaked in 1883, when 40 canneries packed 43 million pounds.[32] Total production of all species reached its maximum in 1911, when almost 50 million pounds of salmon were packed. By 1975, the amount of Columbia salmon that was canned dropped to less than that in 1867, the second year of cannery operations. The industry had collapsed along with the fish runs on which it depended.

The overfishing problem had long been recognized. In 1894, for example, Hollister McGuire, the Oregon fish and game protector, wrote:

> It does not require a study of the statistics to convince one that the salmon industry has suffered a great decline during the past decade, and that it is only a matter of a few years under the present conditions when the chinook of the Columbia will be as scarce as the beaver that once was so plentiful in our streams. Common observation is amply able to apprehend a fact so plain. For a third of a century, Oregon has drawn wealth from her streams, but now, by reason of her wastefulness and a lack of intelligent provision for the future, the source of that wealth is disappearing and is threatened with annihilation.[33]

But McGuire's comments went unheeded—perhaps because the abundance seemed too great, perhaps because the ethos of *Pierson v. Post* was too pervasive. There was money to be made and no one paid undue attention to tomorrow. As Samuel Clarke had remarked in the 1870s: "The immense supply of the chinook salmon that forms the staple of this great commerce is to be had for the taking."[34]

Allocating Habitat to Miners, Farmers, Ranchers, and Loggers

Overfishing was not the only cause of the decline of the runs. River fisheries are fragile because rivers are fragile. The discovery of gold in the Snake River basin, for example, led to the destruction of miles of salmon spawning and rearing habitat by placer and dredge mining.[35] Early loggers often used streams to transport logs to mills, and turning rivers into highways damaged the riverine ecosystems.[36] Irrigation diversions dried up rivers: salmon-producing streams such as the Lemhi River on the Idaho-Montana border and the Umatilla River in Oregon are drained dry by irrigators during all but the wettest years.[37]

Fisheries are also fundamentally fragile because they are inescapably intertwined with the land in the watershed that envelops them. All land uses that affect the physical structure of the riparian environment or the quantity or quality of a river's water affect the fish that live in that water. Although irrigation is the major source of water quantity problems, they may also be caused by land uses that do not directly remove water from streams. Removal of vegetation by grazing or logging, for example, accelerates runoff and increases the likelihood of lower-than-normal flows during the dry summer months.[38]

But the quality of the water is nearly as important as its quantity. Fish kills caused by pollution dramatize the importance of high-quality water. More insidious, because they are less dramatic, are temperature, water chemistry, and erosion problems. A fish's body temperature varies with water temperature, so water temperature is the most important determinant of the type of fish that live in a river. Salmonids are cold-water species that have difficulty surviving in water with temperatures above 62°–68° F.[39] Land uses such as logging, grazing, and farming increase water temperatures when they remove tree cover or riparian vegetation and expose a stream to the heat of direct sunlight.[40]

Salmon are also highly susceptible to changes in water chemistry. Alteration of stream acidity and contamination of the water with heavy metals are particularly injurious. Mining in the Panther Creek watershed in central Idaho, for example, destroyed chinook and steelhead runs when acidic and heavy-metal wastes leached into the stream.[41] The use of herbicides, pesticides, and fertilizers in agriculture and forest and range management is a more common source of water chemistry problems.[42]

Finally, salmon are harmed by silt-laden water. Sedimentation reduces the ability of eggs to hatch and of alevin and fry to survive.[43] Erosion is a frequent result of land-disturbing activities such as grazing, road-building, mining, and many agricultural activities.[44]

The point is simple: salmon require clean, cold, highly oxygenated water. Land uses that alter these critical parameters of the stream ecosystem—and there are a wide variety of such uses—will harm or destroy salmon.

The numerous small insults to the region's streams from mining, logging, grazing, and agriculture took their cumulative toll: habitat loss contributed to the decline of the salmon runs traceable to overfishing. Surprisingly little was done to prevent the loss. In part this reflected a fatalism common to the period. Resource exhaustion was simply an inexorable result of economic development: both gold nuggets and salmon were depleted when transformed into wealth. Although perhaps regrettable, it was nonetheless unavoidable. Spencer Baird, the first United States fish commissioner, wrote in 1878: "It may safely be said that wherever the white man plants his foot and the so-called civilization of a country is begun inhabitants of the air, land, and the water, begin to disappear." Fish were, he noted, particularly susceptible to this "fatal influence."[45] This fatalism reinforced the laissez-faire ideology that governmental regulation was generally inappropriate because landowners should be permitted to do as they wished with their land. Farmers had a right to dry up rivers, loggers should be permitted to clog spawning gravels with silt from denuded hillsides, and ranchers should not be required to keep their cattle out of streams.

Furthermore, a corollary of the decision in *Pierson v. Post* also worked to limit the legal protection of wildlife habitat. The common law's primary method of adjusting conflicting land uses is a body of law known as "nuisance" law. If you pollute the stream running through my land, I can get a court to force you to stop injuring my use and enjoyment of the property. But nuisance requires ownership, and fish and foxes were unowned until captured. Private individuals therefore lacked standing to challenge actions that destroyed habitat. Thus, under the common-law land-use regulation in existence, only the state had the power to prevent destructive conduct. Although some states did enact statutes prohibiting obstruction of passage and destruction of habitat, enforcement was generally underfunded and neglected.[46]

By the end of the nineteenth century, overfishing and habitat destruction had combined to produce a notable reduction in the size of the runs into the Columbia. This was only one of many changes. The post–Civil War period was a time of dramatic economic development and social transformation. The economic free-for-all frequently produced dramatic waste: bison were slaughtered for their hides and left to rot on the Great Plains; passenger pigeons were shipped east as the live precursors of clay pigeons; rivers were destroyed to produce gold nuggets. One response to such waste and inefficiency was the progressive movement, which sought a "new world which conscious purpose, science, and human reason would create out of the chaos of laissez-faire economy where short-run individual interest provided no thought for the morrow."[47]

THE FISHERY AS A SUSTAINED-YIELD MACHINE

As the tenets of progressive conservation became the model for governmental resource management, the passive, hands-off doctrine of *Pierson v. Post* was supplemented by a more interventionist approach. Government responded to declining salmon runs with two types of actions: it enacted statutes imposing restrictions on taking salmon and on polluting waterways, and it sought to increase salmon runs through artificial propagation of fish in hatcheries. These responses reflected the core tenet of progressive conservation—that scientific management of natural resources could produce a sustained high yield of commodities.

Increasingly Comprehensive Harvest Management: The Problem with Monocular Vision

Declining runs prompted states to adopt restrictions on taking salmon. In 1877, for example, Washington prohibited fishing on the Columbia during March, April, August, and September; Oregon instituted a closed season for March, August, and September the following year.[48] Since the closed period lacked any biological basis (neither state employed a biologist), and since the legislatures refused to provide adequate funding for enforcement, the closed seasons did little to halt overfishing.

Declining runs also led to increasing competition among fishers. Gill-netters fought trappers and fish-wheel operators; upriver fishers fought downriver fishers; recreational fishers fought commercial fishers—and everyone employed the rhetoric of conservation.[49] In 1926, for example, gill-netters mounted an initiative in Oregon to ban fish wheels, arguing that the wheels "take practically all the fish that . . . should be entitled to proceed, to the spawning grounds to lay their eggs and perpetuate the industry."[50] The gill-netters chose not to notice that they were simply less efficient at taking "practically all the fish." In fact, the issue was allocation: who would be allowed to catch fish? By imposing gear and catch restrictions, the states allocated salmon among various groups of fishers. During the 1960s and 1970s, the states imposed increasingly severe restrictions on the river fishery while leaving the ocean fishery unregulated—a strategy that effectively allocated the bulk of the catch to ocean sports fishers and commercial trollers.[51]

Indian commercial fishers in particular bore the brunt of this reallocation as the states sought to prohibit their traditional river fishery in the guise of "conservation."[52] The states' assertion of authority to regulate the treaty right to fish "at all usual and accustomed places" led to a series of judicial decisions that reinvigorated Indian rights in the fishery. The federal courts repeatedly held that state regulation, although justified as resource conservation, was in fact an allocation of fish to non-Indians. This reallocation violated treaty rights.[53] The result was a fundamental reordering of management of the fishery: Indian fishers are now entitled to up to 50 percent of the runs.

The reinvigoration of Indian treaty rights and the creation of a legal entitlement not only reallocated fish but also revealed the inadequacies of the existing management structure. One source of difficulty is the species' biology: a salmon's life cycle leads it almost willy-nilly across numerous human boundaries. A chinook salmon hatched in the Salmon River in Idaho grows to maturity in the Alaskan gyre, a rotating ocean environment off the Aleutian Islands and the Gulf of Alaska. On these feeding grounds, salmon are taken by Alaskan fishers. Returning salmon migrate along the coast of British Columbia, where they are intercepted by Canadian fishers. Those fish that cross the bar at the mouth of the Columbia pass through more than a dozen additional jurisdictions before reaching their natal stream: the Washington Department of Fisheries, the Oregon Department of Fish and Game, the Columbia River Compact, the Pacific Northwest Electric Power Planning and Conservation Council, the Fish and Wildlife Service (U.S. Department of the Interior), the National Marine Fisheries Service (U.S. Department of Commerce), the Bonneville Power Administration (U.S. Department of Energy), the Army Corp of Engineers (U.S. Department of Defense), the Bureau of Reclamation (U.S. Department of the Interior), the Federal Energy Regulatory Commission (U.S. Department of Energy), the Columbia River Intertribal Fish Commission, the Warm Springs, Umatilla, Yakama, and Nez Perce Tribes, and the Idaho Department of Fish and Game. No single jurisdiction has the power to conserve the fish.[54] Conservation of the fishery requires agreement among the entire list of managing agencies.

This transboundary problem has been long recognized and has led to a jury-rigged structure of treaties, statutes, and regulations designed to manage the taking of anadromous salmonids. Indeed, a significant thread in the history of the Columbia basin fisheries is their increasing interjurisdictional coordination and management. Interjurisdictional management began in 1918 when Congress approved an interstate compact between Oregon and Washington for joint regulation of the salmon and steelhead runs of the Columbia River.[55] In 1945, Congress extended cooperation to the ocean fishery when it permitted California, Oregon, and Washington to create an interstate compact commission.[56] The commission, however, was restricted to an advisory role and lacked the authority to curb overfishing. Furthermore, the states had no power to regulate fishing beyond their three-mile territorial limits—a regulatory gap that contributed to the growth of the ocean trolling industry during the 1960s and 1970s.

The federal government also lacked power to regulate fishing on the high seas except through international agreements. In attempting to create ever-more-encompassing regulatory systems to manage salmon throughout their migratory course, the United States has negotiated several treaties. The first was the 1930 Convention on the Sockeye Salmon Fisheries that created the joint Canada–United States International Pacific Salmon Fisheries Commission to allocate (and restore) the sockeye runs that spawned in British Columbia's Fraser River.[57] In 1952, treaty restrictions were extended to Columbia basin salmon when the United States entered into the Trilateral Convention with Canada and Japan to restrict Japanese access to North American salmon.[58]

This was the management structure—state and interstate regulation of the harvest in the basin and international agreements covering the high-seas fishery—that was in operation in the 1960s. Its inadequacy became apparent in the 1970s. Driven by the reinvigoration of Indian treaty rights, collapsing fish stocks, the jurisdictional incompetence and general inadequacy of state regulation, and increased foreign competition,[59] Congress enacted the Magnuson Fishery Conservation and Management Act (FCMA) in 1976.[60] The act dramatically reordered both the international and domestic ocean trolling industry by creating a joint federal-state structure to regulate all ocean fisheries within an "exclusive economic zone" that extends 200 miles offshore.[61] In addition, the FCMA asserted management authority over anadromous fish that spawn in the United States throughout their migration—except when such fish are "within any foreign nation's territorial sea or fishery conservation zone."[62]

The FCMA necessitated a new treaty between the United States and Canada to manage runs that crossed the international boundaries between the two nations. The Pacific Salmon Treaty was finally approved in 1985. It created a joint commission to allocate catch in proportion to the salmon produced in each nation's waters.[63] The blockade of an Alaskan ferry during the summer of 1997 by disgruntled Canadian fishers—who felt that the allocational system was inaccurate—demonstrated that implementation problems remain.[64]

As a result of the various statutes and treaties, Columbia River salmon are now subject to some managing authority throughout their migratory wanderings. But the increasingly comprehensive management of taking—

from state to interstate to federal to international—has failed to halt the decline of the basin's salmon runs.

The current regulatory structure is an unfolding of the intrinsic philosophy of progressive conservation. It reflects the simple fact that anadromous fish cross jurisdictional boundaries, so that management structures with more encompassing authority are likely to have the greatest success. This insight was a core proposition of the progressives, whose philosophy began with the belief that centralized management by impartial mechanics could produce an unending sustained yield of desired resources. Increased management has not, however, prevented continuing decline in the runs of salmon or the listing of several species as threatened or endangered under the Endangered Species Act. At least four aspects of this failure are attributable to the philosophy that underpins progressive conservation.

The first is the management model employed to regulate catch: fisheries are managed to achieve "maximum sustainable yield," the number of fish that can be harvested in perpetuity without depleting the stock. This paradigm suffers from the shortcomings implicit in the progressive conservation perspective, the belief that nature is a mechanical equilibrium that can be manipulated to produce our desired outputs. As one ecologist noted, "The formal management of marine species was based on the belief that nature undisturbed is constant and stable."[65] Although fishery managers have become increasingly sophisticated over time, and "sustained yield" has taken on the trappings of science—with managers, fisheries biologists, and economists employing mathematical equations and graphs to explain what they believe is occurring—catch regulations continue to reflect a belief that there is a linear relationship between harvest and productivity: a run's capacity to reproduce itself is a function of the number of breeders allowed to escape to spawn. By determining the number of fish necessary to reproduce—an "escapement goal"—managers could assure runs in perpetuity.[66] Unfortunately, nature is far more chaotic than that. Fish populations have failed to conform to the smooth mathematical curves that science constructs to explain them. There is no place in the maximum sustainable yield universe for interactions among fish species, ocean currents, changing technologies, or weather. Events such as El Niño have no place in the tidy calculations.

The second reason for the failure of increasingly intensive management of catches is the belief that it is possible to have *impartial* experts as decisionmakers. The progressives believed that the mechanics—those who would operate the river to produce a sustained yield of electricity, irrigated crops, and salmon—would remain immune from politics. But the decisions they make involve people's livelihoods: water can be stored to irrigate crops during the summer drought or to generate electricity in the winter or to flush salmon smolts down to the ocean in the spring; it cannot do all three. Thus, there is political pressure on the "expert" decision-makers to overestimate the size of the salmon runs and their ability to enhance them so that it will be unnecessary to choose among the alternatives—and thus to deny some river users.

This political pressure reinforces a third element of the progressive model, the inherent optimism of the belief that nature can be managed to produce a sustained yield—a belief Paul Hirt calls a "conspiracy of optimism." The belief that nature can be manipulated like a machine leads to a technological approach. If dams prove an impediment to downstream smolt migration, the solution is to take the smolts out of the river and barge them downstream to the ocean. One technological fix that has been repeatedly employed has been to supplement natural runs through artificial propagation. Managers have repeatedly relied on hatcheries to produce more fish than nature did, rather than imposing restrictions on habitat destruction, dam-produced mortality, or commercial and sports fishing. Underlying this technological optimism is a fundamental belief about the nature of nature: progressive conservationists were cornucopians; they viewed nature as boundless. "Multiple use" was used to justify additional use regardless of how many uses an ecosystem was already being required to absorb.

Finally, the failure of harvest regulations to stem the decline of the runs is attributable to the narrowness of the regulatory focus: catch restrictions cannot restore runs that have no spawning habitat. The migratory biology of salmon requires a more comprehensive approach.

Pollution Control versus Water Quality Control

Progressive-era conservationists emphasized centralized decision-making by experts, an attempt to impose order on the chaos of the laissez-faire free-for-all. But nature, too, is chaotic. The failure of catch restrictions to stem the decline of the salmon runs reflects in part the narrowness of the progressives' approach. Harvest restrictions do not prevent habitat degradation and destruction—the pervasive water-quality problems caused by land uses such as logging, grazing, and farming. The impacts of heat, siltation, increased nutrients, altered water chemistry, and the like remained outside of the management of the fishery; fishery managers treated habitat loss as a constraint because they were powerless to address such problems directly.

In addition, these problems have proved largely unamenable to the progressives' approach because they do not fit easily within their model of regulation—a "command-and-control" model that relies upon centralized decision-making. This model is most applicable to pollution that pours from a pipe, or "effluents."[67] But the habitat degradation problems facing salmon in the Columbia basin are generally not effluent problems. Instead, in the modern legal jargon, they are "nonpoint source" water-quality problems: land-use activities that produce diffuse problems such as heat and sedimentation.[68] Nonpoint sources are less amenable to command-and-control regulation than are "point" sources because it is difficult to regulate thousands of widely dispersed land uses. Centralized regulation of such pervasive conduct would require too much and too costly regulation.

But the failure of the progressive conservation model to stem habitat loss in the Columbia basin is even more fundamental: the primary cause of the demise of the salmon runs is the basin's hydroelectric system—a system that has its roots in the progressives' view that comprehensive planning could produce the greatest good for the greatest number.

The Hydroelectric System

Hydropower development in the Columbia basin began in 1888 with the construction of the T. M. Sullivan Dam at Willamette Falls.[69] Within 20 years, 14 more hydroelectric facilities had been constructed on the Snake, Boise, and Spokane Rivers in Idaho, on the Rock Creek and the Clackamas and Deschutes Rivers in Oregon, and on the Similkameen, Naches, Spokane, and Wenatchee Rivers in Washington. These early facilities were low dams with relatively small storage capacities. Construction of major dams did not begin until the 1930s with Rock Island, Bonneville, and Grand Coulee Dams. There are now almost 130 hydroelectric or multipurpose dams on the Columbia and its tributaries.

Although this elaborate, highly integrated system has produced an abundance of kilowatts, it has also been the major cause of destruction of the region's once-abundant anadromous fish runs. The hydroelectric system causes two distinct types of problems. First, construction of the system flooded and shut off habitat. Second, the system has been operated to maximize the production of electricity rather than to produce both protein and power.

Habitat loss caused by all other land uses is relatively insignificant in comparison with that caused by damming the region's rivers.[70] In 1941, when the Bureau of Reclamation closed the gates on Grand Coulee Dam, it ended salmonid access to more than 1,100 river miles of spawning habitat. Idaho Power's Brownlee Dam in Hell's Canyon on the Snake River and Portland General Electric's Pelton Dam on the Deschutes River destroyed major runs by closing off access to spawning and rearing habitat. Main-stem Columbia and Snake River dams also drowned spawning and rearing habitat beneath their reservoirs. The cumulative impact is staggering: Columbia basin spawning habitat has been reduced from 163,000 to 73,000 square miles (fig. 12.1).

In addition to reducing the basin's aggregate capacity to produce salmon by closing off access to spawning habitat, dams also kill salmon by changing the river environment. Salmonids evolved in cold, free-flowing streams; dams have replaced such habitat with its antithesis: a series of computer-controlled, slack-water, oxygen-depleted, warm reservoirs. The impoundments increase water temperatures, alter water chemistry, and support predators that feed on migrating juveniles.[71]

Perhaps most fundamentally, however, the minimal current in the impoundments substantially retards downstream migration. Juvenile salmon do not swim downstream; instead, they are flushed to the ocean by spring floods. Prior to the dams, smolts from the Snake, Salmon, and Clearwater drainages reached the estuary at the mouth of the Columbia River in 10 to 14 days; the same trip now requires more than 50 days.[72] The increased time is critical because the salmon are undergoing a physiological transformation that allows them to live in saltwater. These effects are cumulative: each dam exacts its toll in a compounding death count.[73]

The environmental problems caused by dams and their impoundments are exacerbated by the operation of the hydropower system. The devastating mortality rates of salmonids migrating downstream are not an entirely unavoidable by-product of hydroelectricity. It is possible to cogenerate fish and electricity.[74] It is not possible, however, to maximize power production without simultaneously destroying the fishery. As one observer commented, "Dams are good scapegoats. . . . To blame 'the dams,' however, is to miss the point. The dams are merely instruments of a technocratic society. . . . A dam is not a problem because it is a dam. A dam is a problem because it creates benefits for some and hardships for others."[75]

The regional hydroelectric system crossed a critical threshold in the early 1970s. Two structural changes—an increased number of dams[76] and the addition of upstream reservoir storage[77]—allowed the system's managers to

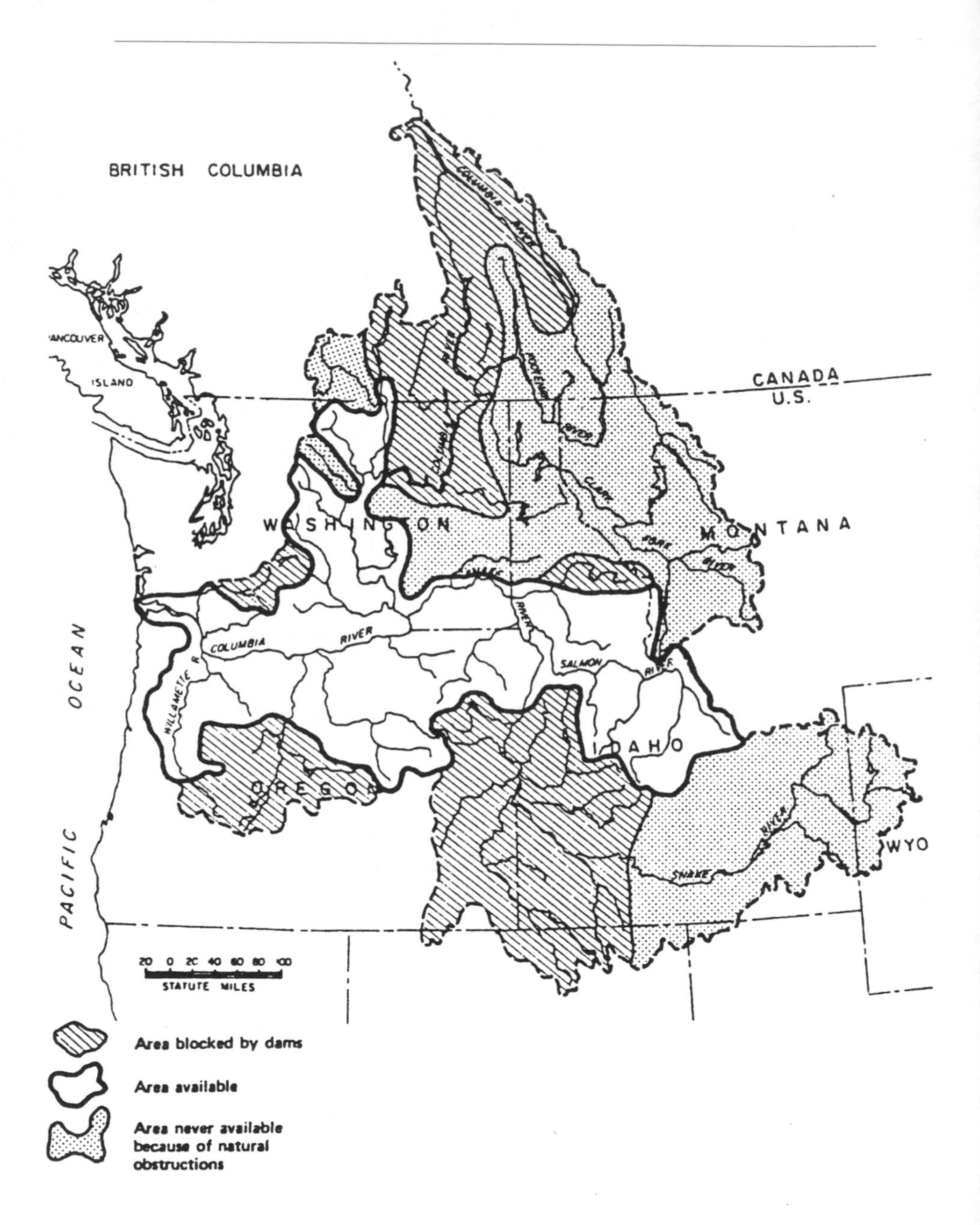

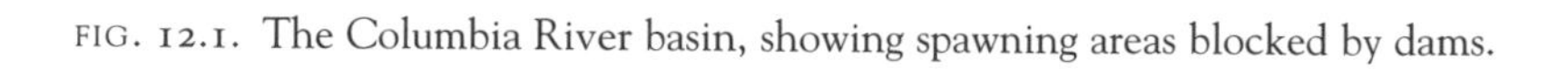
FIG. 12.1. The Columbia River basin, showing spawning areas blocked by dams.

restructure its operation by storing spring runoff for winter's high energy demand. Although this operational approach had significant energy-production advantages,[78] it devastated the basin's fish runs. The change in seasonal flows had two synergistic effects on salmon and steelhead. First, it compounded the general environmental problems caused by the impoundments: delayed out-migration, altered water chemistry, and increased predation. Second, the addition of storage capacity allowed river managers to route a higher percentage of water—and fish—through their turbines. As a result, turbines are now the major cause of juvenile salmon mortality. Turbine mortality is approximately 15 percent per dam, totaling at least 75 percent fatality for Idaho smolts passing through the eight dams.[79]

The decimation of the basin's anadromous fishery is a particularly damning commentary on the failures of progressive conservation. River basin planning was a key policy objective of the progressives, who argued that by treating the basin as a whole, it would be possible to produce a wide range of public benefits: power, flood control, irrigation, navigation, recreation, and fish. The chasm between this noble objective and the tawdry reality is dramatic. The "planned" Columbia River basin is a river basin without the fish that once defined it. The new Columbia River basin is a carp, squawfish, shad, and bass ecosystem rather than a salmon and steelhead ecosystem.[80]

The decision to create this new river was made with knowledge of the probable consequences. The Bureau of Reclamation's design for Grand Coulee Dam, for example, was a conscious decision to close off more than 1,100 river miles of habitat. Although the full effect of the regional system on anadromous fish runs was not revealed until the structural changes of the mid-1970s, there have been those who urged caution. In 1946, for example, the United States Fish and Wildlife Service recommended an indefinite moratorium on the construction of additional Columbia and Snake River dams, arguing that the effect of dams on fish was cumulative. Such objections were drowned in the chorus of power, irrigation, and navigation interests—and in a general exuberance to "tame" and "harness" nature.[81] Project proponents labeled questioners "naysayers," claiming that they simply failed to dream dreams as big as the country. The optimism inherent in the sustained-yield vision reinforced and reflected American pride in Yankee ingenuity, in the belief that there were engineering solutions to every problem.

Congress has had no greater success in forcing the operators of the hydroelectric system to protect anadromous fish runs than did other critics.[82] In 1934, for example, Congress enacted the Fish and Wildlife Coordination Act.

The statute required fish to be accorded "equal consideration" when water projects were planned and operated—a requirement that had no discernible effect on the system's designers.[83] Similarly, the Pacific Northwest Power Planning and Conservation Act of 1980 created a state-appointed regional council to oversee the operation of the system. The council was told to ensure that fish receive "equitable treatment" from all federal agencies with regulatory or management authority over both federal and nonfederal hydroelectric projects. Despite high hopes, the council failed to produce the promised parity of fish and electricity.[84] One measure of the council's failure is the subsequent listing of several runs as threatened or endangered under the Endangered Species Act.

The inability to force the hydropower agencies to protect the fishery is at least partially attributable to the fact that decisions are made by agencies with vested interests in nonfishery uses. Bonneville Power Administration, the Federal Energy Regulatory Commission, the Army Corps of Engineers, and the Bureau of Reclamation exist because of electricity, navigation, or irrigation, not because of fish. Multiple-use requirements have little effect on single-use agencies. One commentator on the politics of the Columbia River has noted that organizational self-interest tends to dominate: "If each organization is free to pursue its own institutional interest, it will do so at a minimum sacrifice to itself, with a concomitant disregard of the costs to others."[85] Fish have been viewed as an impediment to maximizing electricity production, a view captured by a statement attributed to the chief of the Army Corps of Engineers: "We do not intend to play nursemaid to the fish!"[86] Although the statement may be apocryphal, the sentiment is not. Nor is the attitude a historical anomaly: during downstream fish-migration periods over the past dozen years, the corps consistently refused to release water from Dworshak Dam to aid the migrants. Hydropower continues to come before fish—regardless of what the law requires.[87]

Although the Fish and Wildlife Coordination Act did not achieve its goal of guaranteeing fish equal consideration, it did, in conjunction with a subsequent statute (the Mitchell Act),[88] produce a lasting effect: dam builders were required to "compensate" for lost habitat. The compensation concept—which became little more than a hatchery construction program—paved the way for additional hydroelectric development because it allowed dam builders to tell the public that fish problems were being ameliorated. As such, compensation simply reflected the techno-optimism that characterizes progressive conservation: when runs decline due to overfishing or habitat loss, the solution is to supplement the remaining natural run through artificial propagation rather than to confront the cause of the decline. As one early proponent of hatcheries commented, the policy was based on "the idea that it is better to expend a small amount of public money in making fish so abundant that they can be caught without restriction . . . than to expend a much larger sum in preventing the people from catching the few fish that still remain after generations of improvidence."[89]

Hatcheries: Techno-Optimism of the Mechanics

Hatcheries are a paradigm of progressive conservation: it is always possible simply to engineer greater abundance. As one conservation biologist noted: "The notion is that we can right virtually any wrong, given enough money, motivation, and innovation."[90] And salmon hatcheries seemed to offer a cornucopia: by hand-stripping eggs from mature fish and rearing the fry in hatcheries, it was possible to produce a large number of young that could be released into the wild. Natural production—limited by relentless competition—was small; hatcheries removed the competition and produced bountiful numbers.

The first hatchery in the basin was constructed in Oregon on the Clackamas River, a tributary of the Willamette, in 1877; others followed in short order.[91] Interest in hatcheries waned during the 1920s as problems developed and their cost-effectiveness was questioned. Several technical improvements in the 1960s, however, reinvigorated the approach, and the state and federal governments embarked on a massive campaign to restore dwindling runs. In 1974, 40 federal, state, tribal, and private hatcheries released some 155 million juvenile salmon into Columbia basin rivers.

But it was a grand experiment based upon arrogant assumptions that humans could produce more fish than nature.[92] Emerging data indicate that the experiment was largely a failure, and for wild runs, a disaster. Hatchery fish have not appreciably restored the dwindling runs but have introduced diseases into wild populations, increased competition for space and food, contributed to overfishing of wild runs in "mixed-stock" fisheries, and led to a loss of genetic diversity. More importantly, hatcheries served to camouflage the fundamental problems facing the wild runs: they allowed river managers to point to the millions of fish that the hatcheries produced as an excuse for failing to address the underlying problems.

EXTINCTION

The Endangered Species Act (ESA) embodies a simple proposition: all species have a right to continued existence. The act recognizes that the way to ensure this right is by prohibiting the adverse modification of the habitat of species listed as endangered or threatened under the act. Finally, the ESA contains provisions that translate these moral propositions into law: no one may harm a listed species or adversely modify its habitat.[93]

In 1978, the ESA acquired a reputation as a rare thing—a statute that stopped development projects—when the Supreme Court decided that the presence of a small fish, the snail darter, precluded the Tennessee Valley Authority from closing the gates on its completed Tellico Dam. The court reached this conclusion because "the language, history, and structure" of the ESA demonstrated "beyond doubt that Congress intended endangered species to be afforded the highest priorities."[94] "Congress," the Court noted, "viewed the value of endangered species as 'incalculable' "—clearly far more valuable than the $100 million invested in the dam.

Since 1978, the optimistic, we-can-engineer-a-solution-to-any-problem attitude has overwhelmed the moral imperative embodied in the ESA. The story of the Columbia basin salmon runs is an example of the change. The National Marine Fisheries Service (NMFS)—the federal agency with responsibility for anadromous fish under the ESA—took steps to list several upriver populations of salmon and steelhead in October 1978.[95] With the enactment of the Northwest Power Planning and Conservation Act in 1980, NMFS held off listing the populations, because the power act offered an alternative approach to preventing extinction of the runs. The Northwest Power Planning Council, however, failed to fulfill its mandate to ensure that fish received "equitable treatment" from all federal agencies with regulatory or management authority over both federal and nonfederal hydroelectric projects.

With the council's failure to overcome the institutional barriers to fish conservation erected by the Bonneville Power Administration, the Bureau of Reclamation, and the Army Corps of Engineers, Indian tribes and environmental groups filed petitions to list Snake River chinook and sockeye salmon under the ESA; the listings occurred in late 1991 and early 1992.[96] The chinook run was subsequently downgraded from threatened to endangered,[97] and NMFS delay led to the probable extinction of the sockeye salmon run into Redfish Lake in central Idaho.[98]

The listings produced a flurry of administrative actions and litigation.

The Northwest Power Planning Council strengthened its fish and wildlife program; NMFS produced a recovery plan. Both were successfully challenged by fish advocates.[99] The council responded by calling for lowering the water behind the lower Snake River dams during spring to speed the migrating smolts to the ocean. NMFS, on the other hand, advocated continuing to transport smolts past the dams in barges. Despite the use of barging for more than a decade, runs have continued to decline. Barging, however, was the approach least disruptive to the existing river operation regime and was favored by power, navigation, and irrigation interests. Despite the ESA's preference for habitat protection, NMFS chose an engineering solution.

CONCLUSION

Salmon once ascended the rivers of the Pacific Northwest in uncountable abundance. But Euro-Americans brought with them a cultural imagination that saw salmon as commodities to be exploited for a profit without regard for the future—a perspective reflected in the story of Pierson, Post, and the fox. The result was a decline in the number of salmon pushing their way up the Columbia to spawn. As the runs declined under the pressure of unregulated fishing for the canning market, the states in the region attempted to regulate the catch. Initially, these efforts were hampered by inadequate enforcement and by the biology of anadromy that led salmon across jurisdictional boundaries. Governments responded by adopting increasingly inclusive schemes: state regulation was supplemented with interstate and then federal and finally international management structures. But the runs continued to decline.

Part of the difficulty was traceable to the theory that underpinned the regulatory structure—a mechanical, equilibrium view of nature. Salmon failed to behave as the graphs said they should. Part of the decline was also traceable to the narrow focus on catch: restricting the taking of salmon could not compensate for habitat destruction. Other fox hunters—seeking gold or timber or wheat—destroyed river habitat necessary for salmon reproduction.

Although overfishing and habitat degradation depleted the runs, it was the hydropower system that landed the final blow. Dams closed off thousands of miles of spawning and rearing habitat and turned the region's rivers into lethal reservoirs. Finally, the river managers, rather than complying with the legal mandate that they operate the system to cogenerate fish and power, operated the dams to maximize power production.

And throughout the decline of the runs, the various mechanics who operated the systems—whether harvest regulation or hydroelectric production—fell back on technology as a solution. Hatchery production allowed them to argue that they were addressing the problem, that there was no need to address the causes of the decline because it was possible simply to augment supply. The optimism that sustained the belief in hatcheries now sustains the belief that barging smolts downstream will prevent a final collapse.

But the optimism wears thin. Increasingly, it covers a simple unwillingness to acknowledge that the status quo will lead to extinction. However we characterize the failure—as a failure of political will, as an inability to imagine a different river, or as the power of entrenched economic interests to subvert the law—the undeniable result is that the once uncountable salmon teeter on the edge of extinction.

NOTES

1. Washington Irving, *Astoria, or Anecdotes of an Enterprise beyond the Rocky Mountains*, ed. Edgeley W. Todd (1836; reprint, Norman: University of Oklahoma Press, 1964), p. 373. Irving noted that "salmon are taken here in incredible quantities" and reported that one Astorian "had seen several thousand salmon taken in the course of one afternoon."

2. Paul Kane, *Wanderings of an Artist among the Indians of North America from Canada to Vancouver's Island and Oregon through the Hudson's Bay Company's Territory and Back Again* (Toronto: Raddison Society of Canada, 1925), p. 218.

3. Gary E. Moulton, ed., *The Journals of Lewis and Clark* (Lincoln: University of Nebraska Press, 1988), pp. 5: 287–88, 327, 335. See also Anthony Netboy, *The Columbia River Salmon and Steelhead Trout: Their Fight for Survival* (Seattle: University of Washington Press, 1980), p. 13.

4. Prior to approximately 10,000 years ago, at least the upper Columbia basin would have been inhospitable to salmonids—initially because of the glaciers that covered portions of northern Idaho and Washington and later because of the substantial amount of sediment produced by the glacier's retreat. Virginia L. Butler and Randall F. Schalk, *Holocene Salmonid Resources of the Upper Columbia* (Seattle: University of Washington, Institute for Environmental Studies, Office of Public Archeology, Research Report 2, n.d.).

5. The annual precontact catch is necessarily conjectural. The most comprehensive review estimates a total annual catch of between 4.5 and 5.7 million salmon,

weighing nearly 42 million pounds. Northwest Power Planning Council, *Compilation of Information on Salmon and Steelhead Losses in the Columbia River Basin* (Portland: Northwest Power Planning Council, 1986), pp. 71, 75.

6. For example, see the description of New England offered by the first Euro-American settlers, as discussed in William Cronon, *Changes in the Land: Indians, Colonists and the Ecology of New England* (New York: Hill and Wang, 1983), pp. 19–33.

7. James Willard Hurst, *Law and the Conditions of Freedom in the Nineteenth-Century United States* (Madison: University of Wisconsin Press, 1956). For a contemporaneous account of this belief, see Alexis de Tocqueville, *Democracy in America*, ed. Phillips Bradley (1840; reprint, New York: Vintage, 1945), vol. 2, pp. 36–49.

8. *Pierson v. Post,* 3 Caine's Reports 175 (New York Supreme Court, 1805).

9. *Spring Valley Water Works v. Schottler,* 110 U.S. Reports 347, 374 (U.S. Supreme Court, 1884) (Field, J., dissenting).

10. Garrett Hardin popularized the phrase in his article "The Tragedy of the Commons," *Science* 162 (1968): 1243–48; see also H. Scott Gordon, "The Economic Theory of a Common Property Resource: The Fishery," *Journal of Political Economy* 62 (1954): 124–42.

11. Similarly, William Cronon's study of contact between New England's indigenous peoples and the European invaders demonstrates that common property arrangements allowed cyclical use of seasonal resources. Cronon concludes that it was the English conception of property as alienable things to be traded in the marketplace that led to overuse. Cronon, *Changes in the Land.*

12. Donald Worster, *Nature's Economy: A History of Ecological Ideas* (New York: Cambridge University Press, 2d ed., 1994), pp. 26–55.

13. Francis Bacon, *Novum Organum*, bk. 1, aphorism 79, in *Selected Writings of Francis Bacon* 539 (1620; reprint, New York: Modern Library, 1955). Bacon believed that man could "establish and extend the power and dominion of the human race over the universe." *Ibid.* Bacon was not alone. Rene Descartes, for example, argued that the scientific method made it "possible to attain knowledge which is very useful in life . . . and thus render ourselves the masters and possessors of nature." Rene Descartes, *A Discourse on the Method of Rightly Conducting the Reason and Seeking Truth in the Sciences*, bk. 6, in *The Philosophical Works of Descartes*, trans. Elizabeth S. Haldane and G.R.T. Ross (Cambridge: Cambridge University Press, 1931 [1637]), pp. 122–23.

14. See Samuel P. Hays, *Conservation and the Gospel of Efficiency: The Progressive Conservation Movement, 1890–1920* (Cambridge: Harvard University Press, 1959); Arthur F. McEvoy, "Toward an Interactive Theory of Nature and Culture: Ecology, Production, and Cognition in the California Fishing Industry," in *The Ends of the Earth*, ed. Donald Worster (New York: Cambridge University Press, 1988), pp. 211–

29; Donald J. Pisani, "Forests and Conservation, 1865–1890," in *Water, Land, and Law in the West: The Limits of Public Policy, 1850–1920* (Lawrence: University Press of Kansas, 1996), pp. 124–40.

15. See generally Daniel B. Botkin, *Discordant Harmonies: A New Ecology for the Twenty-First Century* (New York: Oxford University Press, 1990); Thomas R. Dunlap, *Saving America's Wildlife: Ecology and the American Mind, 1850–1990* (Princeton: Princeton University Press, 1988).

16. 16 USC secs. 1531–1544.

17. See generally Kenneth W. Cummins, "Structure and Function of Stream Ecosystems," *BioScience* 24 (November 1974): 631–41; Rudolph J. Miller and Ernest L. Brannon, "The Origin and Development of Life History Patterns in Pacific Salmonids," *Proceedings of the Salmon and Trout Migratory Behavior Symposium*, eds. E. L. Brannon and E. O. Salo (Seattle: University of Washington School of Fisheries, 1982), pp. 296–309; Randall F. Schalk, "The Structure of an Anadromous Fish Resource," in *For Theory Building in Archeology: Essays on Faunal Remains, Aquatic Resources, Spatial Analysis, and Systemic Modeling*, ed. Lewis R. Binford (New York: Academic Press, 1977), pp. 207–49; Deanna J. Stouder, Peter A. Bisson, and Robert J. Naiman, eds., *Pacific Salmon and Their Ecosystem: Status and Future Options* (New York: Chapman Hall, 1997).

18. The six species belong to a single genus, *Oncorhynchus:* the chinook (o. *tschawytscha*), sockeye (o. *nerka*), coho (o. *kisutch*), pink (o. *keta*), chum (o. *gorbuscha*), and steelhead trout (*O. mykiss*). There are two other anadromous species in the region: the cutthroat trout (*Salmo clarki clarki*) and the bull trout (*Salvelinus confluentus*), which is actually a char rather than a trout.

19. See Allan C. Hartt, "Juvenile Salmonids in the Oceanic Ecosystem—The Critical First Summer," in *Salmonid Ecosystems of the North Pacific*, eds. William J. NcNeil and Daniel C. Himsworth (Corvallis: Oregon State University Press, 1980), pp. 25–57.

20. John Terbogh, "Preservation of Natural Diversity: The Problem of Extinction-Prone Species," *BioScience* 24 (1974): 715–22, 721.

21. C. J. Cederholm, D. B. Houston, D. L. Cole, and W. J. Scarlett, "Fate of Coho Salmon (*Oncorhynchus kisutch*) Carcasses in Spawning Streams," *Canadian Journal of Fisheries and Aquatic Sciences* 46 (1989): 1347–55.

22. On salmon carcasses as source of nutrients: Robert E. Bilby, Brian R. Fransen, and Peter A. Bisson, "Incorporation of Nitrogen and Carbon from Spawning Coho Salmon into the Trophic System of Small Streams: Evidence from Isotopes," *Canadian Journal of Fisheries and Aquatic Sciences* 53 (1996): 164–73; Thomas C. Kline, Jr., John J. Goering, Ole A. Mathisen, Patrick H. Poe, and Patrick L. Parker, "Recycling of Elements Transported Upstream by Runs of Pacific Salmon: N. and C. Evidence

in Sashin Creek, Southeastern Alaska," *Canadian Journal of Fisheries and Aquatic Sciences* 47 (1990): 136–44; Mary F. Willson and Karl C. Halupka, "Anadromous Fish as Keystone Species in Vertebrate Communities," *Conservation Biology* 9 (1995): 489–97. On Kennedy Creek: Jeff Cederholm, fisheries biologist with Washington Department of Natural Resources, personal communication.

23. These preliminary results from a study by Charlie Robbins, a Washington State University biologist, were reported in *The Spokesman-Review*, 29 November 1998.

24. Quoted in Cederholm et al., "Fate of Coho Salmon," p. 1347.

25. See John N. Cobb, *Pacific Salmon Fisheries* (Washington, D.C.: Government Printing Office, Bureau of Fisheries Document no. 902, 3d ed., 1921); Joseph A. Craig and Robert L. Hacker, *The History and Development of the Fisheries of the Columbia River* (Washington, D.C.: Government Printing Office, Bureau of Fisheries Bulletin no. 32, 1940); Samuel Eliot Morison, "New England and the Opening of the Columbia River Salmon Trade, 1830," *Oregon Historical Quarterly* 28 (1927): 111–32; William G. Robbins, "The World of Columbia River Salmon: Nature, Culture, and the Great River of the West," in *The Northwest Salmon Crisis: A Documentary History*, eds. Joseph Cone and Sandy Ridlington (Corvallis: Oregon State University Press, 1996), pp. 2–24; Courtland L. Smith, *Salmon Fishers of the Columbia* (Corvallis: Oregon State University Press, 1979).

26. Quoted in Smith, *Salmon Fishers of the Columbia*, p. 13.

27. Craig and Hacker, *The History and Development of the Fisheries of the Columbia River*, p. 151.

28. Quoted in Smith, *Salmon Fishers of the Columbia*, p. 21.

29. R. D. Hume, "The First Salmon Cannery," *Pacific Fisherman* (January 1904): 19.

30. Vernon Carstensen, "The Fisherman's Frontier on the Pacific Coast: The Rise of the Salmon-Canning Industry," *The Frontier Challenge: Responses to the Trans-Mississippi West*, ed. John G. Clark (Lawrence: University Press of Kansas, 1971), pp. 57–79.

31. John S. Hittell, *The Commerce and Industries of the Pacific Coast of North America* (San Francisco: A. L. Bancroft and Co., 2d ed., 1882), pp. 372–86. In addition to the canning process and the tin can, the refrigerated railroad car (which allowed the shipping of fresh salmon) also played a significant role marketing salmon. A. T. Pruter, "Review of Commercial Fisheries in the Columbia River and Contiguous Ocean Waters," in *The Columbia River Estuary and Adjacent Ocean Waters: Bioenvironmental Studies*, eds. A. T. Pruter and Dayton L. Alverson (Seattle: University of Washington Press, 1972), pp. 81–120.

32. Kirk T. Beiningen, "Apportionment of Columbia River Salmon and Steel-

head," in *Investigative Reports of the Columbia River Fisheries Project* (Portland: Pacific Northwest Regional Commission, Report R, 1976); Northwest Power Planning Council, *Compilation of Information on Salmon and Steelhead Losses*, App. A.

33. Hollister D. McGuire, "Decadence of the Fisheries," *First and Second Annual Reports of the Fish and Game Protector to the Governor, 1893–94* (Salem, Oreg.: 1894), reprinted in Cone and Ridlington, *The Northwest Salmon Crisis*, p. 75.

34. Quoted in Lisa Mighetto and Wesley J. Ebel, *Saving the Salmon: A History of the U.S. Army Corps of Engineers' Efforts to Protect Anadromous Fish on the Columbia and Snake Rivers* (Seattle: Historical Research Associates, Inc., 1994), p. 24.

35. Barton W. Evermann, "A Preliminary Report upon Salmon Investigations in Idaho in 1894," *United States Fish Commission Bulletin* 15 (1896): 253–84.

36. Rollin R. Geppert, Charles W. Lorenz, and Arthur G. Larson, *Cumulative Effects of Forest Practices on the Environment: A State of the Knowledge* (Olympia: Washington Forest Practices Board, 1984); James R. Sedell and Wayne S. Duval, "Water Transportation and Storage of Logs," *Influence of Forest and Rangeland Management on Anadromous Fish Habitat in Western North America* (Portland: Pacific Northwest Research Station, General Technical Report PNW-GTR-186, 1985), vol. 5.

37. On irrigation's impact on salmonids, see D. W. Rieser and T. C. Bjornn, "Habitat Requirements of Anadromous Salmonids," *Influence of Forest and Rangeland Management on Anadromous Fish Habitat in Western North America* (Portland: Pacific Northwest Research Station, General Technical Report PNW-GTR-96, 1979), vol. 1, pp. 4–5, 9–13, 22–23, 33–34; Q. J. Stober, M. R. Griben, R. V. Walker, et al., *Columbia River Irrigation Withdrawal Environmental Review: Columbia River Fishery Study* (Seattle: University of Washington Fisheries Research Institute, Report no. FRI-UW-7010, 1979).

38. Leonidas G. Liacos, "Water Yield as Influenced by Degree of Grazing in California Grasslands," *Journal of Range Management* 15 (1962): 34–42, pp. 40–41; William R. Meehan and William S. Platts, "Livestock Grazing and the Aquatic Environment," *Journal of Soil and Water Conservation* 33 (1978): 274–78; William S. Platts, "Effects of Livestock Grazing," *Influence of Forest and Rangeland Management on Anadromous Fish Habitat in Western North America* (Portland: Pacific Northwest Research Station, General Technical Report PNW-GTR-124, 1981), vol. 7, pp. 2, 12.

39. Such temperatures retard the growth rate of juveniles, disrupt smolting, contribute to diseases, either halt the upstream migration or seriously deplete the stored body fat of the migrating fish, and adversely affect gonadal development. See Bouck, "The Importance of Water Quality to Columbia River Salmon and Steelhead," in *Columbia River Salmon and Steelhead*, ed. Ernest Schwiebert (Washington, D.C.: American Fisheries Society, Special Pub. no. 10, 1977), pp. 149–51; Reiser and

Bjornn, "Habitat Requirements of Anadromous Salmonids," pp. 27–28; Stober et al., *Columbia River Irrigation Withdrawal Environmental Review*, pp. 163–67.

40. T. W. Chamberlin, "Timber Harvest," *Influence of Forest and Rangeland Management on Anadromous Fish Habitat in Western North America* (Portland: Pacific Northwest Research Station, General Technical Report PNW-GTR-183, 1982), vol. 3; Geppert, Lorenz, and Larson, *Cumulative Effects of Forest Practices*, pp. 155–56; Platts, "Effects of Livestock Grazing," p. 11; William O. Saltzman, "Impact of Streamside Use on Fisheries," in *Columbia River Salmon and Steelhead*, ed. Ernest Schwiebert (Washington, D.C.: American Fisheries Society, Special Pub. no. 10, 1977), pp. 160–63. Irrigation return water is also warmer than stream water. Stober et al., *Columbia River Irrigation Withdrawal Environmental Review*, p. 155.

41. William S. Platts, "The Effects of Heavy Metals on Anadromous Fish Runs of Salmon and Steelhead in the Panther Creek Drainage, Idaho," *Proceedings of the 52d Annual Conference of Western American State Game and Fish Commissioners* (1972): 582–600. See generally Susan B. Martin and William S. Platts, "Effects of Mining," *Influence of Forest and Rangeland Management on Anadromous Fish Habitat in Western North America* (Portland: Pacific Northwest Research Station, General Technical Report PNW-GTR-119, 1981), vol. 8.

42. Charles R. Berry, Jr., "Impact of Sagebrush Management on Riparian and Stream Habitat," in *The Sagebrush Ecosystem: A Symposium* (Logan: Utah State University College of Natural Resources, 1978), pp. 192–209; L. Norris, H. Lorz, and S. Gregory, "Forest Chemicals," *Influence of Forest and Rangeland Management on Anadromous Fish Habitat in Western North America* (Portland: Pacific Northwest Research Station, General Technical Report PNW-GTR-149, 1983), vol. 9; Stober et al., *Columbia River Irrigation Withdrawal Environmental Review*, pp. 185–86. Fertilizers may also cause algal growth that reduces the dissolved oxygen available for fish. Berry, "Impact of Sagebrush Management," pp. 195–96; Chamberlin, "Timber Harvest," p. 19; Reiser and Bjornn, "Habitat Requirements of Anadromous Salmonids," p. 20.

43. Berry, "Impact of Sagebrush Management," p. 195; Platts, "Effects of Livestock Grazing," p. 11; Reiser and Bjornn, "Habitat Requirements of Anadromous Salmonids," pp. 21–22, 29–32; Stober et al., *Columbia River Irrigation Withdrawal Environmental Review*, pp. 169–70.

44. Northwest Power Planning Council, *Compilation of Information on Salmon and Steelhead Losses*, p. 145; Platts, "Effects of Livestock Grazing," p. 11; Stober et al., *Columbia River Irrigation Withdrawal Environmental Review*, pp. 167–68; Carlton S. Yee and Terry D. Roelofs, "Planning Forest Roads to Protect Salmonid Habitat," *Influence of Forest and Rangeland Management on Anadromous Fish Habitat in Western North America* (Portland: Pacific Northwest Research Station, General Technical Report PNW-GTR-109, 1980), vol. 4.

45. U.S. Fish Commission, *Report of the Commissioner for 1878* (Washington, D.C.: Government Printing Office, 1880), pp. xlv, xlvi–l.

46. Lawrence M. Friedman, *A History of American Law* (New York: Simon and Schuster, 2d ed., 1985), p. 187.

47. Hays, *Conservation and the Gospel of Efficiency*, p. 124.

48. Netboy, *Columbia River Salmon and Steelhead Trout*, p. 35.

49. See generally Ivan J. Donaldson and Frederick K. Cramer, *Fishwheels of the Columbia* (Portland: Binfords and Mort, 1971); Courtland L. Smith, *Oregon Fish Fights* (Corvallis: Oregon State University, Sea Grant Publication no. ORESU-T-74-004, 1974); Smith, *Salmon Fishers of the Columbia*, pp. 30–40; Richard White, *The Organic Machine: The Remaking of the Columbia River* (New York: Hill and Wang, 1995), pp. 39–48.

50. Quoted in Smith, *Oregon Fish Fights*, p. 4.

51. In 1976, 66.3 percent of the total chinook and coho catch was taken by ocean sport and commercial fishers. Pacific Fishery Management Council, *Perspective on Management of Ocean Chinook and Coho Salmon Fisheries in the Fishery Conservation Zone off California, Oregon, and Washington* (Portland: Pacific Fishery Management Council, 1982), pp. 22–23.

52. See Beiningen, "Apportionment of Columbia River Salmon and Steelhead," p. 3; Ralph Johnson, "The States versus Indian Off-Reservation Fishing: A United States Supreme Court Error," *Washington Law Review* 47 (1972): 207–36.

53. See, e.g., *Washington v. Washington State Commercial Passenger Fishing Vessel Association*, 443 U.S. Reports 658 (U.S. Supreme Court, 1979). The dispute had strong racist overtones. See United States Civil Rights Commission, *Indian Tribes: A Continuing Quest for Survival* (Washington, D.C.: Government Printing Office, 1981).

54. See Dale D. Goble, "Introduction to the Symposium on Legal Structures for Managing the Pacific Northwest Salmon and Steelhead: The Biological and Historical Context," *Idaho Law Review* 22 (1986): 417–67; Pacific Fishery Management Council, *Perspective on Management of Ocean Chinook and Coho Salmon Fisheries*, p. 13; Charles F. Wilkinson and Daniel K. Conner, "The Law of the Pacific Salmon Fishery: Conservation and Allocation of a Transboundary Common Property Resource," *University of Kansas Law Review* 32 (1983): 17–109.

55. Act of Apr. 8, 1918, Pub. L. no. 65–123, 40 Stat. 515. The compact provided for the joint regulation of seasons, gear, and catches on the Columbia River. See Elmer Wollenberg, "The Columbia River Fish Compact," *Oregon Law Review* 18 (1938): 88–107.

56. Pacific Marine Fisheries Compact, Pub. L. no. 80-232, art. 1, 61 Stat. 419 (1947).

57. Convention on the Sockeye Salmon Fisheries, May 26, 1930, United States–Canada, 50 Stat. 1355, T.S. no. 918. See James A. Crutchfield and Giulio Pontecorvo, *The Pacific Salmon Fisheries: A Study of Irrational Conservation* (Baltimore: Resources for the Future, 1969), pp. 140–46.

58. Trilateral Convention, May 9, 1952, United States–Canada–Japan, 4 UST 380, TIAS no. 2786.

59. During the 1960s, domestic trollers were joined by the fishing fleets of several other countries—the Soviet Union, Japan, North Korea, East Germany, and Poland—which began to harvest fish immediately outside the three-mile territorial limit. See Smith, *Salmon Fisheries of the Columbia*, p. 90; Pruter, "Review of Commercial Fisheries in the Columbia River and in Contiguous Ocean Waters," pp. 116–18.

60. 16 USC secs. 1801–1882.

61. The FCMA established eight regional fishery management councils composed of state and federal representatives. The councils prepare management plans for the harvestable species within their jurisdiction; the plans are subject to federal oversight and enforcement. Two of the regional councils have responsibility for Pacific salmon: the North Pacific Fishery Management Council (with representatives from Alaska, Oregon, and Washington) and the Pacific Fishery Management Council (representing California, Idaho, Oregon, and Washington). In preparing the plans, the councils analyze the status of the various stocks and set both annual and long-term goals intended to provide sufficient fish to maintain a healthy run and to satisfy Indian treaty rights. To achieve these goals, the plan imposes catch, gear, and season restrictions. Pacific Fishery Management Council, *Perspective on Management of Ocean Chinook and Coho Salmon Fisheries*, pp. 14–23.

62. 16 USC sec. 1812(2). Columbia River salmon migrate through Canadian waters.

63. Treaty Concerning Pacific Salmon, Jan. 28, 1985, United States–Canada, TIAS no. 11091.

64. Timothy Egan, "Salmon Wars in Northwest Spurs Wish for Good Fences," *New York Times*, 12 September 1998, p. 1.

65. Botkin, *Discordant Harmonies*, p. 22; McEvoy, "Toward an Interactive Theory of Nature and Culture."

66. For a statement of this perspective, see James A. Crutchfield, "The Fishery: Economic Maximization," in *Pacific Salmon: Management for People*, ed. Derek V. Ellis (Victoria, B.C.: University of Victoria, 1977), pp. 1–33.

67. The progressives paid surprisingly little attention even to effluent regulation—and then only as it applied to public health issues. Earl F. Murphy, *Water Purity: A Study in Legal Control of Natural Resources* (Madison: University of Wisconsin Press, 1961); Donald J. Pisani, "Fish Culture and the Dawn of Concern over

Water Pollution in the United States," *Environmental Review* 8 (Summer 1984): 117–31.

68. Under the Federal Clean Water Act, a National Pollutant Discharge Elimination System (NPDES) permit is required for discharges from "point sources." 33 USC secs. 1362(14), 1342. "Nonpoint source" is the residual category that is not covered by the NPDES permit system.

69. Michael Blumm, this volume; Craig and Hacker, *History and Development of the Fisheries of the Columbia River*, p. 193; Northwest Power Planning Council, *Compilation of Information on Salmon and Steelhead Losses*, App. C, pp. 100–101; George C. Young and Frederic J. Cochrane, *Hydro Era: The Story of Idaho Power Company* (Boise: Idaho Power Co., 1978), pp. 11–33.

70. Ed Chaney and L. Edward Perry, *Columbia Basin Salmon and Steelhead Analysis* (Portland: Pacific Northwest Regional Commission, 1976), p. 3; Ralph W. Larson, "A Few Lessons from the Past," in *Columbia River Salmon and Steelhead*, ed. Ernest Schwiebert (Washington, D.C.: American Fisheries Society, Special Pub. no. 10, 1977), pp. 111–14; Netboy, *Columbia River Salmon and Steelhead Trout*, 94–97.

71. John W. Attey and Drew R. Liebert, "Clean Water, Dirty Dams: Oxygen Depletion and the Clean Water Act," *Ecology Law Quarterly* 11 (1984): 703–29; Gerald B. Collins, "Effects of Dams on Pacific Salmon and Steelhead Trout," *Marine Fisheries Review* (November 1976): 39–46; Wesley J. Ebel, "Major Passage Problems," in Schwiebert, *Columbia River Salmon and Steelhead*, pp. 33–39; Howard L. Raymond, "Effects of Dams and Impoundments on Migrations of Juvenile Salmon and Steelhead from the Snake River, 1966 to 1975," *Transactions of the American Fisheries Society* 108 (1979): 505–29.

72. Rob Lothrop, "The Failure of the Fish Passage Provisions of the Columbia Basin Fish and Wildlife Program and Some Suggested Remedies," *Anadromous Fish Law Memo* 34 (1985): 1–7; Howard L. Raymond, "Migration Rates of Yearling Chinook Salmon in Relation to Flows and Impoundments in the Columbia and Snake Rivers," *Transactions of the American Fisheries Society* 97 (1968): 356–59.

73. Prior to the completion of upper two dams on the Lower Snake River, 89% of the chinook salmon fry made it to Ice Harbor Dam; following completion of these two dams, only 33% of the fingerlings arrived at Ice Harbor. Raymond, "Effects of Dams and Impoundments on Migrations of Juvenile Salmon and Steelhead," pp. 519–20. During low-flow years such as 1973 and 1977, losses of 95% and 99% were recorded for salmon and steelhead as a result of the Snake River Dams alone. Ebel, et al., *The Snake River Salmon and Steelhead Crisis*, p. 9.

74. Ed Chaney, *Cogeneration of Electrical Energy and Anadromous Salmon and Steelhead in the Upper Columbia River Basin: An Economic Perspective on the Question of Balance*, (Eagle, Idaho: Northwest Resources Information Center, 1982), p. 11.

75. Smith, *Salmon Fishers of the Columbia*, p. 4; Michael C. Blumm, "Hydropower v. Salmon: The Struggle of the Pacific Northwest's Anadromous Fish Resources for a Peaceful Coexistence with the Federal Columbia River Power System," *Environmental Law* 11 (1981): 211–300.

76. As recently as 1956, a salmon migrating between the Salmon River and the Pacific faced only two dams, Bonneville and McNary. Then, in rapid succession, six additional dams were added: The Dalles (1956) and John Day (1968) on the Columbia, and Ice Harbor (1961), Lower Monumental (1969), Little Goose (1970), and Lower Granite (1975) on the lower Snake. See Bonneville Power Administration, *Multipurpose Dams of The Pacific Northwest* (Portland: Bonneville Power Administration, 1980).

77. The crucial structural change was the addition of storage capacity. Even with the completion of all the dams, the system's storage capacity was less than one-tenth of the average annual runoff. It thus was not possible to prevent the large spring flows that propelled migrating fingerlings downstream. In 1964, however, the United States and Canada ratified a treaty authorizing the construction of four storage dams: Libby Dam in Montana and three dams in Canada—Keenleyside, Duncan, and Mica. With the addition of Dworshak in Idaho, the system had the capacity to store most of the spring runoff. See Blumm, this volume; Chaney, *A Question of Balance*, pp. 12–13.

78. There are at least two energy-production advantages. First, the firm energy load capability of the regional system—the amount of energy it can reliably produce—is increased because water is not "wasted" by being spilled during the spring runoff, since it can be stored until needed to meet energy demand. Second, the system's flexibility is increased because the stored water can be used to meet demand whenever it occurs. For example, there often is a "peak" in the demand for electricity when people get up in the morning, turn on lights, take showers, and prepare breakfast.

79. Lothrop, "Failure of the Fish Passage Provisions," p. 3. Although actual mortality varies with a number of factors, the combination of predation and turbine losses can be enormous, and the cumulative effect of such per-dam losses is staggering. For example, in the low-flow years of 1973 and 1977, 95–99 percent of all fingerlings leaving the Snake River Basin failed to reach The Dalles Dam. Blumm, "Hydropower v. Salmon," p. 211; Chaney, *A Question of Balance*, p. 7.

80. As Richard White has pointed out, the Columbia River is not "dead." It simply is no longer a river that produces salmon and steelhead. Richard White, "Columbia River History," paper presented at the Pacific Northwest Regional Environmental History Conference, 2 August 1996.

81. As one commenter noted, "These days I look back again at our sorry history

with the dams, thinking that every time we build a dam, somebody assures us that we're going to take excellent care of the fish resource. Yet when there's a problem that demands mitigation, those efforts begin after damages have occurred for two or three years." Larson, "A Few Lessons from the Past," pp. 112–13. See also Netboy, *Columbia River Salmon and Steelhead Trout*, pp. 78–79; Robbins, "The World of Columbia River Salmon"; White, *The Organic Machine*, 59–88.

82. See Blumm, this volume.

83. 16 USC sec. 661; see also Comptroller General, *Improved Federal Efforts Needed to Equally Consider Wildlife Conservation with Other Features of Water Resource Developments* (Washington, D.C.: General Accounting Office, GAO Rep. no. B-118370, 1974), p. 8. See generally Blumm, "Hydropower v. Salmon," pp. 268–276.

84. 16 USC secs. 839–839g. The "equitable treatment" requirement is found at 16 USC sec. 839b(h)(11)(A)(i). See Michael C. Blumm, "Implementing the Parity Promise: An Evaluation of the Columbia Basin Fish and Wildlife Program," *Environmental Law* 14 (1984): 277–358.

85. Joel E. Haggard, "The Columbia River: Protein, Power, Preservation, and Politics," *Environmental Law* 10 (1980): 221–33, p. 231.

86. See Netboy, *Columbia River Salmon and Steelhead Trout*, p. 75 and n. 2.

87. See Michael C. Blumm, "Reexamining the Parity Promise: More Challenges than Successes to the Implementation of the Columbia Basin Fish and Wildlife Program," *Environmental Law* 16 (1986): 461–515.

88. 16 USC secs. 755–757.

89. Hugh Smith, "The United States Bureau of Fisheries: Its Establishment, Functions, Organization, Resources, Operations, and Achievements," *Bulletin of the Bureau of Fisheries* 28 (1910): 1371.

90. Gary K. Meffe, "Techno-Arrogance and Halfway Technologies: Salmon Hatcheries on the Pacific Coast of North America," *Conservation Biology* 6 (1992): 351; see generally David Ehrenfeld, *The Arrogance of Humanism* (New York: Oxford University Press, 1981).

91. See Roy J. Wahle and Robert Z. Smith, *A Historical and Descriptive Account of Pacific Coast Anadromous Salmonid Rearing Facilities and a Summary of Their Releases by Region, 1960–76* (Washington, D.C.: Government Printing Office, NOAA Technical Report NMFS SSRF-736, 1979); James A. Lichatowich and John D. McIntyre, "Use of Hatcheries in the Management of Pacific Anadromous Salmonids," in *American Fisheries Society Symposium* 1 (1987): 131–36; George Straley, "The Growth of Artificial Propagation in Oregon," in Cone and Ridlington, *The Northwest Salmon Crisis*, pp. 38–42.

92. Meffe, "Techno-Arrogance and Halfway Technologies," p. 351.

93. 16 USC sec. 1538(a)(1)–(2).

94. *Tennessee Valley Authority v. Hill*, 437 U.S. Reports 153, 174 (U.S. Supreme Court, 1978).

95. *Federal Register* 43 (October 3, 1978): 45,628.

96. *Federal Register* 56 (November 20, 1991): 58,619 (Snake River sockeye listed as endangered); *Federal Register* 57 (April 22, 1992): 14,653 (Snake River chinook listed as threatened). See generally Michael C. Blumm, Michael A. Schoessler, and R. Christopher Beckwith, "Beyond the Parity Promise: Struggling to Save Columbia Basin Salmon in the Mid-1990s," *Environmental Law* 27 (1997): 21–126.

97. *Federal Register* 59 (August 18, 1994): 42,529 (Snake River chinook downgraded to endangered).

98. For a review of the status of the region's anadromous fish runs, see Willa Nehlsen, Jack E. Williams, and James A. Lichatowich, "Pacific Salmon at the Crossroads: Stocks at Risk from California, Oregon, Idaho, and Washington," *Fisheries* 16 (2) (March–April 1991): 4–21.

99. *Northwest Resource Information Center v. Northwest Power Planning Council*, 35 Federal Reporter 3d 1371 (9th Circuit Court of Appeals, 1994); *Idaho Department of Fish and Game v. National Marine Fisheries Service*, 850 Federal Supplement 886 (District Court for the District of Oregon, 1994).

13 / The Northwest's Hydroelectric Heritage

MICHAEL C. BLUMM

Natural resources law is only beginning to probe its historical roots. Yet it is clear that many of today's resource conflicts are legacies of the policies and programs of earlier eras. This is particularly true in the Pacific Northwest. Although historical perspective has illuminated legal studies of Pacific Northwest timber and fishery resources,[1] no legal analysis of Northwest hydroelectric policies from a historical perspective exists.

This is regrettable, because without an understanding of how and why the present hydroelectric system developed, some emerging issues are likely to be incorrectly characterized as novel, and the lessons of the past may go unheeded. This essay aims to supply a historical dimension to the Pacific Northwest's hydroelectric system.

Today's electric power system evolved over more than half a century of regional development. Some system characteristics are inextricably related to national policies and politics. Indeed, a number of fundamental issues that have confronted every generation of hydroelectric policymakers first arose in the progressive era of the early 1900s. Many issues that dominated the electric power agenda of previous generations recur repeatedly. Tracing the roots of regional hydroelectric policy-making illuminates the forces that influenced the growth of the system in the past and those that will shape its future.

THE PRE–NEW DEAL ERA

Long before it became the driving force behind economic growth in the Pacific Northwest, a federal role in water resources development was widely recognized as a key to settling the nation's western frontier.[2] The motivation for the first national plan for river improvements and canals, developed by treasury secretary Albert Gallatin in 1808, was a desire to facilitate settlement of the vast territory acquired from France in the Louisiana Purchase.[3] After the Supreme Court settled the issue of the federal government's constitutional authority over navigation in 1824,[4] the question of the federal role in developing canals and other water improvements became one of the great political controversies of the pre–Civil War years.[5] Following the Civil War, populist midwestern farmers advocated water development to provide an alternative to high-cost railroad transportation.[6] This antimonopoly sentiment would become a persistent theme when the electric age took shape at the turn of the twentieth century.

The Progressive Era

If the Pacific Northwest and electric power "grew up together,"[7] they were raised on a diet of the reformist zeal that characterized the progressive movement in American politics. Progressives in turn were heavily influenced by the overcrowding, poverty, and crime accompanying a sevenfold increase in the nation's urban population between 1860 and 1910. They found in government-sponsored resource development programs, particularly water projects,[8] a means to preserve the rural, small-town, individualistic life that post–Civil War industrialism threatened with big corporations, big cities, and big political machines.

Although the moral principles of progressive conservationists lay in a simpler America and its agrarian past, the means by which they sought to reach these ends involved harnessing new forces in American life—science, technology, and industrial management—to achieve efficient resource use.[9] The basic tenets of progressive water policies exemplified this attempt to preserve the equities of an older America by employing the efficiencies of corporate management principles. The equity side of progressive thought manifested itself in assertions that streamflows were part of the public domain and that their benefits should be widely shared, not controlled by narrow monopoly interests.[10] The efficiency side was reflected in the belief that multiple uses of streamflows should be maximized through scientific, basinwide planning.[11] In pursuit of both goals, progressive conservationists charted an active role for the federal government as planner, regulator, and developer of water projects.

Belief in the public nature of streamflows led Theodore Roosevelt to appoint a commission to survey prospective waterway improvements and to recommend a suitable role for the federal government. In its 1908 report, the Inland Waterways Commission declared water to be a public resource and recommended basinwide federal water planning to serve multiple purposes, including navigation, flood control, water power, irrigation, and pollution control. The commission also recommended that a central, national water planning agency be established to coordinate the activities of the Army Corps of Engineers, the Bureau of Reclamation, and other federal agencies.[12]

Centralized, national water planning did more than fit the progressive notion of maximizing efficiency. It was also an attempt to reduce the influence of congressional "log rolling," by which individual legislators exercised a disproportionate influence in securing congressional approval of uneconomic, parochial projects. These progressive reforms, however, were stymied by shifting political sentiment and world developments. In 1920, Congress culminated a 15-year progressive struggle to establish a uniform licensing scheme for nonfederal water project development by passing the Federal Water Power Act.[13] But opponents of federal development succeeded in including in that legislation a provision that terminated the Inland Waterways Commission without transferring its planning functions to the newly established Federal Power Commission.

In effect, the 1920 act substituted federal regulation of nonfederal development for the centralized water planning advocated by the progressives. It also rejected conservationist philosophy by providing only relatively low federal charges and failing to earmark those revenues for federal multiple-purpose projects. Nevertheless, a number of progressive notions embodied in the act became enduring principles of national water policy, including a preference for publicly sponsored river development, limited-term licenses that reserved ultimate ownership in the public domain, and license criteria that, while not expressly incorporating the multiple-use concept, were at least amenable to it.

The Twenties

Republican party ascendancy in the 1920 presidential and congressional election considerably altered the federal rôle in water resources development. The Republicans rejected progressive antimonopoly concerns and objected to federal hydropower development, believing it competed with private enterprise.[14] Consequently, in the decade following World War I, private power interests predominated. Multiple-purpose federal water planning nonetheless made significant strides, and the first multiple-purpose federal project was authorized in 1928.[15] Although the 1920s displayed little of the progressive zeal that linked water development with democratic reform, the Republican Congress did initiate the planning of many projects that the federal government would pursue in the next decade.

Although federal development was overshadowed by private projects, the principal federal development agencies—the U.S. Army Corps of Engineers and the Bureau of Reclamation—became advocates of multiple-purpose planning in the 1920s. The corps, which had staunchly resisted the progressives' call for comprehensive basinwide planning, became a convert in 1927 when Congress authorized it to undertake multiple-use plans.[16] The Bureau of Reclamation saw hydropower sales as a means to finance its irrigation projects; it undertook the first federal, multiple-purpose water project when the Boulder Canyon project was authorized in 1928. In the late 1920s, both agencies studied the feasibility of developing the Columbia basin, and the bureau's Grand Coulee project was nearly approved by President Hoover.[17] The Republican attachment to private power was strong enough, however, to prevent federal construction at Grand Coulee until after Franklin Roosevelt's election in 1932.

THE NEW DEAL

The philosophical underpinnings of Pacific Northwest hydropower development are found in the progressive impulse, but it was the economic realities of the Great Depression that transformed that philosophy into action. The New Deal belief that the federal government could stimulate economic recovery through public works projects was soon translated into water project construction, for which the Army Corps' of Engineers' "308 Reports" provided blueprints.[18] Dam-building, however, was not simply a means of unemployment relief; it was also fundamentally related to the notion that inadequate regulation of private utilities led to inequities in the distribution of electric power. New Dealers sought cheaper and wider distribution of electric power through both federal regulation of private power and federal promotion of public power.

The Bonneville Project Act

In the Pacific Northwest, New Deal water planning proceeded quickly because the first of the Corps of Engineers' "308 Reports" on the Columbia River and its tributaries had been completed in 1931. The report called for a series of 10 dams on the main-stem Columbia, including what were to become the Bonneville and Grand Coulee Dams. Construction began on both projects in 1933. After the Supreme Court ruled in 1935 that the two projects had not been properly authorized, Congress quickly reauthorized them, and construction continued.[19] As the Bonneville project drew to completion, distribution of the surplus power produced by the dam became the subject of debate.

Between 1935 and 1937, 38 bills were introduced in Congress to resolve how hydropower from Bonneville Dam would be marketed. The crucial issues in this debate were (1) which agency should have responsibility for project operations and power marketing; (2) what the scope of federal responsibilities regarding construction of transmission lines should be; and (3) whether power should be sold at uniform rates irrespective of geographic distance from the project. Public power advocates generally wanted establishment of a regional Columbia Valley Authority with comprehensive powers similar to those possessed by the Tennessee Valley Authority (TVA). This authority would promote public power and wide distribution of electricity use by constructing transmission lines throughout the region, delivering power at a uniform rate, and giving preference to publicly owned utilities.

Private power interests, on the other hand, argued for a more limited federal role. They favored project operation and power sales by the corps, limited federal construction of transmission lines, and rates that reflected the cost of power transmission.[20] Allied with the private power companies were economic interests in Portland, which wished to maximize their proximity to Bonneville. The debate over the Bonneville legislation thus invoked urban-rural conflicts, private-public power competition, and intergovernmental power struggles.

A compromise was reached in 1937 that preserved the corps's role as project operator but authorized a Bonneville Project Administrator to market power, construct transmission lines, and set electricity rates. Although the legislation did not establish the TVA-like agency that private power interests had feared, its directives to give priority to public power and to foster "widespread use" gave public power advocates much of what they sought. Whether or not rates should be uniform was left to the discretion of the new administrator. Discretionary power over such a vital issue, combined with the broad delegation of authority to the administrator throughout the statute, made the selection of the first administrator a very important appointment.

"Postage Stamp" Rates and Rural Electrification

When FDR dedicated the Bonneville Dam in September 1937, he was met with numerous placards proclaiming "We Want Ross" for Bonneville administrator. J. D. Ross, formerly with Seattle City Light and then a commissioner with the Securities and Exchange Commission, was the choice of public power advocates who believed he would promote uniform, cheap "postage stamp" rates and rural electrification. Opposed to Ross were private utilities, the Portland Chamber of Commerce, and Oregon Republican governor Charles Martin, for whom uniform rates meant higher rates to pay for electrification of rural areas and a consequent loss of Portland's locational advantage. Ross got the job and in 1938 proceeded to institute a uniform rate of two mills per kilowatt hour, a rate that survived for 27 years. Uniform rates were designed to encourage dispersion of industry throughout the region and to induce rural electrification.[21]

Although head of Bonneville for less than a year and a half, Ross left a substantial legacy. In addition to establishing a uniform rate of two mills, he placed a heavy emphasis on construction of transmission lines, actively encouraged formation of public utility districts, and indicated a preference for selling power to industries employing large numbers of workers. After his unexpected death in March 1939, his successor, Paul Raver (who was to serve as administrator for nearly 15 years), pursued Ross's notion of social planning by signing industrial power contracts, although he backed away from openly advocating public power.

By the time the attack on Pearl Harbor thrust the nation into war, Bonneville's temporary status had been made permanent. By constructing a high-voltage line between Bonneville and Grand Coulee Dams, the Bonneville Power Administration (BPA) laid the cornerstone of what would become a regionwide, integrated electric power grid. Moreover, in order to finance its transmission line construction, Bonneville was compelled to lure to the region industries that consumed large amounts of public power.

THE WAR AND ITS AFTERMATH

The war years pushed energy policy into the forefront of the defense effort. In the Pacific Northwest, the prewar antimonopoly concerns that had fueled the often bitter public-versus-private power struggles took a back seat to efforts to mobilize the region behind the war effort. Founded as an agency to promote public power, the BPA evolved into a regional chamber of commerce during the war, helping to attract defense industries with its cheap power. After the war, while not entirely abandoning the public power movement and continuing to foster rural electrification, BPA devoted its primary efforts to increasing the electric capacity of the region, rather than to encouraging additional public takeovers of private utilities. Although it would be inaccurate to describe the war and postwar years as a period in which public and private utilities completely buried their hatchets, the era witnessed the operational integration of the system and laid the groundwork for the regionwide cooperation that would become the dominant characteristic of the Pacific Northwest electric power industry in the 1950s.

In 1938, with war impending in Europe and the power shortages of World War I in mind, FDR asked the Federal Power Commission and the War Department to survey the adequacy of the nation's power capacity. Their report concluded that the country faced serious potential energy deficits in the event of war and recommended that immediate steps be taken to rectify the situation.[22] Their proposal to finance power plant construction through government loans to private power companies was opposed by public power advocates such as interior secretary Harold Ickes, who believed it would undermine the New Deal's public power program. Without such government assistance, private utilities refused to expand. This stalemate persisted until Pearl Harbor. Three weeks after the Japanese attack, the private industry's policy think-tank, Edison Electric Institute, reported a nationwide power shortage.

While the rest of the nation was power-short going into the war, the Pacific Northwest was blessed with a surfeit of power. Given authority by executive order to market Grand Coulee power, BPA moved aggressively to complete the transmission line between Bonneville and Grand Coulee and to enter into contracts with defense-related industries. By 1942, 92 percent of BPA's load was committed to industry.[23] The early war years saw numerous defense contracts awarded to Pacific Northwest industries. BPA embarked on a crash transmission line construction program to meet defense plant deadlines. The agency also stepped up its schedule for installing additional generators at Bonneville and Grand Coulee Dams.

Perhaps the most noteworthy effect of the war on the Northwest's hydroelectric system was the War Production Board's 1942 order to interconnect operations of the region's utilities to maximize efficiency. This order prompted formation of the Northwest Power Pool, a voluntary partnership of the region's utilities designed to increase cooperation and information sharing. Interconnection of utility operations had the short-run effect of enabling the region to meet power demands despite the lowest streamflows in more than 50 years, but its long-run effects were more significant. Interconnection signaled the beginning of an era in which public and private utilities would operate their projects as essentially one utility, which remains the chief characteristic of system operations today.[24]

The economic boom that the war brought to the Northwest spurred interest in additional water projects. Although the Roosevelt administration opposed any new project that was not justified as a war measure, a considerable amount of planning for postwar times was under way. The upper Columbia basin was studied for reservoir sites in 1943, resulting in the authorization of the Bureau of Reclamation's Hungry Horse project a year later. Released from executive control and prodded by fear of postwar depression, Congress authorized an unprecedented number of water projects in the 1944 Flood Control Act and the 1945 Rivers and Harbors Act, including the lower Snake River dams, which were not completed until the mid-1970s.

New water projects could not, however, avoid postwar economic problems without an expanding industrial base to compensate for the slowdown in defense industries. Production curtailments in the shipbuilding and aluminum industries led to layoffs and a power surplus of more than a million kilowatts by 1946. Indeed, this power surplus threatened to undermine attempts to expand the Northwest hydroelectric system, as private utilities, fearing additional public power projects, testified to Congress that no additional federal dams were needed. Clearly, if the Pacific Northwest was to use its water power potential as the springboard to economic development, the region could not afford fractious public-versus-private power disputes.

With its role as power marketer assured by the end of the war, the Bonneville Power Administration sought to expand its role as a regional economic planner. In 1945, the Interior Department authorized BPA to move beyond its limited role of marketing existing power supplies and begin to anticipate future needs and help ensure that supplies were adequate to meet increased future demands. This decision led BPA to sell half of its million-kilowatt surplus to the aluminum industry in 1946.[25] Increased industrial sales in subsequent years evaporated the surplus and created a close relationship between BPA and its industrial customers that persists to this day.

BPA also took steps to overcome private utility resistance to expanding federally produced electricity, playing an instrumental role in the 1946 founding of the Tacoma Conference, later named the Pacific Northwest Utilities Conference Committee (PNUCC), a coalition of public and private utilities and industries. More than simply an advisory committee, PNUCC assembled individual utility forecasts, lobbied in Congress for new dams and increased funding, and played a critical role in expanding the system during the postwar years. By preaching a "gospel of growth" in which additional supplies of hydropower were portrayed as essential to the economic development of the region, PNUCC provided a means by which the region's public and private utilities and electroprocess industries began to cooperate closely with each other. Although their disagreements certainly did not disappear, under BPA's leadership the seeds of regional cooperation were sown.

THE PARTNERSHIP YEARS

The inauguration of Dwight Eisenhower as president in January 1953 had important effects on the development of the regional hydroelectric system. Ideologically committed to private power development, the Republicans authorized no new federal projects and backed off from planned developments on the mid-Columbia and middle Snake Rivers. Despite this, the federal hydroelectric system continued to grow throughout the Eisenhower years, as some previously authorized projects were completed. Moreover, the proliferation of nonfederal projects in the 1950s required greater coordination between public and private power to increase system reliability.

In his 1953 State of the Union message, newly elected President Eisenhower asserted that "the best natural resources program for America will not result from exclusive dependence on Federal bureaucracy. It will involve a partnership of State and local communities, private citizens, and the Federal Government, all working together."[26]

This partnership philosophy was a product of the new administration's close ties with the private utility industry. In June 1953, Eisenhower's interior secretary, Douglas McKay, terminated construction of a BPA transmission line in southwestern Oregon and sold the right-of-way to a private utility. Three months later, in an effort to reduce rate disparities, the Interior Department announced that any federally produced power not committed to public utilities in long-term contracts would be sold to private utilities on a long-term basis. BPA subsequently negotiated 20-year contracts with a number of private power companies, assuring them of access to low-cost hydropower until 1973.

Given the Republicans' antipathy toward federal intervention and public power, it is not surprising that they quickly rejected a 1952 recommendation of the Truman-appointed National Security Resources Board for increased federal funding of new hydroelectric projects. In fact, Secretary McKay blamed electric shortages in the Pacific Northwest on the existence of federal power, not on a shortage of it. The Eisenhower administration chose to let private utilities determine whether new projects should be constructed, restricting new federal developments to situations where power needs were beyond the capabilities of local utilities. Spurred by favorable tax laws and increased demand from defense industries in response to the Korean crisis, nonfederal generating capacity in the Northwest nearly quadrupled during the 1950s.[27]

Two important areas of nonfederal development were on the mid-Columbia above Hanford and on the Snake River in Hells Canyon. In early 1953, the Interior Department dropped its opposition to the Idaho Power Company's application to develop the middle Snake. Although public power advocates challenged the private utilities' plans, three dams were licensed in 1956. On the mid-Columbia, after Congress revoked federal authorization of the Priest Rapids Dam in 1954, the Federal Power Commission (FPC) licensed four nonfederal projects to three Washington public utility districts.

Although no federal projects were initiated during the Eisenhower years, the federal partner was not inactive. Congress continued to appropriate money to construct previously authorized projects and to extend BPA lines to rural areas.[28] The boom in nonfederal projects, however, coupled with an economic recession and a poor aluminum market, reduced BPA power sales in the late 1950s. Reduced power sales produced BPA monetary deficits and a power surplus, a situation that would have an important influence on policymakers in the 1960s.

The power partnership of the 1950s produced a diverse system of federal and nonfederal projects linked by BPA's transmission grid. With increasing amounts of nonfederal power coming on line, the BPA power grid offered utilities a means of transmitting their new power supplies to markets without having to construct new lines. This transmission of power for another party is called "wheeling."[29] By establishing regionwide transfers, wheeling increased regionwide reliability and encouraged nonfederal water project developments by reducing the need to construct transmission lines. As a result, by the close of the Eisenhower "partnership" years, BPA had assumed a central role in both public and private utility planning and operations. This role was to expand far beyond power marketing in the 1960s.

THE GOLDEN AGE

In the late 1950s, a slow economy and a proliferation of nonfederal power projects left BPA with a surfeit of power. At the same time, a series of financial deficits evaporated the agency's accumulated monetary surplus and jeopardized its ability to fulfill its repayment obligations. Administrator Charles Luce, hoping to avoid a rate increase he believed would impair regional economic growth, adopted a new repayment policy and doubled industrial power sales. As a result, Luce was able to defer BPA's first rate increase in 27 years until 1965 and to limit it to less than 3 percent.

While this strategy kept Northwest electric rates far below rates in other regions of the country, it resulted in a repayment policy that the General Accounting Office has criticized as inefficient and probably unfair. Perhaps more significantly, the doubling of industrial power sales made future energy shortfalls appear imminent and helped to spur interest in expanding electrical generating capacity.

The most obvious means of increasing system capacity is to regulate streamflows to maximize power production. Because of the great variation in Columbia basin flows,[30] as well as a lack of significant storage capability, a large percentage of streamflows in high-water years must be spilled over run-of-the-river dams, producing no power. On the other hand, in low-flow years there is too little water to maximize power output. These realities had produced long-standing interest in upper-basin water storage projects. Because 30 percent of the Columbia's streamflows originate in Canada, studies of joint Canadian–United States development had been initiated in 1944. Fifteen years of technical studies and legal and policy debates ensued.[31]

The principal stumbling block concerned whether Canada was entitled to a portion of the downstream power production and flood control benefits produced in the United States, and, if so, what share. United States resistance to the concept of downstream benefits precluded an agreement during the 1950s. When Canada indicated it was prepared to proceed with an alternative development plan on the Peace River, the United States, perhaps also influenced by mounting BPA deficits, agreed to share downstream benefits on an equal basis. This agreement led to the signing of the Columbia River Treaty on January 17, 1961.[32]

The treaty authorized construction of four large water storage reservoirs, doubling the basin's storage capacity. Just as important, it induced a number of agreements that increased interconnected system operations and enhanced BPA's role as regional power marketer.

The first series of agreements was prompted by the fact that British Columbia did not need its share of the power produced by the treaty projects. As a result, the province sold its power rights to a consortium of public utilities to finance its share of the project construction. The consortium sold half of its Canadian entitlement to 4 private utilities and half to 41 public utilities. Each utility, in turn, transferred these variable entitlements to BPA in exchange for firm power. The result was increased pressure on BPA to manipulate streamflows to meet peak power demands.

Maximizing the downstream benefits produced by the treaty storage projects required increased coordination among federal and nonfederal project operators. Marketing the additional power resulting from the treaty also required another agreement. Power sales to California were given impetus by technological innovations and by the Kennedy administration, which saw regional interties as the basis of a fully coordinated national power network. Expanding BPA's service area outside the region alarmed industrial customers, who feared that competition from additional customers would deprive them of cheap power. The result was the enactment of the 1964 Northwest Regional Preference Act, which required BPA to export only surplus power for which there was no demand in the Northwest.[33] Almost simultaneously, Congress appropriated $45 million to fund the federal portion of intertie lines that would in 1967 connect Los Angeles with the Northwest. Thus, the beneficiaries of the Canadian treaty were not confined to the Northwest.

The legacies of the Columbia River Treaty were many and varied. It resulted in greater coordination of system operations, led to California power sales, and enhanced the peaking capacity of the hydropower system. By harnessing the spring freshet upon which anadromous fish depended for transportation of the ocean, the treaty projects and their aftermath contributed greatly to a precipitous decline of upper basin fish runs. Finally, in negotiating the treaty and the subsequent Coordination Agreement, and in lobbying for the Preference Act and intertie authorization, BPA emerged from its Eisenhower-era passivity. Although lacking the authority of TVA, BPA nonetheless possessed the administrative expertise and fiscal resources to function as the region's central planning agency, so long as it had the confidence of its utility and industrial customers. This role would become increasingly important as the region began to look to thermal generated power.

THE RISE AND FALL OF THE HYDRO-THERMAL POWER PROGRAM

As the Canada treaty projects neared completion in the late 1960s, most of the region's large-scale hydroelectric sites had been developed. At the same time, power forecasts projected serious deficits. To avoid shortfalls, the region turned to coal- and nuclear-generated thermal power. The additional costs of thermal power, however, made this transition difficult and controversial. And because of limited BPA authority and limited utility financing, far fewer thermal plants were initiated than were originally proposed. This inability to expand the system prompted the movement to obtain a congressional solution in the late 1970s.

Phase 1

System expansion appeared to be a pressing need in the late 1960s. Beginning in 1967, the long lead time required to obtain siting approval for thermal plants doubled BPA's planning horizon from 10 to 20 years. This expanded time horizon, the doubling of industrial power sales, and forecasting assumptions of steady growth in the use of electricity made power shortfalls seem imminent.[34] Electric demand was expected to rise not only because of regional growth but also because of expected increases in per capita electric consumption. The region's utilities claimed that these demands should be met by expanding the system rather than through conservation measures.

Facing a projected tripling of power demands, the Joint Power Planning Council, a consortium of BPA and more than 100 public and private utilities, formulated a long-range planning document calling for a $15 billion, 20-year program of system expansion. This Hydro-Thermal Power Program was refined a year later into a 10-year program calling for seven thermal plants, new hydroelectric generators to meet peak load demands, and additional transmission facilities. The thermal plants were the fulcrum of the program, and BPA helped spur development of these plants by contracting in advance to purchase their power—in essence, buying electricity "futures" that BPA would later add to the larger pool of electricity it distributed to its customers.

The key to the program was the "net billing" concept. This financing scheme made it possible for public utilities, most of which had little equity, to commit themselves to thermal plant construction.[35] Net billing disguised actual electricity production costs for companies investing in relatively expensive thermal power by spreading wholesale rate increases uniformly among all BPA customers.[36] This hid the full costs of thermal plants, because BPA's rates reflected average system costs—that is, both high-cost thermal power and low-cost hydropower. Such "melded" rates made increments of thermal power appear to be cheaper than in fact they were. By enabling BPA to purchase electric "futures," net billing allowed the agency to expand the power system, an authority that Congress never expressly granted it.

Five years after its initiation, the Hydro-Thermal Power Program came to an abrupt halt. First, skyrocketing construction costs exhausted BPA's net billing capacity as thermal plant costs outstripped its wholesale power rates. Second, a 1972 Internal Revenue Service regulation removed the tax exemption from bonds financing plants if BPA purchased more than 25 percent of the plant's output. Thus, in late 1973 BPA terminated net billing, effectively ending Phase 1 of the program.

Although it was short-lived, the Phase 1 program's legacies proved to be enduring. Construction of seven thermal plants began: Pacific Power and Light's Jim Bridger coal plant in Rock Springs, Wyoming; Portland General Electric's Boardman coal plant in Boardman, Oregon; Pacific Power and Light's and Washington Water Power's Centralia coal plant in Centralia, Washington; Portland General Electric's Trojan Nuclear Plant in Rainier, Oregon; and the first three Washington Public Power Supply System (WPPSS) nuclear plants in Hanford, Washington. The four nuclear plants were net billed, although BPA acquired only 30 percent of Trojan's output and 70 percent of WPPSS No. 3. In addition, there were unanticipated long-run costs to salmon runs from manipulating Columbia and Snake River flows to meet peak loads. These costs would become more apparent in subsequent years.

Phase 2

The demise of the Phase 1 program left the region with projected energy shortages and without a plan to add resources. To avoid power shortfalls, BPA terminated sales of firm (guaranteed) power to private utilities in 1973. Deprived of guaranteed access to low-cost federal hydropower, consumers served by private utilities began to experience significant rate increases.[37]

Late in 1973, BPA and its customers agreed to a new version of the Hydro-Thermal Power Program, quickly christened Phase 2. Designed to produce at least seven additional thermal plants, Phase 2 depended for its success on

BPA's negotiating expertise rather than its deep pockets. Instead of purchasing the planned output of thermal plants, BPA would act as an agent for its preference customers, negotiating purchase agreements for them with thermal plant developers. Although this would guarantee project sponsors markets for their power, the sponsors retained the risks of plant delays or unsatisfactory performance. This financing arrangement resembled that employed to construct the mid-Columbia dams during the Eisenhower years. BPA, however, would exercise some oversight over the performance of project sponsors by acting as "trust agent" for its preference customers.

The federal role in the Phase 2 program, however, was not simply that of a power broker. Installation of hydroelectric peaking generators and construction of additional transmission lines would remain a federal responsibility. Indeed, in order to decrease the financial uncertainties inherent in the congressional appropriation process, BPA sought authority to pay operation and maintenance costs out of its revenues and to finance transmission system additions through bond sales. It received this new authority in the 1974 Columbia River Transmission System Act.[38]

But it was Phase 2's reliance on more thermal plants to meet projected long-term shortages that proved its undoing. Formulated without any significant public review of environmental impacts, the program's first industrial sales contract,[39] and then the program itself,[40] was enjoined for violating the National Environmental Policy Act.[41] Unable to participate in the expansion of the system until it completed a comprehensive environmental impact statement on its role in the Hydro-Thermal Power Program, BPA notified its preference customers in 1976 that after 1983 it no longer would be able to meet their load growth. The agency also informed its industrial customers that it was likely that their contracts would not be renewed when they expired in the 1980s.

The demise of the Hydro-Thermal Program thrust the region into an anticipated power crisis induced not only by impending power shortages but also by rate inequities. Consumers served by private utilities, which had been cut off from BPA firm power in 1973, began to experience steep rate increases due to thermal plant costs.[42] In 1977, the Oregon state legislature attempted to regain access to low-cost hydropower for its private utilities by authorizing the creation of a publicly owned Oregon Domestic and Rural Power Authority if rate disparities remained in 1979. If this authority succeeded in its goal of claiming a share of preference rights for all domestic and rural ratepayers in the state, it would immediately overtax BPA's resources and precipitate a regionwide civil war over federal hydropower entitlements.

Toward a Congressional Solution

In 1977, BPA and its customers, through the Pacific Northwest Utilities Conference Committee, drafted legislation to solve the region's energy problems. Senator Henry Jackson introduced the PNUCC bill in September 1977, but neither it nor a less complex successor drafted a year later managed to progress very far by the time the Ninety-fifth Congress adjourned in late 1978.

When the Ninety-sixth Congress convened in 1979, a coalition of BPA customers was solidly behind a legislative solution to the Northwest's power crisis. Neither BPA nor its customers wanted an administrative allocation of limited power supplies, arguing that such an allocation would be the subject of protracted litigation. Instead, they argued that Congress could avoid the uncertainties by devising a legislative allocation scheme and equipping BPA with the authority to purchase power from nonfederal sources on a long-term basis. Congress apparently agreed. On December 5, 1980, President Carter signed the Pacific Northwest Electric Power Planning and Conservation Act into law.[43]

THE NORTHWEST POWER ACT

The Northwest Power Act became law largely because it seemed to benefit all the interest groups who lobbied for it. First, the act minimized rate disparities through an exchange of power between BPA and private utilities, which effectively reduced the rates paid by residential and small farm consumers served by private utilities.[44] Second, BPA's industrial customers, primarily aluminum companies, which would finance this power exchange, received long-term contracts assuring their continued presence in the region.[45] Third, the statute reiterated the 1937 Bonneville Project Act's preference for power sales to public utilities and promised that public utility rates would increase no more than they would have without the legislation.[46] Fourth, although the act gave BPA new authority to acquire resources to meet the needs of its customers, it subjected the use of that authority to a number of checks, including approval by a new interstate planning council.[47] Fifth, Congress vested this new interstate council, the Northwest Power Planning Council, with the responsibility of devising an environmentally sensitive regional electric power plan favoring conservation programs and renewable resources.[48] Sixth, fish and wildlife interests were promised a new basinwide plan, formulated by the council, to protect and restore Columbia basin fish and wildlife damaged by the hydroelectric system, especially the basin's salmon runs.[49] Finally, while the statute authorized new regional authorities, it reasserted state powers to set consumer electric rates, plan and site new facilities, and control the allocation of water rights.[50]

The Revised Electric "Constitution"

The Northwest Power Act made four significant changes to the electric power system's basic marching orders. It modified the old paradigms of public power preference, widespread use, sound business principles, and lowest possible rates with new directives for regional cooperation, a conservation preference, concepts of shared powers and open processes, and an enterprise theory of liability.[51] The changes are evident in the purposes of the 1980 act.

First, the act recognized that BPA was no longer simply a public power promoter. In fact, the agency's role as an open advocate for public power was short-lived; since World War II it had preached a gospel of growth through electric power use. Electric growth required the cooperation of private utilities, which became BPA customers during the 1950s and which were key partners in the formulation of the Hydro-Thermal Power Program in the 1960s.

Second, although the 1980 act did not expressly repeal the earlier "widespread use" policy, it did give priority to conservation and efficient use of electric power.[52] This change reflected the fact that energy conservation is often the cheapest means to satisfy energy demands.[53]

The third change required BPA to involve the states, local governments, and the public in regional electric power planning.[54] Related to this commitment to open processes was the establishment of the Northwest Power Planning Council, which Congress authorized to develop comprehensive plans for future energy supply and fish and for wildlife conservation in the Northwest. The council, composed of representatives from Idaho, Montana, Oregon, and Washington, would accomplish this through pluralistic administrative processes, shared powers among the four affected states, public participation in planning processes, and the opportunity for judicial review of council and federal agency decisions.[55]

Fourth, the Northwest Power Act modified considerably the Bonneville Power Act's 1937 mandate of "lowest possible rates," particularly in its promise to ensure that all environmental costs were paid by electric users.[56] In fact, the Northwest Power Act made a significant commitment to compensate for past losses sustained by the Columbia basin's fish and wildlife resources at the hands of the hydroelectric system.[57] Fish and wildlife and other environmental resources would no longer be sacrificed to subsidize electric rates.

The Columbia Basin Fish and Wildlife Program

The Northwest Power Act directed the Northwest Power Planning Council to develop a comprehensive program to preserve and rehabilitate fish and wildlife to the extent that their declines were attributable to the development and operation of the hydroelectric system.[58] Although concern over the effects of dam-building on salmon runs was evident during the New Deal,[59] the cumulative effect of the postwar boom in dam-building upon these fish runs was little appreciated. Even less understood was the long-term effect of the events of 1964—the doubling of the basin's storage capacity as a result of the ratification of the Columbia River Treaty with Canada and the formalization of integrated system operations in the Columbia River Coordination Agreement. The increased storage capacity deprived downstream migrating salmon of much of the spring freshet upon which they depended for transportation to the ocean,[60] and integrated operations also promoted the "one utility" concept that enabled the Hydro-Thermal Power Program to rely increasingly on streamflows to generate power to meet peak load demands.[61]

The region's hydroelectric power was long underpriced, partly because the costs in fish and wildlife losses were not accounted for in BPA's rates. In effect, those rates were subsidized by fish and wildlife losses. Accordingly, Columbia basin salmon runs experienced precipitous declines.[62]

The 1980 act made it clear that such losses were unacceptable and ordered the Northwest Power Planning Council to develop a remedial program for BPA, the Army Corps of Engineers, the Bureau of Reclamation, and other federal agencies to implement. The statute's goal was to elevate the status of fish and wildlife to that of a "coequal partner" with hydropower, "on a par" with the other authorized purposes of the dams.[63] The vehicle for achieving this equity was the council's basinwide restoration program, which the statute required to be based on the "best available" scientific knowledge. The law specifically called for river managers to provide water flows sufficient to restore salmon runs[64]—a practice that would result in some loss to hydroelectric generating capacity.

The council promulgated its first Columbia Basin Fish and Wildlife Program in 1982, after the region's fish and wildlife agencies and Indian tribes submitted some 700 pages of recommendations for conservation measures.[65] The program called for increased river flows in the spring to facilitate salmon migration, improvements in the engineering systems to allow fish to bypass the dams, habitat protection and restoration measures, and hatcheries where they would not jeopardize the genetic integrity of wild salmon runs. But rather than the fixed flows in the spring that the fish and wildlife agencies and tribes had recommended, the council instead adopted what it called a "water budget," a dedicated volume of water that would be controlled by representatives of the region's fish and wildlife agencies and Indian tribes and used to augment river flows in the spring to help salmon fry in their downstream migration to the sea. Unfortunately, this water budget contained too little water to produce effective river flows for the smolts, and the water project operators frequently ignored the flows called for by fish and wildlife agencies and tribes. For example, one spring the water budget was met only 6 of 26 days during the critical migration season.[66] Instead of flushing juvenile salmon downstream with increased river flows, federal water managers developed and implemented a system of truck and barge transportation. Both the council and the fish and wildlife agencies acquiesced to this artificial transport program, despite the Northwest Power Act's explicit call for improved river flows.[67]

Despite its limitations and lack of enforcement, the Columbia Basin Fish and Wildlife Program and the statute that created it marked a significant change in hydropower law and policy. The statute established a goal of regionwide salmon restoration, and the Northwest Power Planning Council supplied an open regional forum in which hydropower-salmon tradeoffs and other salmon restoration issues, such as the feasibility of relying on hatcheries to supplement wild fish runs, could be publicly discussed and debated. The fish and wildlife program made important contributions to salmon restoration efforts.

Nonetheless, the statute and the program promised more than they were able to deliver. The act's limited scope—focused exclusively on the hydroelectric system while overlooking salmon harvest regulations, federal land management activities in the watershed, and irrigation diversions—made a comprehensive approach to salmon restoration impossible. Moreover, the act contained weak enforcement language that encouraged federal water management agencies such as the Corps of Engineers and the Bonneville Power Administration to consider the program's measures to be merely advisory, so that they routinely failed to carry out the program's main-stem river flow provisions. The council also showed itself to be a poor overseer of program implementation, as was Congress. Thus, while the program did chart an innovative, systematic approach to salmon restoration, it failed to achieve its basic goal of producing substantial improvements in salmon migration or parity between hydropower and fish and wildlife.[68]

The Northwest Power Plan

In addition to developing the Columbia Basin Fish and Wildlife Program, the Northwest Power Act charged the Northwest Power Planning Council with the responsibility for promulgating an electric power and conservation plan.[69] A fundamental assumption of the statute's drafters was that the region would need new electric resources to meet steadily increasing demand. This proved to be a faulty premise, because increased power rates and a regionwide economic recession dampened power demands during the 1980s and produced an unexpected regional electric surplus. As a result, the acquisition of new power supplies, which had been a paramount concern in the debate over the Northwest Power Act, became a largely forgotten issue after the statute's passage.[70] The overriding question was no longer the kinds of resources the region should acquire to meet projected deficits but how to manage a sustained surplus of power. Not only did this surplus make the contents of the Northwest Power Plan a less pressing issue, but also new cheap sources of electricity such as that generated from gas turbines in the early 1990s helped to create a financial crisis for BPA. It was this fiscal crisis, along with Endangered Species Act (ESA) listings of Snake River salmon, that became the focus of attention in the 1990s, not the council's prescription for the region's electric future.

THE 1990S: CRISIS YEARS

The 1990s began with a crisis precipitated by petitions to list several stocks of Snake River salmon for protection under the ESA. The ESA listing eventually produced competing salmon restoration plans, court decisions suggesting that neither of the plans was sufficient, and revised plans, one of which was largely ignored while the other received judicial sanction. BPA alleged that its financial crisis was exacerbated by the increased costs these plans entailed, and the agency succeeded in brokering an administrative agreement that placed a ceiling on annual salmon restoration expenditures. Looming over both the ESA listings and BPA's financial crisis was electricity deregulation, which threatened to destabilize both BPA's power markets and funding for salmon restoration.

The ESA Listings of Snake River Salmon

Petitions submitted by Indian tribes and environmental groups led to the listing of Snake River chinook and sockeye under the federal ESA by early 1992.[71] These listings induced the Northwest Power Planning Council to amend its program to increase salmon migration measures.[72] Nevertheless, Indian tribes and environmentalists challenged the council's 1992 amendments to the fish and wildlife program as well as the 1993 National Marine Fisheries Service's (NMFS's) salmon recovery plans with surprising success. Two different courts ruled each restoration effort flawed and called for a "major overhaul" in hydroelectric system operations.[73]

The judicial rebukes produced two markedly different administrative responses. The interstate Northwest Power Planning Council promulgated program amendments calling for drawing down Snake River reservoirs during the salmon migration season to speed juvenile salmon past the dams. NMFS, on the other hand, advocated removing more than half of the juvenile fish from the river and transporting them downstream in trucks and barges.[74] This transportation program, as it was called, had been operated by the Corps of Engineers for over a decade with no evident success, but it was favored by power, navigation, and irrigation interests because it was least disruptive to existing river operations. NMFS, surprisingly, embraced this program of artificial transport, alleging that a sound salmon restoration program required a systematic comparison of the efficiency of truck and barge transportation versus in-river migration over a four-year period. Despite protests from the State of Oregon, the tribes, and environmentalists that this approach was insufficient to meet the requirements of the ESA, it received judicial ratification in 1997,[75] allowing NMFS to continue its experiment until 1999. The council's call for seasonal reservoir drawdowns was ignored as an unenforceable advisory opinion, although a panel of independent scientists endorsed an even more radical proposal of permanent drawdowns: breaching the four lower Snake River dams.[76]

BPA's Financial Crisis

The region's long-term electric surplus, coupled with rising BPA expenditures, threatened BPA's fiscal integrity in the mid-1990s. As a consequence,

BPA and its customers lobbied Congress to impose a "cost cap" on BPA's fish and wildlife expenditures.[77] Like the timber salvage rider, this cap would have exempted BPA from compliance with environmental laws, particularly the ESA.[78]

The Clinton administration refused to support such an exemption from environmental laws, so a legislative cost cap was not imposed. The administration, however, helped to broker what amounted to an administrative cost cap that limited BPA's out-of-pocket expenditures to $252 million annually and committed the hydroelectric system to implement the NMFS restoration plan but not the council's reservoir drawdown plan.[79] Whether the cost cap was necessary to ensure BPA's financial solvency is debatable, but it did confirm that, contrary to the directive of the Northwest Power Act, economics, not biology, was the driving force in salmon restoration.[80]

CONCLUSION

Although the region's hydroelectric system was an offspring of the New Deal, its philosophical underpinnings were rooted in the progressive conservation movement. Progressive notions about the public nature of streamflows and the opportunity provided by basinwide waterway developments to promote social equity, prevent economic monopolies, and preserve the rural way of life had enduring influence on New Deal thought. The economic crisis that ushered in the New Deal provided the impetus to marry progressive social philosophy with large-scale federal public works projects to stimulate economic recovery. Water projects not only put people to work but also produced electricity in competition with private utilities, whose excesses in the 1920s resulted in high rates, poor service, and rural areas with no electricity. Federal power as a "yardstick" for private utility rates and service reflected both a distrust of private utilities, which were often controlled by large holding companies far removed from local consumers, and a fundamental lack of faith in the ability of government to control utility excesses through regulation.

The chief regional legacy of the New Deal was the Bonneville Power Administration. BPA was established to market wholesale power from federal dams and to promote public utilities as retailers of federal power. BPA's limited charter was the product of a compromise between New Dealers (who sought a TVA-like authority to plan and operate a basinwide federal water and power system) and private utility interests (which wanted to limit federal power in the region and provide their ally, the Corps of Engineers, with power-marketing authority). The subsidence of the public power movement during and after World War II made an expanded mandate for BPA impractical.

In the postwar era, the hydropower system expanded rapidly through Corps of Engineers planning and congressional appropriations. The rejection of the New Deal paradigm of centralized national water planning left Northwest hydroelectric power in the hands of regional interests. While this arrangement produced more local control, it also allowed key decisions to be made in low-visibility technical reports and appropriations hearings, largely out of the public spotlight. With Congress willing to bankroll a hydroelectric system that surpassed the region's immediate needs, BPA employed its marketing authority to expand its industrial customer base, first lured to the region by defense contracts in World War II. Just as important, the agency forged institutional links with the region's private utilities to coordinate demand forecasts. High forecasts induced more dam proposals; more dams meant that BPA could market more power to industries and private utilities after supplying the needs of its preference customers. Cheap, federally produced power became the engine driving regional economic growth.

The partnership era of the 1950s solidified the role of private utilities as an integral element of the regional power puzzle. The private utilities gained long-term BPA power contracts and took advantage of a moratorium on new federal dams to secure licenses for their own hydroelectric projects. Diversity of project ownership induced BPA to broadly construe its authority to wheel nonfederal power in order to increase regional efficiency. Of even greater long-term significance, private and public utilities collaborated on financing arrangements that enabled the equity-short public utilities to construct a number of new projects, notably on the mid-Columbia. This kind of cooperative financing would become a keystone of the region's approach to thermal plant construction in the 1960s and 1970s.

The hydroelectric system matured in the "golden age" of the 1960s. Ratification of the Columbia River Treaty doubled the basin's storage capacity and promoted a series of contractual arrangements that increased system coordination and interregional power sales. In a classic example of achieving short-term gains while incurring long-term losses, power surpluses were dissipated by a doubling of power sales to industrial customers. Increased industrial power sales produced forecasts of future power shortages. With large dam sites all but exhausted, the region formulated plans to develop thermal power plants.

The transition to an integrated hydro-thermal system proved to be a difficult and controversial one. The initial Hydro-Thermal Power Program foundered when rising construction costs overtaxed BPA's financing scheme and the IRS limited the tax advantages available to project sponsors of federally backed plants. Phase 2 of the program, financed without federal guarantees but still with significant federal responsibility for manipulating streamflows to meet peak power demands, was even shorter lived. Formulated by BPA and its customers without public involvement, the courts halted the program for violating the National Environmental Policy Act, which proved to be perhaps the most cost-effective decision of the decade. In effect, the court rulings reflected the program's lack of political legitimacy. The considerable costs of thermal plants, in terms of both increasing rates and their spillover costs to the environment and the region's salmon runs, made it clear that decisions about expanding the electric system could not be made by technical experts alone. A broader regional consensus was necessary.

That consensus produced the 1980 Northwest Power Act, with its commitments to open processes, shared authority, and fish and wildlife restoration. More fundamentally, the 1980 act represented a dramatic departure from the New Deal model of broad charters to federal administrators. Throughout the postwar era, electric policy was made largely by BPA and its customers, coupled with congressional acquiescence and appropriations. The detailed provisions of the 1980 statute indicated that Congress wished to narrow considerably the agency's statutory mandates. Moreover, in creating the Northwest Power Planning Council and directing it to chart the region's energy future, Congress made a significant reallocation of power to the states. Although the power surpluses of the 1980s made the council's power plan less consequential, its Columbia Basin Fish and Wildlife Program became the centerpiece of salmon restoration during the 1980s.

The failure of that program to restore Snake River salmon led to the ESA listings of the early 1990s. The listings allowed the reassertion of federal authority over river operations, which, surprisingly, produced a salmon restoration effort relying heavily on trucking and barging juvenile salmon in order to minimize disruption to system operations. After initially striking down the federal salmon restoration plan as inadequate, the courts refused to do so twice. The council's more radical prescription for reservoir drawdowns was ignored, and BPA's financial troubles led to a "cost cap" on fish and wildlife expenditures in the mid-1990s. Looming ahead at the time of this writing is NMFS's 1999 decision on whether to continue to truck and barge most juvenile salmon around the dams or to make in-river migration safer for the fish by breaching the four lower Snake River dams to restore natural river flows.

In early 1998, a restructured BPA, with a new independent entity controlling transmission in a new era of electric competition, is on the near-term horizon. The new competitive era makes the need for regional power planning uncertain and could undermine financing of salmon restoration efforts. As the region's power brokers move to restructure the system, however, they should be mindful of the Northwest Power Act's promise of public involvement in all regional power decisions. Although the public nature of streamflows has not been seriously challenged since the progressive conservation movement, regional hydroelectric policymakers have frequently sacrificed public involvement in the name of administrative expertise. Unfortunately, the practical effect of unfettered administrative discretion has been to emphasize the short term at the expense of the long term and to emphasize utility and industrial customers' access to decision-makers at the expense of the general public. The 1980 act's commitment to open processes reflected a recognition that the region no longer could afford to make policies uninformed by public comment and unable to withstand public challenges in the courts. Although public comment and judicial review have been attacked as dilatory and inefficient, it seems clear that the benefits of ensuring sound administrative decision-making far exceed the costs of delays. The lessons of the past indicate that the long-term costs of poor decisions are simply too high for the region not to encourage active, vigorous, and critical public debate on the region's electric future.

NOTES

An earlier version of this chapter was published as "The Northwest's Hydroelectric Heritage: Prologue to the Pacific Northwest Electric Power Planning and Conservation Act," *Washington Law Review* 58 (1983): 175–244. Reprinted with the permission of the *Washington Law Review*.

1. See James L. Huffman, "A History of Forest Policy in the United States," *Environmental Law* 8 (1978): 239–80; Michael C. Blumm, "Hydropower v. Salmon: The Struggle of the Pacific Northwest's Anadromous Fish Resources for a Peaceful Coexistence with the Federal Columbia River Power System," *Environmental Law* 11

(1981): 211–300; Jack L. Landau, "Empty Victories: Indian Treaty Fishing Rights in the Pacific Northwest," *Environmental Law* 10 (1980): 413–56.

2. See, e.g., Frank E. Smith, *The Politics of Conservation* (New York: Pantheon, 1966), chs. 1–5.

3. Beatrice H. Holmes, *A History of Water Resources Programs, 1800–1960* (Washington, D.C.: U.S. Department of Agriculture Economic Research Service, U.S. Department of Agriculture Miscellaneous Publication no. 1233, 1972), p. 3.

4. *Gibbons v. Ogden*, 22 U.S. Reports (9 Wheaton) 1 (U.S. Supreme Court, 1824).

5. See, e.g., Richard Hofstadter, *The Age of Reform: From Bryan to F.D.R.* (New York: Vintage Books, 1955), pp. 38–40.

6. Samuel P. Hays, *Conservation and the Gospel of Efficiency: The Progressive Conservation Movement, 1890–1920* (Cambridge: Harvard University Press, 1959), p. 92.

7. Gus Norwood, *Columbia River Power for the People: A History of the Policies of the Bonneville Power Administration* (Portland: U.S. Department of Energy, 1981), pp. 271–72 (hereinafter cited as *BPA History*).

8. The Progressives' emphasis on water development is reflected in their definition of "conservation": construction of reservoirs to conserve water for use during dry seasons. Hays, *Gospel of Efficiency*, p. 5.

9. Hays, *Gospel of Efficiency*, pp. 265–71 (describing the reliance on efficiency, organization, and scientific technicians to achieve agrarian ideals as seeking Jeffersonian ends through Hamiltonian means).

10. See J. Leonard Bates, "Fulfilling American Democracy: The Conservation Movement, 1907–1921," *Mississippi Valley History Review* 44 (1957): 29–57, 30–31.

11. The best publicized Progressive conservationist, Gifford Pinchot, stated that "every river is a unit from its source to its mouth." Quoted in Bates, "Fulfilling American Democracy," p. 31.

12. Holmes, *History of Water Resource Programs*, p. 6.

13. Pub. L. no. 66–280, 41 Stat. 1063 (1920) (codified at 16 USC secs. 791–823 [1996]).

14. Holmes, *History of Water Resource Programs*, p. 10.

15. Boulder Canyon Project Act, Pub. L. no. 70–642, 45 Stat. 1057 (1928) (codified as amended at 43 USC sec. 617–617t [1996]).

16. Rivers and Harbors Act, Pub. L. no. 69–560, 44 Stat. 1010 (1927).

17. The idea of a high dam at Grand Coulee to irrigate the Columbia basin's "Inland Empire" had been seriously discussed since the end of the war. A plan to irrigate the eastern Washington and Oregon deserts was endorsed by the 1920 Democratic platform. As in the case of the Boulder Canyon Act, however, private

power interests opposed construction of a project that was to be financed out of federal power sales. Instead, they backed construction of an irrigation canal between the Pend Oreille River and the desert. In the mid-1920s, the Bureau of Reclamation felt the project was too expensive and complex to pursue. But when the corps's "308 Report" on the Columbia River included the Coulee Dam among the nine projects it recommended for construction, the bureau had a change of heart. By 1930, both it and the corps advocated the project. When the corps decided to emphasize development of the lower river in 1932, the project became the bureau's. President Hoover (who lent his support to the project as secretary of commerce in 1926) withheld approval of the dam, however, favoring the Pend Oreille canal. Donald C. Swain, *Federal Conservation Policy, 1921–1933* (Berkeley: University of California Press, 1963), pp. 91–93.

18. The Rivers and Harbors Act of 1927 authorized the corps to pursue surveys of river basins that were identified in a joint Corps–Federal Power Commission study authorized two years before and printed in House Document 308. The general investigatory nature of these surveys (referred to as "308 Reports") equipped the corps with the discretion to initiate project planning similar to that enjoyed by the Reclamation Service (now Bureau of Reclamation) since 1902. Although it took more than 20 years for the corps to complete these "308 Reports," they provided the basis of most of the water project development of the New Deal and postwar eras. This is particularly true in the Columbia basin. Blumm, "Hydropower v. Salmon," pp. 225–43.

19. Both projects were funded under the National Industrial Recovery Act of 1933 and later required reauthorization after the Supreme Court's decision in *United States v. Arizona*, 295 U.S. Reports 174, 186–92 (U.S. Supreme Court, 1935).

20. Cost-of-service pricing and few federal lines would maximize the locational advantage of private utilities in Oregon. By reducing their rates, these utilities hoped to keep a lid on the proliferating public power movement. Norwood, *BPA History*, pp. 79–80.

21. Norwood, *BPA History*, pp. 80, 85–86. Two mills meant a rate of $17.50 per kilowatt-year (kwy). A slightly lower rate of $14.50 per kwy was authorized for customers within 15 miles of the dam who provided their own transmission.

22. See Phillip J. Funigiello, *Toward a National Power Policy: The New Deal and the Electric Utility Industry, 1931–1941* (Pittsburgh: University of Pittsburgh Press, 1973), p. 229n30.

23. Norwood, *BPA History*, p. 123.

24. See Bonneville Power Administration, *The Role of the Bonneville Power Administration in the Pacific Northwest Power Supply System, Including Its Participation in a Hydro-Thermal Power Program* I-10 (Final Environmental Impact Statement, December 1980).

25. Norwood, *BPA History*, p. 134. Industrial power sales were seen as means to finance Northwest project development in a series of studies from 1932 to 1937. Paying for projects with industrial sales was, of course, an old Progressive notion.

26. Quoted in Norwood, *BPA History*, p. 189.

27. Nonfederal capacity, which was 2 million kilowatts in 1953, grew by late 1960 to 7.5 million kilowatts completed or under construction. Crauford D. Goodwin, William J. Barber, James L. Cochrane, Neil D. Marchi, and Joseph A. Yager, *Energy Policy in Perspective* (Washington, D.C.: Brookings Institute, 1981), p. 273.

28. Federal capacity, including projects under construction, increased from 2.6 to 8 million kilowatts during the Eisenhower years. *Energy Policy in Perspective*, p. 273. In the Pacific Northwest, federal capacity increased from 2.46 million kilowatts in June 1952 to 6 million kilowatts in June 1960; during this period, the BPA transmission grid expanded from approximately 5,000 to 8,000 miles. Norwood, *BPA History*, p. 200.

29. Wheeling—which is usually accomplished under long-term contracts—was authorized by the Federal Columbia River Transmission Act of 1974, 16 USC sec. 838g–h.

30. The Columbia's record high flows are 34 times higher than its record low flows. Norwood, *BPA History*, p. 180.

31. Ralph Johnson, "The Canada–United States Controversy over the Columbia River," *Washington Law Review* 41 (1966): 676–763, 711–12. For Canadian perspectives, see Neil A. Swainson, *Conflict over the Columbia: The Canadian Background to an Historic Treaty* (Montreal: McGill-Queens University Press, 1979), and Maxwell Cohen, "Some Legal and Policy Aspects of the Columbia River Dispute," *Canadian Bar Review* 36 (1958): 25–41.

32. A dispute between the British Columbia provincial government and the Canadian federal government over who was to pay for the storage projects and where the power produced from them would be marketed prevented ratification of the treaty until 1964.

33. Pub. L. no. 88–552, 78 Stat. 756 (1964) (codified at 16 USC secs. 837–837h (1996)).

34. Among the reasons for increased per capita consumption were (1) increased per capita income, resulting in purchase of additional electric consumptive appliances, (2) development and marketing of such appliances, (3) greater use of electricity for space heating, and (4) increased commercial and industrial consumption to improve productivity.

35. Kai N. Lee, Donna L. Klemka, and Marion E. Marts, *Electric Power and the Future of the Pacific Northwest* (Seattle: University of Washington Press, 1980) (citing H. R. Doc. no. 403, 87th Cong., 2d Sess. [1958]), pp. 65–66; Atomic Energy

Appropriations Act of 1962, Pub. L. no. 87–701, 76 Stat 599; Norwood, *BPA History*, pp. 224–25.

36. Kai Lee pointed out that melded rates produce inequities such as making utilities that are not experiencing significant load growth pay for part of the costs of meeting the incremental loads of those that are. Lee, Klemka, and Marts, *Electric Power and the Future*, pp. 77, 81–82.

37. Henry M. Jackson, "The Pacific Northwest Electric Power Planning and Conservation Act: Solution for a Regional Dilemma," *University of Puget Sound Law Review* 4 (1980): 7–25, 12.

38. Pub. L. no. 93–454, 88 Stat. 1376 (1974) (codified at 16 USC secs. 837–838k [1996]). Self-financing also reduced prospects for congressional review of the Phase 2 program.

39. *Port of Astoria v. Hodel*, 8 Environment Reporter Cases 1156 (U.S. District Court for Oregon, 1975), *aff'd*, 595 Federal Reporter 2d 467 (U.S. 9th Circuit Court of Appeals, 1979).

40. *Natural Resources Defense Council v. Hodel*, 435 Federal Supplement 590 (U.S. District Court for Oregon, 1977), *aff'd sub nom. Natural Resources Defense Council v. Munro*, 626 Federal Reporter 2d 134 (U.S. 9th Circuit Court of Appeals, 1980).

41. 42 USC sec. 4331 (1996) (requiring major federal actions significantly affecting the quality of the human environment to be preceded by an environmental impact statement).

42. For example, Portland General Electric's Trojan Nuclear Power Plant, completed in 1975, resulted in a quadrupling of rates in Portland within three years. Lee, Klemka, and Marts, *Electric Power and the Future*, p. 140.

43. Pub. L. no. 96-501, 94 Stat. 2697 (codified at 16 USC secs. 839–839h [1996]). See Michael C. Blumm, "Columbia River Basin," in *Waters and Water Rights*, ed. Robert E. Beck (Charlottesville, VA: Michie Co., 1994), vol. 6, pp. 57–135.

44. 16 USC sec. 839c(c).

45. *Ibid.*, secs. 839e(c) (DSI rates), 839c(g)(5) (new DSI contracts).

46. *Ibid.*, secs. 839c(b), 839e(b)(2).

47. *Ibid.*, sec. 839d(c) (prescribing hearings prior to BPA's acquisition of large new resources and requiring the council's approval, subject to override by Congress).

48. *Ibid.*, sec. 839b(e).

49. *Ibid.*, sec. 839b(h).

50. *Ibid.*, secs. 839g(a), 839g(h).

51. *Ibid.*, secs. 839c (public preference), 839c(b) (widespread use), 839e(a)(1) (sound business principles), 839e(e) (lowest possible rates).

52. *Ibid.*, secs. 839(1)(A), 839b(e)(1).

53. See Northwest Conservation Act Coalition, A *Model Electric Power and Conservation Plan for the Pacific Northwest* (Summary) (May 5, 1982).

54. 16 USC sec. 839(3).

55. *Ibid.*, sec. 839f(e) (authorizing judicial review of final agency actions).

56. *Ibid.*, secs. 839b(e)(1), (2), 839a(4) (cost-effective resource acquisitions to include quantifiable environmental costs and benefits and consideration of non-quantifiable costs and benefits).

57. *Ibid.*, sec. 839b(h).

58. *Ibid.*, sec. 839b(h)(10)(A).

59. See Blumm, "Hydropower v. Salmon," pp. 228–29 nn. 74, 76 (describing a 1937 report of the commissioner of fisheries and the 1934 Fish and Wildlife Coordination Act, 16 USC secs. 661–666c [1976]).

60. See *ibid.*, p. 244.

61. See "A Reexamination of Columbia Basin Fish and Wildlife Program Issues," *Anadromous Fish Law Memo* 19 (September 1982): 3.

62. For example, in 1995 only 1,116 wild adult spring-summer chinook salmon reached Lower Granite Dam (the last dam salmon must surmount on the Snake River), an all-time low. Estimates of juvenile Snake River spring-summer chinook in 1996 were also at an all-time low. See Michael C. Blumm, Michael A. Schoessler, and R. Christopher Beckwith, "Beyond the Parity Promise: Struggling to Save Salmon in the Mid-1990s," *Environmental Law* 27 (1997): 21–126, 28n33.

63. H.R. Rep. no. 976, 96th Cong. 2d Sess. 49, 56 (1980).

64. 16 USC sec. 839b(h)(6).

65. See Michael C. Blumm, "Implementing the Parity Promise: An Evaluation of the Columbia Basin Fish and Wildlife Program," *Environmental Law* 14 (1984): 277–358; Blumm, "Columbia River Basin," pp. 99–110.

66. Michael C. Blumm and Andy Simrin, "The Unraveling of the Parity Promise: Hydropower, Salmon, and Endangered Species in the Columbia Basin," *Environmental Law* 21 (1991): 657–744, 688.

67. 16 USC sec. 839b(6)(E)(ii) (requiring the council's program to "provide flows of sufficient quality and quantity to improve production, migration, and survival of [anadromous] fish as necessary to meet sound biological objectives").

68. See Joseph Cone and Sandy Ridlington, eds., *The Northwest's Salmon Crisis: A Documentary History* (Corvallis: Oregon State University Press, 1996), pp. 262–64. In 1987, the council set as an interim goal the doubling of the basin's salmon runs, from 2.5 million adult fish annually to 5 million. See Blumm and Simrin, "Unraveling of the Parity Promise," p. 686.

69. 16 USC sec. 839b(e).

70. See James O. Luce, "When the Walls Come Tumbling Down: The Demise of

the Northwest Power Act," *Hastings West-Northwest Journal of Environmental Law and Policy* 3 (1996): 299–326, 309 (reporting that between 1980 and 1996 BPA acquired only 1,385 average megawatts, 580 of which were conservation measures).

71. *Federal Register* 56 (November 20, 1991): 58,619 (Snake River sockeye listed as endangered); *Federal Register* 57 (April 22, 1992): 14,653 (Snake River chinook listed as threatened); *Federal Register* 59 (August 18, 1994): 42,529 (Snake River chinook downgraded to endangered).

72. See Blumm, Schoessler, and Beckwith, "Beyond the Parity Promise," 39–40.

73. *Idaho Department of Fish and Game v. National Marine Fisheries Service*, 850 Federal Supplement 886, 900 (U.S. District Court for Oregon, 1994); *Northwest Resource Information Center v. Northwest Power Planning Council*, 35 Federal Reporter 3d 1371, 1395 (U.S. 9th Circuit Court of Appeals, 1994).

74. See Blumm, Schoessler, and Beckwith, "Beyond the Parity Promise," pp. 49–83 (comparing the restoration programs in detail).

75. *American Rivers v. National Marine Fisheries Service*, Civ. no. 96-384-MA (U.S. District Court for Oregon, Apr. 3, 1997).

76. Independent Scientific Group, *Return to the River: Restoration of Salmonid Fishes in the Columbia River Ecosystem* (prepublication copy, 1996), discussed in Blumm, Schoessler, and Beckwith, "Beyond the Parity Promise," pp. 112–17. On the dam breeching proposal, see Michael C. Blumm, Laird J. Lucas, Don B. Miller, Daniel J. Rohlf, and Glen H. Spain, "Saving Snake River Water and Salmon Simultaneously: The Biological, Economic, and Legal Case for Breaching the Lower Snake River Dams, Lowering John Day Reservoir, and Restoring Natural River Flows," *Environmental Law* 28 (forthcoming 1999).

77. Even including the alleged "costs" of operating the hydroelectric system somewhat more sensitively to facilitate salmon migration (which is a product more of political assumptions than of accounting), BPA fish and wildlife costs were roughly $345 million in 1995, while the carrying costs of the agency's failed nuclear power program were $485 million (a reduction from previous years of $600 million, due to bond refinancing). See Blumm, Schoessler, and Beckwith, "Beyond the Parity Promise," p. 103, n. 592.

78. *Ibid.*," p. 103.

79. See *ibid.*, pp. 104–6.

80. But see 16 USC sec. 839b(h)(6)(C) (authorizing pursuit of "minimum cost alternatives" only where they achieve "the same sound biological alternative").

14 / Fish First!

The Changing Ethics of Ecosystem Management

CAROLYN MERCHANT

"Fish first!"—the slogan has many possible meanings. Is it the most important thing for the individual fisher, for example, to take fish first above every other consideration? Or should fish be caught first for the good of society and only secondarily for the good of the individual? Should the fish themselves come first before all human considerations? Do humans or fish or both have rights? Under what circumstances do fish win by being at the table rather than on the table? Each alternative reading entails a particular approach to management, and each form of management entails an underlying environmental ethic. These approaches may be illustrated by the history of changing policies, ethics, and management strategies for fisheries in the Pacific Northwest from the nineteenth century to the present. By identifying the ethical approaches underlying earlier policies, we can formulate grounds for new ethics to guide future policy and management choices.

The first Euro-American fishery in the Pacific Northwest began in 1823; it was based on the trading and marketing of chinook salmon. Until the 1880s the fishery was marked by the progress of the laissez-faire market economy.[1] Laissez-faire capitalism was rooted in the "egocentric ethic," which pertains to individual fishers' or fishing companies' taking fish from the rivers and seas (see fig. 14.1). An unregulated fishing economy, managed by individual and corporate fishers and based on the freedom of the seas, developed as the West Coast was settled in the nineteenth century. Individual humans had rights of ownership over individual stocks of fish.

FIG. 14.1
Egocentric Ethics: Self

Maximization of Individual Self-Interest:
What Is Good for the Individual Is Good for Society as a Whole
Mutual Coercion Mutually Agreed Upon

Philosophers of Self-Interest	*Religious Traditions*
• Thomas Hobbes	• Judeo-Christian Ethic
• John Locke	• Arminian "Heresy"
• Adam Smith	
• Garrett Hardin	

The basic ethical, economic, and policy assumption behind the egocentric ethic was that what is good for the individual is good for society as a whole.[2]

A second assumption behind the egocentric ethic (and the industrial development and management that reflected it) was that the fisheries were inexhaustible. If one particular fishery lost its productivity and profits declined, fishers could move on to another fishing ground, leaving the first to recover.[3]

A third assumption of the laissez-faire economic approach and its underlying egocentric ethic was that fish were passive objects. Fish were treated like the gold nuggets that had been discovered in California, serving as the coin of trade.[4] As passive resource objects that could be extracted from the state of nature, they could be turned into profit. They were not living beings possessing individual spirits; they were entities of lesser value.

The policy of taking fish from the commons—that is, from the state of nature treated as a commons for everybody, as a free-for-all—has been characterized by environmental historian Arthur McEvoy as the "fisherman's problem."[5] This is the idea of the "tragedy of the commons," popularized by ecologist Garrett Hardin: fishing by individuals for profit degrades the environment.[6] When fishing is done competitively to produce marketable surpluses, there are powerful incentives to overfish, especially under common property regimes. When resources are owned in common but used competitively, each individual fisher gains by taking an additional fish while the degradation of the commons is shared by all. Hardin's characterization of the "tragedy of the commons" led him to propose extremely tight, coercive regulation—"mutual coercion, mutually agreed upon"—as a solution. His solution, based on the assumption that human beings are an economically maximizing species, ignored the cooperative actions of subsistence-oriented peoples both in medieval Europe and in native and colonial America.[7] The indigenous peoples of the Columbia Plateau, for example, took salmon from the Columbia for at least 10,000 years without tragic results.

A fourth assumption of the laissez-faire approach to fisheries management was that the fish themselves, once extracted from the commons, were private property—a bundle of human rights and privileges in relation to a thing. These ideas were given their classical shape by the seventeenth-century political philosophers Thomas Hobbes and John Locke. For Locke, private property was created by mixing one's labor with the soil.[8] Similarly, in the very act of mixing your labor as a fisher with the seas to extract a fish, you create ownership of the fish that are caught. Human property rights take precedence over the rights of fish to continue to exist.

Barbara Leibhardt Wester has proposed an interesting comparison between Western culture's notion of private property as a bundle of human rights and privileges and the Yakama Indian Nation's understanding of a sacred bundle of relationships and obligations between humans and other organisms, such as fish.[9] The Western idea of property stems from the Roman notion of fasces, bundles of sticks carried by Roman lictors as symbols of authority, power, and justice. This notion was exemplified most blatantly in modern times by the use of a bundle of sticks as emblem of the Italian fascist regime of Mussolini.[10] By contrast, the Yakama believed that there were sacred bundles of magical objects given to individuals by guardian spirits. These bundles were defined not as rights and privileges as in the Western system but as relationships and obligations to other human beings, to the tribe, to nature, and to the spirit world. The difference between the Yakama's bundle of relationships and capitalism's bundle of egocentric rights is the basis of the tragedy of the commons. Thus, under laissez-faire capitalism a very different ethic replaced the Native American belief system for managing the commons in the Pacific Northwest.

Nineteenth-century efforts to extract fish from the oceans and rivers and to export them as marketable commodities under the laissez-faire system led to a collapse of the fisheries on the West Coast. In the 1850s, the first gill nets were used on the Columbia River below Portland. Purse seines, traps, and squaw nets were added during the decades of the 1850s and 1860s. In 1879, fish wheels were introduced on the Columbia River; these were like Ferris wheels with movable buckets, attached either to a scow or to rock outcrops along the edge of the river. They operated day and night, scooping fish out of the river and dumping them into large bins on the shore to be packed and salted. By 1899, there were 76 fish wheels on both sides of the river. In 1866, the canning industry began operating on the banks of the Columbia near

Eagle Cliff, Washington; by 1883, there were 39 canneries shipping to New York, St. Louis, Chicago, and New Orleans.[11]

What were the consequences of unregulated fishing? In 1894, the Oregon Fish and Game Protector observed, "It does not require a study of statistics to convince one that the salmon industry has suffered a great decline during the past decades, and that it is only a matter of a few years under present conditions when the chinook of the Columbia will be as scarce as the beaver that was once so plentiful on our streams."[12] In 1917, John H. Cobb of the U.S. Bureau of Fisheries pronounced, "Man is undoubtedly the greatest present menace to the perpetuation of the great salmon fisheries of the Pacific Coast. When the enormous number of fisherman engaged, and the immense quantity of gear employed is considered, one sometimes wonders how any of the fish, in certain streams at least, escape."[13]

The solution of "mutual coercion, mutually agreed upon" (Garrett Hardin's approach) to this overfishing would have required extreme policing and strict laws leveled on the fisheries. The idea of a police state was certainly incompatible with the then-current notion of laissez-faire and with the idea of freedom of the seas. How, then, was the problem of the decline of the fisheries to be resolved? It was approached by the passage of laws and regulations that would help to manage the fisheries and the fluctuating fish populations.

The new approach exemplified a second environmental ethic, the utilitarian or homocentric ethic: human society first and fish second (see fig. 14.2). This ethic arose in the United States in response to problems of resource management. Responding to the fact that forests, along with fish, wild animals, and birds, were in decline during the nineteenth century, a movement developed to manage and conserve these resources for human needs. The homocentric approach stemmed from the utilitarian ethic of nineteenth-century philosophers Jeremy Bentham and John Stuart Mill.[14] It was concerned with the questions, What is the social good, rather than the individual good? What is the public interest, rather than the private interest of the individual or corporation? The utilitarian approach to conservation ethics, as modified by Gifford Pinchot and W J McGee in the early twentieth century, was based on the concept of "the greatest good for the greatest number for the longest time" and on the idea of duty to the whole human community.[15] But like the egocentric ethic, it gave precedence to the rights of the human species over those of nonhuman species. As applied to fisheries, homocentric ethics underlay the regulation of the laissez-faire market.

In the United States, the concept of legal limitation was set out by the Supreme Court, which in 1876 decreed that those businesses "affected with a

FIG. 14.2
Homocentric Ethics: Society

The Greatest Good for the Greatest Number for the Longest Time
Social Justice
Duty to the Human Community

Utilitarian	*Religious*
• John Stewart Mill	• John Ray
• Jeremy Bentham	• William Derham
• Gifford Pinchot	• René Dubos
• Peter Singer	• Robin Attfield
• Barry Commoner	
• Murray Bookchin	

public interest" could be regulated.[16] Regulation entailed the utilitarian idea of cost-benefit analysis—that is, one must weigh both the benefits and the costs resulting from competing interests. For example, in California, the conflict between mining interests and farming interests, two groups that each had a stake in the quantity and quality of the water flowing out of the Sierra Nevada, was resolved by balancing the costs and benefits of the two groups. Natural resources such as fish were simply ignored; they were treated as externalities.

In the 1870s, California made fish and game state property to be regulated for the public good.[17] The State Board of Fish Commissioners was created "to provide for the restoration and preservation of fish in the waters of this state."[18] The U.S. government participated in helping to manage and regulate fisheries through the creation of the U.S. Fish Commission. Its first director, Spencer Fullerton Baird, promoted research and development along the Pacific coast to determine the varieties of fish distributed in coastal waters and to map the places where they occurred in greatest abundance.[19] The premise was that if one knew the numbers associated with particular species in a fishery, then one could manage the fishery according to the idea of "maximum sustainable yield." These ideas were based upon the *logistic curve*, defined by Pierre Francois Verhulst in 1849. The curve was believed to reveal the carrying capacity, the maximum number of individuals that could be sustained without damage to the environment. The *fluctuation point* represented the level of maximum sustainable yield, basically one-half of the number of individuals at carrying capacity.[20] Fishers were to take only as many fish as the fish themselves reproduced in a given season.[21]

The idea of maximum sustainable yield embodied the homocentric ethic and was in accord with the rational balancing of costs and benefits that characterized the conservation movement. During the late nineteenth and early twentieth centuries, fisheries employed the idea of maximum sustainable yield, but fish stocks continued to decline. Regulations were instituted in Oregon and Washington to control the technologies used. In 1877, for example, Washington closed the fisheries in March and April and again in August and September to give the fish a chance to reproduce. Oregon followed suit in 1878. The states also regulated the kind of gear that could be used. The mesh sizes of the nets were specified, and their use was limited to only a third of the width of the river. In 1917, purse seines were prohibited, and in 1948, size regulations were instituted, limiting catchable fish to those above 26 inches in length.[22]

A bigger threat to the fisheries, however, began in the 1930s with construction of the first large dams along the Columbia River and its tributaries. Dams for hydropower and flood control are exemplars of the homocentric ethic with its utilitarian ideal and dedication to the public good. But this public good did not coincide with the good of fish. Fish ladders and elevators had only limited effects in sustaining fish migrations and failed to address the downstream migration.[23] The chief engineer of the Bonneville Dam initially proclaimed, "We do not intend to play nursemaid to the fish."[24] In 1937, George Red Hawk of the Cayuse Indians observed, "White man's dams mean no more salmon."[25] By 1940, the catch of Coho salmon amounted to only one-tenth of that taken in 1890, and Willis Rich of the Oregon Fish Commission noted that "the decline is well below the level that would provide the maximum sustained yield."[26] By 1948, the Army Corps of Engineers reported that although more than 300 dams had been built in the Columbia Basin, "only in a few instances has any thought been paid to the effect these developments might have had on the fish and wildlife."[27] The utilitarian or homocentric ethic was ineffective: the concept of "the greatest good for the greatest number for the longest time" still meant human society first and fish second.

By the 1950s, the homocentric ethic, although still powerful, began to give way to a third approach—the ecocentric ethic (see fig. 14.3). Aldo Leopold first formulated it as the "land ethic" in 1949.[28] The ecocentric ethic is based on the idea that fish are equal to other organisms—including humans—and therefore deserve moral consideration. As Leopold put it: "A thing is right when it tends to preserve the integrity, beauty, and stability of the biotic community. It is wrong when it tends otherwise."[29] We could expand his idea of the land ethic and call it a "land and water ethic." As such, "it enlarges the boundaries of the community to include soils, waters,

FIG. 14.3
Ecocentric Ethics: Cosmos

Rational, Scientific Belief-System Based on Laws of Ecology
Unity, Stability, Diversity, Harmony of Ecosystem
Balance of Nature

Eco-Scientific	*Eco-Religious*
• Aldo Leopold	• American Indian
• Rachel Carson	• Buddhism
• Deep Ecologists	• Spiritual Feminists
• Restoration Ecologists	• Spiritual Greens
• Biological Control	• Process Philosophers
• Sustainable Agriculture	

plants, and animals [including fish] or collectively: the land."[30] It changes the role of *Homo sapiens*, Leopold said, "from conqueror of the land community to plain member and citizen of it."[31] There is an intrinsic value to all living and nonliving things; all have a right to survive. Fish, as well as humans, have rights and can even have standing in a court of law.

The idea that began to emerge in the 1950s and 1960s was that fish themselves had a right to survive and that one should cooperate with each stock's own strategy for survival. The interaction between harvesting, environmental change, and cooperation with the species' own survival strategy reflected the new ecocentric approach to management. The conclusion that arose from such ecological considerations was that "the benefit to the nation occurs by leaving the fish in the ocean."[32] This was a policy of fish first and people second, or fish for the sake of the fish.

The idea of "optimum sustainable yield" (a modification of maximum sustainable yield) was developed in conjunction with the ecocentric approach to management. The optimum level of harvest was the level that could be obtained indefinitely without affecting the capacity of the population or the ecosystem to sustain the yield. The optimum yield was the maximum sustainable yield as modified by any relevant economic, social, or ecological factor.[33] Endangered species must be taken into consideration, and there must be limited entry to the fisheries. The idea of freedom of the seas was challenged. Both the Marine Mammal Protection Act of 1972 and the Fisheries Conservation and Management Act of 1976 were based on the idea of maintaining the health and stability of marine ecosystems with the goal of obtaining an optimum sustainable population.[34]

What problems arise from this ecocentric approach? One is that the idea of optimum sustainable yield retains certain assumptions. It is based on the idea, current in the 1960s and 1970s, that ecology reflects the balance of nature.[35] It retains the assumptions that the fish population will follow the classical logistic curve, that there is a fixed carrying capacity, that there is an absolute maximum sustainable level, and that nature left undisturbed is constant and stable. These are the classical assumptions of the concept of the balance of nature, which was the motivating inspiration behind the ecocentric ethic and the environmental movement of the 1970s.[36]

But the notion of the balance of nature has recently been questioned by ecologists (particularly population ecologists) and by proponents of the ideas of chaos theory and complexity theory.[37] Chaos theory questions the idea of the constancy and stability of nature, the idea that every organism has a place in the harmonious workings of nature, the idea that nature itself is fixed in time and space—like the environment in a petri dish in a modern scientific laboratory—and the idea that the logistic curve is a permanent and final explanation.

The ecologist Daniel Botkin has proposed the concept of discordant harmonies as an alternative to the idea of the balance of nature. Botkin says that we must move to a deeper level of thought and "confront the very assumptions that have dominated perceptions of nature for a very long time. This will allow us to find the true idea of a harmony of nature, which, as Plotinus wrote so long ago, is by its very essence discordant, created from the simultaneous movements of many tones, the combination of many processes flowing at the same time along various scales, leading not to a simple melody, but to a symphony sometimes harsh and sometimes pleasing."[38]

The idea of discordant harmonies, the notion of the complex and chaotic behavior of nature, and the conclusion that natural disturbances are equally and in many cases more important than human disturbances have led to a wholesale questioning of earlier approaches—not only the egocentric and homocentric but even the ecocentric approach—to environmental ethics and ecosystem management.[39]

The three dominant forms of environmental ethics, therefore, all have conceptual shortcomings. Egocentric ethics—reflecting narcissistic, cutthroat individualism and competition among individual and corporate fishers—privileges not only people over fish but also, among humans, the few at the expense of the many. Homocentric ethics privileges human majorities at the expense of minorities, to the extent of encompassing environmental racism

FIG. 14.4

Partnership Ethics: People and Nature

The Greatest Good for the Human and Nonhuman Communities
Is in Their Mutual Living Interdependence

- Equity between the human and nonhuman communities
- Moral consideration for both humans and other species
- Respect for cultural diversity and biodiversity
- Inclusion of women, minorities, and nonhuman nature in the code of accountability
- Ecologically sound management is consistent with the continued health of both the human and nonhuman communities

and the subjugation of Indian fishers. Ecocentric ethics privileges the whole at the expense of the individual; it has even been called a sort of holistic fascism that gives precedence to fish over people.[40] Moreover, egocentric and homocentric ethics are often lumped together (by deep ecologists, for example) as anthropocentrism. But the anthropocentric/nonanthropocentric approach masks the role of economics, particularly the role of capitalism, placing the onus on human hubris and domination rather than on the capitalist appropriation of both nature and labor. Similarly, this approach fails to recognize the positive aspects of the social justice approach of homocentric ethics. The ecocentric approach of many environmentalists, however, suggests the possibility of incorporating the intrinsic value of nature into an emancipatory green politics.[41]

In attempting to resolve some of these dilemmas, I propose that we consider a new kind of ethic, which I call a partnership ethic—a synthesis between the ecocentric approach and the social justice aspects of the homocentric approach (fig. 14.4).[42] A partnership ethic holds that *the greatest good for the human and nonhuman communities is to be found in their mutual living interdependence*. It is based on the idea that people and nature are equally important. Both people and fish have moral standing.

For most of human history, up to at least the seventeenth century, nature had the upper hand over human beings. Humans fatalistically accepted the cards that nature dealt. Harvests, famines, and droughts were considered God's way of punishing humans for acting unethically. Since the seventeenth century, however, the pendulum has swung the other way. Western culture has developed the idea that humans are more powerful than nature and that as European Americans we can dominate, control, and manage it.[43]

Because humans are above nature, we can control fisheries, for example, through ideas such as logistic curves and maximum or optimum sustained yields. We need to bring the pendulum back into balance so that there is greater equality between the human and nonhuman communities.

The partnership ethic that I propose is a synthesis of the ecocentric approach, based on moral consideration for all living and nonliving things, and the homocentric approach, based on the social good and the fulfillment of basic human needs. All humans have needs for food, clothing, shelter, and energy, but nature also has an equal need to survive. The partnership ethic recognizes the interdependence of these two spheres. It questions the notion of an unregulated market, eliminating the idea of the egocentric ethic, and instead proposes a partnership between nonhuman nature and the human community.

A partnership ethic is based on the following precepts: first, equity between human and nonhuman communities; second, moral consideration for both humans and other species; third, respect for both cultural diversity and biodiversity; fourth, inclusion of women, minorities, and nonhuman nature in the code of ethical accountability; and fifth, the belief that an ecologically sound management is consistent with the continued health of both human and nonhuman communities.[44] We might come back to the notion that Barbara Wester proposed in her comparison of Native and European Americans—the idea of the "sacred bundle." Like the Native American sacred bundle of relationships and obligations, a partnership ethic is grounded in the notions of relation and mutual obligation.[45]

What would a partnership ethic mean for ecosystem management? How would it be implemented in the fisheries professions? Each stock of fish has a home spawning stream and an ocean habitat connected with it over many miles of river. Each stock has a season for returning to its primal ecological community to reproduce. Seasonal changes, as well as chaotic disturbances in ocean currents, temperature changes, and predation affect return rates. So do human disturbances, such as timber removal, erosion, watershed pollution, dams, and fishing quotas and regulations.

A partnership ethic means that in each linked human and nonhuman biotic community, all the parties and their representatives must sit as partners at the same table. This includes fishers (individual, corporate, and tribal), foresters, dam builders, conservation trusts, soil and fishery scientists, community representatives, and spokespersons for each stock of fish affected. The needs of fish and the needs of humans should both be discussed. Examples of efforts at such partnerships include resource advisory committees, watershed councils, self-governing democratic councils, collaborative processes, and cooperative management plans.

Consensus and negotiation should be attempted as partners speak together about the short- and long-term interests of the interlinked human and nonhuman communities. Seated at the table, participating in the discourse, are not only representatives of human concerns but also representatives of nonhuman entities. The meetings will be lengthy. As in any partnership, there will be give and take as the needs of each party are expressed, heard, and acknowledged. If the partners identify their own egocentric, homocentric, and ecocentric ethical assumptions and agree to start anew from a partnership ethic of mutual obligation and respect, there is hope for consensus. A partnership ethic does not mean that all dams must be removed, that electric production must be forfeited, or that irrigation must be curtailed for the sake of salmon and redwoods. It does mean that the vital needs of humans and the vital needs of trees and fish, along with their mutually linked terrestrial and aqueous habitats, must be given equal consideration. Indeed, there is no other choice, for failure means a regression from consensus into contention, and thence into litigation.

Many difficulties exist in implementing a partnership ethic. The greatest challenge is the free-market economy's growth-oriented ethic, which uses both natural and human resources inequitably to create profits. The power of the global capitalist system to remove resources—especially those in Third World countries—without regard to restoration, reuse, or recycling is a major roadblock to reorganizing relations between production and ecology. Even as capitalism continues to undercut the grounds of its own existence by using renewable resources, such as fish, faster than species can replenish themselves, so green capitalism attempts to Band-Aid the decline by submitting to some types of regulation and recycling. Ultimately, new economic forms need to be found that are compatible with sustainability, intergenerational equity, and a partnership ethic.

A second source of resistance to a partnership ethic is the property rights movement, which in many ways is a backlash against both environmentalism and ecocentrism. The protection of private property is integral to the growth and profit-maximization demands of capitalism and egocentrism and to their preservation by governmental institutions and laws. Although individual, community, or common ownership of "appropriate" amounts of property is not inconsistent with a partnership ethic, determining what is sustainable and hence appropriate to the continuation of human and nonhuman nature is both challenging and important.

As a start, we might propose an ethic for the American Fisheries Society, inspired by that proposed for the Society of American Foresters: partnership with the land and the aquatic habitat is the cornerstone of the fisheries profession; compliance with its canons demonstrates respect for the land and waters and for our commitment to the wise management of ecosystems.

As we move forward in the twenty-first century, we must also move into partnerships between human beings and the nonhuman community in which both are equal and share in mutual relationships and obligations. A partnership ethic will not always work—but it is a beginning, and with it there is hope.

NOTES

An earlier version of this essay was presented at the Pacific Northwest Regional Environmental History Conference, August 2, 1996, and was published in *Human Ecology Review* 4(1) (1997): 25–30. This revised and expanded version is reprinted with the permission of *Human Ecology Review*.

1. Anthony Netboy, *Salmon of the Northwest: Fish versus Dams* (Portland: Binfords and Mort, 1958).

2. Carolyn Merchant, "Environmental Ethics and Political Conflict: A View from California," *Environmental Ethics* 12 (1990): 45–68.

3. Arthur McEvoy, *The Fisherman's Problem: Ecology and Law in the California Fisheries, 1850–1980* (New York: Cambridge University Press, 1986), p. 6.

4. *Ibid.*

5. *Ibid.*

6. Garrett Hardin, "The Tragedy of the Commons," *Science* 162 (1968): 1243–48.

7. Susan Jane Buck Cox, "No Tragedy on the Commons," *Environmental Ethics* 7 (1985): 49–69; McEvoy, *The Fisherman's Problem*, pp. 11–12; Richard White, *The Organic Machine* (New York: Hill and Wang, 1995), pp. 39–40.

8. John Locke, *Second Treatise of Government*, ed. R. H. Cox (1690; reprint, Arlington Heights, Ill.: Harlan Davidson, 1982), p. 21; C. B. MacPherson, *The Political Theory of Possessive Individualism: Hobbes to Locke* (New York: Oxford University Press, 1962).

9. Barbara Leibhardt, "Law, Environment, and Social Change in the Columbia River Basin: The Yakima Indian Nation as a Case Study, 1840–1933" (Ph.d diss., University of California, Berkeley, 1990).

10. *Oxford English Dictionary; Scott's Standard Postage Stamp Catalogue* (New York: Scott Publications, 1948), p. 510.

11. Netboy, *Salmon of the Northwest;* Courtland L. Smith, *Salmon Fishers of the Columbia* (Corvallis: Oregon State University Press, 1979), p. 104.

12. Hollister D. McGuire, "Decadence of the Fisheries," *First and Second Annual Reports of the Fish and Game Protector to the Governor*, 1893–94 (Salem, Oreg.: 1894), quoted in Anthony Netboy, *The Salmon: Their Fight for Survival,* (Boston: Houghton Mifflin, 1974), p. 282.

13. Quoted in Netboy, *Salmon of the Northwest*, p. 29.

14. Merchant, "Environmental Ethics and Political Conflict."

15. Gifford Pinchot, *Breaking New Ground* (New York: Harcourt Brace, 1947), p. 326.

16. *Munn v. Illinois*, 94 United States Reports 113 (U.S. Supreme Court, 1876); McEvoy, *The Fisherman's Problem*, p. 117.

17. McEvoy, *The Fisherman's Problem*, p. 118.

18. *Ibid.*, p. 101.

19. *Ibid.*

20. Daniel Botkin, *Discordant Harmonies: A New Ecology for the Twenty-First Century* (New York: Oxford University Press, 1990), pp. 20–21.

21. McEvoy, *The Fisherman's Problem*, p. 6.

22. Netboy, *Salmon of the Northwest*, pp. 28–30.

23. *Ibid.*, pp. 44–46.

24. Quoted in Netboy, *The Salmon: Their Fight for Survival*, p. 287.

25. Quoted in Netboy, *Salmon of the Northwest*, p. 48.

26. *Ibid.*, p. 39.

27. *Ibid.*, p. 34.

28. Aldo Leopold, *A Sand County Almanac, and Sketches Here and There* (New York: Oxford University Press, 1949), pp. 201–04, 221–25.

29. *Ibid.*, p. 225.

30. *Ibid.*, p. 204.

31. *Ibid.*

32. McEvoy, *The Fisherman's Problem*, p. 227.

33. Botkin, *Discordant Harmonies*, pp. 22–23.

34. *Ibid.*, p. 160; McEvoy, *The Fisherman's Problem*, p. 237.

35. Botkin, *Discordant Harmonies*, pp. 18, 145–47.

36. *Ibid.*, pp. 21–23.

37. *Ibid.*; James Gleick, *Chaos: The Making of a New Science* (New York: Viking, 1987); Robert T. Lackey, "Pacific Salmon, Ecological Health, and Public Policy," *Ecosystem Health* 2 (1996): 1–8, 4; Mitchel Waldrop, *Complexity: The*

Emerging Science at the Edge of Order and Chaos (New York: Simon and Schuster, 1992).

38. Botkin, *Discordant Harmonies*, p. 25.

39. Robert T. Lackey, "Ecological Risk Assessment," *Fisheries* (September 1994): 14–18.

40. On the land ethic as a case of "environmental fascism," see Tim Regan, *The Case for Animal Rights* (Berkeley: University of California Press, 1983), p. 262. For a response, see J. Baird Callicott, *In Defense of the Land Ethic* (Albany: State University of New York Press, 1998), pp. 92–94, and J. Baird Callicott, "Moral Monism in Environmental Ethics Defended," *Journal of Philosophical Research* 19 (1994): 51–60, p. 53.

41. On ecocentrism as the ground for an emancipatory green politics, see Robyn Eckersley, *Environmentalism and Political Theory: Toward an Ecocentric Approach* (Albany: State University of New York Press, 1992), p. 47.

42. Carolyn Merchant, *Earthcare: Women and the Environment* (New York: Routledge, 1996), pp. 216–19.

43. *Ibid.*, p. 218.

44. *Ibid.*, p. 216–17.

45. *Ibid.*, p. 217.

PART IV / AGRICULTURE

Agriculture plays a major role in the environmental history of the Pacific Northwest. Farms, both irrigated and dry-land, make up 29 percent of the Northwest's land base and a large portion of its economy. Farmers have cleared millions of acres of forest and grassland for commercial crop production. Irrigation has turned millions more acres of desert and arid scrubland into greenbelts. Indeed, agricultural irrigation has been responsible for many of the dramatic changes to the Northwest's rivers discussed in part three. Grand Coulee Dam on the Columbia River, for example, was constructed to provide irrigation water for the Columbia Basin Project in central Washington.

Agriculture is more than crop farming. It also includes livestock ranching—primarily beef and dairy cattle, and sheep. There are 4.5 million cattle and nearly 1 million sheep in the region; some 51.7 million acres is rangeland. As the authors of the first essay in part four note, grazing is the most widespread land use in the interior Northwest.

Of the four chapters in this section, two examine irrigated agriculture and two discuss livestock grazing. Their authors are biologists, geographers, and historians. The essays emphasize ecological conditions, cultural perceptions, and historical changes in local environments. Part four thus represents the sciences as well as the humanities, providing a useful diversity of perspectives—and once again exemplifying the interdisciplinary nature of environmental history.

We begin with a chapter by a team of biologists and ecologists who detail the large-scale effects of livestock grazing on Northwest ecosystems. As Kathleen A. Dwire, Bruce A. McIntosh, and J. Boone Kauffman point out, northwesterners, including Native Americans, have grazed horses, sheep, and cattle on the grasslands, river bottoms, deserts, and mountain meadows of the region for over 300 years. Although the effects of livestock grazing thus are both ubiquitous and long-standing, few people are aware of the significance of this land use because its impacts are less apparent to the untrained observer than are the environmental changes caused by crop farming. Dwire, McIntosh, and Kauffman briefly survey the history of grazing in the Northwest before examining the environmental consequences of grazing on different types of land—forestland, grassland, arid steppeland, and riverine (riparian) land. They conclude with an assessment of the trend in ecological conditions on land damaged by grazing.

The second chapter is by two geographers, William Wyckoff and Katherine Hansen, who provide a historical case study of ranching in the Madison Valley near Yellowstone National Park. Where the first chapter provides breadth, this one provides depth: it is an example of the kind of bioregional history that Dan Flores argued was the essence of place. Beginning with a discussion of the environmental setting, Wyckoff and Hansen trace the historical settlement of the valley through four eras that reflect shifts in the economy and in environmental conditions. Access to markets in the late 1800s brought rapid development of the livestock industry. The consequent increase in cattle caused environmental changes that culminated in deterioration of range conditions. People engaged in the ranching economy attempted to resolve these problems by imposing controls on livestock—regulations and active management—as well as through efforts to better understand range resources. Some responses were successful, others were not. Wyckoff and Hansen conclude with an assessment of contemporary environmental conditions and economic trends, including the shift toward a tourism-based economy that now poses challenges for those engaged in traditional resource-based economies. Here the authors echo Ewert's earlier discussion of the "New Northwest."

The essay by Mark Fiege addresses irrigated agriculture on the arid but fertile Snake River Plain of southern Idaho. Fiege deftly combines environmental and social analysis, paying close attention to the place as well as to the people inhabiting it. His "hybrid landscape" theme is a remarkably apt characterization of the part-natural, part-artificial ecosystems that result when humans occupy a place. Fiege reminds us that the grandiose systems we design to control nature to make it produce our desired commodities are always only partly successful—nature never quite conforms to our engineering designs. In his words, Idaho's irrigated landscapes represent "a synthesis of human designs and recalcitrant natural processes." The reciprocal interactions between human institutions and nature create a hybrid landscape that reflects an ambiguous mixture of artifice and nature.

The final chapter, by Dorothy Zeisler-Vralsted, emphasizes cultural perceptions of the irrigated landscape and serves as a counterbalance to the ecological emphasis of the first chapter. Like Wyckoff and Hansen, Zeisler-Vralsted offers an insightful case study—of the Yakima Irrigation District in Kennewick, Washington, and the Bitterroot Valley Irrigation District in western Montana. Both areas are experiencing a slow but steady shift from rural-agricultural use to suburban and urban habitation. Zeisler-Vralsted focuses on how this shift affects communities, people's perceptions of the environment—particularly water resources—and the institutions that control the water. Her essay recalls William Lang's and reminds us of the importance of human perceptions in the creation of western landscapes.

15 / Ecological Influences of the Introduction of Livestock on Pacific Northwest Ecosystems

KATHLEEN A. DWIRE, BRUCE A. MCINTOSH,
AND J. BOONE KAUFFMAN

Livestock grazing is the most widespread land use throughout the interior Pacific Northwest. With the exception of inaccessible locations, most of the forests, grassland steppes, shrub-steppes, and riparian zones have been or are currently grazed by livestock. Relict areas that have not been influenced by livestock are among the rarest of ecosystems. In addition to private land, most public land is grazed, including the majority of federal land administered by the United States Bureau of Land Management and the United States Forest Service. Livestock grazing also occurs in national wildlife areas, in some national parks and wilderness areas, and on state land.[1]

Intense grazing by cattle and sheep has occurred since Euro-Americans first settled in the Pacific Northwest. Although native ungulates were present and Native Americans acquired horses in the eighteenth century, large numbers of herbivores were not present in the Pacific Northwest prehistorically. Cattle and sheep were introduced during the period of homesteading and settlement, beginning approximately 150 years ago. As the most pervasive land use, livestock grazing has had widespread and dramatic ecological impacts, including loss of native species, changes in species composition, soil deterioration, degradation of fish and wildlife habitat, and changes in ecosystem structure and function.[2] Livestock grazing is one of several land uses that have worked in concert to change the ecosystems of the Pacific Northwest. Other factors with impacts that are hard to separate from those of livestock grazing include logging, mining, suppression of natural fires, elimination of beaver, agriculture, urbanization, and grazing by native ungulates.

This essay describes the ecological changes resulting from livestock grazing that have occurred in the interior Pacific Northwest (that is, east of the Cascade Mountains) over the past 150 years. We review the history of livestock grazing, describe a simple model of the primary, secondary, and tertiary influences of livestock grazing, and discuss ecological changes that have occurred in forested, steppe, riparian, and stream environments.

GRAZING HISTORY OF THE PACIFIC NORTHWEST

The historical record indicates that large herbivores such as bison, deer, and elk were uncommon in the steppe grasslands of the Pacific Northwest. Early explorers rarely reported seeing bison in eastern Oregon and Washington, although small herds were frequently noted in the upper Snake River Plain until the mid-nineteenth century.[3] The fossil record also confirms the rarity of bison. Whereas deer and elk were uncommon in grasslands, they were probably more common in forested regions. Richard Mack and John Thompson argued that limited sources of permanent summer water and the phenology of grasses adapted to long dry summers made this region unsuitable for large numbers of herbivores.[4] The small herds that persisted utilized areas of permanent water and remained at low densities. The dependence of Native Americans in the Pacific Northwest on fish as their primary source of food, unlike the Plains Indians, who depended on large ungulates, further supports this argument. Mack and Thompson concluded that the grasslands of the Pacific Northwest did not evolve under the influences of large herbivores and that they were therefore extremely vulnerable to the effects of overgrazing by horses, cattle, and sheep.

Large, non-native herbivores were first introduced into the Pacific Northwest by Native Americans who brought Spanish horses into the region in the early 1700s.[5] By the early 1800s, Native Americans had bred large bands of horses; early Euro-American explorers of the region regularly noted large herds of horses around indigenous settlements. Although the impact of the introduction of horses is unknown, grazing by horses was an anthropogenic disturbance to native ecosystems of the Pacific Northwest. Given the relatively small populations of Native Americans and their scattered distribution in the Pacific Northwest, the impact of horses was probably localized.

The first cattle were brought to a settlement on Vancouver Island, British Columbia, by Spaniards in 1789.[6] By 1825, cattle were found in isolated settlements on Puget Sound and the lower Columbia River. In 1850, livestock were common in the scattered Euro-American settlements throughout the region. These early settlements were concentrated in the Willamette Valley and along the lower Columbia River; they expanded inland with the discovery of gold in the 1860s.[7] Settlers soon recognized the economic potential for livestock, given the seemingly abundant forage and water resources to support them. As immigration increased on the Oregon Trail in the mid-1800s, the livestock industry grew rapidly in the Pacific Northwest.

To analyze historical trends in livestock grazing, we assessed cattle and sheep data for Oregon, Washington, and Idaho from the United States Census of Agriculture.[8] Data have been collected every 10 years since 1850 on a county-by-county basis; livestock data contain the type of livestock and the number of animals. We analyzed trends in data for sheep and cattle, excluding dairy cows. Animal numbers were converted to number of "animal unit months" (AUMS)—representing the amount of forage required by one cow and calf for one month—in order to standardize the relative effects of the different livestock classes.[9] For this analysis, five sheep were considered to be equivalent to one cow and calf, that is, one AUM.

Our analysis indicated that although livestock use increased rapidly before 1900, the highest levels of livestock use in the Pacific Northwest have occurred since World War II. From 1850 to 1900, livestock numbers in the Pacific Northwest grew in response to growing populations of recent immigrants (fig. 15.1). By 1900, livestock were found throughout the Pacific Northwest. During this settlement period, the numbers of AUMS of cattle and sheep were about equal.

From 1850 to 1905, grazing on unclaimed public land was unregulated. There was intense competition among users, with little consideration for controlling the number of animals, duration of grazing, or season of use.[10] The lack of grazing management led to severe degradation of rangeland and forest ecosystems. An 1883 report noted the almost immediate effect of the large-scale introduction of non-native herbivores.[11] Rangeland throughout the interior Columbia basin was described as badly damaged by overgrazing. Early stockmen were accustomed to the productive pastures of the eastern and southern United States; they did not understand the limits of western rangeland and the vulnerability of native plants to overstocking. Although cattle ranchers fought among themselves for preferred range, most of their anger was directed at sheep ranchers.[12] Cattle ranchers believed that the millions of sheep hooves destroyed the grass and that the smell of sheep was a deterrent to cattle and horses using the range.

The apparent and widespread decline of Pacific Northwest rangeland due to overgrazing caused concern among both ranchers and government offi-

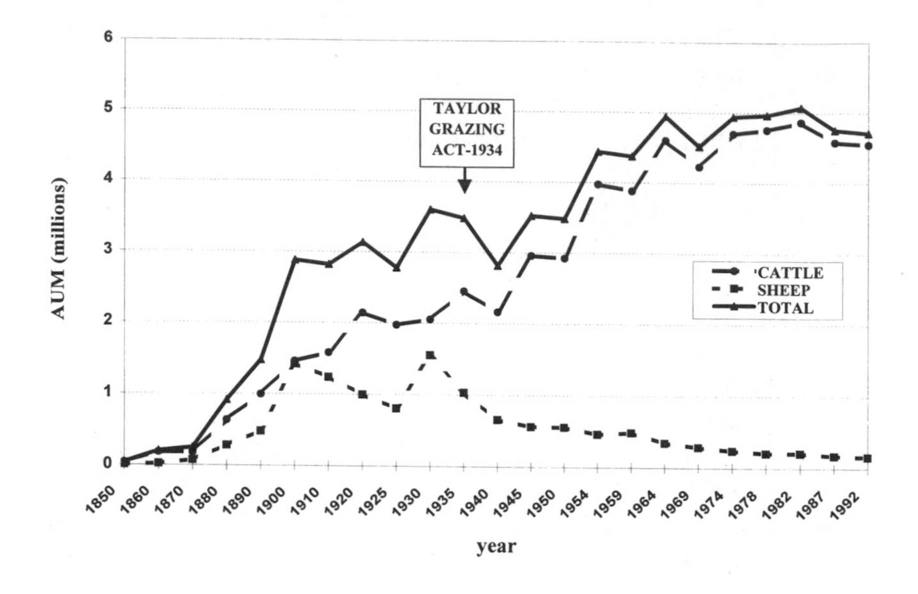

FIG. 15.1. Changes in animal unit months (AUMs), by livestock class, in the Pacific Northwest from 1850 to 1992.

cials. Despite these concerns, livestock grazing continued to increase until the turn of the century, when forest reserves (predecessors of today's national forests) were established. An 1898 National Academy of Sciences report focused national attention on overgrazing in the forest reserves.[13] With the formation of the United States Forest Service in 1905, grazing on forest reserves began to be actively managed. The number of AUMs decreased by 23 percent (2.6 million to 2.0 million) from 1900 to 1925 (fig. 15.1). The decrease was due primarily to a decline in sheep numbers while cattle numbers increased slightly. From 1925 to 1930, livestock AUMs increased by 45 percent to a historical high of 2.9 million. In 1934, passage of the Taylor Grazing Act brought the remainder of public land under supervised grazing. The intent of the act was to "stop injury to the public grazing lands by preventing overgrazing and soil deterioration, to provide for the orderly use, improvement and development [of public grazing land], and to stabilize the livestock industry dependent upon the public range."[14] The Grazing Service (which later became the Bureau of Land Management) was formed to create grazing districts and manage public rangeland.

From 1930 to 1940, the number of livestock AUMs in the Pacific Northwest decreased by approximately 30 percent. But the decline in total live-

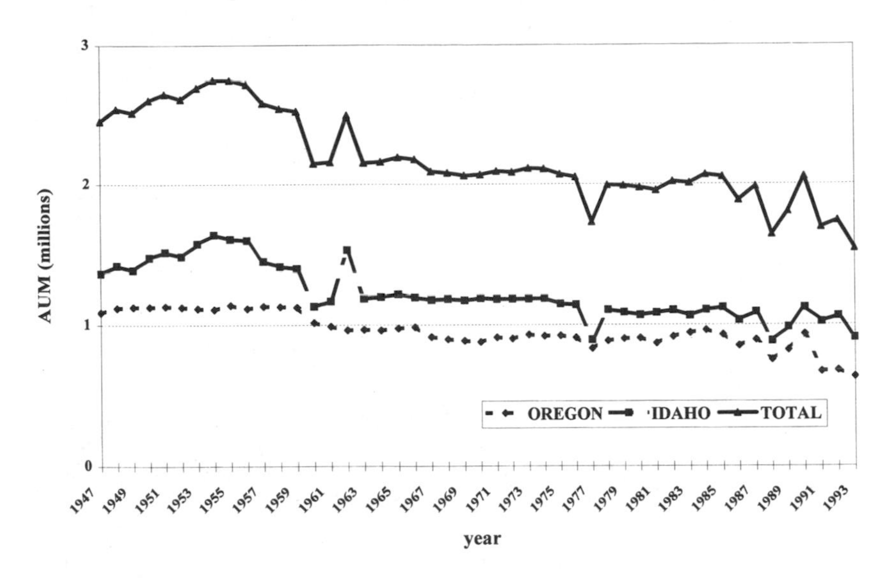

FIG. 15.2. Changes in animal unit months (AUMS) on land administered by the Bureau of Land Management (BLM) in Oregon and Idaho from 1947 to 1993. Data for Washington are not included because the BLM administers very little acreage in that state.

stock numbers was short-lived. In 1940, a significant change began in the livestock industry of the Pacific Northwest. From 1940 to 1992, the number of cattle more than doubled while the number of sheep fell to a fraction of the 1930 high; total livestock AUMS increased by 110 percent. Currently, livestock numbers in the Pacific Northwest are near the highest levels since Euro-American settlement began. Cattle surpassed sheep in 1935 and currently account for 96 percent of the AUMS in the Pacific Northwest.

Livestock grazing has been increasing and has shifted from public to private land. Data for land administered by the Bureau of Land Management show that the number of AUMS in Oregon and Idaho has steadily decreased since 1947 (fig. 15.2).[15] Limited data indicate a similar trend for land administered by the U.S. Forest Service.[16] Because overall numbers of livestock increased from 1850 to 1992, this means that more private land is being grazed.

The livestock history of the Pacific Northwest can be characterized in terms of three distinct periods. The settlement period, from 1850 to 1900, exhibited exponential growth in both cattle and sheep numbers. From 1900 to 1940, livestock grazing came under increased regulation and livestock

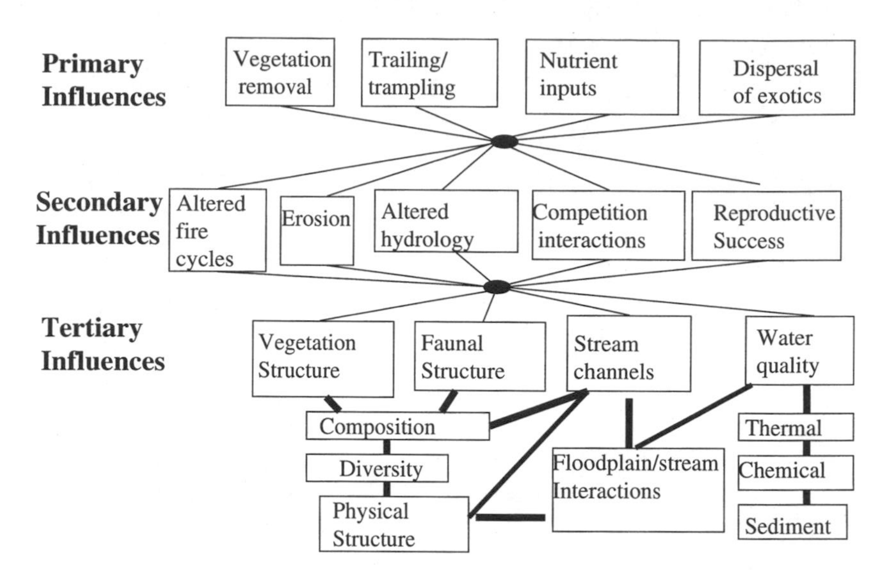

FIG. 15.3. A hierarchical model of the influences of livestock grazing on terrestrial and aquatic ecosystems. Primary influences are direct impacts; secondary influences are the ecosystem responses to the direct impacts; tertiary influences are the long-term, cumulative impacts that can ultimately occur.

numbers did not change substantially. Since 1940, livestock numbers have increased steadily, and cattle have become the dominant grazing animal in the region.

ECOLOGICAL INFLUENCES OF LIVESTOCK GRAZING

The introduction of livestock has dramatically affected Pacific Northwest ecosystems. Several reviews of grazing impacts have described the major ecological changes that have occurred in forested,[17] riparian, and meadow areas[18] and in nonforested rangelands[19] of the western United States. These reviews describe both immediate, local impacts and the long-term, cumulative effects of livestock grazing. As shown in figure 15.3, livestock directly influence ecosystems in four ways: (1) by removing vegetation through grazing; (2) by creating trails and trampling soils and vegetation; (3) by redistributing nutrients via defecation and urination; and (4) by dispersing exotic organisms, including noxious plant species and pathogens. These influences are readily measured and can have significant impacts on a particular site. The interaction of these four primary influences leads to physical and biotic responses in different ecosystems; these responses are labeled "secondary influences" in figure 15.3. The long-term cumulative impacts of the primary and secondary influences are shown as "tertiary influences" in figure 15.3.

Primary Influences

The first direct effect of livestock is simply the eating of vegetation. Grazing by livestock can decrease the quantity and quality of available forage for wild grazers such as deer, elk, bighorn sheep, geese, and rodents. Vegetation removal also decreases the protective litter layer and the quantity of organic matter that is available to be incorporated into soils. Along streams, consumption of riparian vegetation by livestock decreases the amount of organic matter that enters the stream. For many streams, riparian vegetation is the primary energy and nutrient source for instream biota, and altered inputs may affect productivity of the aquatic ecosystem.

The second direct influence of livestock is damage to soil and vegetation caused by trail-making and trampling. Damage occurs from the mechanical compression and compaction of soil and the physical breakage of vegetation. On established cattle trails, old driveways, and stock concentration areas, the surface horizons of some soils have not recovered or become revegetated.[20] Trails along and through streams have caused bank erosion and sedimentation.

The third direct impact of livestock is localized nutrient and bacterial inputs resulting from defecation and urination. These impacts can have serious consequences for water quality and aquatic organisms. Concentrated nutrient inputs can increase bacterial and protozoan pathogens, promote algal growth, and alter water chemistry—particularly concentrations of dissolved oxygen, which affect fish and invertebrate populations.

The fourth direct influence, dispersal of exotic organisms, works with the other direct effects. Plant seeds may be dispersed via transport on the fur and hooves of livestock and in fecal deposition. As animals move through an area, they serve as dispersal agents for exotic species while simultaneously preparing a seedbed through grazing and trampling. Many of these undesirable plant species are adapted for dispersal by animals and successfully establish themselves and spread in disturbed soil. Livestock can also be vectors of disease that harm wildlife. Herds of bighorn sheep have been locally extirpated because of diseases spread through contact with domestic sheep.[21]

Secondary Influences

The primary influences of livestock grazing in turn promote three types of secondary influences: changes in species composition, altered soils and hydrology, and altered fire cycles. Livestock are selective in their diet and preferentially graze certain palatable species while avoiding others. Excessive removal of plant material weakens the preferred plants and alters their physiology,[22] causing them to adjust their patterns of growth and reproduction. Livestock grazing can thus affect competitive interactions between different plant species that are using the same limited resources. Selective removal of preferred species gives a competitive advantage to unpalatable species; with continued grazing over a number of seasons, the most palatable species will decline in abundance while less palatable species increase.[23] The continued overgrazing of palatable plants by livestock can also decrease seed production, thus limiting potential for recovery. As a result, the dynamics of plant succession can be altered. In many areas, palatable native bunchgrasses, weakened by heavy grazing pressure, have been outcompeted and displaced by more aggressive, less palatable exotic species.

Numerous exotic plant species have become widely distributed since their introduction into North America. No other species has increased so dramatically as cheatgrass (*Bromus tectorum*), an annual grass that was accidentally introduced from Eurasia in the mid-1800s. By the turn of the century, cheatgrass had become widespread. It is now ubiquitous throughout the western United States at low to middle elevations and dominates much of the steppe region.[24] The enormous increase in the abundance of cheatgrass has coincided with a marked decrease in native bunchgrasses. What were the reasons for the rapid and extensive spread of cheatgrass? In a series of field and greenhouse studies, Grant Harris demonstrated that cheatgrass develops an extensive root system earlier and faster than the native bluebunch wheatgrass, giving cheatgrass a competitive growth advantage in spring.[25] As cheatgrass increased in abundance, competition for soil moisture increased, putting even greater stress on bluebunch wheatgrass. In addition, cheatgrass can produce large numbers of seeds, which germinate under favorable conditions or remain dormant in the soil if conditions for growth are unfavorable. Livestock facilitated the spread of cheatgrass by avoiding it and preferentially grazing the native bunchgrasses, thus increasing the proportion of cheatgrass every year. The dominance of cheatgrass has reduced the biodiversity of plant species in semiarid regions of the Pacific Northwest.

Another secondary influence of livestock grazing is the alteration of local soils and hydrology. When plant cover and the protective litter layer are removed by grazing, the proportion of bare ground is increased, making the soil more susceptible to freezing and erosion.[26] Plant cover and litter are important sources of nutrients and organic matter for the soil; their removal, coupled with changes in plant species composition, can cause changes in nutrient cycling and soil fertility. Extensive trampling and trail-making by livestock cause the surface horizons of soils to compact, reducing aeration and the water-holding capacity of the soil and thus decreasing the rate of infiltration of rain and snow. During storm events such as summer thundershowers, low infiltration results in greater rates of surface runoff, which can increase soil erosion. Increased runoff also means that less water is retained on-site, so that soil moisture reserves are diminished and water is unavailable for plant growth. Streamflows in overgrazed watersheds are typified by higher peak flows in spring and lower summer base flows (the relatively constant low flow to which the stream returns after storms or snowmelt periods).[27]

Finally, livestock grazing secondarily influences natural fire cycles. Prior to settlement by Euro-Americans, fire (either lightning-caused or set by Native Americans) was an important disturbance that influenced the vegetation structure and species composition of ecosystems throughout the inland West. Low-elevation forests typically consisted of open stands of fire-adapted species. Steppe grasslands were primarily herbaceous communities; shrub-steppes were codominated by grasses and shrubs. Frequent fires maintained these ecosystems by killing shrub and vulnerable tree seedlings and favoring the growth of herbaceous species. Native plant and animal species were adapted to the frequency and severity of natural fires; several conifer species actually require periodic fire in order to persist at a site.

Livestock grazing reduced the fine fuels in both forested and steppe environments, which diminished the occurrence of fires and thus led to substantial changes in plant species composition. For sagebrush steppe, Boone Kauffman and David B. Sapsis reported dramatic differences in fire behavior between ungrazed and lightly spring-grazed areas.[28] Flame lengths were greater than 18 meters in areas protected from grazing, whereas fire would not carry in grazed areas. Long before federal and state agencies implemented active fire-suppression policies, the frequency of fire decreased in the western United States, coincident with the introduction of livestock.[29]

Tertiary Influences

The cumulative effects of livestock grazing in the interior Pacific Northwest have altered the structure of ecosystems. Physical and biological aspects of ecosystem structure strongly influence ecosystem functions—how energy and nutrients are processed. Because livestock grazing affects both the physical and biological, its longterm effects have dramatically altered ecosystem function. In this section we describe the changes in ecosystem structure and function that have occurred in forests, steppes, and riparian areas of the interior Pacific Northwest.

Many low- to mid-elevation forests of the region are composed of stands dominated by ponderosa pine (*Pinus ponderosa*) or a mix of conifers including ponderosa pine, lodgepole pine (*Pinus contorta*), Douglas fir (*Pseudotsuga menziesii*), and western larch (*Larix occidentalis*).[30] Over the last century, the structure and composition of these forests have changed substantially because of logging, fire suppression, and grazing. These land-management practices have resulted in dramatic shifts in species composition and increased tree density. Historically, these low-elevation forests were composed of an uneven-aged mosaic of even-aged stands. Frequent (every 8 to 10 years), low-intensity surface fires killed the youngest trees, reduced the shrub cover, and maintained old-growth stands that were open and parklike, consisting of large, widely spaced, fire-tolerant trees—mostly ponderosa pine and western larch.[31] In many areas, the open forests have been replaced by dense thickets of fire-sensitive, shade-tolerant, and disease-susceptible species such as grand fir (*Abies grandis*). These changes have been attributed to forest management practices, notably the selective logging of the larger, more fire-tolerant trees and the active suppression of fire.[32] Joy Belsky and D. M. Blumenthal have argued, however, that livestock grazing has also been a factor in altering the stand dynamics of these forests.[33]

The removal or reduction of grass cover by livestock fostered the establishment of less palatable tree seedlings, which previously had been outcompeted by bunchgrasses or eliminated by low-intensity fires. Livestock grazing also eliminated the protective litter layer and disturbed the soil, creating a seedbed for establishment of tree seedlings. Reduction of the understory cover also meant reduction in the quantity of fine fuels. Prior to

Euro-American settlement, the dense herbaceous understory facilitated and carried frequent surface fires. After livestock were introduced and active fire suppression efforts began in the early 1900s, the natural fire regime of interior forests was greatly altered.[34] Rather than frequent low-intensity surface fires, there now are less frequent but more extensive high-intensity, stand-replacing fires in which all above-ground vegetation may be killed. Severe fires, such as those that swept through low-and mid-elevation forests of central and eastern Washington and Oregon during the summers of 1994 and 1996, are partially the result of overgrazing, logging, and fire suppression. The interactive effects of these practices have also created forest stands particularly susceptible to insects and disease.[35]

In the rain shadow east of the Cascade Mountains, the lowlands and basins are covered by grassland steppe and shrub-steppe: vegetation characterized by perennial bunchgrasses such as bluebunch wheatgrass (*Pseudoroegneria spicatum*), Idaho fescue (*Festuca idahoensis*) and Sandberg's bluegrass (*Poa secunda*), sagebrushes (*Artemisia tridentata*, *A. arbuscula*, and *A. rigida*), and other shrubs. The steppes of the interior West, the largest semidesert of North America, have undergone dramatic changes since the introduction of livestock.

Livestock grazing has affected the steppes in ways fundamentally similar to the forests, the major results being shifts in species composition, soil degradation, altered fire regimes, and changes in ecosystem structure and function. Overgrazing caused a rapid decrease in the abundance and distribution of native perennial bunchgrasses. As the native grass cover declined and the frequency of fire diminished, the cover and extent of woody species—notably sagebrush and juniper species—increased markedly. Richard Miller and coauthors described this sequence as follows (we omit their footnotes):[36]

> The increase in woody vegetation during the late 1800s and early 1900s was due to reduced fire frequency, intensive season-long grazing, and climatic change. Fire, which maintained the dominance of herbaceous vegetation on many areas, declined with the reduction of fine fuels through grazing, building of roads, and fire suppression. The influence of North American Indians on fire frequency was also reduced in the 19th century due to the decline of American Indian populations caused by European diseases and movement of Indians to reservations in the early 1870s. The overall trend on a large portion of the sagebrush steppe has been a reduction in palatable grasses and forbs and an increase in sagebrush species.

Although livestock grazing has undoubtedly played an important role, Miller and colleagues suggested that the encroachment of juniper into shrub-steppe cannot be attributed entirely to overgrazing.[37] Recent changes in climate, pollen production, and reproductive success may also be operating.

Livestock grazing has also facilitated the introduction, establishment, and rapid spread of numerous exotic, invasive, and undesirable plant species. Many of these weedy species were introduced from the steppe regions of central Asia, where climatic conditions are similar to those of semiarid regions of western North America. The increase in cheatgrass is an example of the effect of exotics. Other notable exotic species include red brome (*Bromus rubens*), medusahead (*Taeniatherum caput-medusae*), Russian thistle (*Salsola tragus*), halogeton (*Halogeton glomeratus*), and several knapweeds (*Centaurea* spp.)[38] These exotic species evolved under continuous grazing pressure and, unlike the bunchgrasses native to the inland western United States, were well adapted to the disturbed conditions caused by livestock grazing. They now dominate large areas of the western United States, reducing the diversity and abundance of native plant species and impoverishing forage for native ungulates.

Livestock have also influenced nonvascular plants of the steppe region, including mosses, lichens, and microphytic crusts. The impacts come primarily through trampling and soil degradation. Microphytic crusts consist of a complex matrix of lichens, mosses, liverworts, algae, fungi, and bacteria. These organisms are important in cycling nutrients, controlling soil erosion, reducing overland water flow, and increasing water infiltration.[39] They are particularly susceptible to physical disturbance and have been eliminated from large areas due to livestock grazing and agricultural cultivation.[40]

Changes in the vegetation structure and plant species composition of the steppe region have also impacted animal populations, primarily through alteration of habitat and prey or forage availability. For example, the density and diversity of small mammals and birds have been shown to be greater on ungrazed or lightly grazed sites than on heavily grazed ones.[41]

Riparian areas are the zones of contact between land and water ecosystems—the mesic, productive environments that border streams, rivers, lakes, and springs. In semiarid regions, riparian areas provide critical habitat for numerous species, including plants, insects, mammals, and migratory and resident birds. More than 85 percent of the wildlife species in the Great Basin region of Oregon depend on riparian areas for all or part of their life cycle.[42] In addition to providing animal habitat, riparian areas also perform important ecosystem functions. Through interactions of soil, vegetation, and water, riparian areas retain and filter sediments, stabilize stream banks, and moderate stream and groundwater flows through storage, flood attenuation, and maintenance of base flow to adjacent streams.[43] Riparian vegetation also plays a crucial functional role in shaping the quality of the aquatic habitat by providing shade and inputs of organic material and nutrients into the stream and by structuring channel morphology.[44]

Overgrazing by cattle has caused substantial damage to stream and riparian ecosystems throughout the Pacific Northwest.[45] Cattle tend to congregate in riparian areas because of the abundant forage, proximity to water, relatively level terrain, and favorable microclimate.[46] The shift from sheep to cattle as the dominant domestic herbivore in the 1930s likely increased the use of riparian areas.[47] Because cattle are difficult to manage in riparian zones and have high preference for these areas, riparian and stream habitats are highly vulnerable to their impacts.

Livestock grazing has negatively affected riparian areas by changing, reducing, and, in some areas, eliminating vegetation. Woody species—including cottonwoods (*Populus* spp.), willows (*Salix* spp.), and alders (*Alnus* spp.)—have declined in density and cover in riparian areas in response to livestock use. Wayne Elmore, who has worked in central Oregon for more than two decades, wrote: "Fur trappers and settlers during the early 1800s reported extensive stands of willows and wide wet meadows along stream systems throughout the western rangelands. By the early 1900s, many of these stream systems were severely damaged or eliminated because of improper livestock use."[48] He noted that the Indian word *ochoco*, the name of a mountain range in central Oregon, means "streams lined with willows."[49]

Other factors, such as use of cottonwood as a wood source, removal of woody species as "phreatophyte control," and hydrologic modification of streams and rivers for agriculture, have contributed to the decrease in the distribution and abundance of woody riparian species. The widespread decline of cottonwood and willows, however, has been largely attributed to land management practices associated with livestock production.[50] Some floodplains formerly dominated by woody riparian species, such as the portion of the John Day River, Oregon, shown in figures 15.4 and 15.5, were converted to cattle pastures or hay fields. In studies comparing the plants growing within and outside exclosures (fenced areas that exclude cattle and wild ungulates), Richard Case and Boone Kauffman found dramatic increases in the growth, reproduction, and establishment of willows, cottonwoods, and alders in the ungrazed exclosures.[51] In heavily grazed riparian areas, these woody species are unable to establish or reproduce. As individu-

FIG. 15.4. The John Day River, Oregon, near Picture Gorge, prior to extensive grazing. The photograph shows a gallery forest of black cottonwood (*Populus trichocarpa*). The main channel of the John Day River runs a sinuous course through the center of the floodplain.

als and stands senesce and die, they are not replaced. Along many western streams, one can observe dead cottonwood or aspen snags where extensive stands of gallery forests once existed. When important woody species such as cottonwoods decline, so do the numerous wildlife species that depend on them for habitat.

Meadows dominated by grasses, sedges (*Carex* spp.), and rushes (*Juncus* spp.) occur along headwater streams, in areas of groundwater seepage, and in the floodplains of rivers and streams. These areas are characterized by high water tables, hydric soils, and diverse herbaceous plant communities.[52] The impacts of overgrazing by sheep from the late 1880s through the 1920s can still be observed in some high-elevation meadows. Sheep were kept in the same meadow for up to four months during the growing season and caused severe soil deterioration and dramatic shifts in plant species composition.

In floodplain meadows at low to middle elevations, livestock impacts on vegetation and soils have resulted in stream channel widening and incision and the lowering of groundwater tables. Hydrologic connectivity between groundwater flow and streams has been altered. In these situations,

FIG. 15.5. A contemporary photo of the same location. The river has been channelized to the edge of the floodplain. The cottonwoods and other woody riparian species were removed when the site was converted to cattle pasture. Black cottonwood forests are among the most valuable wildlife habitats in North America, and their loss signals a decline in wildlife diversity.

mesic plant species may be replaced by species that typically occur in more upland or drier environments.[53] Extensively rooted sedges and rushes may be replaced by shallow-rooted pasture grasses and exotic annual species, leading to reduction in stream-bank stability with subsequent sedimentation and loss of pool habitat. Conversion of herbaceous plant community types can also result in reduction of stream-bank cover and shade and decreased organic matter inputs into the stream.[54]

With changes in hydrology, the areal extent and duration of flooding and soil saturation in riparian areas has decreased, thus altering the cycling of nutrients and organic matter. In many cases this has resulted in spatial reduction of the riparian area itself. Water flows more rapidly through such degraded stream sections, which can reduce nutrient uptake and storage by vegetation and result in greater quantities of dissolved nutrients being exported from the watershed.[55] Changes in streamwater quality, such as higher stream temperatures, increased turbidity and sediment loads, and lowered

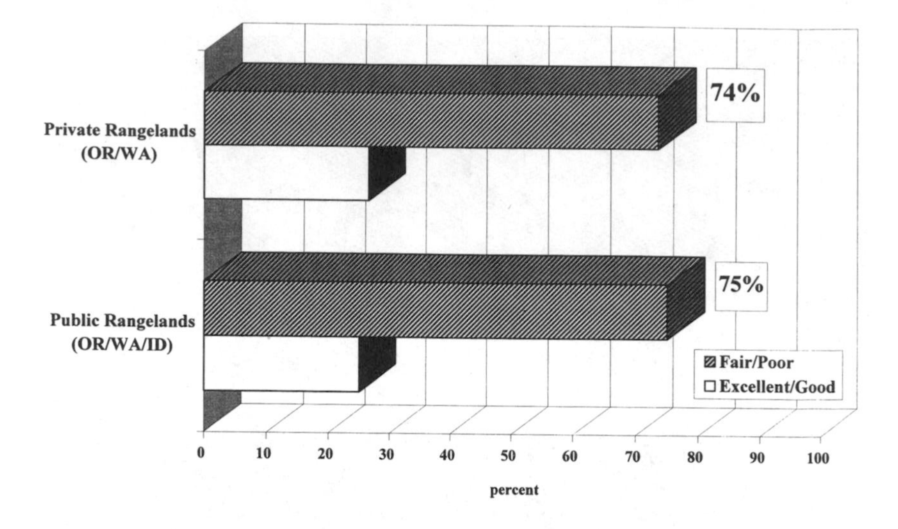

FIG. 15.6. Range condition on public and private rangelands in the Pacific Northwest.

oxygen content, can also occur. The cumulative impact of these changes on streams and riparian areas degrades spawning and rearing habitat for salmon and other native, cold-water fish species.

CURRENT CONDITION

Traditionally, range managers have rated range conditions as excellent, good, fair, or poor. These classes indicate the productivity of the vegetation and the condition of the soil at a particular site relative to its ecological potential. Sites in excellent or good condition typically have adequate plant and soil cover, exhibit little or no soil erosion, and are producing at or near their expected ecological potential. Sites with low plant cover and eroding soils are rated as being in fair condition. Poor condition indicates that the site is producing only a fraction of the vegetation produced on similar land in good or excellent condition. Poor or fair condition indicates a degraded landscape where human-caused desertification is occurring.

Although there are few trend data, the literature suggests that large-scale range degradation was halted by the 1930s and that range conditions have improved since then.[56] Although range condition may be improving, the most recent assessment shows that 75 percent of federal rangeland in the Pacific Northwest remains in fair to poor condition (fig. 15.6).[57] Another recent assessment found that 74 percent of private rangeland in eastern Oregon and Washington was also in fair to poor condition.[58] Additional documentation suggests that most of the improvements have been in the uplands, and riparian areas remain in poor condition.[59]

SUMMARY

Rangeland in the Pacific Northwest did not evolve under intensive grazing by large herbivores as it did on the Great Plains of North America. As a result, Northwest rangeland was extremely vulnerable to the effects of overgrazing by introduced livestock. Although Native Americans had acquired horses, it was not until Euro-Americans began migrating to the Pacific Northwest that domestic herbivores were introduced there on a widespread basis. Since Euro-American settlement began in the mid-1800s, the livestock industry has grown steadily in the Pacific Northwest to the present. The industry can be characterized in terms of three distinct periods. The settlement period (from 1850 to 1900) was the era of the open range, when livestock numbers grew exponentially and grazing was unmanaged. The regulation of livestock grazing began with the formation of the United States Forest Service in 1905. From 1900 to 1940, cattle numbers increased steadily, but overall grazing use was relatively constant. During this period, the livestock industry shifted from a mixture of cattle and sheep to one dominated by cattle. During the third period (1940 to the present), the livestock industry grew steadily, with current grazing use near the highest levels since Euro-American settlement began. In addition, grazing use appears to be shifting from public to private land since the 1940s.

The introduction of livestock has had widespread and dramatic impacts on the ecosystems of the Pacific Northwest. Primary influences, such as vegetation removal and trampling, have immediate, local impacts that can be readily measured. Secondary influences, which include changes in species composition, altered soils and hydrology, and altered fire cycles, have local as well as regional manifestations. The long-term, synergistic, and cumulative impacts of the primary and secondary influences have resulted in fundamental changes in the structure and function of forest, steppe, riparian, and stream ecosystems. Livestock grazing has contributed to soil loss, declines in plant and animal biodiversity, and loss of ecosystem integrity.

Although range trend data indicate that range conditions have been improving since the 1930s, most of the improvement has occurred in upland areas rather than in meadow and riparian ecosystems. These improvements have been slow, requiring in excess of 70 years in most cases. In addition, current range condition data indicate that the majority of public and private rangeland in the Pacific Northwest is in poor to fair condition. Current range conditions reflect the fact that livestock management over the past 100 years has focused primarily on improved forage production rather than on the sustainability of ecosystem processes, functions, and biodiversity.

Throughout the Pacific Northwest, stream and riparian environments have been degraded by grazing and other land management practices.[60] The recent listing under the Endangered Species Act of several salmon species that rear or spawn in streams and rivers east of the Cascade Mountains has focused attention on the condition of eastside streams. Of the 214 stocks of anadromous fish that occur in the Pacific Northwest, 47 percent are considered to be at high risk of extinction.[61] Numerous factors, including dams, the introduction of exotic fish species, and fish harvesting, have contributed to the decline of cold-water fish communities. The degradation of freshwater habitats that are critical for spawning and rearing has also severely affected the reproductive success of salmonid species. Restoration of riparian habitats has become a major priority for federal and state land management agencies and tribal governments. Because of the many impacts of grazing, exclusion of livestock from many streamside areas is necessary for the recovery of riparian and aquatic habitats.

Livestock management policies are just beginning to address restoration and sustainability of ecosystems. We know of no comprehensive research that has taken a holistic approach to understanding the influences of livestock grazing on the processes and functions that shape and maintain entire ecosystems. Most ecologists would agree that it is impossible to return to a presettlement ecological condition on many rangelands. We can begin, however, to integrate our understanding of ecosystem structure and function with what is known about the influences of natural disturbance (such as fire) and anthropogenic disturbance (such as livestock grazing), creating strategies for restoration and management of Pacific Northwest ecosystems.

NOTES

1. T. L. Fleischner, "Ecological Costs of Livestock Grazing in Western North America," *Conservation Biology* 8 (1994): 629–44.

2. J. Belsky and D. M. Blumenthal, "Effects of Livestock Grazing on Stand

Dynamics and Soils in Upland Forests of the Interior West," *Conservation Biology* 11 (1997): 315–27.

3. Gerald F. Schroedl, "The Archeological Occurrence of Bison in the Southern Plateau" (Ph.D. diss., Washington State University, 1973); B. R. Butler, "Bison Hunting in the Desert before 1800: The Paleoecological Potential and the Archeological Reality," *Plains Anthropologist* 23 (1978): 106–12.

4. Richard N. Mack and John N. Thompson, "Evolution in Steppe with Few Large, Hooved Mammals," *American Naturalist* 119 (June 1982): 757–73.

5. Francis Haines, "The Northward Spread of Horses among the Plains Indians," *American Anthropologist* 40 (1938): 434–35; William G. Robbins and Donald W. Wolf, *Landscape and the Intermontane Northwest: An Environmental History* (Portland: Pacific Northwest Research Station, General Technical Report PNW-GTR-319, 1994).

6. William A. Galbraith and E. William Anderson, "Grazing History of the Northwest," *Rangelands* 13 (October 1991): 213–18.

7. Robbins and Wolf, *Landscape and the Intermontane Northwest.*

8. U.S. Department of the Interior, Census Office, *Census of the United States* (Washington, D.C.: Department of the Interior, 1850, 1860, 1870, 1880, 1890, 1900); U.S. Department of Commerce, Bureau of the Census, *Census of the United States* (Washington, D.C.: Department of Commerce, 1910, 1920); U.S. Department of Commerce, Bureau of the Census, *Census of Agriculture* (Washington, D.C.: Department of Commerce, 1930, 1940, 1950, 1954, 1959, 1964, 1969, 1974, 1978, 1982, 1987, 1992).

9. Harold F. Heady, *Rangeland Management* (New York: McGraw-Hill, 1975).

10. Laurence A. Stoddart, Arthur D. Smith, and Thadis W. Box, *Range Management* (New York: McGraw-Hill, 3d ed., 1975).

11. C. Gordon, C. W. Gordon, and J. C. McCoy, *Report on Cattle, Sheep, and Swine: Supplementary to Enumeration of Livestock on Farms in 1880* (Washington, D.C.: Department of the Interior, 1883), vol. 3, pp. 953–1116.

12. Galbraith and Anderson, "Grazing History of the Northwest."

13. Larry L. Irwin, J. G. Cook, R. G. Riggs, and J. M. Skovlin, *Effect of Long-Term Grazing By Big Game and Livestock in the Blue Mountains Forest Ecosystems* (Portland: Pacific Northwest Research Station, General Technical Report PNW-GTR-325, 1994).

14. National Research Council, *Rangeland Health: New Methods to Classify, Inventory, and Monitor Rangelands* (Washington, D.C.: National Academy Press, 1994).

15. U.S. Department of the Interior, Bureau of Land Management, *Public Land Statistics* (Washington, D.C.: Department of Interior, 1947–93).

16. Bruce A. McIntosh, "Historical Changes in Stream Habitats in the Columbia River Basin" (Ph.D. diss., Oregon State University, 1995).

17. Fleischner, "Ecological Costs of Livestock Grazing"; J. Belsky and D. M. Blumenthal, "Effects of Livestock Grazing on Stand Dynamics and Soils in Upland Forests of the Interior West," *Conservation Biology* 11 (1997): 315–27.

18. J. Boone Kauffman and W. C. Krueger, "Livestock Impacts on Riparian Ecosystems and Streamside Management Implications . . . A Review," *Journal of Range Management* (September 1984): 430–38; William S. Platts, "Livestock Grazing," *American Fisheries Society Special Publication* 19 (1991): 389–423.

19. R. F. Miller, T. J. Svejcar, and N. E. West, "Implications of Livestock Grazing in the Intermountain Sagebrush Region: Plant Composition," in *Ecological Implications of Livestock Herbivory in the West,* eds. M. Vavra, W. A. Laycock and R. D. Pieper (Society of Range Management, 1994), pp. 101–46.

20. Jon M. Skovlin and Jack Ward Thomas, *Interpreting Long-Term Trends in Blue Mountain Ecosystems from Repeat Photography* (Portland: Pacific Northwest Research Station, General Technical Report PNW-GTR-315, 1995).

21. William L. Robinson and Eric G. Bolen, *Wildlife Ecology and Management* (New York: Macmillan, 2d ed., 1989).

22. D. D. Briske and J. H. Richards, "Physiological Responses of Individual Plants to Grazing: Current Status and Ecological Significance," in Vavra, Laycock, and Pieper, *Ecological Implications of Livestock Herbivory*, pp. 147–76.

23. Stoddart, Smith, and Box, *Range Management*.

24. R. N. Mack, "The Invasion of *Bromus tectorum L.* into Western North America: An Ecological Chronicle," *Agro-Ecosystems* 7 (1981): 145–65.

25. G. A. Harris, "Some Competitive Relationships between *Agropyron spicatum* and *Bromus tectorum*," *Ecological Monographs* 37 (1967): 89–111.

26. J. M. Facelli and S.T.A. Pickett, "Plant Litter: Its Dynamics and Effects on Plant Community Structure," *Botanical Review* 57 (1991): 1–32.

27. W. Elmore, "Riparian Responses to Grazing Practices," in *Watershed Management*, ed. Robert Naiman (New York: Springer-Verlag, 1992), pp. 442–57.

28. J. Boone Kauffman and D. B. Sapsis, "The Natural Role of Fire in Oregon's High Desert," in *Oregon's High Desert: The Last 100 Years* (Corvallis: Oregon State University Agricultural Experiment Station, Special Report no. 841, 1989).

29. M. Savage and T. W. Swetnam, "Early Nineteenth-Century Fire Decline following Sheep Pasturing in the Navajo Ponderosa Pine Forest," *Ecology* 71 (1990): 2374–78.

30. Jerry F. Franklin and C. T. Dyrness, *Natural Vegetation of Oregon and Washington* (Corvallis: Oregon State University Press, rev. ed., 1988).

31. James K. Agee, *Fire Ecology of Pacific Northwest Forests* (Washington, D.C.: Island Press, 1993).

32. Nancy Langston, *Forest Dreams, Forest Nightmares: The Paradox of Old Growth in the Inland West* (Seattle: University of Washington Press, 1995).

33. J. Belsky and D. M. Blumenthal, "Effects of Livestock Grazing on Stand Dynamics and Soils in Upland Forests of the Interior West," *Conservation Biology* 11 (1997): 315–27.

34. Franklin and Dyrness, *Natural Vegetation of Oregon and Washington*.

35. Boyd E. Wickman, *Forest Health in the Blue Mountains: The Influence of Insects and Disease* (Portland: Pacific Northwest Research Station, General Technical Report PNW-GTR-295, 1992).

36. Miller, Svejcar, and West, "Implications of Livestock Grazing."

37. *Ibid*.

38. *Ibid*.

39. N. E. West, "Structure and Function of Microphytic Soil Crusts in Wildland Ecosystems to Arid and Semi-arid Regions," *Advances in Ecological Research* 20 (1990): 179–223.

40. Miller, Svejcar, and West, "Implications of Livestock Grazing."

41. T. D. Reynolds and Charles H. Trost, "The Response of Native Invertebrate Populations to Crested Wheatgrass Planting and Grazing by Sheep," *Journal of Range Management* 33 (1980): 122–25; D. M. Taylor and C. D. Littlefield, "Willow Flycatcher and Yellow Warbler Response to Cattle Grazing," *American Birds* 40 (1986): 1169–73.

42. J. W. Thomas, C. Maser, and J. E. Rodiek, "Riparian Zones," in *Wildlife Habitats in Managed Forests: The Blue Mountains of Oregon and Washington*, ed. Jack Ward Thomas (Washington, D.C.: Government Printing Office, USDA Forest Service Agricultural Handbook no. 553, 1979), pp. 40–47.

43. W. Elmore, "Riparian Responses to Grazing Practices," in Naiman, *Watershed Management*, pp. 442–57; S. V. Gregory, F. J. Swanson, W. A. McKee, and K. W. Cummins, "An Ecosystem Perspective of Riparian Zones," *Bioscience* 41 (1991): 540–51.

44. R. L. Beschta, "Riparian Shade and Stream Temperature: An Alternative Perspective," *Rangelands* 19 (1997): 25–28; J. D. Allen, *Stream Ecology* (London: Chapman and Hall, 1995); C. Clifton, "Effects of Vegetation and Land Use on Channel Morphology," in *Practical Approaches to Riparian Resource Management: An Educational Workshop*, eds. Robert E. Gresswell, Bruce A. Barton, and Jefferey L. Kershner (Billings, Mont.: Bureau of Land Management, 1989), pp. 121–29.

45. Kauffman and Krueger, "Livestock Impacts on Riparian Ecosystems"; Platts, "Livestock Grazing."

46. L. R. Roath and W. C. Krueger, "Cattle Grazing and Behavior on a Forested Range," *Journal of Range Management* 35 (1982): 332–38.

47. Platts, "Livestock Grazing."

48. W. Elmore, "Riparian Responses to Grazing Practices," in Naiman, *Watershed Management*, pp. 442–57.

49. Here Elmore was citing Lewis A. McArthur, *Oregon Geographic Names* (Portland: Press of the Oregon Historical Society, 5th ed. 1982).

50. K. G. Busse, *Ecology of the Salix and Populus Species of the Crooked River National Grassland* (M.S. thesis, Oregon State University, 1989).

51. R. L. Case and J. Boone Kauffman, "Wild Ungulate Influences on the Recovery of Willows, Black Cottonwood, and Thin-leaf Alder following Cessation of Livestock Grazing in Northeastern Oregon," *Northwest Science* 71 (1997): 115–26.

52. E. A. Crowe and R. R. Clausnitzer, *Mid-Montane Wetland Associations of the Malhuer, Umatilla, and Wallowa-Whitman National Forests* (Portland: Forest Service, R6-NR-ECOL-TP-22–97, 1997).

53. J. B. Kauffman, W. C. Krueger, and M. Vavra, "Effects of Cattle Grazing on Riparian Plant Communities," *Journal of Range Management* 36 (1983): 685–91.

54. N. J. Otting, "Vegetation and Environmental Gradients in Riparian Meadows of the Upper Grande Ronde Watershed, Oregon," (M.S. thesis, Oregon State University, 1998); J. Boone Kauffman, unpublished data.

55. D. M. Green and J. Boone Kauffman, "Nutrient Cycling at the Land-Water Interface: The Importance of the Riparian Zone," in Gresswell, Barton, and Kershner, *Practical Approaches to Riparian Resource Management*, pp. 61–68.

56. Thadis W. Box, "Rangelands," *Natural Resources for the Twenty-first Century* (Washington, D.C.: Island Press, 1990); U.S. Department of the Interior, *State of the Public Rangelands* (Washington, D.C.: Government Printing Office, 1990).

57. Johanna Wald and David Alberswerth, *Our Ailing Public Rangelands: Condition Report—1985* (n.p.: National Wildlife Federation and Natural Resources Defense Council, 1985).

58. Grant A. Harris and Martha Chaney, *Washington State Grazing Land Assessment* (Spokane: U.S. Department of Agriculture, 1984); U.S. Department of Agriculture, Soil Conservation Service, *Oregon Soil: A Resource Condition Report* (Portland: Department of Agriculture, 1985).

59. Ed Chaney, Wayne Elmore, and William S. Platts, *Livestock Grazing on Western Riparian Areas* (Washington, D.C.: Environmental Protection Agency, 1990); Comptroller General, Government Accounting Office, *Public Rangelands: Some Riparian Areas Restored but Widespread Improvement Will Be Slow* (Washington, D.C.: Government Printing Office, GAO/RCED-88–105, 1988).

60. B. A. McIntosh, J. R. Sedell, J. E. Smith, R. C. Wissmar, S. E. Clarke, G. H. Reeves, and L. A. Brown, "Historical Changes in Fish Habitat for Select River Basins of Eastern Oregon and Washington," *Northwest Science* 68 (1994): 36–52; R.

C. Wissmar, J. E. Smith, B. A. McIntosh, H. W. Li, G. H. Reeves, and J. R. Sedell, "A History of Resource Use and Distribution in Riverine Basins of Eastern Oregon and Washington," *Northwest Science* 68 (1994): 1–35.

61. W. Nehlsen, J. E. Williams, and J. A. Lichatowich, "Pacific Salmon at the Crossroads: Stocks at Risk in California, Oregon, Idaho, and Washington," *Fisheries* 16(2) (1991): 4–21.

16 / Environmental Change in the Northern Rockies

Settlement and Livestock Grazing in Southwestern Montana, 1860–1995

WILLIAM WYCKOFF AND KATHERINE HANSEN

Changes in the global resource economy, the technologies of natural resource exploitation, and attitudes toward nature have transformed the environment of the western United States in often dramatic ways over the past 150 years. These large-scale changes have shaped local patterns of settlement and land-use management in southwestern Montana's Madison Valley and have altered the region's biogeography (fig. 16.1). The Madison Valley and its traditional livestock economy can serve as a case study for many areas of the interior West.

Our study of the Madison Valley has been shaped by the work of several other scholars. Donald Meinig examined general processes of regional development in which previously isolated areas were increasingly affected by and connected with outside economic, political, and cultural influences.[1] Among the key forces of change he identified were evolving transportation technologies and political institutions that bound the regional and national economies ever more closely together. Meinig demonstrated that these external influences affected the human geographies of areas undergoing change.

Patricia Nelson Limerick and William Robbins have each emphasized the unequal power relationships between isolated western settings and the larger institutional forces that mold their landscapes.[2] They argue that the West was not the promised land of frontier myth but was instead a region controlled and often abused by an expanding global economy and by imperialistic political and cultural institutions. Donald Worster and Susan Neel each argue further

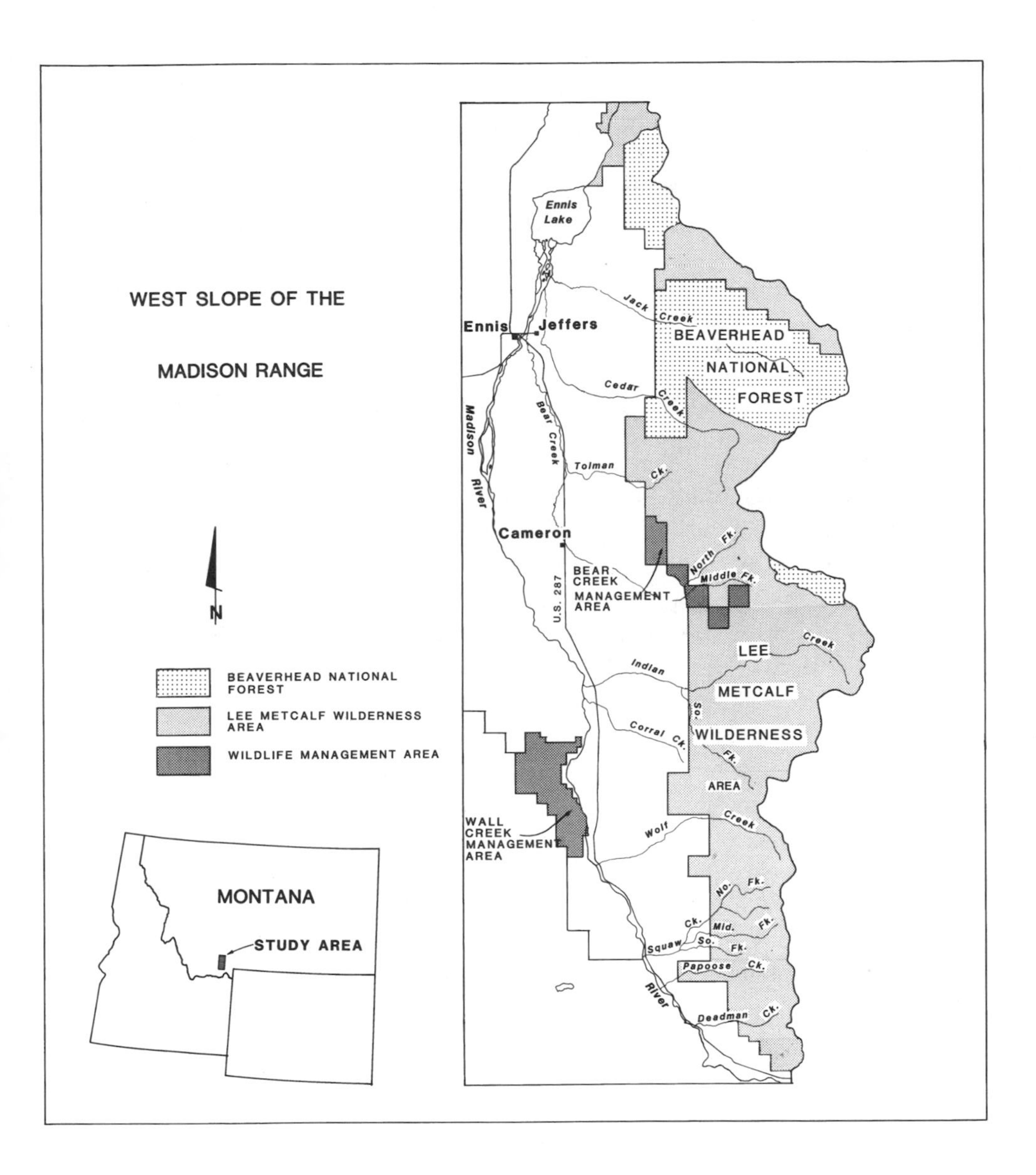

FIG. 16.1. The study area in southwestern Montana.

that the West's distinctive character was an expression of how these outside forces interacted with an often-capricious western environment.[3]

Finally, other scholars have emphasized the region's internal geographical diversity, arguing that this diversity requires an examination of a variety of settings to appreciate the West's larger environmental history.[4] All of these perspectives are useful in understanding the Madison Valley, because they suggest insights into how an isolated, geographically complex part of the West came to be dominated by larger economic, political, and cultural forces and how those forces shaped a specific locality in particular ways.

THE WESTERN MONTANA SETTING

Characteristic of much of the intermountain West, southwestern Montana's topography alternates between long and broad, generally northwest-southeast trending valleys and a succession of adjacent mountain ranges that define those valleys. The Madison Valley, southwest of Bozeman, Montana, has an average elevation of 5,000 to 6,000 feet (1,525 to 1,830 meters). It lies immediately west of the Madison Range, which extends from northeast of the town of Ennis southward to near the Idaho border. A series of elevated, gently sloping benchlands extends from the river to the mountain front. These benches are frequently terraced, intermittently cut by streams, and composed of large alluvial fans. They often extend directly to the abrupt and usually steeply sloping, west-facing foothills that form the western edge of the Madison Range.

The regional climate is cool and continental. Annual precipitation (spring maximum) ranges from less than 12 inches (30 centimeters) at lower valley locations to more than 25 inches (65 centimeters) in the mountains, much of it falling as snow between October and May.[5] Temperatures also vary with elevation. In the valley locations, highest averages occur in July (average of 65°F; 18.4°C) and lowest in January (average of 23°F; −5.2°C). The entire region is subject to large diurnal changes in temperature, and extreme variations can occur in winter when arctic air masses depress temperatures or when chinook winds produce abnormally high temperatures.

Vegetation varies with microclimatic and geographic conditions.[6] Below the upper tree line at approximately 9,500 feet (2,900 meters), a mixture of coniferous trees including whitebark pine, Englemann spruce, and subalpine fir alternates with meadows. At lower elevations the forest is dominated by lodgepole pine, limber pine, and Douglas fir, with scattered stands of Rocky

Mountain juniper and aspen. The lower forest–valley grassland ecotone generally occurs in the foothills at 6,500 to 7,500 feet (1,920 to 2,287 meters). Valley grasslands vary widely and include a mixture of Idaho fescue, needlegrass, wheatgrass, and grama grass that alternates with sagebrush communities, especially in the southern portions of the valley.[7] Riparian vegetation along tributary stream courses and adjacent to the Madison River includes stands of willow, cottonwoods, and aspen.

The Madison Valley thus is a high, semiarid valley surrounded by wetter mountain ranges. When Euro-Americans began to move into the valley, it was largely a grassland that gave way to forests at higher elevations. These resources became attractive after the discovery of gold in the vicinity in the early 1860s.

SETTLEMENT AND ENVIRONMENTAL CHANGE

The impact of external economic, political, and cultural influences on the Madison Valley landscape grew dramatically once precious metals were discovered. Because the valley lent itself particularly to grazing domestic livestock, its economy became increasingly intertwined with regional, national, and global demands for cattle, sheep, and horses. Settlement in the Madison Valley can be divided into five major periods.

The initial period (pre-1870) was characterized by very little permanent Euro-American occupancy. Seasonal grazing by domestic livestock, however, marked the arrival of a commercial agricultural system that persistently pushed into the isolated valleys of the northern Rockies as nearby mining communities developed and transportation infrastructure improved. A second period (1870–90) began with the arrival of permanent settlers. A mosaic of private landholdings developed, and ranchers increasingly identified regional and national rather than local markets for their livestock. Changes to the region's biogeography quickened during this period but were minor compared with the dramatic and deleterious impacts of the subsequent era (1890–1934). During those years, large increases in livestock populations combined with periodic drought to accelerate human-induced changes that led to the first, though ineffective, set of grazing controls. Thereafter (1934–60), local landowners and public land managers acknowledged the area's limited carrying capacity. The recent past (post-1960) has been a period of resource reallocation. Traditional land uses have been challenged by new interpretations of the region's resources.

The Period of Seasonal Use, Pre–1870

Before 1870, human impacts on the Madison Valley were quite limited. The high semiarid valleys of southwestern Montana were not traditionally home to large numbers of Native Americans. Native inhabitants, however, periodically journeyed through the region and often established seasonal hunting camps in the valleys. They may have done some burning, but they did not maintain permanent populations in the valley. At the time of early American exploration and settlement, Shoshoni and Bannock Indians were most common in the region.[8]

Southwestern Montana's human geography began to change with the discovery of placer gold in the early 1860s at Alder Gulch, about 20 miles west of the Madison Valley. The Alder Gulch placers generated tremendous excitement in 1863. The boomtown of Virginia City was soon established, and within a year approximately 10,000 people lived in the vicinity.[9] This influx of miners set the stage for the American occupation and settlement of the Madison Valley beginning in the middle 1860s.

Although permanent Euro-American settlement in the Madison Valley was limited before 1870, commercial stockraisers increasingly made seasonal use of the region's pastures (fig. 16.2).[10] They grazed cattle and horses along the Madison River's riparian habitats and nearby benchlands. In the Madison Valley, especially in its upper southern reaches, this pattern of use began in the middle 1860s and early 1870s as nearby mining settlements created new local demands for beef and horses. Although this was public land, the federal government did little to manage it. The land was simply left open to entrepreneurs who seasonally herded stock into the cooler, upper sections of the valley. Cattle and horses were herded south in late spring and then allowed to scatter between the river and the mountain front as they drifted north down the valley before being rounded up in the late fall. Overall biogeographic impacts were minimal, because herds remained widely dispersed. Still, as symbolized by the arrows in figure 16.2, a radically new commercial economic system penetrated the area and presaged the more dramatic alterations to come.

The Early Settlement Period, 1870–1890

The transition to more permanent settlement in the valley after 1870 produced important changes. New ranches on private land (fig. 16.3) increasingly were commercial enterprises with links beyond the local economy. These ranches were boosted by growing regional and national demand for

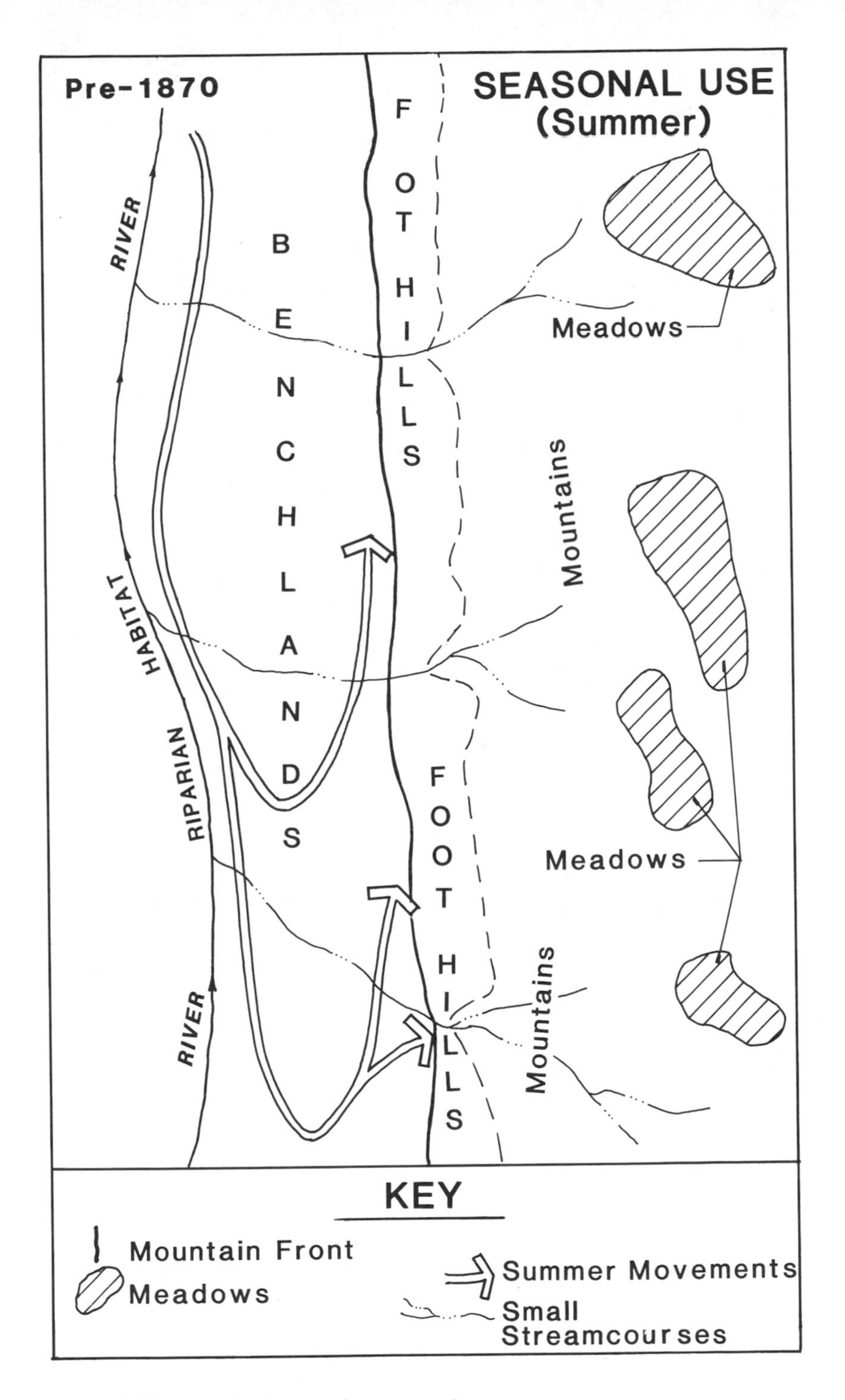

FIG. 16.2. The period of seasonal use, pre–1870.

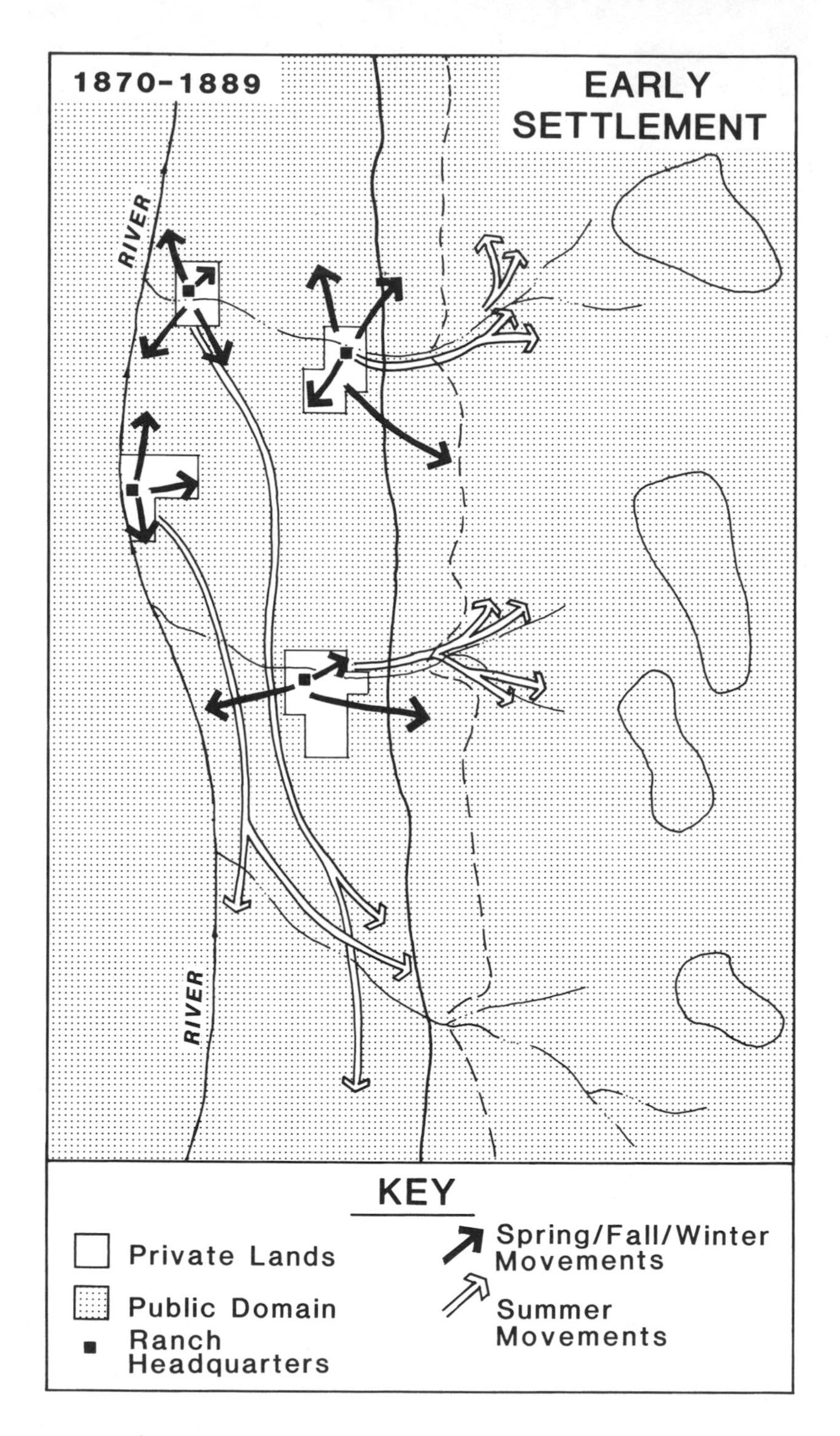

FIG. 16.3. The early settlement period, 1870–1889.

livestock products. After 1880, demand was further stimulated by the completion of railroad lines connecting the region to northern Utah and to the northern Midwest.[11] Although human population remained low in these widely scattered and remote operations, animal numbers increased markedly. By 1890 there were approximately 16,000 horses and 38,000 cattle in Madison County, many of them in the Madison Valley.[12] With the introduction of sheep in the 1870s from Oregon, California, and Idaho, an additional ranching economy took root.[13] By 1890, there were over 25,000 sheep in the county, including several large flocks that were headquartered or grazed seasonally in the Madison Valley. The impacts of these new settlements and economic activities varied across the region (fig. 16.3).

The Homestead, Timber Culture, and Desert Land Acts offered easy methods of acquiring ranches.[14] Usually settlers located their parcels near the river or with access to tributary streams. In addition to experiencing seasonal impacts from grazing, these lands were also put into cultivation as ranchers grew wheat, oats, barley, potatoes, and peas.[15] Ranchers also made extensive use of the surrounding public domain.[16] Several local sawmills harvested timber from nearby mountain slopes, although commercial operations never became a major factor in the region's economy. Most of the land in the valley continued to be publicly owned, though little was done to limit the free use of this public land or its resources.

A common method of transhumance involved herding livestock onto the open range of nearby foothills or south into the cooler upper valley during May and June. Pastures nearer ranch headquarters were used after the stock was herded home in October or November. This system of use meant that summer impacts were focused in the mountains and southern portions of the valley (especially south of Indian Creek), whereas nonsummer impacts were more concentrated on the benches and riparian habitats nearer ranch headquarters.

Overall, the early settlement period was a time of moderate, though sharply increased, use of the region's grazing resources. The period was also characterized by the continuing incorporation of the valley into the national and global economic system, since much of the livestock was shipped to markets outside the immediate region.

The Period of Rapid Settlement and Initial Controls, 1890–1934

The pace of settlement in the Madison Valley accelerated greatly at the end of the nineteenth century. The years between 1890 and the early 1930s were marked by fluctuating global markets for livestock products that increasingly shaped the local economy and resulted in a shifting mix of cattle and sheep, depending on prices and demand. Overall, demand rose sharply though unevenly during this period, placing ever-greater pressure on grazing resources. When combined with the unpredictable local climate, the result was a general deterioration in the area's grasslands.

Significant changes in ownership and control of land also occurred during this period. After 1890, most of the valley passed into private ownership; the mountains, however, remained in public ownership in a newly established forest reserve (national forest) (fig. 16.4).[17] Neither set of land managers effectively reconciled an expanding commercial grazing economy with an unpredictable and limited physical environment.

In the 1890s, cattle and sheep grazing increased, although the ratio of cows to sheep fluctuated. Low wool prices in 1892 and 1893, for example, favored cattle production, but by the end of the decade, improving demand for mutton and wool led to increases in sheep numbers.[18] Overall, cattle production over the decade rose about 40 percent while sheep numbers increased by 400 percent. The Jeffers family, for instance, after an early history of cattle ranching, trailed several flocks of sheep into the northern part of the valley from eastern Oregon in the middle 1880s; by the turn of the century, they were grazing more than 17,000 sheep.[19]

Residents noticed the first signs of overgrazing in the 1880s.[20] The decline in available open range in the valley increased fall-winter-spring impacts on both privately held ranches and the valley's dwindling public domain. An ever-increasing need for forage put more pressure on these valley bottoms and benches (as suggested by fig. 16.4). During the summer, ranchers were compelled to send their herds into the mountains. Despite this increased pressure on public forest land, the managers of the federal forest reserves had not yet prescribed any limits on grazing.

Three significant changes occurred in the first third of the twentieth century. First, United States Forest Service grazing management began in 1906. Until the severe drought of 1919, however, Forest Service grazing policies strongly favored increased use of public rangelands.[21] Even after 1919, a continuing rise in demand for grazing permits pressured the Forest Service to increase use on many allotments despite the increasingly apparent overgrazing.[22] The federal agency responsible for managing the area's public land accordingly used its authority for the benefit of local ranchers as the commercial livestock economy grew and the demand for pastureland expanded.

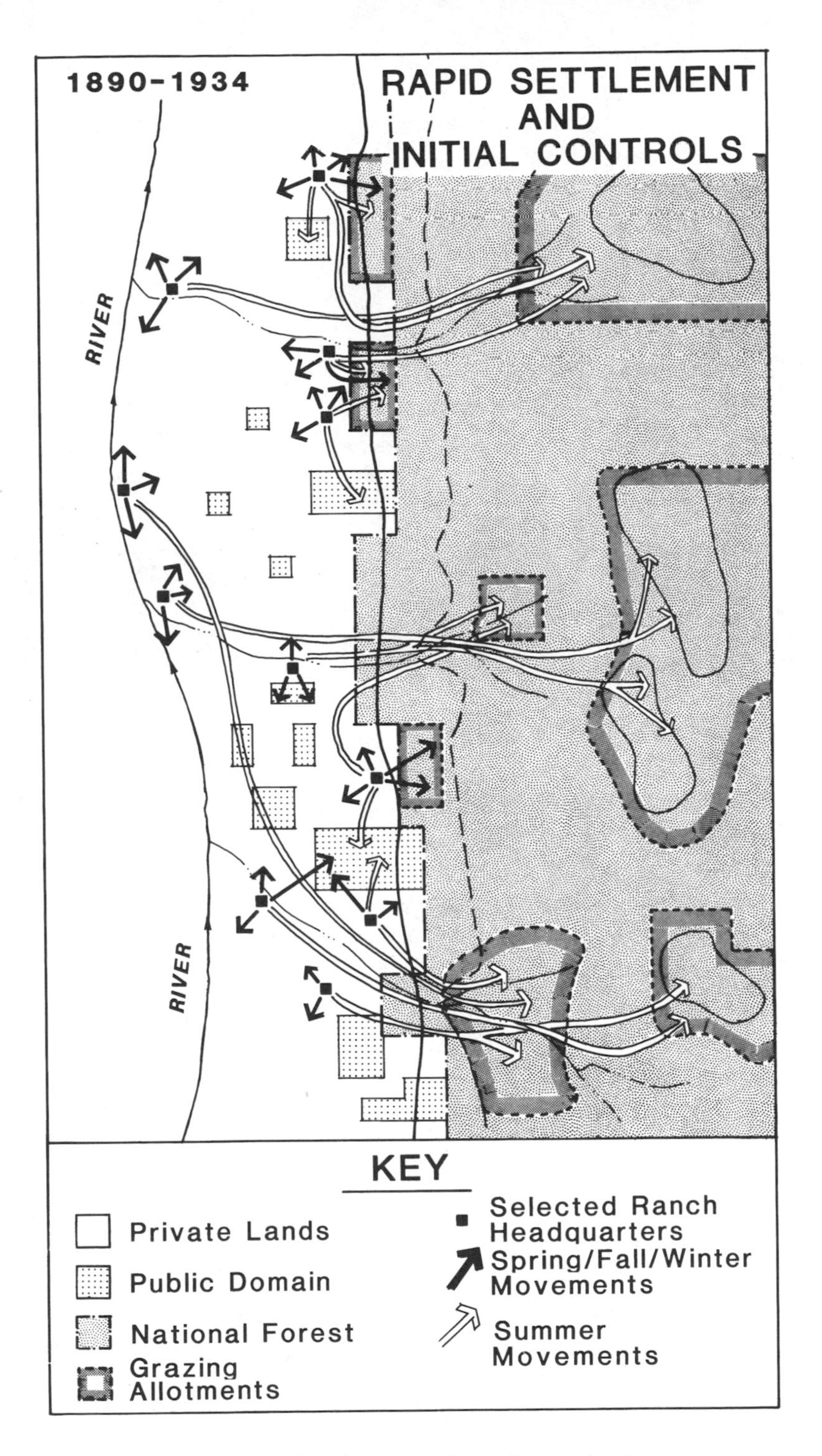

FIG. 16.4. The period of rapid settlement and initial controls, 1890–1934.

FIG. 16.5. The number of sheep increased tremendously throughout southwestern Montana between 1890 and 1930. This is the Madison National Forest Office, Sheridan, Montana, circa 1915 (source: Beaverhead National Forest).

National grain markets caused a second change. High grain prices and a shortage of available dryland farm acreage encouraged farmers to move temporarily into grain farming in portions of the Madison Valley between 1909 and 1920.[23] This had two effects on the ranching economy: some land in the long-settled northern part of the valley was converted to grain operations, and the open ranges of the higher, southern part of the valley were quickly homesteaded and removed from the public domain.[24]

Finally, despite the decline in available rangeland, there was a tremendous increase in sheep grazing throughout southwestern Montana between 1890 and 1930 (fig. 16.5). High wool prices between 1915 and 1926 encouraged a significant shift away from cattle.[25] Many of the valley's longtime cattle ranchers converted to sheep, and new ranchers set up sheep operations during the period.

By the early 1930s, the conditions for disaster were in place. County sheep numbers totaled more than 210,000 by 1930, almost double the total from a decade earlier. In 1919, a Forest Service memo acknowledged "the

FIG. 16.6. The gently sloping benchlands in the central and northern part of the Madison Valley suffered the worst abuse by livestock. This view shows the east side of the Madison Valley in 1927 (source: USGS Photo Archives, Denver Federal Center, Denver, Colorado).

overstocked conditions of many of the Madison ranges," and overgrazing was noted with increasing frequency after 1923.[26] Price declines for livestock also accelerated by the late 1920s. As a result, many ranchers were reluctant to reduce the size of their herds.[27]

Until 1934, the region had experienced nearly half a century of dramatically increasing livestock use as a result of a vigorously expanding economic system that encouraged ranchers to maximize herd sizes and to make use, wherever possible, of public resources controlled by the generally accommodating officials in the Forest Service. What were the local impacts of these institutional forces on flora and fauna? By most accounts, the gently sloping benchlands in the central and northern part of the valley suffered the worst abuse (fig. 16.6).[28] Private land in this part of the valley was fully utilized, especially in the fall, winter, and spring. The most abused acreage before 1934 was undoubtedly the scattered remaining parcels of open range. Accessible grassy foothills on both private parcels and loosely regulated public grazing allotments were also often overused, especially in spring and fall.[29]

Summer grazing patterns shifted substantially between 1890 and 1934. With less open range available in the valley, ranchers herded their animals into the mountains (see fig. 16.4). Impacts were greatest along stock herding routes and major stream courses, as well as in mountain meadows made increasingly available under a liberal U.S. Forest Service policy.[30]

All of these changes had profound consequences for the local environment. As range conditions deteriorated on the benchlands, fescue and wheatgrasses decreased in importance and were replaced by rabbitbrush, grama grasses, and sagebrush. Cattle and sheep especially damaged delicate riparian and meadow habitats where seasonal grazing concentrated. These degraded environments reduced the quality of downstream fisheries,[31] as well as terrestrial wildlife, particularly competitive grazers such as elk and antelope.[32]

A fundamental shift in the valley's livestock grazing history occurred in 1934. That year marked the worst drought in memory across much of the West. Emergency conditions existed in the valley as the range was ravaged from drought and overgrazing and as low livestock prices produced huge losses for ranchers liquidating their herds. Emergency programs established in the New Deal era provided government disaster relief for those who remained.[33] That year marked a turning point in the management and use of the region's resources.

The Period of Adjustment and Accommodation, 1934–1960

The quarter century following the pivotal 1934 drought was a period of ongoing demographic, economic, and environmental adjustment in the Madison Valley. The population of the region declined more than 20 percent between 1935 and 1960. Most of the valley acreage remained in private hands, although turnover of parcels continued to alter patterns of ownership, especially in the lean times of the 1930s. As a result of the Taylor Grazing Act of 1934, federal land managers assumed greater regulatory control over the previously isolated blocks of open range in the valley that had been grievously overused. Most of the mountain country remained in Forest Service hands, where stricter grazing policies gradually contributed to improving range conditions. Private landowners also modified their management practices to accommodate their herds to the environmental realities of the region.

The changing national and global economy continued to shape local livestock production (fig. 16.7). A dramatic shift in the region's livestock industry came with the pronounced decline in sheep grazing as cattle ranching once again assumed a marked dominance across the region.[34] Sheep numbers in the county declined from 200,000 in the early 1930s to less than

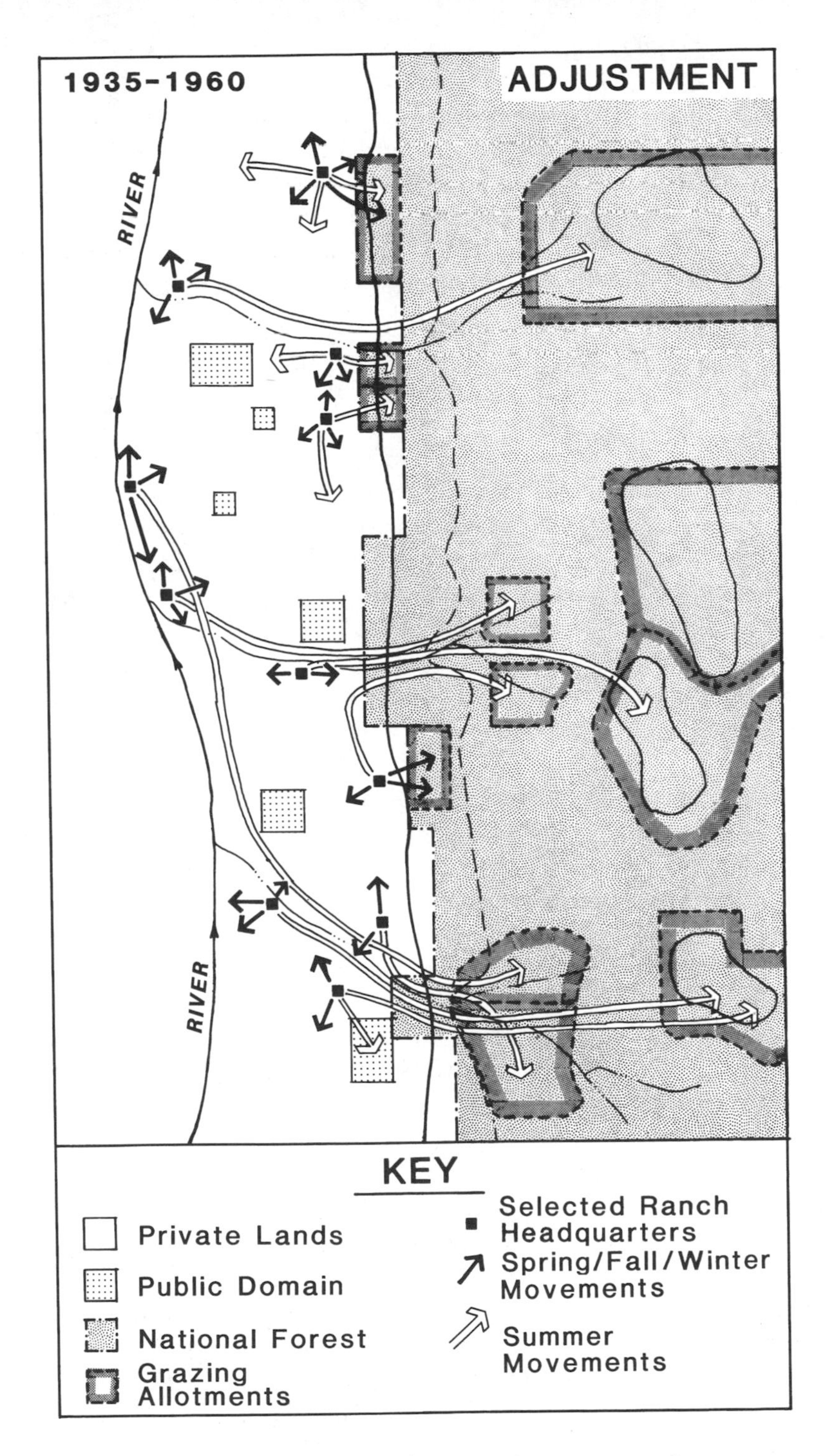

FIG. 16.7. The period of adjustment and accommodation, 1935–1960.

80,000 by 1960. Cattle numbers, on the other hand, more than doubled during the period. Numerous factors contributed to the shift. Local ranchers reported poor demand and low prices for sheep in the 1940s.[35] Cattle ranchers, on the other hand, enjoyed increased demand. Even before the end of World War II, ranchers reported beef cattle numbers at an "all time high."[36] Rising postwar beef prices continued the strong incentive to shift to cattle.[37] Sheep raising was also more labor intensive than the cattle business, and numerous operations lost their herders during and after World War II.[38]

Changes in livestock management also typified the period. Improved grazing practices, combined with adequate precipitation, contributed to overall range improvement after 1934. On public land, the Taylor Grazing Act and more intensive management of U.S. Forest Service grazing allotments contributed to improvement in public rangeland between 1936 and 1940.[39] Forest Service policy encouraged shorter grazing seasons and fewer animals. More stringent enforcement practices further minimized abuse.[40] Finally, increased use of fencing subdivided selected grazing allotments into more appropriate management units.

On private land, range conditions also improved as ranchers learned from the bitter lessons of poor grazing practices and the drought of the early 1930s.[41] Overuse still occurred, most noticeably during the dry years of the mid-1950s, but overall, operators adjusted their expectations to a more realistic understanding of the land's carrying capacity.[42] Cooperative range conservation programs that emphasized improving water resource access, fence building, and grass reseeding of abused pastures also had a positive impact.[43]

Different local environments responded to these changes in the global economy and in management practices in distinctive ways. Meadows experienced a selective recovery of certain grasses and forbs but were invaded by trees as fire suppression, enhanced spring precipitation, and reduced livestock grazing encouraged tree survival.[44] Open foothill slopes also experienced widespread tree and shrub invasion, especially by Douglas fir.[45] Lower grasslands on the foothills and benchlands recovered selectively, depending on levels of continued use and specific management practices.[46] In overgrazed zones, sagebrush, rabbitbrush, and blue grama often increased, although sagebrush invasions were often thwarted by burning or spraying as early as the 1930s.[47] In areas of recovering range (especially where reseeding occurred), wheatgrass, Idaho fescue, and needlegrass flourished. Riparian habitats experienced selected increases in wet meadow bluegrasses, sedges, rushes, brome grasses, and Idaho fescue. Elk and antelope benefited from decreased utilization of benchland grasses by livestock and increased winter storage of hay, while mule deer numbers increased in response to the larger number of palatable trees and woody shrubs, including sagebrush.[48]

The Resource Reallocation Period, Post-1960

Since 1960, the balance of forces shaping the Madison Valley region has shifted once again. The area's agricultural economy still reflects national economic factors that favor cattle over sheep production.[49] Range management techniques, however, have changed significantly, and new cultural attitudes toward the environment have redefined the value of the local resource base.

Four specific changes in cattle management can be noted since 1960. First, the overall utilization of Forest Service land has declined significantly.[50] Grazing seasons have been shortened, and allowable numbers of animals have often been reduced. Similarly, fewer grazing allotments are available for use (fig. 16.8). These changes reflect a gradual shift in the Forest Service's multiple-use policy from one that favored local grazing interests to one that recognizes larger national priorities emphasizing recreation and ecological health. In the context of the Beaverhead National Forest, the process was accelerated by a 1993 lawsuit filed by the National Wildlife Federation and its Montana chapter. The lawsuit alleged that the Forest Service had allowed local ranchers to overgraze their allotments causing harm to wildlife. An out-of-court settlement forced the Forest Service to reevaluate the ecological health of dozens of grazing allotments, to establish a 10-year plan of range review and improvement, and to focus particular attention on protecting and restoring heavily impacted riparian areas.[51]

Second, where grazing persists, management strategies on many public and private parcels have shifted to a greater use of grazing rotation techniques under which stock use a given allotment or pasture for a sequence of different grazing seasons. The new techniques include fallow periods when no grazing is permitted, to allow for range recovery.[52] Rotation strategies have been combined with additional fencing and water storage facilities and with a more careful consideration of the size and extent of allotment and pasture boundaries. These strategies have altered livestock impacts in some areas (fig. 16.8).

Third, a number of ranches have become highly specialized breeding operations. On these ranches, feeding is tightly controlled and cattle are less likely to be turned loose on the range.[53]

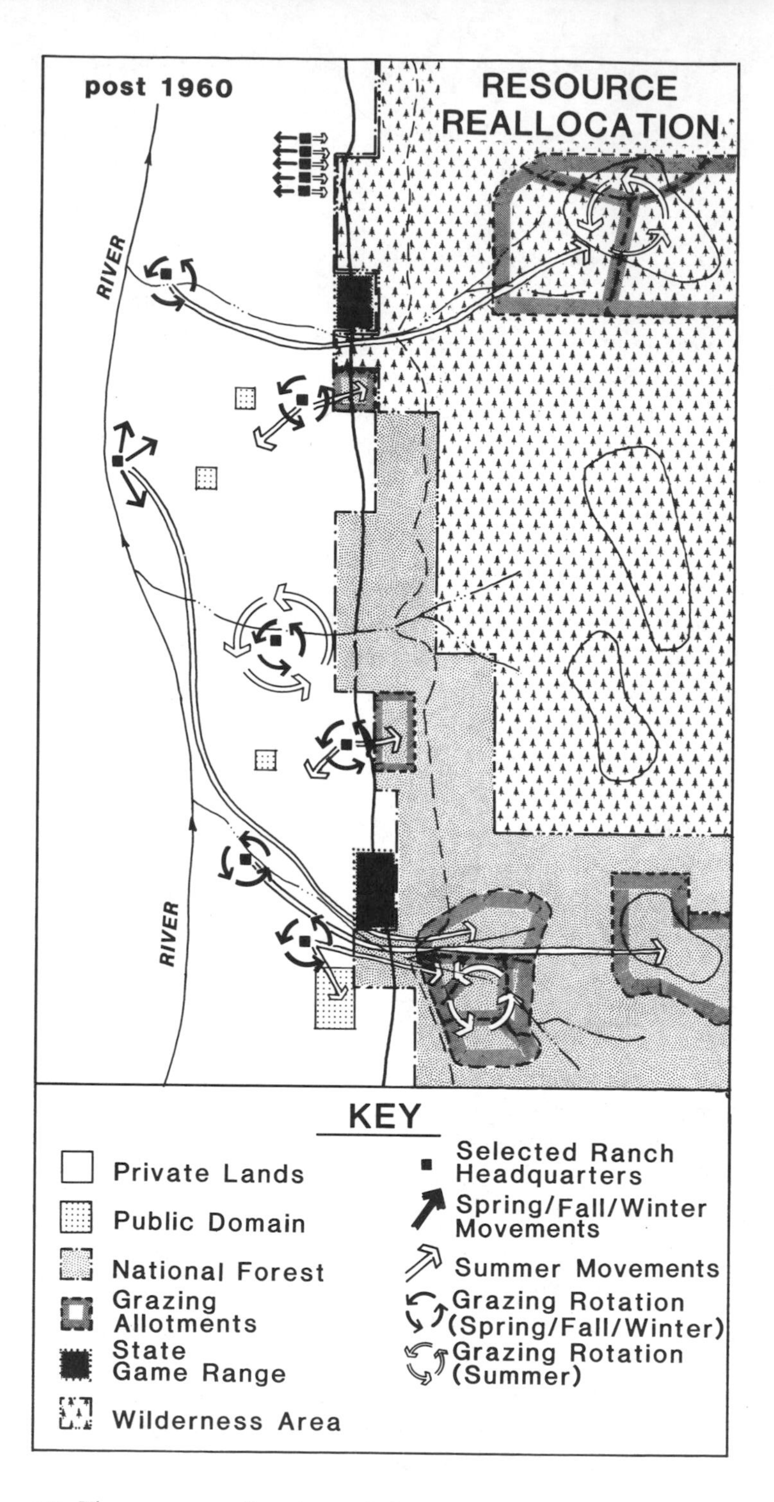

FIG. 16.8. The resource reallocation period, post-1960.

Finally—and with an effect opposite to those of the first three changes—an increasing number of private landowners have quit the cattle business entirely, choosing to lease their land to other local and nonlocal cattle owners. Not surprisingly, these leased acres are the most prone to be overutilized, because returns to the landowner and to the lessee are based on the number of cows the range supports.[54]

In addition to these changes within the livestock industry, broader national trends emphasizing the value of recreational and environmental amenities have contributed to the post-1960 reallocation of resources in the region.[55] Like earlier economic variables impacting the valley, the new forces have been largely external, reflecting the ongoing vulnerability of the area to outside influences. Improved highway and air links to the region hastened the transformation. As before, the character of specific localities was changed as large-scale evolutionary processes manifested themselves. Public land-use policies echoed new calls for more preservation—demands that led to the creation of the nearby Lee Metcalf Wilderness in the 1980s. A number of grazing allotments in the wilderness were closed to promote increased numbers of big game, especially elk.[56] In addition, beginning around 1960 the State of Montana accumulated former private land along the range front near Bear Creek to form a state game refuge.[57] No domestic livestock are allowed there. Instead, the primary goal has been to increase elk, deer, and antelope in the Madison Range (see fig. 16.8).

The importance of recreational fishing has also increased greatly. By 1990 more than 50 professional guide services per day floated the Madison River. Anglers visiting the county poured more than $30 million annually into the state's economy.[58] Ironically, another exogenous force, an introduced parasite that caused whirling disease in the area's rainbow trout, decimated the local economy in the mid-1990s. Guide service revenues, real estate prices, and local businesses in nearby Ennis were adversely affected by a malady that may have been introduced through the illegal stocking of contaminated hatchery fish.[59] If anything, the battle with the disease has increasingly sensitized many to the value of the area's riparian environments, thus keeping the pressure on the ranching community to minimize its impacts on stream waters.

Tourism is important to the local economy in other ways as well. Some ranches have become dude operations, catering to visitors who buy a week or two in the saddle to enjoy a taste of the West.[60] These opportunities are enhanced by the area's proximity to Yellowstone and Grand Teton National Parks, where recreational activities have grown immensely since 1960.

Horses have become the primary stock grazed on these dude ranches, although cattle may also be part of the scene, perhaps as much for appearances as for necessity.

Another sign of shifting priorities in the region is the increasing subdivision of large ranches into smaller parcels for resale, often to wealthy out-of-state residents.[61] By the mid-1990s more than 20 percent of the valley's open land had been subdivided into parcels of 40 acres or less. A growing number of local residents have become alarmed at the area's population growth and are lobbying for more county-level land-use controls to halt the spread of ranchettes.[62] Often these small units support a few cattle and horses, reflecting the appeal of classic cowboy culture. One such example shown near the top of figure 16.8 is a cluster of small ranches located in an area previously dominated by a single large landowner who decided to subdivide and sell his acreage. Typically, the new owners are retirees or other persons who are financially free to live where they wish and are not dependent on livestock operations.

The flora and fauna of the region continue to respond to these larger economic and cultural transformations. Impacts in the higher elevations reflect the decline of domestic grazing and the increasing role of recreation. Tree invasion and meadow recovery continue in areas once heavily grazed that are today in non-use status or only lightly grazed by cattle.[63] On the other hand, rising recreational use has increased human impacts in camping areas, including the introduction of exotic weeds and the failure of conifer seedling survival around camp edges.[64] Excessive use of pack stock by recreational outfitters has resulted in a decrease in vegetal cover and a reduction in both grass and forb stems, particularly in the vicinity of high-elevation camps.[65] Strategies designed to encourage big game have also proven successful, and elk and antelope numbers have increased markedly on both state-controlled and federally controlled land.[66] Further downslope, the foothills zone also continues to see new tree invasion as favorable climatic factors combine with reduced grazing and effective fire suppression policies to enable the establishment of trees in previously open, nonforested sites.[67] Finally, benchland grass environments have recovered somewhat in areas of declining livestock numbers, on ranches where rest-rotation grazing is used, and on land that is now subdivided for recreational and second home use. Poor range conditions remain in areas with leased land that is fully utilized over a number of years, in areas with smaller, more marginal ranching operations, and in areas periodically overgrazed by rising numbers of wildlife, particularly elk.[68]

CONCLUSION

After 1860, the Madison Valley found itself variably shaped by national and even global economic, technological, political, and cultural changes. As Meinig suggested in his assessment of the larger American West, these changes were manifested in several ways. Evolving transportation technologies spelled the end of the isolation of the Madison Valley. In the nineteenth century, improved road and rail links were pivotal in opening the region to larger markets for its livestock products. In the twentieth century, a network of highways and airline routes further opened the region to national influences, including increased numbers of visitors who value the area for its environmental amenities. Meinig's framework also reminds us of the escalating grasp of external political and cultural institutions on these isolated western settings. For the Madison Valley, public land managers represent larger governmental institutions responding to national pressures that have shaped the local environment. The rising impact of recreational activities and the preservation movement also illustrates the increasing significance of outside influences across the region.

Limerick's and Robbins's emphasis on the unequal nature of power relationships in the West is also borne out in the Madison Valley. Its livestock economy is at the mercy of global price and demand trends that have alternatively encouraged first cattle and then sheep raising and then cattle again. For example, strong sheep markets after 1915 led ranchers to increase their numbers of sheep, whereas the bust of the 1930s dramatically reduced sheep numbers and compelled ranchers to return to cattle. The livestock markets were unpredictable and inherently beyond the control of local ranchers. Livestock markets are examples of a resource-hungry economic system that established a colonial relationship with many areas of the American West. The broken lives and failed promises that are a part of the Madison Valley's history (especially during the lean times of the early 1920s and middle 1930s) illustrate how such forces affected particular persons and how the dream of an American frontier with unlimited opportunity often remained a dream unfulfilled.

Finally, Worster's and Neel's perspective reminds us that the western environment was both a key agent and a victim of this development process. The Madison Valley reflects how resource-based capitalism was not always able to adjust to the unpredictability and constraints of the resources being exploited. When forage conditions deteriorated, local land-use managers were not necessarily able to adjust their livestock numbers in ways that contributed to the rehabilitation of the landscape. Often environmental problems persisted or worsened as overworked, drought-ridden ranges continued to be grazed.

The changing settlement eras diagramed in figures 16.2–16.4 and 16.7–16.8 emphasize how the economic, political, and cultural changes impacting the Madison Valley shaped a diverse local environment in distinctive ways. The biogeographies of riparian habitats, benchlands, foothills, and mountain meadows all responded differently, depending on the impacts in those localities and the types of plants and animals being altered. In that sense, our study is but a small-scale reminder of the larger environmental diversity that characterizes the West. The region remains an intricate mosaic of plant and animal communities, a mosaic often defined by subtle variations in slope, elevation, drainage, and precipitation.

Today, conflicts between traditional ranchers, forest service managers, recreational enthusiasts, and wilderness preservationists remind us that natural resources in the West remain finite and vulnerable to forces far removed from the quiet valleys of southwestern Montana.

NOTES

An earlier version of this essay was published as "Settlement, Livestock Grazing, and Environmental Change in Southwest Montana, 1860–1990," *Environmental History Review* 15 (Winter 1991): 45–72.

1. Donald W. Meinig, "American Wests: Preface to a Geographical Interpretation," *Annals of the Association of American Geographers* 62 (1972): 159–84.

2. Patricia Nelson Limerick, *The Legacy of Conquest: The Unbroken Past of the American West* (New York: W. W. Norton, 1987); William Robbins, *Colony and Empire: The Capitalist Transformation of the American West* (Lawrence: University Press of Kansas, 1994).

3. Susan Neel, "A Place of Extremes: Nature, History, and the American West," *Western Historical Quarterly* 25 (1994): 488–505; Donald Worster, "New West, True West: Interpreting the Region's History," *Western Historical Quarterly* 18 (1987): 141–56; Donald Worster, *An Unsettled Country: Changing Landscapes of the American West* (Albuquerque: University of New Mexico Press, 1994).

4. Dan Flores, "The Rocky Mountain West: Fragile Space, Diverse Place,"

Montana: The Magazine of Western History 45(1) (Winter 1995): 46–56; William Wyckoff and Lary M. Dilsaver, *The Mountainous West: Explorations in Historical Geography* (Lincoln: University of Nebraska Press, 1995).

5. Montana Department of Natural Resources and Conservation, Water Resources Division, *The Madison River Basin: A Resource Review* (Helena: Department of Natural Resources and Conservation, 1972), pp. 1–3; D. T. Patten, "Vegetational Pattern in Relation to Environments in the Madison Range, Montana," *Ecological Monographs* 33 (1963): 376.

6. Patten, "Vegetational Pattern," pp. 375–406.

7. Julia Wells Samuel, "Effects of Tree Invasion on Understory Vegetation in Areas Grazed by Livestock, Madison Range, Montana" (M.S. thesis, Montana State University, 1992).

8. Michael P. Malone and Richard B. Roeder, *Montana: A History of Two Centuries* (Seattle: University of Washington Press, 1988), pp. 3–16; Minnie Paugh Papers, Special Collections, Montana State University Library, Bozeman, Montana; Jimmie Spray, "Early Days in the Madison Valley: A History Written in Commemoration of Our Honored Pioneers" (n.p., 1937).

9. John W. Reps, *Cities of the American West: A History of Frontier Urban Planning* (Princeton: Princeton University Press, 1979), pp. 497–502.

10. Paugh Papers; Works Projects Administration (WPA), Livestock History Papers, Special Collections, Montana State University Library, Bozeman, Montana.

11. Paugh Papers; WPA Livestock History Papers.

12. U.S. Bureau of the Census, *The Eleventh Census of the United States, 1890.*

13. Jeffers Papers, Special Collections, Montana State University Library, Bozeman, Montana; Spray, "Early Days." For purposes of comparison, current range management considers five sheep to be the equivalent of one cow-calf unit. See Harold F. Heady, *Rangeland Management* (New York: McGraw-Hill, 1975).

14. Madison County History Association, *Pioneer Trails and Trials: Madison County, 1863–1920* (Virginia City: Madison County History Association, 1976); Paugh Papers.

15. George E. Cope, *Statistical and Descriptive Report upon the Mines, Farms and Ranges of Madison County, Montana* (Virginia City: George Cope, 1888); Jennie Ennis Chowning Papers, Special Collections, Montana State University Library, Bozeman, Montana; James Handly, *The Resources of Madison County, Montana* (San Francisco: Francis and Valentine, 1872); Jeffers Papers.

16. *Pioneer Trails*; M. H. Saunderson, *Readjusting Montana's Agriculture: Economic Changes in Montana's Range Livestock Production* (Bozeman: Montana State College, Agricultural Experiment Station Bulletin no. 311, 1936), pp. 3–5.

17. Paugh Papers; U.S. Forest Service, Madison National Forest records, 1906–23, boxes 8638 and 8639, Supervisor's Office, Beaverhead National Forest, Dillon, Montana.

18. *The Madisonian*, 29 October 1892, 1 July 1893; U.S. Bureau of the Census, *Eleventh Census*, 1890; U.S. Bureau of the Census, *Twelfth Census*, 1900.

19. Interview with Shirley Gustafson, ranch owner, by Wyckoff, Jeffers, Montana, July 1988 (tapes of all interviews are in Wyckoff's possession); WPA Livestock History Papers.

20. U.S. Department of Agriculture, *The National Forests of the Northern Region: Living Legacy* (College Station, Tex.: Intaglio, Inc., 1993), pp. 70–72; *The Madisonian*, 10 February 1894, 9 November 1895; Paugh Papers.

21. Madison National Forest records, boxes 8638 and 8639.

22. *Ibid.*; interview with Parm Hacker, retired sheep rancher, by Wyckoff, Alder, Montana, August 1988; interview with Randall Storey, retired ranch owner, by Wyckoff, Ennis, Montana, July 1988; interview with Lloyd Wortman, retired ranch owner, by Wyckoff, Ennis, Montana, July 1988.

23. Paugh Papers; *Pioneer Trails;* Saunderson, "Readjusting Montana's Agriculture," pp. 3–5; Spray, "Early Days," pp. 194–96; Lloyd Wortman interview.

24. Madison Valley Publicity Club, *Madison County, Montana: Its Resources, Opportunities, and Possibilities* (1911), pp. 35–38.

25. Montana Agricultural Extension Service, Annual Reports of Agents, Madison County, 1921–30, Special Collections, Montana State University Library, Bozeman, Montana (hereafter cited as MAES); interview with Minnie Paugh, retired ranch owner and historian, by Wyckoff, Bozeman, Montana, July 1988; *Pioneer Trails;* M. H. Saunderson, *A Study of the Trends of Montana Livestock Numbers, Prices and Profits* (Bozeman: Montana State College, Agricultural Experiment Station Bulletin 329, 1936).

26. Madison National Forest records, boxes 8638 and 8639.

27. Lloyd Wortman interview.

28. Interview with Chuck Aaberg, retired ranch employee, by Wyckoff, Ennis, Montana, July 1988; Randall Storey interview; Lloyd Wortman interview.

29. Parm Hacker interview.

30. Madison National Forest records; Parm Hacker interview; Randall Storey interview; Lloyd Wortman interview.

31. U.S. Department of the Interior, Bureau of Land Management, *Draft Environmental Statement on Grazing Management in the Mountain Foothills of Montana* (1980), pp. 95–97.

32. C. E. Brooks, *The Living River* (New York: Doubleday, 1970).

33. MAES annual reports, 1934; Minnie Paugh interview; Richard Lowitt, *The*

New Deal and the West (Bloomington: Indiana University Press, 1984). Donald Worster examines the broader economic and political context of New Deal planning ethics in *Dust Bowl: The Southern Plains of the 1930s* (New York: Oxford University Press, 1979), pp. 181–230.

34. U.S. Bureau of the Census, *Fifteenth Census, 1930*; U.S. Bureau of the Census, *Eighteenth Census, 1960*.

35. Interview with Lyn Hoig, retired ranch owner, by Wyckoff, Bozeman, Montana, July 1988; interview with Garnet Oliffe, cattle rancher, by Wyckoff, Cameron, Montana, July 1988; Minnie Paugh interview.

36. MAES annual reports, 1944.

37. MAES annual reports, 1946, 1947; Minnie Paugh interview; Lloyd Wortman interview.

38. Parm Hacker interview; Garnet Oliffe interview; Minnie Paugh interview.

39. U.S. Forest Service, Beaverhead and Gallatin National Forest records, Supervisor's Office, Beaverhead National Forest, Dillon, Montana.

40. Parm Hacker interview; Lloyd Wortman interview.

41. Chuck Aaberg interview; interview with Lloyd Coleman, ranch employee, by Wyckoff, Cameron, Montana, July 1988; interview with Erwin Werner, retired ranch owner, by Wyckoff, Ennis, Montana, July 1988; Lloyd Wortman interview.

42. Parm Hacker interview; Minnie Paugh interview; Erwin Werner interview; Lloyd Wortman interview.

43. Interview with Catherine Armitage, retired ranch owner, by Wyckoff, Cameron, Montana, July 1988; MAES annual reports, 1937.

44. David R. Butler, "Conifer Invasions of Subalpine Meadows, Central Lemhi Mountains, Idaho," *Northwest Science* 67 (1977): 28–45; Lara Dando and Katherine Hansen, "Tree Invasion into a Range Environment near Butte, Montana," *Great Plains Rocky Mountains Geographic Journal* 18 (1990): 65–76; Katherine Hansen, William Wyckoff, and Jeff Banfield, "Shifting Forests: Historical Grazing and Forest Invasion in Southwest Montana," *Forest and Conservation History* 39 (1995): 66–76; Samuel, "Effects of Tree Invasion"; Alan H. Taylor, "Tree Invasion in Meadows of Lassen National Park, California," *Professional Geographer* 42 (1990): 457–70; Thomas R. Vale, "Tree Invasion of Montane Meadows in Oregon," *American Midland Naturalist* 105 (1981): 61–69.

45. S. F. Arno and G. F. Gruell, "Douglas-Fir Encroachment into Montana Grasslands in Southwestern Montana," *Journal of Range Management* 39 (1986): 272–76; Dando and Hansen, "Tree Invasion"; Hansen, Wyckoff, and Banfield, "Shifting Forests"; Samuel, "Effects of Tree Invasion"; B. W. Sindelar, "Douglas-Fir Invasion of Western Montana Grasslands" (Ph.D. diss., University of Montana, 1971).

46. Bureau of Land Management, *Draft Environmental Statement*, pp. 49–64; Montana Department of Natural Resources, *Madison River Basin*, pp. 4–5.

47. Beaverhead National Forest records; Minnie Paugh interview.

48. S. Bayless, "Montana Deer: A History of Change," *Montana Outdoors* (Sept.–Oct. 1975): 8–10; George E. Gruell, *Post-1900 Mule Deer Irruptions in the Intermountain West: Principle Cause and Influences* (Ogden: Intermountain Research Station, General Technical Report INT-206, 1986); V. B. Richens, "Characteristics of Mule Deer Herds and Their Range in Northeastern Utah," *Journal of Wildlife Management* 31 (1967): 651–66.

49. Parm Hacker interview; Garnet Oliffe interview; Minnie Paugh interview; Bret Wallach, "Sheep Ranching in the Dry Corner of Wyoming" *Geographical Review* 71 (1981): 51–63.

50. Beaverhead and Gallatin National Forest records; interview with Sonny Smith, ranch manager, by Wyckoff, Cameron, Montana, August 1988; Lloyd Wortman interview.

51. *Bozeman Chronicle*, 23 December 1993; *Bozeman Chronicle*, 18 June 1995; *Great Falls Tribune*, 4 February 1994.

52. Beaverhead and Gallatin National Forest records; interview with Dale Black, ranch manager, by Wyckoff, Cameron, Montana, August 1988.

53. Interview with Pauline Nelson, retired ranch owner, by Wyckoff, Cameron, Montana, July 1988; Minnie Paugh interview; Randall Storey interview.

54. Interview with Greg Brooks, ranch employee, by Wyckoff, Cameron, Montana, August 1988; interview with Joe Gecho, retired ranch employee, by Wyckoff, Cameron, Montana, July 1988.

55. David A. Clary, *Timber and the Forest Service* (Lawrence: University Press of Kansas, 1986), pp. 169–94; Roderick Nash, *Wilderness and the American Mind* (New Haven: Yale University Press, rev. ed., 1973), pp. 228–62; U.S. Department of Agriculture, *National Forests of the Northern Region*, pp. 201–99.

56. For example, North Bear and Burger Creek allotments are no longer open to domestic livestock grazing. See Beaverhead and Gallatin National Forest records.

57. For land acquisitions, see Land Records and Plat Books, Madison County Clerk's Office, Virginia City, Montana.

58. *Bozeman Chronicle*, 22 December 1994.

59. *Bozeman Chronicle*, 25 August 1996; *Great Falls Tribune*, 25 February 1995.

60. Greg Brooks interview; Lloyd Coleman interview; Sonny Smith interview; Randall Storey interview.

61. Lloyd Coleman interview; Pauline Nelson interview; Garnet Oliffe interview; interview with Ross Stalcup, rancher, by Wyckoff, Ennis, Montana, July 1988; Randall Storey interview.

62. *Bozeman Chronicle*, 19 January 1992.

63. Lloyd Coleman interview; Ross Stalcup interview; Katherine Hansen, Todd Kipfer, and Charles Worth, "Whitebark Pine and Effects of Development," *Proceedings of the International Stone Pine Conference, St. Moritz, Switzerland* (Bozeman: U.S. Forest Service, 1993).

64. K. K. Allen, "The Distribution of Exotic Plants Adjacent to Campgrounds in Yellowstone National Park, USA" (M.S. thesis, Montana State University, 1996); J. Y. Taylor, "Impact of Backcountry Campsite Use on Forest Structure, within Yellowstone National Park, USA" (M.S. thesis, Montana State University, 1995).

65. K. M. Olson-Rutz, C. B. Marlow, K. Hansen, L. C. Gagnon, and R. J. Rossi, "Recovery of a High-Elevation Plant Community after Packhorse Grazing," *Journal of Range Management* 49 (1996): 541–45.

66. Parm Hacker interview; Minnie Paugh interview; Sonny Smith interview; interview with Ken Storey, rancher, by Wyckoff, Cameron, Montana, July 1988.

67. Arno and Gruell, "Douglas-Fir Encroachment"; Hansen, Wyckoff, and Banfield, "Shifting Forests"; Sindelar, "Douglas-Fir Invasion."

68. Chuck Aaberg interview; Dale Black interview; Greg Brooks interview; Joe Gecho interview; Shirley Gustafson interview; Lyn Hoig interview; Sonny Smith interview; Ross Stalcup interview; Ken Storey interview.

17 / Creating a Hybrid Landscape

Irrigated Agriculture in Idaho

MARK FIEGE

Americans have celebrated and decried irrigated landscapes. To some observers, irrigation has represented the attainment of an agricultural ideal in a harsh, forbidding environment. Often with the assistance of engineers and the federal government, farmers "conquered" or "reclaimed" arid "wastelands" and made them "blossom as the rose."[1] Still popular, such sentiments reflect a deeply entrenched national predisposition toward farmers, agricultural landscapes, and utilitarian uses of nature.[2] In recent years, however, historians, geographers, and other commentators have attacked irrigation for economic excesses and environmental damage.[3] The work of journalist Marc Reisner sums up much of this critical viewpoint. To Reisner, the American West's extravagant use of water has produced a "Cadillac desert," an environmentally costly and ultimately unsustainable hydraulic order.

Despite their differences, both celebrants and critics overlook the ways in which irrigated landscapes are the products of an ongoing, dynamic relationship between human systems and land, water, and biota. Irrigated landscapes are neither just controlled, verdant gardens nor simply degraded environments. Idaho's irrigated landscape, for example, represents a synthesis of human designs and recalcitrant natural processes. Irrigators built dams, excavated canals, plowed the earth, and planted crops, but their systems did not function precisely. Streams flowed erratically, canals leaked, and fields and waterways provided habitat for organisms that often disrupted farming and irrigation operations. In response, farmers and engineers adjusted their agricultural practices and renewed their efforts to shape the land to their liking.

Over time, this pattern of reciprocal interactions between human systems and water, landforms, soil, plants, animals, and climate brought into being a hybrid landscape that reflected an ambiguous mixture of artifice and nature. Indeed, in this place "nature" at times could be a nebulous category, as indeterminate as the landscape itself.[4]

TRANSFORMING THE PRE-IRRIGATION LANDSCAPE

To build irrigation systems, farmers and engineers in Idaho's Snake River valley altered existing geological formations, streams, flora, and fauna. Between the 1860s and 1920s, small groups of farmers, private corporations, and the federal Reclamation Service (renamed the Bureau of Reclamation in 1923) gradually covered the region with some 13,000 miles of canals that carried water to approximately 2 million acres.[5] Small groups of settlers developed the first systems. These usually consisted of temporary rock, brush, and canvas diversion dams and relatively short canals (often less than 10 miles) that conveyed water to small tracts of irrigated land totaling from a few hundred to several thousand acres. After 1900, private corporations and the Reclamation Service built large dams and developed huge canal networks hundreds of miles long that encompassed up to 200,000 acres. To lay out their fields, farmers plowed the grassy, brushy bottomland along creeks and rivers, but mostly they tore up the vast stands of big sagebrush (*Artemisia tridentata*) and associated grasses in the arid soils above the streams.

Construction of irrigation systems destroyed plant and animal habitat, but the engineers and farmers did not simply replace "pristine" ecosystems with artificial agriculture. Most landscapes are dynamic, are shaped by humans, and do not exhibit a simple "balance of nature."[6] Even before irrigation, human actions altered the stream bottoms and sage-covered land of Idaho's Snake River valley. Indians burned the land and hunted on it for millennia. From the 1810s to the 1830s, fur trappers extracted beaver and other animal skins from it. During the overland migration of the 1830s–1860s, horses, mules, and oxen consumed grasses that flourished in the shadow of the sagebrush and in the moist stream bottoms. Between the 1860s and the early 1900s, stock growers herded cattle and sheep through the area. Irrigators thus did not impose their systems on a static, undefiled environment. They contributed another level of change to an already changing land.[7]

For irrigators, changing the land required manipulating the water that flowed through it. Idaho irrigators tapped streams that in turn were part of a larger process operating everywhere on the planet—the hydrologic cycle.[8] As snow and rain, water falls to earth. Plants absorb and transpire it, and it evaporates into the dry air. Some water infiltrates the soil and percolates downward into gravel and volcanic rock, forming underground aquifers. In Idaho, most water falls on mountains adjacent to the arid Snake River valley: the Boise, Sawtooth, Pioneer, Lost River, Lemhi, and Beaverhead ranges to the north; the Owyhee, Albion, Deep Creek, and other mountains to the south; the Caribou, Snake and craggy Tetons to the east. Streams pour from these mountains and converge on the Snake, which emerges from the eastern ranges and flows due west in a huge arc across southern Idaho.[9]

Idaho's early irrigators sought to intercept the hydrologic cycle and make it serve their agricultural objectives. As they put it, they endeavored to "tame" or "harness" obstreperous rivers such as the Snake and make them "do duty" for "man." To irrigators, rivers were akin to wild horses that needed to be broken and drafted into productive labor. So to break and harness the rivers, the irrigators built dams: Milner, Minidoka, Jackson Lake, American Falls, Palisades, and many others.[10]

HARNESSING A RIVER?

Today this metaphorical harnessing is most dramatic at Milner Dam on the south-central reach of the Snake. Completed in 1904, the rock-fill structure diverts water into irrigation canals on both sides of the river. The Twin Falls Main Canal runs west about 85 miles to the vicinity of Twin Falls, where it serves 200,000 acres of farms. The North Side Canal runs northwest roughly 35 miles to 130,000 acres around Shoshone.[11] When the canals are full, the water flows with menacing speed and power. But below Milner Dam, save for a miniscule trickle of water, the Snake is usually dry and empty. Only in spring, when warm weather melts the accumulated mountain snow, is there enough water in the river to flow over Milner Dam.

Early irrigators viewed the dry Snake as evidence of their ability to control the natural environment. On March 1, 1905, a crowd gathered as Twin Falls Land and Water Company officials first closed the gates at the dam and stopped the river. "As the gates dropped lower and the roar of the river grew fainter," one storyteller recounted, "the voices of the people dropped almost to a whisper. There was a stillness in the canyon such as it had probably never known before." About 20 miles downstream, another crowd watched the dramatic recession of water flowing over Shoshone Falls, the "Niagara of the

West." To this audience, the irrigators' ability to shut off the falls had superseded the falls itself as an example of sublime power. The next morning, irrigation company officials staged one more ceremony when they let water into the Twin Falls Main Canal: "As the water flowed into the big canal to be carried out to transform and redeem a desert, cheer after cheer was given and bottles of wine were broken over the gates."[12]

Irrigators certainly had every reason to celebrate. They had mastered the Snake; they had won control of the water. Or had they? However dramatic, Milner Dam and the dry bed constituted but one place on a long, complex stream, itself only one element in a still larger pattern of water movement. Those early irrigators mastered the Snake well enough to create a productive agriculture, but they could not precisely control the entire river or the hydrologic cycle. Dams such as Milner intercepted and appropriated the rivers; in turn, rivers, the hydrologic cycle, and even the natural characteristics of water itself preempted and compromised dams and irrigation systems. The result of this interplay was a hybrid river—a blend of human systems and natural hydrological processes—that caused the irrigators a great deal of trouble.

Consider the early history of Jackson Lake Dam, several hundred miles above Milner in Wyoming, within the shadow of the Tetons.[13] The U.S. Reclamation Service built the structure in 1907 to store water for the federal government's Minidoka irrigation project and for farmers downstream at Twin Falls. The government's plan of operation seemed simple enough: impound the spring runoff in a reservoir, then release it in the summer and let it flow down the Snake to the Minidoka and Twin Falls irrigators. The Minidoka and Milner dams would then catch and divert this "stored water."

In practice, this system was complicated by a second group of irrigators and hydrological conditions. Irrigators who lived midway between Jackson Lake and Minidoka–Twin Falls, and who antedated Jackson Lake Dam, held rights to the undammed, "natural flow" of the Snake. Since the 1880s, these farmers had simply diverted the Snake and applied the water to their land. But with construction of Jackson Lake Dam, they had to differentiate between "natural flow" and "stored water," which now blended in the Snake channel. Doing so was not easy, given the complexity of the river.

Over its length and over time, the Snake fluctuated in volume. In some reaches, it lost water to evaporation and a gravel bed. In other places, it gained from tributaries, rainfall, a high water table, springs, or the "return flow" that seeped in from canals and fields. Even these groundwater inflows varied. When the Snake swelled with spring runoff or stored water, the river's mass blocked groundwater entry; when the river receded, groundwater filtered in. Moreover, the volume of the reservoir itself fluctuated. Pooled in Jackson Lake, varying amounts seeped into the ground; when the Reclamation Service drew the reservoir down, different quantities of "bank storage" percolated back in. Under such conditions, the exact quantity of stored water or natural flow at any given point on the Snake River was difficult to determine. To which category did a loss or gain apply?

Public officials who administered the Snake—locally elected watermasters, U.S. Geological Survey employees, and deputies of the state engineer—measured the river at numerous places and then tried to calculate the losses and gains that each category sustained. That their calculations were informed estimates, not precise science, did not necessarily bother the irrigators. In years when the Snake River provided abundant water, it essentially washed away the irrigators' need to know the exact amounts of stored water and natural flow. But during the 1910s and early 1920s, increasing use of water and then drought made loss and gain critical to the irrigators. Conflict developed. Both natural flow and stored water users complained that public officials shorted them; accusations, lawsuits, and threats of lawsuits abounded.

The fighting consumed time and money but resolved nothing. As one Idahoan said, lawsuits enriched lawyers but never made more water.[14] Consequently, the two groups of irrigators gradually turned away from conflict and began to search for ways to overcome their differences. One possible solution was technical: if they could know the river scientifically and precisely, the irrigators surmised, they could accurately and justly divide the water. In 1923, they decided to hire an engineer of "disinterested expert judgement" to help them comprehend the Snake. They sought someone with "broad experience in the control of 'intermittent' rivers—rivers that disappear and rise again; rivers that belong to different states or counties or valleys thru which they flow—through which they flow and sink and rise and sink again, serving perhaps different settlements over and over as they travel on and rise as return flow in the parent stream."[15] After a brief search, they asked Ralph Meeker of Denver to study the Snake and recommend a water distribution plan.

Meeker's conclusions ultimately disappointed the irrigators. The engineer found that the problem of loss and gain to stored water and natural flow had no clear technical answer. "Flowing water," he wrote, could not "be measured with the same precision" as "an acre of land . . . or a ton of wheat";

it could not be divided "with the same accuracy as a quart of milk."[16] Meeker easily could have added that a dam and a river did not function like plumbing, like a stoppered bathtub that engineers turned on and off like a tap.[17] Irrigation systems—Jackson Dam and water that seeped from irrigation canals—complicated an already complex river, making it even more difficult for irrigators to control.

With no precise technical solution—and with drought imminent—the irrigators decided to negotiate and compromise. By 1923, the two sides already had made tentative efforts to formulate methods of allocating the river without fighting. Irrigators met and devised a somewhat arbitrary plan for determining losses and gains to stored water and natural flow. In essence, quantities such as stored water loss and return flow served as bargaining chips in the negotiations that produced the plan. In 1924, for example, the natural flow users accepted a minimal percentage loss to stored flow and agreed that bank storage at Jackson Lake reservoir equaled the bank storage prior to the dam's construction.

The opposing groups quickly institutionalized the new allocation methods in an extralegal organization called the Snake River Committee of Nine. Each subsequent year, nine representatives met and hammered out a schedule of water distribution. Ironically, the irrigators' attempts to alter and control nature had trapped them in conditions that in turn forced them to change. They had to establish a new social institution—the Committee of Nine—through which they adjusted to the hybrid river that they had helped to create. The ritual at Milner Dam celebrated the ideal and promise of environmental control; the Committee of Nine represented a more ambiguous reality.

RIVERS, FISH, AND BIRDS

Irrigation altered the hydrology of rivers such as the Snake. It also transformed the streams ecologically. These changes further exemplified the ways in which human systems blended with streams to produce disorderly new environments that did not always accord with the irrigators' plans.

The ecological transformation of streams began with habitat destruction. When engineers and farmers intercepted the hydrologic cycle, they ruined the aquatic habitat of cold-water fish such as trout and salmonids (*Onchorynchus* spp.). Some dams slowed migration and spawning runs, and they also slowed water and raised temperatures to intolerable levels. Warm water, suffused with nitrogen and phosphates from agricultural runoff, promoted the growth of plants such as algae that in turn deprived fish of oxygen. Irrigation diversions, especially during drought, exacerbated the harsh conditions, as did farm pesticides that filtered into streams. Irrigation systems also trapped fish. A trout that swam into a canal might never get out; death in a farmer's field might be its fate. And dams inundated the riparian habitat of much terrestrial fauna.[18]

Irrigation systems, however, did not just destroy habitat; dams paradoxically enhanced habitat for other fish. Indigenous species such as Utah chub (*Gila atraria*) and suckers (*Catostomus* spp.) proliferated in the warm, deoxygenated, muddy water. Introduced fish, notably the carp (*Cyprinus carpio*), also flourished there.[19] During the 1880s and 1890s, Idaho farmers acquired carp from the U.S. Fish Commission and brought them to the irrigated landscape, intending to raise the species in ponds for food. By accident and intent, the fish spread through drainage ditches and streams. Not only was the carp ideally suited to a "degraded" aquatic environment, but it also helped to create its own habitat. A voracious feeder, it edged out other fish, and by rooting in the mud for plants, it stirred up sediments that made the water intolerable for species needing cleaner conditions. Much to their dismay, Idahoans later discovered that they could not eliminate carp from their rivers. In its own way, the carp reflected the larger transformations that were creating the irrigated landscape. Manipulated and introduced by irrigators, the fish became part of a hybrid environment over which humans exercised incomplete control.[20]

Reservoirs also exemplified the ways in which human systems and wildlife habitat intermingled.[21] Reservoirs inundated habitat, but many birds flocked to these bodies of water and there found resting and feeding sites and, on occasion, places to nest. During the early twentieth century, thousands of birds congregated around American Falls, Lake Walcott, Lake Lowell, and other reservoirs. Taking note of the birds, government officials decided that the reservoirs would serve a secondary role as wildlife refuges. In 1909, for example, executive order established the Minidoka National Bird Reservation (later national wildlife refuge) at Lake Walcott, the reservoir behind Minidoka Dam. The USDA's Bureau of Biological Survey managed the site, which attracted an array of avian species including white pelicans (*Pelecanus erythrorhynchos*), great blue herons (*Ardea herodias*), cormorants (*Phalacrocorax auritus*), avocets (*Recurvirostra americana*), ducks (*Anas* spp.), and numerous other birds that frequented lakes and wetlands.[22] Nothing could have been more ironic: a "wildlife refuge" on a body of water that many people would have considered "man-made."

NOT LIKE PLUMBING

The Snake certainly did not function like plumbing; neither, for that matter, did canals. As much as dams, canals have symbolized the quest to dominate and control nature.[23] Idaho farmers and engineers, however, found that they could not precisely regulate these conduits. Uncontrollable, disruptive geomorphic and hydrological processes affected the operation of the canals almost as much as the irrigators themselves.

Canals and ditches are constantly exposed to the action of weather and climate. Virtually all Idaho irrigation conduits are earthen structures, either excavated from the ground or built up from soil and rock. As soil alternately freezes and thaws, banks gradually slump. Runoff from rainstorms on frozen or steep ground deluges canals and ditches, erodes their surfaces, and fills them with silt. Even the creeks and rivers that feed irrigation systems carry silt into them. Maintaining the physical condition of the conduits absorbed a great deal of the early irrigators' resources, especially in the largest systems. Irrigation companies and districts retained workers whose duties included dredging and reshaping canals and ditches, not simply shunting water through them.[24]

Irrigators, moreover, sought to regulate the hydrologic cycle, but that same natural process in turn influenced the operation of the canals themselves. Some of the water evaporated, but even more—as much as 30 to 60 percent—seeped into the ground, particularly where porous soils underlay the canals, ditches, and fields. In 1912, irrigators on the north side of the federal government's Minidoka project lost almost 50 percent of the 314,000 acre feet of water that they diverted from the Snake River. This seepage had important consequences for the evolution of the irrigated landscape.[25]

To stop the seepage and to retain the shape of canal and ditch banks, Idaho irrigators might have lined their canals with concrete. Except in a few places, they did not. The cost of construction stopped them, but more importantly, climate discouraged the practice. Idaho's frosty winters cause concrete to flake and crumble, much more so than in the warmer irrigated landscapes of California or Arizona. Maintaining concrete canals under such harsh conditions would only have added to the heavy installation expense.[26]

A BIOLOGICAL PROBLEM

Earthen canals not only leaked, they also provided a habitat for plants and animals that in turn hampered irrigators' ability to manipulate water. Although not usually recognized as such, canals became a new aquatic and riparian habitat that consisted of earthen structures, water, and a biota that irrigators struggled to control.

Exposed soil on the canal banks provided outstanding conditions for weeds, plant species that thrive on disturbed land and that are the bane of all farmers. Tumbleweeds (*Salsolsa kali* L. and other species), for example, spread through canals and flopped into the water, inhibiting flow and thus increasing seepage. Other "weeds" included cultivars such as alfalfa (*Medicago* spp.), sweet clover (*Melilotus* spp.), and barnyard grass (*Panicum crus-galli*) that moved in from adjacent fields and pastures. Perhaps the most troublesome "weeds" were those that irrigators deliberately introduced. During the early 1900s, irrigation system managers planted willow trees (*Salix* spp.) along ditches and canals in an effort to stabilize the soil. Their dense, tough roots held the banks in place, but at the price of tremendous quantities of water that were transpired back into the air through the leaves. Irrigators eventually concluded that they could no longer afford the price and attempted to destroy the willows. But when they cut down the trees, vigorous roots sent up fresh new shoots.[27]

Aquatic plants spread as well. These flora included submerged plants such as pondweed (*Potamogeton* spp.) and algae, which irrigators lumped in the category of "moss." Canals and ditches also provided habitat for cattails (*Typha* spp.) and other species whose stalks grew above the water. As prolific and tenacious as their terrestrial cousins, aquatic weeds posed major problems for irrigation system operation. They trapped silt, further reducing canal capacity; they slowed water movement, increasing ground absorption and seepage; they interfered with the measurement of water in flumes and weirs; and they transpired water.[28]

Terrestrial and aquatic flora in turn provided food and cover for small mammals that undermined canal banks. Pocket gophers (*Thomomys* spp.) inhabited the margins of the waterways, as did the badgers (*Taxidea taxus*) that preyed upon them. Marmots (*Marmota flaviventris*) and ground squirrels (*Citellus richardsoni*) also lived there. Aquatic mammals likewise found the canals and ditches to their liking. Beavers (*Castor canadensis*) spread from creeks and rivers into irrigation systems, where willow and other plants provided food and material for lodges and dams. Muskrats (*Ondatra zibethicus*) took nourishment from the fleshy, starchy roots of cattails and built homes in the banks.[29]

The result was sometimes devastating. Seeping water penetrated mammal burrows and saturated surrounding soil; canal banks gradually weakened and collapsed. Difficult and costly to repair, canal breaks could spell disaster for farmers. Breaks flooded fields and could leave irrigators without water for days. In July 1923, ground squirrels caused a serious washout on the main conduit of the Aberdeen-Springfield Canal Company (ASCC) system. On one section of canal consisting of fill, the "squirrel holes had been giving some trouble and had been the occasion for anxiety." When the bank finally collapsed, water surged out and cut roads and railroad grades and flooded several farms. Workers had difficulty closing the headgates at the Snake River, and water continued to flow from the canal breach for 24 hours. The destruction halted water delivery for about a week, until the canal company rebuilt the structure. The break also interrupted railroad service on the Oregon Short Line's Aberdeen branch for six days. When laborers repaired the railroad bed, they missed a section of embankment that the flood had weakened. As the first train passed over it, the bed crumbled and the locomotive tumbled into the water.[30] Collapsing technology—canal and railroad—underscored irrigators' sometimes fragile control of water.

In their effort to tame rivers and master nature, engineers and farmers had inadvertently provided a habitat for organisms that in turn threatened to overwhelm irrigation systems. To retain control, irrigators adopted various techniques to rid the canals of plants. They sometimes burned and mowed bank vegetation. Irrigation districts and companies invited the pasturing of sheep flocks along ditches and canals because the grazers were so effective at cropping unruly plants. Hydraulic technology thus provided a common pasture. Irrigators even erected fences along the waterways to help confine the animals that farmers and stockmen brought in to graze.[31]

While irrigators battled terrestrial weeds, they carried on another fight against the aquatic plants that grew rank each summer. When the "moss" proliferated in the hot sun, irrigation system managers drained the canals to deprive the plants of water. They used exotic technologies such as the Ziemsen submarine saw to clean weeds out of canals and ditches. In 1918, Twin Falls Canal Company (TFCC) workers modified disc harrows so that horses could drag the implements through the canals. One of the most popular techniques was "chaining." Horses or tractors on opposite sides of a canal each pulled one end of a heavy chain, dragging it through the water and tearing the "moss" loose. Workers then heaped the cut vegetation in huge piles to putrefy in the desert heat.[32]

Irrigation companies and districts routinely attempted to clear small mammals from their canals. Much to their delight, irrigators discovered that sheep also kept rodents in check. By keeping the vegetation down, the flocks deprived small mammals of cover, and their hooves trampled and destroyed burrows. In 1922, TFCC manager J. C. Wheelon expressed hope that with grazing, the irrigation system "would become a model of neatness and convenience instead of an unsightly jungle of willows and a harboring place for all kinds of rodents and noxious weeds."[33] Hunting and trapping became a regular part of canal and ditch maintenance. Irrigation districts and companies authorized farmers and trappers to run trap lines, and they often paid small bounties for ground squirrels and gophers. In 1932, ASCC managers sent form letters to local school principals, asking them to inform students that the company would pay 10 cents for gophers. Public officials assisted irrigators with trapping. In 1945, the Idaho Fish and Game Department divided the state into "beaver allotments" in which "caretaker" trappers removed beaver from "problem areas" such as canals.[34]

With chemicals, irrigators turned the full force of their modern technology against the pests. Idaho farmers and agronomists experimented with herbicides during the early 1900s but did not make widespread use of the chemicals until 1920, when they applied sodium arsenite to perennial weeds. Later in the decade, farmers and their allies began to use carbon bisulfide, sodium chlorate, and other treatments; they even used plain salt. After World War II, irrigators doused weeds with still more potent herbicides, such as 2,4-D, and they poured compounds with names like xylene, aqualin, and acrolein into canals and ditches.[35]

They also poisoned burrowing mammals. During the early twentieth century, irrigators laced oats and carrots with strychnine and scattered the bait through their canal systems. After World War II, they adopted a more powerful poison: sodium fluoroacetate, commonly known as compound 1080 or just "ten-eighty." A host of government and public agencies—such as the U.S. Department of Agriculture's Bureau of Biological Survey—assisted irrigators in poisoning mammals.

A SOCIAL PROBLEM

Weeds and small mammals were not only biological problems; they were also social problems. As these organisms spread through canals, ditches, and fields, they complicated an important human institution—private property—that structured the irrigated landscape. Weeds cast seeds into irrigation conduits; the water then carried the seeds across property boundaries onto farms and, ultimately, into creeks and rivers and still other irrigation systems. Borne on the breeze, the seeds passed from one piece of land to another. Small mammals traveled through canals and ditches and then scurried across property lines onto farms; in turn, the animals moved from farms and public land into other irrigation systems.

The passage of weeds and small mammals across boundaries led to controversy: who was responsible for these mobile organisms? Some irrigators criticized their careless neighbors who, they claimed, did nothing to destroy weeds that launched seeds onto surrounding farms. Irrigation system managers complained that while they struggled to destroy burrowing mammals, the animals kept coming in from adjacent farms, other private holdings, and public land.[36]

A social problem now required a social solution. Many irrigators contended that to counter pests effectively, landowners had to work together. They also argued that community good must take precedence over individual rights: it was wrong, they said, for some landowners to allow weeds and rodents to proliferate on their property. Accordingly, during the early 1900s, the Idaho legislature established a series of weed and rodent laws that qualified the private property right in the interest of agricultural communities.[37] Under the statutes, county officials could require landowners to destroy weeds and small mammals within designated areas, often called weed or rodent districts. The measures also authorized officials to control pests on the property of uncooperative landowners and to collect payment for this work. In addition, the laws allowed officials to use county funds for cooperative, public control programs carried out under the auspices of government agents.

A RESISTANT BIOTA

Given the irrigators' enormous killing apparatus—ranging from deadly poisons to coercive laws—one might expect that the flora and fauna inhabiting the canals were easily wiped out. In fact they were not. Weeds proved amazingly tenacious. Mechanical removal of aquatic plants sometimes only hastened their spread. Cutting or chaining shook loose seeds that landed in sediment and sprouted. Fragments of roots and leafy stems regenerated. Powerful herbicides could not completely stop either aquatic or terrestrial weeds. As the chemicals wore off—as plants absorbed them and water washed them away, or as they percolated deeper into the ground—weeds again moved into the exposed soil. In 1935, the TFCC manager reported that "the progress made in killing weeds on canal banks is not very encouraging." Each year, he said, the company killed one patch of weeds only to find a new patch growing somewhere else. "The total area" infested with weeds, he stated, remained "about the same from one year to another."[38]

Irrigators noticed, too, that weeds developed resistance to chemicals and that new, unusual species often took the place of old ones. Sometimes these new weeds resisted herbicides, which allowed them to thrive in places once occupied by species that chemicals had destroyed. In effect, the killing of one weed opened space for another. Irrigators thus looked on as a strange and bewildering array of weeds continued to parade through their canals, ditches, and fields. In 1954, the TFCC manager observed the appearance in the system of "new types of moss," including "a very bushy type" and "other types" that grew "as much as 3 feet long," clung tenaciously to rocks, and required "more frequent mossing."[39]

Extermination methods could not stop rodents and other small mammals from moving into irrigation systems either. In addition to the ability of these animals to find secure places to eat and reproduce, control programs contained inherent limits. Economics conscribed trapping and hunting in irrigation systems; when fur prices fell, trapping waned. Once trappers had reduced a population to a certain level, they moved on. Broad economic trends, not just the economy of trapping, also restricted hunting and trapping. During the 1940s, gasoline rationing, more remunerative jobs elsewhere in the economy, and the movement of young men into the military lessened the intensity of trapping and hunting.

Moreover, the rodent control act was at times inefficient and cumbersome: implementation of the law required petitions from farmers, identification of problem areas, coordination between the public and government agencies, and the allocation of limited funds. Human values, too, inhibited control efforts. Because many Idahoans valued muskrat and beaver for their furs and as important wildlife symbols, laws regulated trapping. Occasionally the legislature relaxed the law to allow irrigators to take beaver and muskrat, but ultimately the statutes still provided the animals with a measure of security in and around irrigation systems. Regulations on the use of poison similarly influenced the survival of rodents and other small mammals in irrigation systems. In 1972, public opposition to poisoning compelled government restrictions on the use of compound 1080, making destruction of ground squirrels and pocket gophers more difficult.

Thus, even after years of effort, irrigators still found mammals in their canals. "After ten years of elimination work on the ground squirrels they seem to be as numerous as ever," one irrigation system manager stated in 1938, "but . . . we believe that if this extermination work [were] not kept up . . . these rodents would soon destroy the canal system."[40] Engineers and farmers could not regulate their conduits as precisely as a machine. A dynamic mixture of flora and fauna, much of it unwanted, became integral to the entire hydraulic system.

FAMOUS POTATOES

And how well did farmers control their crops, the ultimate purpose of all that irrigation water? Irrigated fields, too, stood in apparent testimony to the human ability to transform and control the natural environment, a view vividly expressed by the *Idaho Republican* in 1907 when it announced the arrival of a settler who had come to take up farming on an irrigation project near American Falls. "Loomis," the paper stated, "left a nice home in Illinois to make a new home on this manless, homeless, weedless, bugless tract where he can put everything on the land just to his liking."[41] Even today, irrigated cropland gives the impression of a precisely arranged environment—subdued, static, controlled. But the same reciprocal interaction between human systems and soil, water, flora, and fauna going on in the canals also affected the crops that farmers grew.

Consider Idaho's most famous crop: the potato. "The Idaho" was the product of a complex interplay between agriculture and the land. In the early 1900s, Idahoans grew no single potato but rather numerous varieties: rural New Yorker, cobbler, bliss triumph, bull moose, Idaho rurals, peachblow, russet Burbank, and others. The farmers achieved remarkable success with these spuds, which encouraged them to intensify production. Year after year they turned out one crop after another. Such concentrated, repetitive cropping, however, made the potatoes increasingly vulnerable to microorganisms. Idaho's desert soils contained, for example, indigenous fungi (*Fusarium* spp., *Rhizoctonia* spp.) that parasitized the tubers; with prolonged production, the fungi increased and gradually overwhelmed the fields. By 1914, fungi and other diseases had destroyed up to one-half of some crops.[42]

One potato variety, the russet Burbank, resisted the attack of microorganisms, especially "scab" fungi, and so disaster yielded the spud that would become "the Idaho." Because of its hardiness, and because consumers liked it, Idaho farmers raised more and more of the russet Burbank. By 1929, it constituted roughly 60 to 70 percent of all potatoes grown in Idaho. This figure rose to 90 percent by 1939; by 1989, the russet Burbank amounted to 95 percent of the entire crop.[43]

Thus Idaho's best known and one of its most lucrative crops owes its existence, at least in part, to the influence of disease. Farmers had not simply put everything on the land just to their liking; rather, the microorganisms that ravaged their fields had limited the range of potatoes they could grow profitably. True, the triumph of the russet Burbank showed the farmers' ability to make a wise choice for the region's growing conditions, but they had hardly imposed their agricultural agenda on the land. The contingency of the irrigated landscape, not just the farmers' intelligence and competence, had yielded "the Idaho" potato.

THE CONSEQUENCES OF SEEPAGE

For linear-minded engineers and farmers, potatoes and other crops marked the end of the passage of water through the irrigated landscape. But the hydrologic cycle, not the linearity of irrigation systems, ultimately governs the movement of water everywhere. Water that seeps from canals and fields does not simply disappear into the ground. In Idaho, much of it has remained in the irrigated landscape and created new problems—and opportunities—for engineers and farmers.

In certain places, seepage raised groundwater levels and saturated farm soils. When water seeped out of canals, ditches, and fields, it encountered impervious layers of volcanic rock or subsoils such as clay that restricted its downward movement. In addition, seepage from irrigation on high ground percolated along gradients to lower land, especially along river bottoms. After prolonged irrigation, groundwater rose and created springs, filled depressions, formed pools on the surface, and in general made the soil soggy. Crops grew poorly or not at all on such land.[44]

Waterlogging, moreover, is not the only condition seepage creates. Seepage water can concentrate salts in the soil, making it somewhat toxic to cultivated plants. Water dissolves and leaches out salts that occur naturally in arid soils. As the water evaporates, the salts precipitate and form patches of thin, poisonous crust on the land.[45]

An area on the south side of the Boise River west of Caldwell shows how the process of seepage and salinity gradually damaged farmers' ability to carry on agricultural production. Originally, this bottomland was the scene of intense farming. In 1877, journalist W. A. Goulder described a portion of this area, the Dixie Valley, also known as Dixie Flats. Goulder observed farmers raising hay, grain, and vegetables in the rich alluvial soil. "The settlers here are all delighted with the country," he wrote, "each one thinking that this valley is the best portion of Idaho, and that his own home occupies the best place to be found."[46] Over the next decades, however, irrigation and seepage from higher ground waterlogged the soil and concentrated salt in it. In 1909, Henry A. Wallace (later secretary of agriculture and vice president) traveled through the area and reported "a good many ragged spots in the grain fields along the river bottom land" caused by patches of salt. By 1940, around 35,000 acres were wet and salty. One can hardly overlook the irony that waterlogged land posed: the water that the irrigators coveted and purported to control had become a serious nuisance.[47]

Waterlogged land provoked engineers and farmers to initiate a new phase in the development of the irrigated landscape: the construction of drainage systems. Drainage ditches, dug below the level of the fields and irrigation canals, would collect the seepage water and carry it back to creeks and rivers. By 1920, Idaho irrigators had excavated roughly 290 miles of drains that served a total of about 65,000 acres. Ten years later, they had excavated 651 miles of ditches to serve approximately 375,000 acres.[48]

Drainage was a complex matter: it not only required costly labor and expertise, but more importantly, it was a controversial social problem that demanded collective responses. Like weeds or rodents, seepage water ignored private property lines. Often, water from canals and farms on high ground seeped down and waterlogged neighboring low land. Farmers who lived on the sodden ground thus were not entirely responsible for seepage, nor did they alone have the resources to construct drains. Consequently, all irrigators on a project or within the same canal system usually paid for drainage, even farmers on high ground whose fields were not waterlogged or salty.

Not all farmers on high ground accepted this responsibility. Drainage systems thus became a dividing line in political and legal controversies. During the 1910s and 1920s, groups of farmers on high ground contended that they should not have to pay for drainage. The seeped land was not theirs, they argued, and they received no direct benefits from drain construction. In 1917, about 150 farmers in the Nampa-Meridian Irrigation District, a project in the Boise Valley, objected to a drainage assessment on their high land. The irrigation district's board of directors then petitioned for a judicial ruling on the legality of the levy. In 1918, the court approved the blanket assessment, confirming the district's power to apply the same per-acre charge to both seeped land and high ground. The irrigated landscape was not just the sum of the private holdings that constituted it; seepage turned it into a hydrological whole, and the law at least tacitly recognized this fact.[49]

Seepage, salinity, and drainage posed great problems for irrigated agriculture, but these developments did not simply represent environmental degradation. Seepage and salinization altered but did not destroy the land's capacity to sustain plant and animal life. Where farmers stopped planting crops, or where crops simply would not grow because of salt or waterlogging, uncultivated flora moved in and flourished. These included greasewood (*Sarcobatus vermiculatus*), rabbitbrush (*Chrysothamnus* spp.), foxtail barley (*Hordeum jubatum*), and saltgrass (*Distichlis spicata*), species that tended to tolerate salty, soggy ground.[50] Farmers then released livestock to forage in this new floral assemblage; what had been cropland became pasture. To be sure, it was not the best of pastures. Cows, sheep, and horses found many of its plants unpalatable. Mostly the animals grazed on the saltgrass, which Henry Wallace called "a short, wiry grass of but little account."[51] Yet this agricultural poverty did not necessarily indicate a correspondingly impoverished environment; quite the contrary. Waterlogged, salty land produced plenty of plants, just not the sorts of species from which farmers reaped large surpluses and profits.

Drainage systems similarly provided more open ground for uncultivated plants as well as additional aquatic and riparian habitat for animals. Cattails spread through the drains. As cattails proliferated, mammals such as beaver and muskrat moved in and then migrated from the drains into irrigation systems. When irrigators tried to drive the animals from canals and ditches, they retreated back into drains. Drains became a refuge because farmers and engineers did not maintain drainage ditches as thoroughly as actual irrigation canals.[52]

Seepage not only resulted in new habitat for certain flora and fauna but in some places proved beneficial for irrigators and other people. Some irrigation systems collected seepage in wells and ditches and then channeled or pumped it back into canals for reuse. Seepage filled creeks and coulees that had once run dry in the summer; irrigators then took the water and applied it to their fields. As early as 1909, aquaculturists established commercial fish farms at Thousand Springs and other places where seepage burst from the wall of the Snake River canyon. Irrigation water that percolated through the ground also made its way back into rivers as return flow. Irrigation depleted stretches of rivers; return flow filled them again downstream.[53]

Irrigators on the Egin Bench in the upper Snake River valley actually made a practice of manipulating seepage to irrigate their crops. During the 1880s and 1890s, settlers moved onto benchland on the west side of Henry's Fork of the Snake River near St. Anthony, an area about 3 miles wide and

14 miles long. Farmers there eventually cultivated about 28,000 acres. The first irrigation water applied to the sandy loam of the bench quickly drained away; the soil would not hold water. But the farmers found that this situation worked to their advantage.[54]

Water percolated down until it reached a layer of volcanic rock; with continued irrigation, groundwater rose until it reached the root zone of crops. Farmers realized that they could actually manage the groundwater level, keeping it within 6 to 18 inches of the surface. They developed a system called "subirrigation" in which they ran water through canals into borrow pits along roads, swales, and special ponding areas. From these sources, water seeped down, raising the groundwater level. Irrigators channeled water through their own ditches to adjust the groundwater level to meet the exact conditions of their individual farms. The Egin Bench farmers used a tremendous amount of water, but subirrigation worked reasonably well for them. They still practice it today.

Egin Bench subirrigation had other, unforeseen consequences for nearby groundwater formations. Water from Egin Bench eventually seeped west about 20 miles until it reached intermittent Mud Lake, which began to fill with Egin Bench seepage around 1900. By 1927, Mud Lake was a permanent lake covering 15,000 acres to an average depth of seven feet. The lake then attracted irrigation companies and projects.[55]

Seeping irrigation water also augmented aquifers that farmers eventually tapped. Idaho's greatest aquifer lies under the Snake River Plain, and it exemplifies the entangling of human purposes and natural processes characteristic of the irrigated landscape as a whole. Twenty-five to 50 miles wide and nearly 200 miles long, the Snake Plain aquifer extends from the Ashton vicinity on the northeast to the Hagerman Valley on the southwest. It consists of loose volcanic and sedimentary rocks interlayered between porous lava flows, the whole permeated with water trickling down from the surface. Rain, snow, and streams such as Henry's Fork, the Big and Little Lost Rivers, and the Snake itself all contribute water to this aquifer. More than a century ago, water diverted through canals and ditches at places such as Egin Bench also began to seep into it. By the late 1940s, high crop prices and electric pumps invited irrigators to sink wells into the aquifer, which by then contained between 60 and 150 million acre feet and which irrigation seepage had boosted to within 70 feet of the surface in places. Over the next three decades, farmers and agricultural corporations used the aquifer to expand the irrigated landscape by about 1 million acres.[56]

Seepage not only expanded the volume of the Snake Plain aquifer; in what amounted to a massive "return flow," it increased the aquifer's discharge into the Snake River. Flowing roughly southwest, the aquifer pours from the north side of the Snake River canyon, between Milner Dam and King Hill. Bubbling from cracks in the volcanic rock wall, the magnificent springs have always fascinated travelers who stop to look. But visitors to Thousand Springs and other founts delight in a seemingly "natural" phenomenon that irrigation systems helped create. In 1902, the Snake Plain aquifer discharged about 3,800 cubic feet per second (cfs); by 1917 the figure had risen to more than 5,000 cfs; by 1956 roughly 6,000 cfs gushed from the springs. In the early 1960s, geologists estimated that irrigation had expanded the flow approximately 60 percent, increasing the discharge by about 1.8 million acre feet per year.[57] Irrigators had intercepted and diverted the hydrologic cycle, but the Snake Plain aquifer eventually brought much of the water back to its river of origin, helping to fill the stream again after it ran dry at Milner Dam.

EVALUATING IDAHO'S IRRIGATED LANDSCAPE

Springs pouring irrigation water back into the Snake do more than just replenish the river. Much like other features of the irrigated landscape, they also challenge popular notions that the river and the adjacent farmlands represent a nature that is controlled (the developer's view) or a nature that is increasingly artificial and in jeopardy of dying (the environmentalist's stance). Neither of these understandings is entirely wrong; irrigators certainly have shaped their environment well enough to make it economically productive, and in the process the river and the valley indeed have experienced ecological losses. But neither view adequately explains the environmental history of this place. The river and the irrigated landscape that it sustains are much more complicated ecologically and socially than popular understandings admit. Irrigation has not so much controlled or ruined the environment as transformed it into something new, a landscape that we as a people have not yet fully comprehended.

Our tendency to view the irrigated landscape in simplistic terms derives in part from the myths that shape our understandings of the places we inhabit. Myths reduce the complexity, ambiguity, and uncertainty of human experience to simple, timeless, compelling stories.[58] Myths of American places, especially irrigated landscapes, frequently revolve around a biblical archetype deeply embedded in the culture of European civilization: Eden. For farmers and engineers, irrigation has often symbolized the redemption of a fallen world. This is the meaning of "reclamation," a popular euphemism for irrigated agriculture: irrigators have reclaimed a wasteland and restored it to its original perfection. The Edenic myth has also influenced the critics of irrigation. For these people, the irrigated landscape is not a garden but a manifestation of the fall from grace, a paradise ruined.

Indeed, idealized and even mythologized notions of pristine nature and a "balance of nature" have influenced attitudes toward the irrigated landscape. To many people, the irrigated landscape deviates from what "nature" actually should be. No doubt this at least partly explains why the Idaho Department of Fish and Game, in an official description of wetlands, includes no mention of reservoirs, irrigation canals, or drainage ditches.[59] Because humans have played such a powerful role in creating these landscapes, they are not worthy of attention as sites where wetland flora and fauna congregate. According to this perspective, "man-made" bodies of water are not part of stable, harmonious ecosystems; they may not even be part of "nature" at all.

Despite their power and importance, such assumptions should not stand in the way of a more realistic view of Idaho's irrigated landscape. This place is more than a restored or destroyed Eden. It is more than simply a disturbed, unbalanced environment. The irrigated landscape is a product of contingent, interacting, and ultimately historical forces. Some of these forces have been human, others natural. Engineers and farmers tried to impose their designs on water and land, but at each stage water and land to a certain degree circumvented their plans. Over time, this interplay of hydrologic cycle and hydraulic system, of agriculture and uncultivated biota, produced a new, hybrid landscape that irrigators could not regulate as much as they desired. Technology and nature pushed against and yielded to one another, forming a dynamic, evolving, often ambiguous synthesis.

Of course, viewing the irrigated landscape as a historical creation and a place of great ecological fluidity does not mean that all changes are desirable or acceptable. Environmental changes, especially those to which humans have contributed, must be judged right or wrong. Yet how the irrigated landscape is judged depends on values and ethics; nature itself provides no solid standards for evaluating this place. Even in the absence of humans, environments undergo change over time; it is difficult if not impossible to find a stable "balance of nature" that can provide a standard for evaluating the human impact.[60] We deem changes wrong or unhealthy because of our own principles, not so much because landscape alterations actually deviate from a fixed natural standard.

A better understanding of Idaho's irrigated landscape, and indeed most landscapes, will emerge when Americans look beyond mythic stereotypes and pat assumptions about the human relationship to nature. Most landscapes are the products of human actions combined with physical and biological processes that outstrip human objectives. Like Idaho's irrigated landscape, most landscapes are hybrid creations, offspring of human dreams and a dynamic natural world.

NOTES

1. William E. Smythe, *The Conquest of Arid America* (1899; reprint, Seattle: University of Washington Press, 1969); James L. Wescoat, "Challenging the Desert," in *The Making of the American Landscape*, ed. Michael P. Conzen (Boston: Unwyn Hyman, 1990); John Rennie Short, *Imagined Country: Environment, Culture, Society* (London: Routledge, 1991); Leonard Arrington, "Irrigation in the Snake River Valley: An Overview," *Idaho Yesterdays* 30 (Spring–Summer 1986): 3–11.

2. Henry Nash Smith, *Virgin Land: The American West as Symbol and Myth* (1950; reprint, New York: Vintage Books, 1957); John R. Stilgoe, *Common Landscape of America, 1580 to 1845* (New Haven: Yale University Press, 1982); John Brinckerhoff Jackson, *Discovering the Vernacular Landscape* (New Haven: Yale University Press, 1984).

3. William L. Preston, *Vanishing Landscapes: Land and Life in the Tulare Lake Basin* (Berkeley: University of California Press, 1981); Donald Worster, *Rivers of Empire: Water, Aridity, and the Growth of the American West* (New York: Pantheon, 1985); Marc Reisner, *Cadillac Desert: The American West and Its Disappearing Water* (New York: Viking Penguin, 1986); Wallace Stegner, *The American West as Living Space* (Ann Arbor: University of Michigan Press, 1987).

4. For a full version of my argument and the evidence on which I base it, see my *Irrigated Eden: The Making of an Agricultural Landscape in the American West* (Seattle: University of Washington Press, 1999).

5. For overviews of irrigation development in the region, see Arrington, "Irrigation in the Snake River Valley"; William Darrell Gertsch, "The Upper Snake River Project: An Historical Study of Reclamation and Regional Development, 1890–1930" (Ph.D. diss., University of Washington, 1974); W. G. Hoyt, *Water Utilization in the Snake River Basin* (Washington, D.C.: U.S. Government Printing Office, U.S. Geological Survey Water Supply Paper 657, 1935). Figures on canals and irrigated acreage were derived from U.S. Department of Commerce, Bureau of the Census, *Fourteenth Census of the United States, 1920, vol. 7: Irrigation and Drainage* (Washing-

ton, D.C.: Government Printing Office, 1922), pp. 164, 170; and the map of G. F. Lindholm and S. A. Goodell, *Irrigated Acreage and Land Uses on the Snake River Plain, Idaho and Eastern Oregon* (n.p.: Department of the Interior, U.S. Geological Survey, Atlas HA-691, 1986).

6. Numerous ecologists, historians, anthropologists, and geographers have made this point. See, e.g., Andrew Goudie, *The Human Impact on the Natural Environment* (Cambridge: MIT Press, 4th ed., 1994); Daniel B. Botkin, *Discordant Harmonies: A New Ecology for the Twenty-First Century* (New York: Oxford University Press, 1990); Richard White, "Environmental History, Ecology, and Meaning," *Journal of American History* 76 (1990): 1111–16; William M. Denevan, "The Pristine Myth: The Landscape of the Americas in 1492," *Annals of the Association of American Geographers* 82 (1992): 369–85; Steven R. Simms, "Wilderness as a Human Landscape," in *Wilderness Tapestry*, eds. Samuel I. Zeveloff, L. Mikel Vause, and William H. McVaugh (Reno: University of Nevada Press, 1992).

7. For early land use in the pre-irrigation Snake River valley, see P. J. Mehringer, "Prehistoric Environments," and K. T. Harper, "Historic Environments," both in *Handbook of North American Indians, vol. 11: Great Basin*, ed. W. L. D'Azevedo (Washington, D.C.: Smithsonian Institution, 1986); Carlos Schwantes, *In Mountain Shadows: A History of Idaho* (Lincoln: University of Nebraska Press, 1991), pp. 13–16, 25–34, 49–53; Leonard Arrington, *History of Idaho*, vol. 1 (Moscow: University of Idaho Press/Boise: Idaho State Historical Society, 1994), pp. 87–113, 141–60, 492–94; E. B. Bentley and Glenn Oakley, "A Climate of Change," in *Snake: The Plain and Its People*, ed. Todd Shallat (Boise: Boise State University, 1994), pp. 75–77; F. Ross Peterson, "Confronting the Desert," in *ibid.*, pp. 124–35; U.S. Department of the Interior, Census Office, *Report on the Productions of Agriculture as Returned at the Tenth Census*, 1880 (Washington, D.C.: Government Printing Office, 1883), pp. 140–47. See also James A. Young and B. Abbott Sparks, *Cattle in the Cold Desert* (Logan: Utah State University Press, 1985); Donald K. Grayson, *The Desert's Past: A Natural Prehistory of the Great Basin* (Washington, D.C.: Smithsonian Institution Press, 1993).

8. Luna B. Leopold, *Water: A Primer* (San Francisco: Freeman, 1974), describes the hydrologic cycle.

9. Idaho geology and hydrology are described in Israel C. Russell, *Geology and Water Resources of the Snake River Plains of Idaho* (Washington, D.C.: Government Printing Office, U.S. Geological Survey Bulletin 199, 1902); Hoyt, *Water Utilization in the Snake River Basin*; Harold T. Stearns, Lynn Crandall, and Willard G. Steward, *Geology and Ground Water Resources of the Snake River Plain in Southeastern Idaho* (Washington, D.C.: Government Printing Office, U.S. Geological Survey Water Supply Paper 774, 1938).

10. Examples of these kinds of metaphors were widespread in the early 1900s. On dams and reservoirs, see Hoyt, *Water Utilization in the Snake River Basin*, pp. 36–46.

11. For the early history of Milner Dam and the Twin Falls canals and irrigation projects, see Gertsch, "The Upper Snake River Project"; James Stephenson, *Irrigation in Idaho* (Washington, D.C.: Government Printing Office, U.S. Department of Agriculture, Office of Experiment Stations, Bulletin 216, 1909); Hoyt, *Water Utilization in the Snake River Basin*, pp. 45, 125–26.

12. Jessie Warrington, "The Wonderful Redemption of a Desert: A Story of the Past, Present, and Future of the Famous Twin Falls Country," *See Idaho First* 5 (September 1914): 3–6, 20. See also *Twin Falls News*, 17 March 1905.

13. See, e.g., "Distribution of Water during Greatest Drought," *Engineering News-Record* 85 (11 November 1920): 927–31; G. Clyde Baldwin, "Transmission and Delivery of Reservoir Water," *American Society of Civil Engineers Transactions* 94 (1930): 296–300, 325; Lynn Crandall, "Crandall on Administrative Water Problems," *American Society of Civil Engineers Transactions* 94 (1930): 318–21; T. R. Newell, "Newell on Administrative Water Problems," *American Society of Civil Engineers Transactions* 94 (1930): 321–24.

14. *Farmer's Friend*, 27 March 1906.

15. *Idaho Republican*, 15 March 1923.

16. Ralph I. Meeker, "Report: Snake River Water Problems, Concerning Operation, Jackson Lake Reservoir, Transmission Losses Reservoir Water [*sic*], Water District 36, Idaho," File 310.2, Minidoka Project Records, Record Group 115, Records of the Bureau of Reclamation, National Archives, Washington, D.C. (hereafter cited as RG 115).

17. For such a characterization of dams, see Stegner, *The American West as Living Space*, p. 50. See *Twelfth Census of the United States, 1900, Agriculture, part 2: Crops and Irrigation* (Washington, D.C.: United States Census Office, 1902), pp. 804–5, for official acknowledgement that rivers do not function like pipelines.

18. The impact of dams on preexisting flora and fauna is discussed in U.S. Fish and Wildlife Service, "A Preliminary Survey of Fish and Wildlife Resources, Upper Snake River Basin," in U.S. Bureau of Reclamation and U.S. Army Corps of Engineers, *Upper Snake River Basin, vol. 3, part 1: Coordination of Reports of Cooperating Agencies* (n.p.: n.p., 1961); and Mark Fiege, "Wildlife and Irrigation Systems along the Snake River, Idaho," *Transactions of the Fifty-Seventh North American Wildlife and Natural Resources Conference* (1992): 724–32.

19. Fiege, "Wildlife and Irrigation Systems along the Snake River"; James Simpson and Richard Wallace, *Fishes of Idaho* (Moscow: University Press of Idaho, 1982); Allan D. Linder, "Idaho's Alien Fishes," *TEBIWA: The Journal of the Idaho State Museum* 6 (1963): 12–15.

20. *Idaho Register*, 20 August 1912; *Idaho Statesman*, 14 October 1979.

21. Fiege, "Wildlife and Irrigation Systems along the Snake River."

22. F. M. Dille, "The Minidoka National Bird Reservation, Idaho," *Reclamation Record* 7 (December 1916): 564–65; T. S. Palmer, "The Deer Flat National Wildlife Refuge," *Reclamation Record* 7 (March 1916): 221–23.

23. Worster, *Rivers of Empire*, pp. 3–15.

24. See, e.g., "Farmers View Aging Canal System," *Idaho State Journal*, 8 November 1983.

25. A. P. Davis, "Why Some Irrigation Canals and Reservoirs Leak," *Engineering News-Record* 80 (4 April 1918): 663–65; F. N. Cronholm, "Drainage System for the North Side Minidoka Irrigation Project," *Engineering News-Record* 69 (25 April 1914): 468–69.

26. Newell, *Report on Agriculture by Irrigation in the Western United States: Eleventh Census, 1890* (Washington, D.C.: Government Printing Office, 1894), p. 19.

27. See, e.g., L. F. Henderson, *Twelve of Idaho's Worst Weeds* (Moscow: University of Idaho, Agricultural Experiment Station, Bulletin 14, 1898); Barry Dibble and T. W. Parry, "Control of Moss, Weeds, and Willows on the Minidoka Project," *Reclamation Record* 8 (April 1917): 192–93.

28. Dibble and Parry, "Control of Moss, Weeds, and Willows on the Minidoka Project."

29. Fiege, "Wildlife and Irrigation Systems along the Snake River."

30. *Idaho Republican*, 19 July 1923.

31. See, e.g., Dibble and Parry, "Control of Moss, Weeds, and Willows on the Minidoka Project."

32. R. M. Adams, "Removal of Vegetation from Twin Falls Irrigation Canals," *Engineering News-Record* 85 (12 August 1920): 319–21; Hugh L. Crawford, "Report on Cleaning Canals with Modified Disk Harrows, Minidoka Project, Idaho," *Reclamation Record* 13 (February 1922): 22–24.

33. Dibble and Parry, "Control of Moss, Weeds, and Willows on the Minidoka Project"; Twin Falls Canal Company, Twin Falls, Idaho, "Annual Report of Canal Operations for Season of 1922."

34. Fiege, "Wildlife and Irrigation Systems along the Snake River, Idaho"; E. H. Neal to Principal, Pingree School, 31 March 1932, File 4.07, Rodent Control, Aberdeen-Springfield Canal Company, Aberdeen, Idaho (hereafter cited as ASCC); Idaho Fish and Game Warden, *Twenty-First Biennial Report, Fish and Game Warden, State of Idaho* (Boise: Idaho Fish and Game Department, 1946), pp. 37–39.

35. Lester Vance Benjamin, "Experimental and Historical Development of Weed Control in Idaho" (M.S. thesis, University of Idaho, 1931); Ralph Scott Bristol, "Chemical Weed Control in Idaho" (M.S. thesis, University of Idaho, 1932).

36. Henderson, "Twelve of Idaho's Worst Weeds," pp. 99–100, 112–14; O. M. Osborne, *Weed Pests of Idaho and Methods of Eradication* (n.p.: n.p., Idaho Agricultural Experiment Station Bulletin no. 71, 1911), p. 12; E. H. Neal to Chase Clark, 13 January 1923, ASCC.

37. *General Laws of the State of Idaho* (Boise: Syms-York, 1907), pp. 24–25; *General Laws of the State of Idaho* (Boise: Syms-York, 1911), pp. 381–83; *General Laws of the State of Idaho* (n.p.: n.p., 1919), pp. 86–88; Paul Wenger, *Pure Seed Law and Weed Control Act* (Moscow: University of Idaho, Agricultural Experiment Station Circular no. 8, 1919).

38. Twin Falls Canal Company, "General Manager's Annual Report for the Year 1935."

39. Twin Falls Canal Company, "Twin Falls Canal Company Annual Report, 1954."

40. Twin Falls Canal Company, "Annual Report for 1938."

41. *Idaho Republican*, 22 February 1907.

42. C. E. Temple, "Potato Diseases and Insect Pests," in *Potato Culture* (n.p.: Idaho Agricultural Experiment Station, Bulletin no. 79, 1914); O. A. Pratt, "Soil Fungi in Relation to Diseases of the Irish Potato in Southern Idaho," *Journal of Agricultural Research* 8 (April 1918): 73–99.

43. J. C. Ojala, S. L. Love, J. J. Pavek, and D. L. Corsini, *Potato Varieties for Idaho* (Moscow: University of Idaho, College of Agriculture, Cooperative Extension System and Agricultural Experiment Station, Current Information Series no. 454, 1989).

44. James Thorp and C. S. Scofield, "Drainage in Arid Regions," in *Soils and Men: United States Department of Agriculture, Year Book of Agriculture, 1938* (Washington, D.C.: Government Printing Office, n.d.).

45. O. C. Magistad and J. E. Christiansen, *Saline Soils: Their Nature and Management* (Washington, D.C.: Government Printing Office, United States Department of Agriculture Circular no. 707, 1944); Milton Fireman and H. E. Hayward, "Irrigation Water and Saline and Alkaline Soils," in *Water: United States Department of Agriculture, Yearbook of Agriculture, 1955* (Washington, D.C.: Government Printing Office, n.d.).

46. *Idaho Statesman*, 31 May 1877.

47. Richard Lowitt and Judith Fabry, *Henry A. Wallace's Irrigation Frontier: On the Trail of the Corn Belt Farmer* (Norman: University of Oklahoma Press, 1991), pp. 127, 129; Robert S. Snyder, Mark R. Kulp, G. Orien Baker, and James C. Marr, *Alkali Reclamation Investigations* (Moscow: University of Idaho, Agricultural Experiment Station Bulletin no. 233, 1940), p. 3.

48. U.S. Department of Commerce, Bureau of the Census, *Fourteenth Census of*

the United States, 1920, vol. 7: Irrigation and Drainage, p. 433; U.S. Department of Commerce, Bureau of the Census, *Fifteenth Census of the United States, 1930: Drainage of Agricultural Lands* (Washington, D.C.: Government Printing Office, 1932), pp. 95, 96.

49. An account of this case is recorded in "Annual Project History of Boise Project for 1917," 165–66, Roll 9, and "Annual Project History of Boise Project Idaho for 1918," 162–64, Roll 10, both Microcopy M96, Project Histories and Reports of Reclamation Bureau Projects, 1905–25, RG 115.

50. Charles A. Jensen and B. A. Olshausen, "Soil Survey of the Boise Area, Idaho," in *Field Operations of the Bureau of Soils* (Washington D.C.: Government Printing Office, 1902), pp. 429–45, plates 65–68; Milton Fireman, C. A. Mogen, and G. Orien Baker, *Characteristics of Saline and Alkaline Soils in the Emmett Valley Area, Idaho* (Moscow: University of Idaho, Agricultural Experiment Station, Research Bulletin no. 17, 1950); Osborne, *Weed Pests of Idaho and Methods of Eradication*, pp. 12–13; George Stewart, "Pastures and Natural Meadows," in *Grass: United States Department of Agriculture, Yearbook of Agriculture, 1948* (Washington D.C.; Government Printing Office, 1948), p. 548.

51. Lowitt and Fabry, *Henry A. Wallace's Irrigation Frontier*, p. 127.

52. Cronholm, "Drainage System for the North Side Minidoka Project"; E. H. Neal to Owen W. Morris, 2 December 1941, ASCC.

53. F. T. Crowe, "Draining the Pioneer Irrigation District," *Reclamation Record* 5 (October 1914): 373–74, provides an example of developed drains supplying water for irrigation. George W. Klontz and John G. King, *Aquaculture in Idaho and Nationwide* (Boise: Idaho Department of Water Resources, 1975), discusses the origins of aquaculture in the Snake River Canyon. For examples of return flow from irrigation, see Stearns, *Geology and Ground-Water Resources of the Snake River Plain*.

54. D. W. Ross, *Biennial Report of the State Engineer to the Governor of Idaho for the Years 1899–1900* (Boise: Capitol Printing Office, n.d.), pp. 17–18; Kate B. Carter, ed., *Pioneer Irrigation: Upper Snake River Valley* (Salt Lake City: Daughters of Utah Pioneers, 1955), p. 249.

55. Julius Hornbein, "Seepage Water, Formerly Wasted, Now Pumped for Irrigation," *Engineering News-Record* 78 (26 April 1917): 192–93.

56. Stearns, *Geology and Ground-water Resources of the Snake River Plain*, pp. 136–202; Lynn Crandall, "Ground Water Flows of the Snake River Plain," 31 March 1953, Idaho–Reclamation of Land File, Idaho Room, Idaho Falls Public Library; B. B. Bigelow, S. A. Goodell, and G. D. Newton, *Water Withdrawn for Irrigation in 1980 on the Snake River Plain, Idaho and Eastern Oregon* (Washington, D.C.: U.S. Department of the Interior, U.S. Geological Survey Atlas HA-690, 1986); Lindholm and Goodell, *Irrigated Acreage and Land Uses on the Snake River Plain*.

57. Stearns, *Geology and Ground-Water Resources of the Snake River Plain*, pp. 142–66; M. J. Mundorf, E. G. Crosthwaite, and Chabot Kilburn, "Ground Water for Irrigation in the Snake River Basin in Idaho," in U.S. Bureau of Reclamation and U.S. Army Corps of Engineers, *Upper Snake River Basin, vol. 3, part 2: Coordination and Reports of Cooperating Agencies: U.S. Geological Survey, Ground Water Branch* (U.S. Department of the Interior, Geological Survey, Open-File Report), pp. 1–4, 10–12, 58–59, 63, 72, 131–34, 149, 151.

58. Smith, *Virgin Land;* Leo Marx, *The Machine in the Garden: Technology and the Pastoral Ideal in America* (New York: Oxford University Press, 1964); Patricia Nelson Limerick, *The Legacy of Conquest: The Unbroken Past of the American West* (New York: W. W. Norton, 1987); Richard Slotkin, "Myth and the Production of History," in *Ideology and Classic American Literature*, eds. Sacvan Bercovitch and Myra Jehlen (New York: Cambridge University Press, 1986).

59. Idaho Department of Fish and Game, *Between Land and Water: The Wetlands of Idaho* (Boise: Idaho Department of Fish and Game, Nongame Wildlife Leaflet no. 9, 1990).

60. Botkin, *Discordant Harmonies*.

18 / Cultural Perceptions of the Irrigated Landscape in the Pacific Northwest

DOROTHY ZEISLER-VRALSTED

Culture shapes perceptions of the environment. To many early settlers, despite evidence to the contrary, the West represented a garden because that was the landscape they anticipated. Coupled with the perception of the West as garden was an image of majestic western landscapes that rivaled European scenic landscapes. Artists such as Albert Bierstadt and Thomas Moran lent credence to the image. Because of these perceptions, Euro-Americans sought to transform the West to meet their expectations. Other early inhabitants of the West, including Native American and Hispanic people, perceived the West differently and transformed it accordingly.

The Euro-American drive to transform is reflected in the rhetoric of resource use. The Walker River in northern Nevada offers a humble example that reveals differences between Northern Paiute and Euro-American rhetoric in the form of different names for the river. The Northern Paiute knew the river as "Agai Hoop" (Trout River), a name indicating their relationship with it as a source of sustenance. In the nineteenth century, the explorer John Fremont renamed it the Walker River in honor of the man he credited with its discovery. The "connection" between the people and the resource that was evident in the Paiute language had been broken.[1]

The history of the Columbia River also reveals changes in people's perceptions of nature. In Richard White's words: "For much of human history, work and energy have linked humans and rivers, humans and nature."[2] Early-nineteenth-century explorers described the Columbia River in terms of the effort it took to travel or the lashings their ships endured. By the twentieth century, this had changed as engineers quantified the resource and began to measure the river by its force or energy. Again, a connection between humans and the environment had been severed.[3]

The irrigated landscape of the Pacific Northwest illustrates these themes—the power that perception and culture have on shaping human interactions with nature; the role of rhetoric in illustrating the link between humans and their resources; and the changed relationship between humans and nature. Beginning in the late 1800s, scores of Euro-Americans traveled west to cultivate land in the arid West. The river bottoms were settled first, leaving the higher, drier parcels for later homesteaders. By the 1890s, it had become obvious that most of the region was unsuited for agriculture without the aid of irrigation. Boosters, settlers, and politicians extolled the region's potential; with reclamation, they promised, the West could "blossom like the rose." No one questioned whether the myth of the yeoman farmer was unrealistic in such a dry and fragile environment. Instead, the agrarian ideal required that the area be populated by agriculturalists in a manner closely resembling that of land east of the Mississippi River.

The federal Reclamation Act of 1902 provided much of the technology, expertise, and money to water millions of acres in the West. To secure passage of the legislation, supporters argued that they were sustaining America's homestead tradition. The western landscape was perceived or "read" in agricultural terms: the numerous valleys lying between mountain ranges and the almost desert strips were seen as potential breadbaskets. The only obstacles were technological or economic. An American culture imbued with notions of the agrarian ideal dictated this response.

Western populations are again changing. In contrast to the earlier pastoral image, many western locales that were once settled by agriculturalists and sustained by federal reclamation projects are being transformed into "Western cityscapes."[4] Cities such as Las Vegas, Phoenix, and Albuquerque expand as newcomers arrive in search of a quality of life against a western backdrop. On a lesser scale, subdivisions and ranchettes are carved out of former agricultural land. Eighty-six percent of the residents of the contemporary West live in urban areas; 60 years ago only a little over 50 percent of the western population did so.[5]

Alarmed over diminishing water resources relative to the burgeoning population, some scholars are examining the effects of urbanization. Some applaud the transformation, arguing that urban users consume less water than agricultural users and that the delivery system for residential use is usually more efficient. Others, notably political scientist Helen Ingram, lament the shift to urbanization because of the impact on rural communities and the loss of a rural lifestyle. To Ingram, the losers are the rural poor, including Native American and Hispanic water users.[6] Yet a question remains: How does the shift from an agricultural to an urban landscape affect humans' perception of the irrigated landscape, and how do these perceptions affect the West's most precious resource, water?

I examine these questions through two case studies. One focuses on the Mid-Columbia Plateau, with particular emphasis on the southern Yakima Valley, where the city of Kennewick is located and provided with irrigation water by the Kennewick Irrigation District. The other examines the Bitterroot Valley, particularly the area irrigated by the Bitterroot Irrigation District. Both areas have witnessed an urban transformation as subdivisions replaced orchards and other agricultural enterprises. Once portrayed as an agricultural mecca, the Tri-Cities—Kennewick, Richland, and Pasco—are now depicted as an urban oasis where people live in quiet, suburban bedroom communities. In the Bitterroot Valley, once touted as the land of the MacIntosh apple, the only visible farming is pasture land; instead of the lure of agriculture, houses are sold for their panoramic views or access to the golf course.

In the nineteenth century, developers and settlers perceived the land in agricultural terms, envisioning fertile, thriving farms through irrigation. Realtors in the Bitterroot Valley now advertise ranchettes and subdivision lots with irrigated lawns. In the Tri-Cities, developers and residents project a landscape dotted with industrial parks, commercial and professional sites, and residential dwellings surrounded by green, manicured lawns. This change in perception has reinforced another belief that existed before urbanization—the potential of technology to change the landscape. The agricultural paradise that settlers first predicted for the area could be realized only through twentieth-century technology. The same holds true today—but instead of building open ditches and small earth-filled dams, contemporary technology includes PVC-lined canals, underground pressurized pipelines, and elaborate sprinkler systems. The manipulation and delivery of finite water resources remains the basis of much western settlement.

THE MID-COLUMBIA BASIN

Of the two areas, the Mid-Columbia Basin has experienced the most dramatic transitions in land use. Located in southern Washington, the basin surprises the first-time visitor with its sharp contrasts: a desert landscape, yet a riverine environment. The arid plateau is a flat plain interrupted by basalt terraces or plateaus that rise out of the river canyons and dry coulees. The Columbia is the major river to cut through this landscape, draining the Yakima, Snake, Spokane, Walla Walla, and Okanogan Rivers.

Rainfall is minimal; annual precipitation ranges from seven to nine inches. The climate varies from high, desertlike temperatures in the summer to cool winters. The area has "arid steppe" vegetation. Cottonwoods and willows grow along the river bottoms; tumbleweed, rabbitbrush, cheatgrass, and sagebrush dominate the drylands above.

The seminomadic cultures of the region—the Yakama, Nez Perce, Palouse, and Klickitat tribes—lived on camas, bitterroot, and berries, on deer, elk, sharp-tailed grouse, and large quantities of salmon. In this riverine environment, water played an important role in their cosmology: "The sun (*an*) is Father; water (*cuus*), the first sacred food." Salmon runs were an integral part of a 12,000-year-old culture.[7] Their lives changed dramatically after they were forced to sign the 1855 Isaac Stevens Treaty. In it, the tribes ceded much of the land that would become Washington, Oregon, and Idaho.

But even with the acquisition of land, Euro-American settlement hesitated until the arrival of the Northern Pacific Railroad in 1883. Early settlers—envisioning an agricultural economy—expressed disappointment when they first encountered the region that became the Tri-Cities. Newcomers described the area surrounding Kennewick, in the southern end of the Yakima Valley near the confluence of the Yakima and Columbia Rivers, as "nothing but hot winds, and dust and sagebrush," "a bunchgrass waste." Later the townsite of Kennewick was depicted as "one of long waiting and hope deferred."[8] Irrigation was the only way to transform what these settlers believed to be a barren, useless land.

After private developers started work on an irrigation project to reclaim up to 50,000 acres, one local newspaper reported in 1893 that "two years ago if a man had settled on a piece of that land he would have been advised to emigrate to Steilacoom or Medical Lake [area mental institutions]. Water makes the difference. The people residing in the district bounded by the canal are putting in the crops, some setting out orchards, others seeding their land to alfalfa, potatoes or oats."[9] The phrase "water makes the difference" underscored the history of Euro-American settlement in the arid West.

The project changed the history of the Mid-Columbia Plateau. Throughout the region, private developers, followed by state and federal agencies, embarked on a series of irrigation enterprises—tapping water supplies in the Yakima and Columbia Rivers—that still shape the political, social, and economic well-being of the area. After forming the Kennewick Irrigation District in 1917, water users began their quest for federal funding. Local boosters extolled the fertility of area farms; the county agriculturalist stated that "there is no other body of land in the Northwest where the per acre yield of those food products so much needed at this time, would be as great as here." The local media portrayed the Kennewick Highlands, a 4,000-acre tract within the irrigation district, as a Garden of Eden with unlimited potential.[10]

By the 1920s, federal officials received alarming reports from farmers in the Kennewick Highlands that if the United States Bureau of Reclamation (USBR) did not offer the settlers hope in the form of irrigation works, their only option would be abandonment.[11] Other reports described the condition of area families as "pitiable," without an adequate water supply.[12] But when the USBR did survey the district to determine whether to award federal monies for irrigation, the engineers concluded: "The unimproved land on the upper part of the division is rough and broken. A relatively long canal will have to be constructed for a narrow fringe of land until it reaches the main body of better land in the vicinity of Kennewick. The land is quite sandy and will require a large amount of water to grow crops under irrigation. This land is not regarded as having the fertility or the water holding capacity of much of the land of the Yakima-Benton Division."[13] Despite the disparaging comments, Kennewick boosters persisted in their efforts with a never-say-die attitude. The persistence of a cultural perception of the land—the vision of the land made bountiful through irrigation—has a long history, persisting despite evidence to the contrary.[14]

Bad luck plagued the district during the 1930s, prompting one area resident to remark that "they have three different classes of people, one here, one coming, and one going."[15] Many of those departing left abandoned homesteads, and the irrigation district assumed ownership of the land. As the town grew, this abandoned land became valuable real estate, but in the 1930s, owners considered the land too worthless to even pay taxes on it.

Kennewick's destiny changed with World War II and the establishment of the nearby Hanford Nuclear Reservation, which expropriated 6,000 to 7,000 acres of irrigated agricultural holdings. Construction of an irrigation project—the Kennewick Division—was a way to replace the lost agricultural land. Federal officials also reasoned that the division would provide homesites for returning veterans and jobs for displaced wartime construction workers.[16] In 1948 the district received congressional authorization for the project. Underlying the report recommending authorization was the assumption that the land developed in the Kennewick Division would be used for agricultural purposes. Although a small portion of the land that had previously been farmed was being converted into residential lots, the USBR planned to divide the newly irrigated land into full-time farms or acreage large enough to sustain a family at "an adequate level of living."[17] Experts provided estimates of incomes from farming operations on various soil classifications, demonstrating that families could earn a livelihood by cultivating the designated 160-acre tracts. This planning demonstrated a continuing belief in the agrarian myth of the yeoman farmer.[18] Despite the misfortunes suffered by many Kennewick agriculturalists in the 1920s and 1930s, the conviction persisted that Kennewick land could and should support a farming community.[19]

Residents in the community shared these sentiments, perceiving their environment to be an agricultural landscape. The only obstacle to realizing this landscape had been a lack of water, and this obstacle was now being overcome through technology and federal investments. Thus with great pride one observer remarked that "engineering history is being made" and that the construction firms were "staging a pitched battle with nature, matching engineering skill against a rampaging river and solid basalt rock to win more than 19,000 acres of raw, arid land over to peaceful usefulness."[20]

The USBR also provided a settler assistance program. Settlers were helped in establishing themselves and in completing homestead entries by bringing new land under irrigation. Federal officials hoped the "reclaimed land" would be devoted to the raising of high-income crops such as mint, tree fruits, and asparagus. To ensure that the newcomers would be farmers, potential settlers who applied for homesteads had to have worked in agriculture for at least two years. In addition, applicants were required to submit five letters of recommendation, of which four had to be from fellow farmers. Once chosen, homesteaders were assured of help by the USBR, with technical information on soil classification and crops. By 1957, the new irrigation and hydro-power works were in place, ready to serve a predominantly agricultural population.[21]

Ironically, the area began to urbanize in the late 1960s, spurred by the sparkplug of the Tri-Cities' economy, the Hanford nuclear works. Urbanization produced one of the first problems the district encountered: delivering water to residential lots (in a growing number of subdivisions) that had once been agricultural land. When the Bureau of Reclamation planned the irrigation works, the procedure for water service was for irrigation district employees—ditch riders—to deliver water to the high point on each 40-acre parcel.[22] When land was sold and broken into subdivisions, the district staff was obligated to supply water to each 40-acre tract. But the Kennewick district was unprepared to deal with the growing number of suburban users, all tied into the one main hookup that had earlier served the 40-acre parcel. Thus, in the area once touted as a Garden of Eden and held out as promising the highest per-acre yield in the Pacific Northwest, ditch riders were unable to deliver an adequate amount of water because of the increased number of residential water users.

These changes precipitated a second problem. Not only did ditch riders have to contend with more irrigators for each 40-acre tract, but the type of water user also changed. Where the agriculturalist had irrigated a precise number of acres at a time and given the ditch rider advance notice of his schedule, the urban user applied water to the lawn at different peak hours with no advance warning. The demand of the residential client was instantaneous, making it difficult for the ditch rider to determine how much to turn into the canal. Moreover, many residential water users who lived near the canal installed pumps that took more than their share of water and left other users without enough. In short, the ditch rider had to contend with unregulated water and unanticipated flows.

The establishment of local improvement districts (LIDs) resolved part of this problem but created its own difficulties.[23] Local improvement districts created a system of pressurized underground pipes that provided water directly to the individual homeowner. Praised for their efficiency, LIDs did cut down on water loss due to seepage and evaporation. Previously, however, seepage from the open canals maintained groundwater level. In areas where LIDs replaced open ditches, springs dried up and any plant or wildlife supported by canals and seepage was eliminated. Furthermore, the seepage had helped to dilute the nitrates (a product of fertilizers used for crops) in surrounding agricultural fields.

Finally, the general assumption that residential water users consume less water than agricultural users has not been the case in the Kennewick district. Some residential users consume as much water, if not more, than farmers. Whether this is a result of the low water rates the district can offer (assessments for residential users are approximately $70 a season) or of a lack of conservation on the part of residents, or both, is unclear, but urban users do not view water as a scarce resource. Regardless of the cause, district personnel contend that residential users are the least conservation-minded customers.[24]

Illustrating this indifference to the area's resources are comments made at a public hearing on water shortages during the summer of 1994. The Yakima Valley was experiencing its third successive drought year, and the USBR decided to maintain a flow of 64,500 gallons of water per minute in the Yakima River to sustain the anadromous fish population, which was in sharp decline. Displeased over a reduced water supply, one residential water user remarked, "Dead grass makes people very unhappy." Other users grumbled about "dry flower beds and yellow grass withering in the desert heat." To someone unaccustomed to the marvels of modern-day irrigation technology, "yellow grass and desert heat" are compatible images, whereas flower beds, green grass, and desert heat are not.[25]

Despite these conflicting images of desert and green lawns, and despite probable future disagreements over the water supply, the Kennewick Irrigation District has increasingly become an urban district. The district is actually promoting the transition. In a recent public-relations newsletter the district published, one headline proclaimed, "Irrigation: A Bargain for Everyone." Under the headline were photos of upscale suburban homes surrounded by immaculate green lawns; a brief paragraph touted the use of inexpensive irrigation water for residential dwellings. Ironically, one reason irrigation water is considered a "bargain," in addition to its low cost, is its environmental advantage. By using district water, "energy, chemicals, storage facilities, normal components of treated domestic water are avoided."[26]

Under a second headline, "KID Looks to the Future," the district's newsletter discussed the Tri-Cities' population growth and the district's role in facilitating that growth by providing water to newly developed residential areas. In guaranteeing growth "at a sustained rate," the district perceived its role as similar to that of a utility company furnishing a service, like the electric or phone company. Thus, not only has district land been urbanized but the water resource itself has undergone a process of urbanization as new technology replaces open canals and drainage ditches. Public awareness of the scarcity of the resource is also diminished. Compounding the situation are the low rates that users pay for irrigation water, which further foster an ignorance of its worth.[27]

In that same newsletter article appeared photos of a residential area being developed in a classic Mid-Columbia landscape—dry, broken land, with sagebrush the only ground cover. The author described the landscape, remarking that "water is the basic unit from which the success of the Tri-Cities is built upon [*sic*]. One need but look at the sage covered slopes of Horse Heaven hills to realize that without water the Tri-Cities would be a dreary, unproductive setting. The district may soon be servicing these areas, since feasibility studies are being conducted to determine whether to construct new LIDs."[28]

Sprinkled throughout the newsletter are pictures of the engineering feats—the new pump stations, canal linings, and drum screens—necessary for the district's increasingly urban complexion, a testimony to the experts' confidence in facilitating the transition. More than 100 years have passed since the first settlers arrived in the Yakima Valley, and although the economy has changed, cultural perceptions regarding the efficacy of technology persist.[29]

Coinciding with district predictions about the future, however, is the issue of water rights. The Yakima River has many water users, and in low water years shortages occur. Controversies including the loss of anadromous fish and the rights of recreational water users plague the valley. In an effort to address these competing claims, the Washington Department of Ecology filed a lawsuit in 1977 against 5,000 defendants representing wildlife, environmental, recreational, agricultural, residential, and Indian interests.[30] It is still unsettled and is unlikely to be for many years.

A finite water supply is not the only obstacle the Kennewick Division is encountering in its transition to an urban district. The booster newsletter fails to describe the problems caused by the development of subdivisions in an area that still services agricultural land. Although the district provides water to more than 14,000 residential users in 140 local improvement districts, 60 percent of district land remains agricultural. Open, earth-lined canals still wind their way throughout the division. Most irrigation canals break at some point, because of burrowing rodents, excessive rains, or soil composition; the canals of the Kennewick project are no exception. In the late 1950s, the losses that were sustained were minimal because the washouts affected only farmland.

As the number of subdivisions multiplied, however, district officials grew concerned about the locations of new subdivisions. Locating a subdivision below an irrigation canal could jeopardize the lives and property of hundreds of urban users if the canal should break and flood lower-lying land. Because of such concerns, the board decided that district officials should work with the Benton County engineer and planning commission regarding the approval of subdivisions within district boundaries. Nonetheless, developers have still located subdivisions below canals. The district took the only step open to it: it placed a caveat on the plats for these subdivisions to warn potential buyers of the possibility of washouts. As land becomes more valuable, developers seem less mindful than ever of the risks involved. In the end, the district is likely to replace open canals with piping, with related environmental changes. Water-table levels will drop, and any habitat supported by the canal system will be eliminated.[31]

Adding to these concerns, the Tri-Cities of the 1990s were in the midst of another boom. With district land increasing in value, the district hired a full-time resource manager to lease, sell, and plan residential development by setting up subdivisions. The resource manager does not oversee actual construction but does set up the infrastructure for the designated subdivision. Profits will reduce assessments for district water users. Yet the subsidy, particularly to residential users, who pay a flat fee for water regardless of how much they use, removes any economic incentive to encourage conservation; water users persist in their unrealistic expectations of area resources.[32]

Despite the trend toward urbanization, district board and staff continue to emphasize that the district still serves a large agricultural population. The remaining orchard and farming interests are productive and successful. But the statistics favor the residential user. In 1994, urban users made up 14,137 accounts, or 96 percent of all district water accounts, while there were 560 agricultural accounts, or 4 percent. In terms of dollars and cents, residential users returned a greater profit to the district. In a recent preliminary study, residential users were found to pay 70 percent of collected assessments while accounting for 66 percent of costs. There is also a trickle-down effect: as a district gains more urban users, the demand for water increases.[33] This growing demand could in turn cause such a rise in assessment (when subsidies from real estate are curtailed) that agricultural use of the water would become impractical. The economic advantages of serving urban consumers can be seen in the experience of the Provo River Project in Utah, where a major share of the water goes for city and industrial use. As the authors of one study concluded, "if reclamation projects such as [the Provo River Project] are to be viewed on a purely dollars-and-cents basis, increasing consideration is apt to be given to proposals to turn over a large share of the water to urban and industrial use."[34]

Regardless of the outcome, the Kennewick district is increasingly urbanized. Today a Wal-Mart, an Aamco Transmission shop, assorted motels, and subdivisions replace a vista that was once farms and dryland. The growth shows no sign of declining, despite recurring droughts and the impending adjudication of water rights on the Yakima River. The irrigated landscape, a product of American culture, continues to serve cultural expectations of the land.

THE BITTERROOT VALLEY

Residents of the Bitterroot Valley have witnessed several cultural transformations of the landscape. Located in western Montana, the valley is 100 miles long with an average width of 8 miles; it comprises some 327,000 acres. It is a place of scenic beauty: the west side of the valley is lined with the impressive and craggy Bitterroot Mountains, and the east side is marked by benchlands that blend into the less-imposing Sapphire Range. The Bitterroot River is the principal waterway; it drains numerous mountainous streams including Skalkaho, Sleeping Child, Lolo, and Roaring Lion Creeks. Rainfall is greater than in the Mid-Columbia Basin, with an average annual precipitation of 10–12 inches. At one time wildlife was plentiful, with elk, moose, deer, grouse, and pheasants living in the valley. Berries such as huckleberry and roots including bitterroot grew in abundance, and aspens and cottonwoods filled the river bottoms.

Despite the valley's impressive vistas, Euro-Americans were skeptical of its agricultural potential. One of the first Euro-American residents in the area, the Jesuit Father DeSmet, arrived in 1841, intending to settle permanently and establish an Indian mission. He found the land cold and harsh. He remarked that the valley had "but one defile, which serves as the entrance to, and issues from the valley. The mountains which terminate it on both sides appear to be inaccessible, they are piles of jagged rocks, the base of which presents nothing but fragments of the same description, while the Norwegian pine grows on those that are covered with earth, giving them a very somber appearance, particularly in the autumn when snow begins to fall."[35] Father Mengarini, a companion of DeSmet's, retained vivid memories of the valley's long, cold winter. Thus the initial Euro-American perception of the Bitterroot Valley was of an uninviting, inhospitable landscape.[36]

Despite these characterizations, DeSmet introduced agriculture to the valley in 1842. He sought to interest the valley's inhabitants, the nomadic Flathead Indians, in farming. From the outset DeSmet and his successors realized that successful cultivation depended on irrigation. Early records indicate that the priests irrigated their wheat crop and vegetable gardens. DeSmet's perception of the landscape became more optimistic:

> St. Mary's, or Bitter Root Valley, is one of the finest in the mountains, presenting, throughout its whole extent of about 200 miles, numerous grazing, but few arable tracts of land. Irrigation, either by nature or artificial means is absolutely necessary to the cultivation of the soil, in consequence of the long summer

> drought that prevails in the region, commencing in April and ending only in October. This difficulty, however, if the country should ever be thickly settled, can be easily obviated, as the whole region is well supplied with numerous streams and rivulets.[37]

Still, farming was difficult, and Father Mengarini complained that "the soil is naturally dry and filled with large rocks . . . and we cannot find arable spots except along the creeks which are often located at great distances from each other."[38]

Soon after the arrival of the Jesuits, the legal status of the valley changed. By 1850, the United States had resolved the Oregon question with Great Britain, and Congress divided the Oregon territory by establishing Washington territory. The Bitterroot Valley was part of the area designated Indian territory. Isaac I. Stevens, governor of Washington territory, was also the territorial superintendent of Indian affairs, a position that included jurisdiction over the Bitterroot Valley.

One of Stevens's first tasks was to explore and evaluate the territory. In 1854, he and a party of scientists and explorers traveled from Fort Owen (the mission site of the Jesuits in the Bitterroot Valley) to Vancouver. One member of the party recalled the scenic mountains and fertile land. Another reported a lack of snow at the valley's periphery and noted that in some places the valley was without snow, whereas outside of it, snow had fallen. Stevens's impressions were also favorable, and he began a campaign to acquire the land for white settlement.[39]

Before settlement could occur, Stevens had to secure congressional approval. In the interim, he asked Dr. Richard Lansdale to inspect the Jocko and Bitterroot Valleys for a proposed Flathead Indian reservation. Lansdale found the Bitterroot Valley "poor in quality and intersected by moraines." On the west side of the valley, he reported, "fertile spots are numerous but of small extent. . . . The amount of fertile, arable lands on the east side of the river [is estimated] at twenty-five sections or square miles. . . . The whole of the open lands in the valley both wet and dry, afford good grazing and may be estimated at three hundred square miles."[40] With federal approval, Stevens selected the Jocko Valley for the reservation, but the Flatheads considered the Bitterroot Valley their home. Many Flatheads remained in the valley up until 1891, despite the presence of white settlers after the early 1860s.[41]

Newcomers settled only where water was readily available, either on the west side of the valley with its numerous mountain rivulets or in scattered spots on the east side near one of its streams. The west side was the more popular in the early years because of the availability of water aided by spring runoff, despite the fact that the soil there was rockier. Although the east side had a rich, fertile layer of topsoil, it had fewer streams with smaller amounts of water. Furthermore, precipitation occurred mostly during the spring, leaving the summer months dry. Without irrigation, the Bitterroot Valley would not support a successful agricultural community.

Early residents recognized the importance of irrigation, and by the 1880s small irrigation ditches laced the central valley. In 1880 a visitor to the valley, Abe Williamson, wrote to his family in the East about irrigation and the possibility of beginning an apple industry.[42] More settlers planted orchards, and in 1893 the valley held its first fruit fair.

Perceptions now changed again. Orcharding became a viable livelihood. In the words of one resident, prior to the fair "those who had planted out large numbers of trees were looked upon as cranks and people in the fruit belt did not realize they were making money from them."[43] After the fruit fair, a boom in orcharding led local newspapers to emphasize the agricultural potential of the valley. The *Bitter Root Times* regularly contained a column titled "Little Pointers on Ways of Doing for the Farm, the Orchard and the Live Stock."[44] In the same paper, one writer boasted that Montana had the second largest orchard in the United States. The Bitterroot Valley, he wrote, had 65,000 trees, and many orchards in western Montana had 6,000 to 10,000 trees. The railroads joined in the clamor. One promoter informed Bitterroot farmers that they would "reap fortunes through the enterprise of the Northern Pacific."[45]

The hoopla reflected changed perceptions. Promoters and settlers now believed that the valley's resources offered the possibility of more than a moderate living. If growers used correct varieties, cultivation, and marketing methods, they could derive large profits from the land. In 1905 a Chicago financier, W. I. Moody, visited the valley and came to share these views. An early-twentieth-century booster, Moody said he was

> charmed with the luxurious scenic beauty of the valley and instantly impressed with the wonderful latent resources abounding all about. Thoroughly conversant with the productivity of the great agricultural states of the Mississippi valley, he was amazed by the prolific yields of the Bitter Root farms and orchards. It seemed incredible that the bench lands should have been permitted to lie unreclaimed. True, it was a big undertaking and would cost millions to provide an irrigation system to water the 40,000 acres, but that thought only served to spur him on—provided the lure of big and successful achievement! He

was determined to apply the magic touch of water, to cause that barren waste to "blossom as the rose."[46]

Moody, with local support, embarked on a large undertaking to water the eastern benchlands. His project, later known as the Bitter Root Valley Irrigation Company, planned to build an 80-mile-long canal to deliver the equivalent of 30 inches of rainfall to 70,000 acres planted in orchards. The company's aim was to "expand the valley's McIntosh apple orchards" and make the Bitterroot Valley the center of the apple industry.[47] Although they did build the "big ditch," Moody and his company failed—but not before they contributed to another shift in cultural perceptions of the irrigated landscape.

Moody found immediate local support for his ideas. Joseph M. Dixon, owner of the *Daily Missoulian*, promoted the proposed irrigation system, which mirrored his view of the state's potential. Another local editor, J. C. Conkey of the *Ravalli Republic*, stated, "We want eastern people to know what we raise out here and join us in making one of the richest valleys in the whole West the ideal home spot."[48]

Hundreds bought tracts from Moody's company and newly formed irrigation enterprise. Land values rose dramatically during the period from 1905 to 1915. In 1909, at the twelfth annual session of the Horticultural Society, the society's president stated that land recently had sold for $25 an acre, but now general commercial orchards charged from $200 to $1,000 per acre.[49]

Valley promoters promised a robust lifestyle, near-perfect climate, and the possibility of becoming rich with little labor or initial investment. "It is the land of homes where people breathe pure air and imbibe the spirit of the mountains. The nerve-racking struggle of existence of the cities is not to be found in these everlasting hills."[50] In promotional literature for another orcharding concern, the company alleged that

> a holder of ten acres of choice bearing orchard can depend upon net profits of from $2,000 to $5,000 a year, according to the age of trees. . . . Sunshine may be experienced approximately 300 days in the year. . . . More than one-half of the average annual rainfall comes in April, May and June, a condition that approaches the ideal in orchard economy. . . . It is a blizzardless country. . . . There has not been a killing frost in the growing season on the bench lands where the company's tracts are located in their orchard history.[51]

Still another advertisement boosted the valley as "The Garden Spot of the West, where Mother Earth shall yield her fruits and grains in abundance and neither extreme heat nor cold shall annoy." In yet another testimony to the profits that could be realized, the Bitterroot Fruit Growers Association provided a sampling of incomes derived from various orchards: Thomas Padden's orchard realized $1,600 from each acre of McIntosh Reds during one season; Gus D. Gorus's orchard on the Lake Como "Bench" yielded $800 per acre of McIntosh Red apple trees.[52]

An agricultural transformation of the valley did occur during the boom years. For example, in 1900 Ravalli County (where the Bitterroot Valley is located) claimed 177,652 acres for farming; in 1910 this figure jumped to 209,266 acres, and by 1920 it had risen to 245,965 acres. Similarly, in 1898 Ravalli county produced an estimated 20,000 bushels of apples; by 1919 the number of bushels rose to 400,000.[53]

Despite impressive statistics and boisterous fanfare, the scheme to reclaim the eastern benchlands failed. Financial difficulties plagued the Bitter Root Valley Irrigation Company from the start. Construction of the ditch and its subsequent breakdowns drained much of the company's capital. Insects, the incompatibility of some valley soils with orcharding, and the company's inability to deliver water were additional problems. One of the most damaging blows was a lawsuit filed against the company by Hans B. Knudsen, a Bitter Root orchard owner acting on behalf of 100 fellow orchardists. The suit charged that the company was insolvent and had accumulated debts of $5 million, and it questioned what the company did with funds collected from annual assessments. In September 1915 the Bitter Root Valley Irrigation Company's board of directors decided to file for bankruptcy. The Bitter Root Irrigation District was formed out of the company's wreckage. Today it irrigates approximately 17,000 acres.[54]

The Bitter Root Irrigation District persisted, despite costly breaks, drought years, and economic downturns in the valley. In 1922, Chester C. Davis, state commissioner of agriculture, wrote an article entitled "Western Montana State's Garden Spot." He praised the valley's scenic beauty but differed from early promoters in his assessment of the valley's agricultural potential. Rather than emphasizing the commercial fruit industry, Davis advocated "dairying and a diversified agriculture." The fruit industry should compose only a part of this diversified agriculture, because "the fruit grown in the western part of Montana is not of the same basic importance as alfalfa as a foundation for farm homes." Although he advocated diversification, Davis nevertheless praised the valley's fruit industry in a manner similar to promotions for the earlier boom. He blamed the previous boom's failure on the "promotion company, which committed numerous blunders." In 1926, the Montana Department of Agriculture, Labor, and Industry published a publicity tract on the state's fruit industry touting the "Montana grown Macintosh apple." In the department's eyes, the Bitterroot Valley remained an agricultural landscape.[55]

But the orchard industry endured more setbacks in the 1920s. In 1922 and 1923, the valley suffered damaging hailstorms combined with late frosts and extreme temperature changes. Orchardists suffered major losses. Nevertheless, the University of Montana Extension Service published a bulletin to encourage small-scale orcharding. It was not until the onset of the Great Depression and a continuing rivalry that allowed eastern buyers to accept only the highest quality fruit that the futility of large-scale orcharding in the valley was finally accepted. In 1930, when the extension service published the bulletin "How to Remove Apple Orchards," it was obvious that the orchard industry had ended.[56]

After the collapse of the orchard industry, the irrigation district continued to serve farmers who had diversified their crops. It remained economically unstable, however, for many years. Canal breaks and washouts were common. Instead of offering an "unfailing water supply," the district could not guarantee delivery of water in drought years; the big ditch had gone dry. The district's water supply was limited. Once the spring runoff had been utilized, the reservoir on Lake Como could supply users for only 45 days. Still, the district persisted in serving predominantly agricultural water users.

In recent years the Bitterroot Valley has undergone a shift similar to that in the Yakima Valley. Subdivided farms and suburban land now characterize the Bitterroot Irrigation District. Residential users comprise approximately two-thirds of the system's water users. Within the last 20 years, the number of water users has increased from 650 to 16,665. Realtors pitch lots with "beautiful 360 degree views" and water rights from the Bitterroot Irrigation District. Instead of economic incentives, landownership offers "endless sights" or the "Montana Dream"—a log home with 10 acres producing 25 tons of hay in the heart of the Bitterroot Valley. Finally, the landscape provides tranquillity. One advertisement for a seven-acre parcel with water rights suggested that potential buyers "walk the land and feel the peacefulness."

But the transition to residential water use entails more than a change in livelihoods. Attitudes toward water have also changed, with parallels to the Kennewick Irrigation District. Bitter Root Irrigation District suburban water users view irrigation water as a municipal supply. Turning water out of an irrigation ditch is comparable to turning on the faucet: the supply is similarly instantaneous. In the past, agricultural water users planned which times to irrigate a field and notified the ditch rider. Now the suburban user comes home from work and turns on the sprinkler. Moreover, when the district was primarily agricultural, conservation efforts were greater; if they used sprinkler systems, farmers moved the sprinklers daily. Today the ranchette owner leaves the sprinkler in one place, which is an inefficient use of the resource. Finally, an understanding of the resource, its place in a semiarid region, and the intricate laws governing water use is lacking. It is not uncommon for a realtor, when showing a property with a ditch running through it, to tell clients that this is their water, when in fact that property might have no water rights in the irrigation ditch.[57]

Clearly, precedents were established with the first irrigation enterprises in the nineteenth century. Euro-Americans did not have to come to terms with an arid environment; technology accommodated settlement and still does. But in our time there are differences. With urbanization, the water user is more removed from the resource base. In the Kennewick district in particular, pressurized, underground water pipelines almost deny the reality of living in an arid environment. But the history of these projects—from entrepreneurial schemes followed by eager, inexperienced farmers to the residential user—is not unique. The communal consensus to cultivate the land through irrigation recurred throughout the western half of the United States, and the transition to urban water user is also occurring throughout the West. Thus it is important to understand the history of these areas, as fragile landscapes are continually asked to make room for more people. If culture dictates one response to the environment, perhaps we as a people need to understand and evaluate that culture.

NOTES

Portions of this essay were published previously as "Hydraulic Cities: The Urbanization of the Mid-Columbia Plateau," in *Change in the American West: Exploring the Human Dimension*, ed. Stephen Tchudi (Reno: University of Nevada Press, 1996), pp. 106–27.

1. For a discussion of the relationship between culture and nature, see Anne F. Hyde, "Cultural Filters: The Significance of Perception in the History of the American West," *Western Historical Quarterly* 24 (1993): 351–77; Wesley Reid, "Rhetoric of a River: Tracing Language and Change," in *Change in the American West: Exploring*

the Human Dimension, ed. Stephen Tchudi (Reno: University of Nevada Press, 1996), pp. 87–106.

2. Richard White, *The Organic Machine: The Remaking of the Columbia River* (New York: Hill and Wang, 1995), p. 4.

3. *Ibid.*

4. John Findlay, in *Magic Lands: Western Cityscapes and American Culture after 1940* (Berkeley: University of California Press, 1992), offers one of the best studies of urbanization in the West, demonstrating not only how long urban centers have been a part of the American West but also their impact on a national culture.

5. *New York Times*, 20 December 1996.

6. Helen Ingram, *Water Politics: Continuity and Change* (Albuquerque: University of New Mexico Press, 1990); Helen Ingram, *Water and Poverty in the Southwest* (Tucson: University of Arizona Press, 1987).

7. Eugene S. Hunn, *Nch-i-Wana, "The Big River": Mid-Columbia Indians and Their Land* (Seattle: University of Washington Press, 1990), p. 91. Hunn provides one of the best ecological and ethnographic treatments of the Mid-Columbia Indians.

8. Interview with W. F. Sonderman by Augusta Eastland, in U.S. Public Works Administration, Federal Writers Project, *As Told by the Pioneers* (Washington, D.C.: Government Printing Office, Washington Pioneers Project, 1937–38, 1936); *An Illustrated History of Klickitat, Yakima and Kittitas Counties* (n.p., 1904), p. 227. For a general history of the Columbia basin, see D. W. Meinig, *The Great Columbia Plain: A Historical Geography, 1805–1910* (Seattle: University of Washington Press, 1968); see also Carlos A. Schwantes, *The Pacific Northwest: An Interpretive History* (Lincoln: University of Nebraska Press, 1989).

9. *Yakima Herald*, 25 May 1893.

10. *Kennewick Courier-Reporter*, 6 December 1917.

11. John J. Rudkin to Senator Wesley Jones, Wesley L. Jones Papers, folder 6, box 74, University of Washington Manuscripts Section, Seattle.

12. M. M. Moulton to Wesley L. Jones, 20 November 1926, Jones Papers, folder 5, box 74.

13. Commissioner to Chief Engineer, Denver, Colorado, 1 August 1923, Yakima Kennewick Project 1919–29, Record Group 115, file 30l, National Archives, Washington, D.C.

14. *Yakima Daily Republic*, 13 February 1930.

15. Mrs. Robert H. Denhoff to President Franklin D. Roosevelt, 12 April 1935, Yakima Kennewick Project 1930–45, RG 115, file 100.2.

16. U.S. Department of Interior, Bureau of Reclamation, *Appendix Report of the Kennewick Division, Yakima Project, Washington* (Boise, 1947), p. 19; Harold G. Fyfe

to C. E. Crownover, 20 June 1945, "Washington State Reclamation Associations file," Kennewick Irrigation District Archives, Kennewick, Washington.

17. *Authorizing the Construction, Operation, and Maintenance, under Federal Reclamation Laws, of the Kennewick Division of the Yakima Project, Washington*, Senate Report 1404, 80 Cong. 2 Sess. (Serial 11207); *Appendix Report of the Kennewick Division:* pp. 1, 19.

18. *Appendix Report of the Kennewick Division*, pp. 116–17.

19. In Bureau of Reclamation, *Definite Plan* (n.d.), pp. 1–9, government officials were more adamant about the primacy of agriculture and cautioned against a local economy without a firm agricultural base.

20. *Tri-City Herald*, 7 March 1954.

21. "Kennewick Irrigation District Minutes," 15 June 1954; "Settler Assistance File," Public Notice no. 1, 23 February 1956, "USBR Correspondence Files pre-1958"; "Kennwick Irrigation District Minutes," 25 October 1954; W. L. Karrer to Van E. Nutley, 12 September 1956, "USBR Correspondence Files pre-1958," Kennewick Irrigation District Archives.

22. "Kennewick Irrigation District Minutes," 5 May 1970.

23. *Ibid.*; Interview with Paul Chasco, Kennewick Irrigation District manager, by author, 15 April 1986, Kennewick, Washington. In a study commissioned by the Kennewick district, the Century West Engineering Corporation concluded that the canals for the old land were "designed to deliver at a constant flow, typical of a farm's demand for water," but with residential users the engineers found "considerable fluctuations" and that "operation difficulties reflected the daily guessing-game of estimating user demand, weather conditions and delivery lagtime." Century West Corporation, "Urban Canal Fluctuations—Kennewick, WA" (1982), Kennewick Irrigation District Archives.

24. Interview with Ben Volk, Kennewick Irrigation District engineer, by author, 6 February 1995. The manager of the Bitter Root Irrigation District reached the same conclusion regarding residential water use. Interview with Mike Shatzer, Bitter Root Irrigation District manager, by author, 10 January 1997.

25. *Tri-City Herald*, 26 July 1994.

26. Kennewick Irrigation District, *Farm and Community Canal* (Kennewick, Wash., 1991).

27. *Ibid.*

28. *Ibid.*

29. *Ibid.*

30. The lawsuit is *Department of Ecology v. James J. Acquavella, et al.* See "Kennewick Irrigation District Minutes," 2 January 1968, 18 May 1977, 1 March 1977, 9

September 1982; *Tri-City Herald*, 22 July 1986; *Spokane Spokesman-Review*, 28 September 1986. I

31. "Kennewick Irrigation District Minutes," 2 March 1971, 6 February 1973, 3 June 1975, 12 October 1978, 11 April 1980; *Tri-City Herald*, 7 May 1979.

32. Interview with Michael Macon, Kennewick Irrigation District resource manager, by author, 20 August 1991; Chasco interview, 21 August 1991. In a 1993 court ruling it was determined that irrigation districts were not statutorily authorized to do economic development, but the resource manager still assists developers in facilitating their needs.

33. Chasco interview, 15 April 1985; interview with Chuck Garner, Kennewick Irrigation District manager, by author, 19 January 1995.

34. Leonard J. Arrington and Thomas G. Alexander, *Water for Urban Reclamation: The Provo River Project* (Logan: Utah State University, 1966).

35. Hiram M. Chittenden and Alfred T. Richardson, eds., *Life, Letters and Travels of Father DeSmet* (1905; reprint, New York: Arno Press, 1969), vol. 1, p. 344.

36. Albert J. Partoll, ed., *Mengarini's Narrative of the Rockies: Memoirs of Old Oregon, 1840–1850, and St. Mary's Mission* (Missoula: Montana State University, Historical Reprints in Sources of Northwest History Series no. 25, 1930), p. 7.

37. Chittenden and Richardson, *Life, Letters and Travels of Father DeSmet*, vol. 1, p. 571.

38. Lucylle H. Evans, *St. Mary's in the Rocky Mountains* (Stevensville, Mont.: Montana Creative Consultants, 1976), p. 69.

39. Seymar Dunbar, ed., *The Journals and Letters of Major John Owen* (New York: Edward Eberstadt, 1927), p. 275; John Fahey, *The Flathead Indians* (Norman: University of Oklahoma Press, 1974), p. 80.

40. Fahey, *The Flathead Indians*, p. 99.

41. *Ibid.*

42. Merrill Burlingame and K. Ross Toole, *History of Montana* (New York: Lewis Historical Publishing Co., 1957), vol. 1, p. 344.

43. *First Biennial Report of the Montana State Board of Horticulture* (1899).

44. *Bitter Root Times*, (n.d. [1894?]).

45. *Bitter Root Times*, 6 May 1898.

46. "Bitter Root Valley Illustrated," *Western News* (special edition), 1910, p. 8.

47. "Hamilton FOCUS," *Ravalli Republic*, January 1979.

48. *Ibid.*

49. *Northwest Tribune*, 12 February 1909.

50. Bitter Root Valley Irrigation Company, *Bitter Root Valley*, p. 11.

51. *Ibid.*

52. "Bitter Root Valley Illustrated," *Western News* (special edition), 1910, p. 41.

53. Burlingame and Toole, *History of Montana*, vol. 1, p. 294.

54. Sherman E. Johnson, *The McIntosh Apple Industry in Western Montana* (Helena: Agricultural College of Montana, Montana Agricultural Experiment Station Bulletin no. 218, 1924), p. 17; Bitter Root Valley Irrigation Company, Bankrupt, Case no. 1390 (1st Montana Judicial District, 1916).

55. *Daily Missoulian* (souvenir edition), 22 July 1922, p. 19; Montana Department of Agriculture, Labor, and Industry, *Montana: Resources and Opportunities Edition* (Helena: 1926), p. 85.

56. *Missoulian*, 29 March 1979, p. 17.

57. Interview with Elaine Culletto, Bitter Root Irrigation District office manager, by author, 6 January 1997; interview with Gary Shatzer, Bitter Root Irrigation District manager, by author, 10 January 1997; Bitter Root Valley Board of Realtors, *Real Estate Directory* (June 1996).

PART V / FORESTS

Although forests make up just 46 percent of the total land base in the states of Washington, Oregon, and Idaho, they largely define the region in the popular imagination. Washington, for example, advertises itself on its license plates as the "Evergreen State." The condition and fate of the Northwest's forests are central to its sense of place, politics, economics, tourist industry, and wildlife resources. This diversity of interests mirrors the wide variety of forest landowners: the timber industry owns about 11 million of the 53 million acres of forest land in the region, and 3.5 million acres are owned by farmers. Other private landowners own an additional 5 million acres; the federal government owns 27.5 million acres; and state, local, and tribal governments own about 6 million acres. Each owner manages the land to achieve a different objective; each parcel has a different land-use history. The forest history of the Northwest is thus complex.

Despite the diversity, industrial timber production is the major use of forests in the region—particularly in the more productive and accessible lower-elevation forests. Although the forest industry owns approximately 20 percent of all Northwest forest land, it owns 55 percent of the most productive timberland—areas supporting more than 120 cubic feet of timber growth per acre per year. This highly productive land is intensively managed for timber production. The national forests have also been extensively logged over the past century to support the wood products industry—despite the fact that much of the federal government's forest land is in remote, steep, less-productive mountainous areas. Although the federal government owns extensive tracts of forest, only 8 percent of its forest land falls into the highest productivity class.

Forest management has been among the most visible and hotly contested environmental issues in the Northwest during the past two decades. Industrial timber production, with its large clear-cuts, erosion, and reductions in biological diversity—on federal, state, and tribal land as well as on industry land—energized a forestry reform movement in the Northwest in the 1980s and 1990s. At the same time, the rapid liquidation of old-growth forests in the region led to a preservation movement that resulted in significant reductions of logging on federally owned land in the 1990s. These reductions were achieved primarily through lawsuits filed under the Endangered Species Act.

Nancy Langston's chapter offers an excellent introduction to these controversies. She provides an ecological history of the inland Northwest forests, focusing on the national forests of the Blue Mountains region of northeastern Oregon. She traces the evolution of the science of forestry, showing how cultural ideas about nature, embedded in United States Forest Service notions about proper forestry, actually helped create the undesirable ecological conditions that foresters were trying to avoid. Finally, Langston ties in the political and economic forces at work on both the forests and the federal foresters, linking ecology, culture, economics, and politics into a clear explanation of the "forest health crisis" of the 1990s.

Paul Hirt focuses on the national forests of northern Idaho and western Montana, examining the boom in timber harvesting that followed World War II. He shows how economic and political pressures moved the Forest Service away from an older, more conservative approach to forest management and toward an aggressive program of intensive management for maximum production of timber—a shift that had disastrous consequences for the forests and led to a dramatic boom and bust in the timber economy. Both Hirt and Langston show that although the Forest Service has been nominally dedicated to the principles of "sustained yield" and "multiple use" for nearly a century, neither ideal has been realized.

In the final chapter of part five, Thomas Cox looks at the forests of the "Westside"—the dense, wet, evergreen jungle west of the Cascades. He considers the entire landscape—both forest and meadow, public and private land—in a sweeping survey tracing Native American occupation, Euro-American agricultural settlement, and the industrial age to the modern period. Cox notes how technologies such as the railroad, the steam donkey, trucks, and tractors increasingly expanded the timber industry's ability to exploit forest resources. This led to the growth of the industry as well as to increasing conflicts between those who valued forests for their wood fiber and those who valued them for biological, recreational, or aesthetic reasons. Remoteness and rugged terrain "once provided protection for vast tracts of land," he says. But now such protection "must be extended through human wisdom, which must at the same time devise means of meeting the very real material demands on the forest that come with a growing regional and world population."

19 / Human and Ecological Change in the Inland Northwest Forests

NANCY LANGSTON

When whites first came to the Blue Mountains of eastern Oregon and Washington in the early nineteenth century, they found a land of lovely open forests full of yellow-bellied ponderosa pines five feet across. These were stately giants the settlers could trot their ponies between, forests so promising that people thought they had stumbled into paradise. But they were nothing like the humid forests to which easterners were accustomed. Most of the forest communities across the inland West were semiarid and fire-adapted, and whites had little idea what to make of these fires.

After a century of trying to manage the forests, what had seemed like paradise was irrevocably lost. The great ponderosa pines were gone, and in their place were thickets of fir and lodgepole. The ponderosas had resisted most insect attacks, but the trees that replaced them were the favored hosts for defoliating insects such as spruce budworm and Douglas-fir tussock moth. As firs invaded the old ponderosa forests, insect epidemics swept the dry western forests. By 1991, on the 5.5 million acres of United States Forest Service land in the Blue Mountains, insects had attacked half the stands; in some stands, nearly 70 percent of the trees were infested.[1]

Even worse, in the view of foresters and many locals, was the threat of catastrophic fires. Although light fires had burned through the open pines every 10 years or so, few had exploded into infernos that killed entire stands of trees. But as firs grew beneath the pines and succumbed to insect damage, far more fuel became available to sustain major fires. Each year the fires seemed to get worse. By the beginning of the 1990s, one major fire after another swept the inland Northwest, until it seemed as if the forests might go up in smoke entirely.

In an unusual admission of guilt and confusion, the Forest Service stated in its 1991 *Blue Mountain Forest Health Report* that the forest health crisis was caused by its own past management practices—and that therefore it would have to change those practices.[2] But no one could agree on exactly which management practices had caused the problems, much less on how to restore the forests. The Forest Service nevertheless felt compelled to do something about the perceived crisis. Because many earlier forests seemed to have been more open, many people within the Forest Service and the timber industry argued for the so-called forest health solution: log the forests in order to restore them to their "original condition" and simultaneously keep the mills running. This was a solution that to many environmentalists seemed worse than the problem.

In this essay I offer an environmental history of the inland Northwest forests, focusing on the Blue Mountains of eastern Oregon, where the crisis first broke.[3] I trace the links between culture, ecology, and politics that led to the current problems. Finally, I show how many of the proposed solutions will only exacerbate the problems because they ignore the root causes of forest decline.

FORESTS: ECOLOGY, CULTURE, AND POLITICS

Forests change not just because people cut down trees, but because they cut down trees in a world where nature and culture, ideas and markets tangle together in complex ways. On one level, landscape changes in the Northwest forests resulted from a series of ecological changes. Heavy grazing removed the grasses that earlier had suppressed tree germination, allowing dense thickets of young trees to spring up beneath older trees. When federal foresters suppressed fires, the young firs grew faster than pines in the resultant shade, soon coming to dominate the forest understory. High-grading—removing valuable ponderosa pine from a mixed-conifer forest—helped change species composition as well. But the story is much more complex than this outline suggests. Changes in the land are never just ecological changes: people make the decisions that lead to ecological changes, and they make those decisions for a complex set of motives.

There are two basic perspectives on the causes of forest decline in the Blue Mountains—both of them inadequate. Many environmentalists argue that federal foresters promoted excessive harvests to feed the demands of the timber companies. These harvests led to soil compaction, removal of valuable pines, and even-aged management, creating a simplified ecosystem that is increasingly susceptible to epidemics.

Many other people believe exactly the opposite: they argue that the Forest Service bowed to the demands of sentimental preservationists and refused to harvest the forests intensively enough to save them from disaster. These people put the blame on past decades of light, selective cutting, arguing that only intensive management can save the forests. Since ponderosa pine is shade intolerant, they reason that clear-cutting, even-aged harvests, and intensive management are the only ways to assure that fir stands do not replace pine stands. In their view, the best way to eliminate insects, disease, and fire from a stand is to manage it as intensively as possible.

These two perspectives on the history of forest health problems obviously lead to radically different management prescriptions. The two stories of what went wrong are, at least in part, shaped by the different ideologies of their proponents. A traditional forester who sees an old-growth forest as a place of decay, waste, and decadence tends to believe that the human role in forestry is to prevent waste and promote a clean, productive, growing stand. Conversely, someone trained in an ecosystem perspective tends to see insects and disease as having a role in a forest and to value old-growth stands more than single-species stands under even-aged management. But ideology, though it shapes one's view of who is to blame, is not everything. The story is more complicated than either perspective suggests.

The early foresters' training was heavily influenced by European silviculture, which had as its ideal a waste-free, productive stand: nature perfected by human efficiency. Early Blue Mountains foresters believed that to make forests sustainable, they needed first to transform decadent old growth into vigorous, regulated stands. Yet until the First World War, they did not try to implement these ideals, largely because there were few markets for the trees. It was neither economically nor technologically feasible to cut the forests heavily enough to bring about intensive sustained-yield forestry. After World War I, however, the Forest Service established extremely high rates of ponderosa pine harvests, creating the ecological and economic conditions that led directly to the forest health crisis of the 1990s.

Why did the Forest Service promote such large harvests? Desire for profit, power struggles, and bureaucratic empire-building all played institutional roles, but none of them can explain the motivations of individual foresters. To understand their decisions, we need to examine the links between ideals and material reality in American forestry. Federal foresters shaped western landscapes according to a complex set of ideals about what the perfect forest ought to be. In turn, these visions were shaped by available logging technology, developing markets for forest products, the costs of silvicultural practices, and what the historian Rich Harmon has called "unrelenting pressures . . . aimed at government officials to make public resources available for private profit."[4]

To understand forest history, we need to pay attention to at least three interwoven categories. First, there is the cultural: How did cultural ideals affect the ways different groups of people changed the forests? What kinds of visions of the relationship between humans and nature did foresters bring to the forests? How did these visions shape the land? In particular, what scientific visions of the forest shaped the foresters' work? In which political and cultural contexts did these scientific theories develop? How did foresters' scientific ideas and technology shape the forests?

Second, there is the ecological category: What were the biotic and abiotic factors that shaped forests? Plant communities, disturbance processes such as fires, floods, and insect epidemics, soil processes, and hydrological processes are major (but often overlooked) players in forest history. How did ecological constraints affect the land's response to management? How did management change the paths of forest history?

Third, there is the political: Over the course of a century, forest communities in the Blue Mountains were transformed into a collection of resources exported out of the region to feed the demands of distant markets. How did different groups negotiate conflicts over who should determine the rate and extent of logging? How did government officials respond to congressional, corporate, and local pressures to make public resources available for private profit? Whose vision of the land determined how the land was shaped?

Most forest histories have focused on this final category of political, administrative, and institutional relationships, treating ecological factors as unchanging givens rather than as dynamic players in the story.[5] In this analysis, I pay less attention to political forces and focus instead on the links between these categories.

ECOLOGICAL FRAMEWORK

To understand what went wrong in the Blue Mountains, we need to consider first the complexity of the forests, particularly the ways in which they differed from humid forests of the eastern United States. The Blue Mountains were disturbance-prone landscapes marked by varying patterns of change. Fire, wind, insect epidemics, and droughts shaped a shifting mosaic in which patterns of forest types and grassland and sagebrush were anything but stable.

Water and fire—and the changes they brought—were at the heart of the differences between western and eastern forests. Much of what went wrong in the Blues was a failure to pay attention to the land's signs. Trees made the Blues appear a fertile, promising, easy place to the first whites who came, but that perception proved to be an illusion. The forests' fertility was based on ash soils that were quite different from eastern soils; when managers tried to apply forestry techniques based on eastern forests, those techniques decimated the soils.

The constraints that aridity imposed were unfamiliar to people who knew forests only in moister places. The critical resource in the Blue Mountains—the resource on which the rest depended—was not trees or grass or even soil but water. Water, the lack of water, the distribution of water, the storage of water affected every aspect of forest and grassland ecology in the Blues. Because most of the Blues' precipitation came in the form of snow, the water that trees needed to grow came not from rain that fell during the period of growth but from stored water.[6] Any activity that decreased the ability of forest soils to hold snowfall had magnified effects on the forests themselves.

Whites who first arrived in the Blue Mountains found a land completely unlike the humid forests of home. What seemed familiar at first glance proved to be different, and this was unsettling. People expected forests to be moist and fertile, but these forests seemed too dry, too open, and not very fertile. Fires burned much more often than people thought was normal or desirable, and no one understood how forests could survive them. Sagebrush typically indicated poor soil, but the soil under this sagebrush seemed better than much of the forest soil. Rivers normally drained to the sea, but many of these rivers drained into the Great Basin and never flowed out. The canyons were far too steep; people could not believe the evidence their eyes gave them. Trees grew on top of these steep canyons but not down by the water, where trees were supposed to grow. It seemed as if someone had turned the world upside down.

Part of what seemed strange to easterners, especially to those used to the vast stretches of climax forests in Maine and the Midwest, was the diversity of vegetation types within the Blues. When government forest inspectors came to classify the Blue Mountains forests at the turn of the century, the variety of trees, habitats, and forest types astonished them. The inspectors walked through steep treeless grasslands covered with sagebrush and bunchgrass and then crossed into juniper woodlands. These gave way to stately ponderosa forests spaced in as open and pleasing a pattern as any that the inspectors knew from landscaped parks in eastern cities. Along the creeks, strips of lush cottonwood forest shadowed the waters, and these cool riparian zones offered shelter from the brutal summer sun.

When the inspectors crossed from the south face of a canyon to the north face, they moved out of the ponderosa forest into much denser forest dominated by Douglas fir, grand fir, and larch—communities they called the north-slope type or fir-larch forests. At first glance, these north-slope forests appeared uniform, but when the foresters looked more closely, they realized that there were many small patches of larch, fir, spruce, and pine jostled together. The inspectors climbed higher into the hills, finding themselves in thickets of lodgepole pine. Their way became nearly impassable as piles of dead wood and tangles of wind-thrown lodgepole blocked the route. If the men kept climbing, they entered high, eerie forests filled with the stumps of subalpine fir and contorted, wind-twisted whitebark pines. Where fires had burned in small, hot patches, lush meadows interrupted the high forests, and finally the forests gave way to mats of wind-cropped fir and then rock and snow.[7]

Out of all these forests, it was the ponderosa pine that caught men's eyes. These were forests on the edge of forestedness—open and easy, not claustrophobic and frightening. Whites loved the ponderosa forests because they seemed like wilderness tamed and made easy—tamed not by a gardener wielding a hoe and trowel but by nature. Ironically, however, the pinelands were a managed landscape. Indian burning, not just natural processes, had shaped them. But whites saw the frequent Indian-set fires as a threat to what they loved rather than as an essential part of it, and so did their best to protect the forests from fire—a decision that radically backfired.

Fire, set both by lightning and by people, had shaped the pine parklands for thousands of years, and without it the forests changed to something utterly different. In the presettlement ponderosa forests, light fires swept through about every 8 to 10 years.[8] They kept the stands open, killed off the youngest trees, reduced fuel loads, and released a surge of nitrogen into the soil that stimulated the growth of grasses and nitrogen-fixing shrubs such as ceanothus. The fires were frequent enough to keep the stands well thinned but variable enough that a few young trees survived to form future stands. This was not at all evident to early foresters, however. They saw reproduction dying in the frequent fires, and they saw the effects of intense slash fires on soil and water. They believed that suppressing fires was the only way to save the forests.

Early foresters recognized that fires were common in the open pine forests, but they were unable to imagine how trees could survive them. It appeared logical that frequent fires would kill young growth; therefore, after a century, all that would remain was forests made up entirely of old trees. Although light fires might maintain nice open stand conditions for old trees, they seemed disastrous for the future, and future forests were the Forest Service's responsibility.

Early foresters recognized that the thickets of young trees that fire suppression and logging created were not an ideal situation. Yet they argued that allowing fires to thin the stands would lead to problems, because thinning would lessen competition, and competition was what led to manly, vigorous trees. An antifire Forest Service report from the 1930s argued that trees would be "hardier, taller, straighter and cleaner for having this heavy competition in their early lives. Dense young thickets of forest seedlings or saplings are what make our tall, straight, clean-limbed forest giants. Did anybody ever see a big, bushy-topped, limby tree that grew in the open, all by itself, that was worth a whoop as a sawlog?"[9] Stand development theory stated that in a forest, some seedlings should grow faster, become dominant, suppress, and then eliminate their neighbors, thus allowing a few trees to grow rapidly into big trees. Competition was a natural and virtuous process ensuring that only the best would survive. Overcrowded stands were a good thing because they would lead to strong trees. But in reality, western conifers—ponderosa, western larch, Douglas fir, and lodgepole in particular—do not obey stand development theory. They do not self-suppress, which is one reason why fire suppression was such a problem in mixed-conifer stands. In the absence of fire, no trees became dominant; instead, the undergrowth became a thicket of same-aged trees whose growth rates had slowed almost to a halt.[10]

Across the Blue Mountains, after years of fire suppression, high-grading, and heavy grazing, millions of acres changed from pine to fir and then from healthy fir to drought-stressed, insect-defoliated fir. In 1912, in what is now the Wallowa-Whitman National Forest, 71 percent of the stands were open and full of old pine; in 1991 only 10 percent fit this description. In the Malheur National Forest in 1938, 78 percent of the forests were open pine stands; by 1980 less than half of those were still pine forests. In the Umatilla National Forest in 1905, 43 percent of the forest area was dominated by open pine stands; in 1991 only a seventh of those forests remained pine.[11]

Old growth faced the most drastic losses: probably less than a tenth of the presettlement old-growth forests remained by the early 1990s. Surviving old growth had become extremely fragmented; by 1993, 91 percent of the old-growth forests existed in patches of fewer than 100 acres. When the Forest Service had first arrived, those mature forests had stretched for hundreds of thousands of acres. For example, the Malheur National Forest once contained some of the finest stands of open ponderosa in the nation. In 1906, one report stated that an open, old-growth ponderosa-larch forest covered 800,000 acres south of the Strawberry Mountains. Fewer than 8,000 acres of the same forest remained in 1993—less than 1 percent of what was present before the Forest Service began management. In 1912, the forester R. M. Evans wrote that in the Wallowa and Minam Forests, there were nearly 600,000 acres of uneven-aged forests containing more than 80 percent ponderosa. About half a million acres—or 85 percent—of these ponderosa-dominated forests were old growth. By 1991, only 18 percent of the forests in the Wallowa-Whitman National Forest were mature or old-growth stands, and very few of those were ponderosa. The rest of the forests were dominated by young fir trees.[12]

VISIONS OF THE FORESTS

The story of these drastic landscape changes is, in its simplest version, a story of the land's transformation into a set of resources that could be removed from one landscape and transported to another. Indians had certainly altered the landscape, but when whites showed up they set into motion changes that far outpaced previous ones. The critical difference was that the Blues finally became a source of commodities—timber, gold, meat, and wool—to feed the engines of market capitalism.

Before whites came, the Blue Mountains were certainly connected to markets outside the region. Local tribes had an extensive set of ties to trading networks that spread west to the Pacific Ocean and east to the Great Plains.[13] Indians did extract elements from the local ecosystem, and in the process they changed the local ecology to meet their needs. But their needs did not include removing large quantities of wood fiber for fuel, fertilizer, or construction. Indian land use was not necessarily sustainable, nor was it in any kind of inherent balance with the land's limits. Yet it was fundamentally different from the land use that whites instituted because it did not include the large-scale extraction of resources and their export elsewhere. Indians who made the Blues their home did not see the land as a set of distinct, extractable resources, the way most whites would come to see it even when they had strong emotional connections to the place.

Euro-American settlement in the Blue Mountains, as in the West at large, had been driven by a vision of limitless abundance. The forests seemed endless, the land in need of improvement, the world available for the taking. But as the timber industry reached the Pacific, people began to fear that the end might be in sight. Many worried that if the nation continued to deplete its forests without thought of the future, it might one day it find itself without the timber upon which civilization depended. Federal scientists in particular were certain that because of wasteful industrial logging practices, a timber famine was about to devastate the United States. By the last decade of the nineteenth century, the Blues seemed to be in serious trouble. The bunchgrass was largely gone, depleted by intense grazing. Wars between small cattle ranchers, itinerant sheepherders, and large cattle operations from California had left thousands of sheep and several sheepherders dead. Timber locators and speculators were taking up the best timberland; small mills and miners were illegally cutting throughout the watersheds; irrigators feared that their investments in water projects would be lost.[14] It was in this context that federal foresters came west in 1902—to save the Blues from unrestricted abuse fostered by the desire for short-term profits.

At the turn of the century, the United States was in a furor over land management. Disposal of the vast tracts of western land was an enormously corrupt process, and to many Americans the federal government seemed more corrupt than anybody else when it came to managing land. The new forest scientists stepped in and said, Science will show us a way out of this chaos of political corruption, if you just leave it to us. By turning to the clear, calm, seemingly universal rules of science, foresters tried to avoid the mess of contentious politics and the contingencies of history. They felt they could introduce science into the muddle of land management, making a better society as well as a better nature. As scientists who had the interests of America and American forests at heart, they felt they were beyond criticism. They alone could serve the public good, they felt, because efficiency rather than short-term profit was their goal.

The Forest Service came to the Blues with the best of intentions: to save the forest from the scourges of industrial logging, fire, and decay. When they looked at the Blues, they saw two things: a "human" landscape in need of being saved because important watersheds seemed to have been ravaged by companies and the profit motive, and a "natural" landscape that also needed saving because it was decadent, wasteful, and inefficient. Not only were federal foresters going to rescue the grand old western forests from the timber barons, they were also going to make them better. Using the best possible science, foresters felt they were going to make the best possible forests for the best of all possible societies: the United States in the brand new twentieth century.

In the eyes of the early conservationists, western old-growth forests and the loggers who denuded them shared a basic flaw: both were wasteful. As the historian Samuel T. Hays argued, the point of American conservation was to reduce waste and increase efficiency. A Umatilla National Forest press release dated September 12, 1906, put it well: the intent of conservation was to "hunt down waste in all its varied forms" in water, forests, land, and minerals. Waste existed not when people overused resources but rather when they failed to use them fully enough. Any water that was not used for irrigation was lost forever; any grass that was not eaten by a cow went to waste—or so the press release insisted. If people did not put everything to full use, it was a moral failure, not just an economic loss.[15]

Early government foresters had several strong articles of faith. First, the point of forestry was to reduce waste and make forests more efficient—which meant making them produce more timber more quickly. Second, the United States needed wood, and demands would continue to increase as quickly as they had increased after the Civil War. Third, if harvesting continued as usual, the country would run out of timber in 25 to 30 years. Fourth, forests ought to be used, but scientifically, to assure a perpetual supply of timber for a growing nation. Finally, scientists were best at solving land management problems, and so scientists, rather than politicians, ought to control the Forest Service.[16] Trained scientists—professional foresters—would redesign the old-growth forests to improve them and so assure a perpetual supply of timber for a growing nation.

The basic premise of the new Forest Service was simple: if the United States was running out of timber, the best way to meet future demands was to grow more timber. More than 70 percent of western forests were old-growth stands—what foresters called "decadent and overmature"—which meant forests that were losing as much wood to death and decay as they were gaining from growth. Because young forests put on more volume per acre faster than old forests, foresters believed that old-growth forests needed to be cut down so that regulated forests could be grown instead. Regulated forests were young and still growing quickly, so they added more volume in a year than they lost to death and decay. The annual net growth could be harvested each year without depleting the growing stock. Scientific forestry seemed impossible until the old growth had been replaced with a regulated forest. This logic shaped a Forest Service that, in order to protect the forest, believed it necessary first to cut it down.

As the emphasis on eliminating inefficiency and waste increased, Gifford Pinchot, the founding chief of the Forest Service, and his young foresters grew increasingly impatient—and then hostile—toward decadent old growth. The best way to free up the land to grow better forests was to sell off the old-growth timber as soon as possible, which meant pushing sales, even in a slow market. On their first day in forestry school, foresters learned how to think about old growth. In 1911, C. S. Judd, the assistant forester for the Northwest region, told the incoming class of forestry students at the University of Washington that a timber famine was on its way unless the Forest Service did something quickly. Since the forest was running out of trees, the way to fix the problem was to get National Forest land to grow trees faster. As Judd put it, "the good of the forest . . . demands that the ripe timber on the National Forests and above all, the dead, defective, and diseased timber, be removed."[17] The way to accomplish this was to "enter the timber sale business" and heavily promote sales. This would get rid of the old growth, freeing up land to "start new crops of timber for a future supply."[18] Foresters saw old growth not as a great resource but as a parasite, taking up land that should be growing trees.

Henry Graves, the second chief of the Forest Service, continued the emphasis on removing old growth to reduce waste and inefficiency. For Graves, forests were just storehouses for public commodities. As he stated in his *Annual Report of the Forester*, "virgin forests are merely reservoirs of wood." Production was the main thing foresters needed to consider. And production, Graves added, "can be secured only by converting them [old-growth forests] into thrifty, growing stands through cutting." The Forest Service's sales policy was clear: "The timber sale policy aims, therefore, as the first requirement of good management, to work over the old stands on the National Forests, utilizing mature timber . . . and putting the ground in such condition that forest production will be renewed at a much faster rate."[19]

The unregulated forest was something to be altered as quickly as possible for moral reasons—to alleviate what one forester, Thorton Munger, termed "the idleness of the great areas of stagnant virgin forest land that are getting no selective cutting treatment whatsoever."[20] The problem was not just with old growth or dying timber; it was with a forest that did not produce precisely what people wanted. The problem was recalcitrant, complex nature, marked by disorder and riot.

Foresters believed that disease, dead wood, old growth, and fire all detracted from efficient timber production. In other words, they assumed that the role of the forest was to grow trees as fast as it could, and anything that did not contribute directly to that goal was bad. Whatever was not producing timber competed with trees that could be producing timber, foresters believed. Any space that a dead tree took up, any light that a fir tree used, any nutrients that an insect chewed up—those were stolen from productive trees. If timber trees did not use all the available water, that water was wasted. If young, vigorous pine did not get all the sun, that sun was lost forever. These assumptions made it difficult for foresters to imagine that insects, waste, disease, and decadence might be essential for forest communities—indeed, that the productive part of the forest might *depend* on the unproductive part of the forest.

LIQUIDATING OLD GROWTH

What effect did all these grand visions of scientific forestry have on the public forests themselves? Until World War I, very little. For all the foresters' desire to cut old growth, the Forest Service sold little in the Blue Mountains until after the war.[21] Forest Service timber was inaccessible, prices were set so high that few contractors were willing to invest, and the industry still had enough private stock to make sales of federal timber unattractive. After the war, markets for national forest timber opened up, and the Forest Service began to push sales of ponderosa pine heavily in the Blues. This in turn enabled them seriously to begin the campaign to regulate the forests by liquidating old growth.[22]

The Forest Service believed that to ensure local prosperity, old-growth forests had to be converted to regulated forests that could produce harvests forever. But to regulate the forests, planners needed markets for the timber, and they needed railroads to get it out to markets. Railroads were extraordinarily expensive, particularly after the First World War. Financing them required capital, which often meant attracting investments by midwestern lumber companies. But these companies were interested in spending money on railroads only if they were promised sales large enough and rapid enough to repay their investments. The results in the Blues, as across the West, often damaged both the land and the local communities who depended on that land.

Throughout the Blue Mountains in the 1920s, Forest Service planners encouraged the construction of mills that had annual milling capacities well above what the Forest Service could supply on a sustained-yield basis. In the Malheur Forest alone, for example, two large sales during the 1920s offered more than 2 billion board feet of pine, out of only 7 billion in the entire forest. Two mills followed—one capable of processing 60 million board feet a year and another that could process 70 to 75 million board feet a year.[23] With mill capacities reaching 135 million board feet a year, it would have taken only 15 years—not the 60 years of the cutting cycle—to process the 2 billion board feet in these sales, and only 52 years to process all the ponderosa in the entire forest.

As the Forest Service tried to get contractors to buy their timber, they made extensive compromises in the hope of attracting business and liquidating the old-growth ponderosa as fast as possible. The Forest Service initially had a strong policy against high-grading—the practice of cutting out only the most valuable ponderosas from a mixed stand and leaving behind the less-valuable trees to form the basis of the future forest. But correspondence between sales planners makes it clear that contractors were refusing to cut inferiors. Rather than force them to meet their contracts, the Forest Service quietly looked the other way. Soon contractors began demanding a reduction in stumpage prices, claiming that they should be given a discount on ponderosa if they had to cut any other species at all. In 1922, the Forest Service decided that continuing to insist on cutting fir would mean that no one would buy ponderosa, and so the policy against high-grading was dropped.[24]

In 1922, the Forest Service also abandoned its light selective cutting ideal and began allowing contractors to remove 85 to 90 percent of the mature forest in each sale, leaving only 10 to 15 percent as a reserve stand for the next harvest cycle. Loggers were allowed to skid out the timber with Caterpillar tractors, even though before 1922 the Forest Service had discouraged tractor skidding because it damaged young growth. And finally, the Forest Service abandoned its policy of requiring that big pines be left in the reserve stand as seed trees; in the 1920s, it told contractors to cut all the pine over 15 inches in diameter in a sale area. Although many foresters now argue that light selective cutting destroyed the ponderosa forests by encouraging fir, it was these silvicultural compromises of the 1920s, rather than light selective logging itself, that helped ensure that fir would take over the mixed-conifer forests.[25]

Concern about the effects of intense harvests on local communities began to emerge in working circle plans during the late 1920s, although foresters did not allow this concern to decrease their recommended harvests. In a Malheur River working circle plan from about 1927, the planner attempted to calculate the annual yield that would be available for local mills beginning in the

1980s, during the second cutting cycle. He realized with dismay that harvests would drop by at least 40 percent in the 1980s if cutting continued at current rates.[26] The planner consoled himself with the thought that because his calculations of growth rates were only rough estimates, perhaps they would turn out to be underestimates, and then there would more timber than anyone expected. He also hoped that "utilization efficiency will greatly increase"—less waste would mean more wood for future mills.

For the Whitman Forest, letters between sales planners, the forest supervisor, and the regional district forester show that by 1927 the Forest Service was worried about the mill capacities it had encouraged. E. A. Sherman, the acting forester in Portland, criticized a draft of the management plan for the Baker Working Circle, complaining that the mill at Baker was too large and was using up too much wood, in excess of annual allowable cuts. He wrote that "the present milling capacity at Baker of between 40,000,000 and 50,000,000 feet annually . . . greatly exceeds the possible sustained yield from the Government lands in this working circle. . . . It does not look as if a reduction in the milling capacity at Baker sooner or later could be avoided."[27] The sales planner who had prepared the plan Sherman was criticizing agreed that harvest reductions would certainly come by the 1980s. Nevertheless, he argued to Sherman, they should do their best to meet the mills' current demands in order to avert possible immediate closures, even though such harvests would come at the expense of the next cutting cycle. Sherman reluctantly agreed, and large harvests continued.

Even though the Forest Service sales program started out conservatively, it quickly gained a momentum that seemed to overwhelm the good sense of foresters. Perhaps the most surprising thing about this history is that in the 1920s, foresters set up plans knowing that harvests would drop by at least 40 percent, leading to probable mill closures in the 1980s. This, unfortunately, is exactly what happened. Harvests collapsed at the beginning of the 1990s, not because of environmentalists or spruce budworm but because planners set it up that way in the 1920s, figuring it was a reasonable price to pay for getting forests regulated as fast as possible.

RESTORING THE FORESTS

After World War II, managers became ever more enamored of intensive forestry. No one had yet proven any of its claims; no one had managed to regulate a western old-growth forest. But the Forest Service was optimistic all the same—surely, someday soon, with the help of loggers, silviculturists would be able to transform all the western forests into vigorous young stands growing at top speed.[28] And when that day came, the Forest Service estimated that loggers would be able to harvest 20 billion board feet a year forever.[29] There seemed hardly to be an end to what managers thought forests could eventually produce. The forest health crisis changed all this. Just before the Forest Service published *The Blue Mountain Forest Health Report* in 1991, loggers had harvested over 860 million board feet a year of timber from the Blues—nearly 600 million of this from federal land. But by 1993, harvests had slowed to a trickle. A lot of money, a lot of timber, and a lot of jobs were at stake.

What, if anything, can we do now to fix the forests? The restoration of forest health has become intensely politicized since 1995, when President Clinton signed the Budget Rescissions Act, setting into motion a salvage logging program. The salvage rider suspended the Endangered Species Act, the Clean Water Act, the National Forest Management Act, and a host of other environmental laws across millions of acres—all under the guise of forest health. Proponents of the rider argued that heavy salvage logging would fix the forest health crisis and restore ponderosa pine to the inland Northwest. The effect, in just the first few months, was to triple or quadruple logging in many areas, returning harvests to the inflated levels of the late 1980s.

Because many presettlement mixed-conifer communities were open and parklike, proponents of salvage logging have argued that we should log out the dense understory of fir now present in these forests. As Eric Pryne wrote in the *Seattle Times*, "careful logging and burning would help return the forests to their original condition, and reduce the scope of future wildfires."[30] Many managers and industry representatives immediately assured the public that ecosystem management meant intensive management. Environmentalists were dismayed when Steve Mealey, a former forest supervisor in Idaho with a reputation for pushing aggressive harvests, was appointed head of the Upper Columbia Basin Ecosystem Management Project, the government team developing a plan for ecosystem management in the inland West. Mealey voiced the feelings of many traditional foresters when he said, "In the inland West, ecosystem management may mean more management than before."[31] Representative Larry LaRocco, the Democrat from Idaho who pushed hard in Congress for the salvage rider to "save" the forests, agreed with Mealey, adding that "the scientific consensus is going to carry the day."[32] Fire and science, taken together, were suddenly providing managers the justification for something that looked very much like business as usual.

Definitions of forest health are at the root of these justifications for salvage logging. These definitions reflect long-held cultural ideals of what a virtuous forest should look like. According to the Idaho Policy Planning Team, the best measure of forest health is when mortality is 18.3 percent of gross annual growth—the definition offered by the Society of American Foresters.[33] By this definition, intensively managed industrial forests in Idaho are in a much healthier condition than nonindustrial forests, and old growth is in the worst condition of all, because its mortality and growth are nearly equal. Therefore, the Idaho report concludes, intensive, industrial management is what keeps forests healthy.

Early foresters justified liquidating old-growth pine forests for exactly this reason—so that young, healthy, rapidly growing forests could take their place. This led us into our current troubles. When human desires for commodities become the definition of health, managers must eliminate anything that detracts from high annual growth rates: insects, disease, decadent trees, fire, anything that does not produce commodities. This definition of health is based on human conceptions of efficiency, not on an understanding of ecological processes of mortality and disturbance.

Traditional management aimed for full efficiency, a notion close to the heart of proponents of salvage logging. Efficiency, however, is a problematic way to manage semiarid forests. The monocultures that are planted after clear-cutting may be economically more efficient than uneven-aged forests, but as the history of the Blue Mountains illustrates, short-term economic efficiency can lead to long-term ecological problems.

At the heart of the desire to save the forests with intensive management is a critical assumption that remains untested. The hope is that by making current forest overstories look the way they used to look, we will make fires behave the way they used to behave. But the world has changed: simply rearranging the trees will not return a forest to its earlier condition.

A hundred years ago, when light fires burned frequently in some mixed-conifer forests, those forests were open, with minimal fuel loads, little organic matter on the ground, and few firs in the understory.[34] But after years of fire suppression and intensive management, the forest is a different place. Even light fires may now have surprising effects.[35] For example, after decades without fire, increased litter has led to cooler microclimates near the forest floor and increased soil moisture. Root structures have changed in response, with more roots clustering close to the surface. In those conditions, even a very light fire may singe tree roots, killing old ponderosas if the soil moisture is low.[36]

Because so many additional changes have radiated out from fire suppression, we need to treat prescribed fire with caution. This does not mean that we should not try to restore light fires to sites historically dominated by ponderosas. In many sites, we have little choice—fire is going to be a part of those ecosystems regardless of our decisions.

Some environmentalists argue that the only alternative to intensive management and salvage logging is to back off and let the forests heal themselves. But we have altered these landscapes so thoroughly that this is no longer a realistic option if we want forests to persist. After the past 70 years of forest management, we have changed the landscape enough that letting the forests burn may prevent the reestablishment of ponderosa pine forests for centuries. Much of the Blue Mountains is already covered with shrub-dominated communities; the current forest soils may have become so depleted that without attempts to lessen fuel loads, additional shrubs might replace many forest stands after a high-intensity fire.[37]

The problem with using salvage logging to restore forest health is not that salvage is always wrong. Sometimes the technique can help heal a particular forest stand. But salvage logging has become a political tool that tries to fix forests by focusing on just one element—changes in overstory composition—while ignoring the policies and cultural ideals that led to the changes: the transformation of forest communities into storehouses of commodities to feed distant markets and fill distant pockets. By ignoring these links, it gets the ecology wrong, because it fails to recognize the ways in which ideology and politics shape definitions of forest health.

Foresters tried to maximize efficiency—by removing snags, clearing out deadwood, getting rid of all insects, stopping fires, simplifying stand structure, and replacing slow-growing trees with faster trees—but efficiency backfired on them. When a manager tried to fix one problem, the solution often created a worse problem elsewhere. Fire managers tried to prevent catastrophic fires by suppressing all little fires; insect managers tried to control insect damage by killing all insects as soon as they appeared or by simplifying individual stands so that insects could not survive. In spite of managers' best efforts, attempts at fire and insect control tended to intensify other problems. Suppressing fires led to fuel accumulation, slowed the growth of many forests, and made future fires more intense. Changing old-growth stands to even-aged stands in order to control insects only eliminated insect predators and contributed to the catastrophic insect damage now apparent in the Blue Mountains. But failures of fire and insect control generally led not to a reevaluation of the approach but to more engineering to fix the problems engineering had created.

Attempts to engineer nature assume that the land is a predictable machine made up of disconnected parts. Any pieces we think wasteful—insects, fires, vermin, deadwood, weed trees such as alders—can simply be eliminated. A few years later, when the trees start dying, we wonder what went wrong. Only then do we notice that those wasteful parts had a critical function in the ecosystem, so we try to restore a few pieces, removing other weeds and vermin to make room for the ones we want to return. Managers call for more science, more experiments, more information about the functional roles those parts play. But more information will not save us from our errors, for we can never learn all there is to know. The world is far too complex. We need to change the way we think about the land, not just change the number of little parts we study, label, and preserve.

Given the limits of our present understanding of forest complexity, forest health problems cannot become the justification for wholesale applications of thinning, burning, and salvaging. We know little about how these forests function now and much less about how they functioned in the past; we need to try to recognize the limits to our knowledge and control. We should try active restoration, but we need to treat it as an experiment, not as doctrine. No one can restore the Blue Mountains back to their original state. We can restore them to an inevitably altered, but not inevitably impoverished, biota—by giving up our ideals of efficiency and maximum commodity production and instead allowing for complexity, diversity, and uncertainty.

NOTES

1. William Gast et al., *The Blue Mountain Forest Health Report* (Washington, D.C.: U.S. Department of Agriculture [USDA], Forest Service, 1991). Following the practice of this report, I use the term Blue Mountains to include the Wallowa-Whitman National Forest (WWNF), Umatilla National Forest (UNF), and Malheur National Forest (MNF). The Wallowa Mountains are therefore included.

2. *Ibid.*

3. Although I restrict my analysis to a 5.5-million-acre region, similar environmental problems have developed from American transformations of forests throughout the West. In this essay, I summarize arguments I developed at greater length in *Forest Dreams, Forest Nightmares: The Paradox of Old Growth in the Inland West*

(Seattle: University of Washington Press, 1995). A longer version of this essay has also been published in Char Miller, ed., *The Greatest Good* (Lawrence: University Press of Kansas, 1997).

4. Rich Harmon, "Unnatural Disaster in the Blue Mountains," Portland *Oregonian*, 24 December 1995, p. E-6 (Review of *Forest Dreams, Forest Nightmares*).

5. There are a large number of excellent forest histories, including those of David Clary, *Timber and the Forest Service* (Lawrence: University Press of Kansas, 1986); Thomas R. Cox, ed., *This Well-Wooded Land: Americans and Their Forests from Colonial Times to the Present* (Lincoln: University of Nebraska Press, 1985); Paul Hirt, *A Conspiracy of Optimism: Management of the National Forests since World War II* (Lincoln: University of Nebraska Press, 1994); Lawrence Rakestraw, *A History of Forest Conservation in the Pacific Northwest, 1891–1913* (New York: Arno Press, 1979); William Robbins, *American Forestry: A History of National, State, and Private Cooperation* (Lincoln: University of Nebraska Press, 1985); and Michael Williams, *Americans and Their Forests: A Historical Geography* (Cambridge: Cambridge University Press, 1989).

6. Information on climate and physical geography comes largely from the three forest plans: *Umatilla National Forest Land and Resource Management Plan, Final Environmental Impact Statement* (Washington, D.C.: USDA Forest Service, 1990), III-30; *Malheur National Forest Land and Resource Management Plan, Final Environmental Impact Statement* (Washington, D.C.: USDA Forest Service, 1990); *Wallowa-Whitman National Forest Land and Resource Management Plan, Final Environmental Impact Statement* (Washington, D.C.: USDA Forest Service, 1990).

7. Jack Ward Thomas, ed., *Wildlife Habitats in Managed Forests: The Blue Mountains of Oregon and Washington* (Washington, D.C.: USDA Forest Service, 1979), p. 20.

8. These fires included both anthropogenic and natural ignitions; distinguishing between the two has proved difficult. See Langston, *Forest Dreams, Forest Nightmares*, pp. 246–63; James K. Agee, *Fire Ecology of Pacific Northwest Forests* (Washington, D.C.: Island Press, 1993), pp. 320–38; Frederick Hall, "Ecology of Natural Underburning in the Blue Mountains of Oregon" (USDA Forest Service, Pacific Northwest Region, R-6, Regional Guide 51-1, August 1977), pp. 1–13. See also Kathleen Maruoka, "Fire History of *Pseudotsuga menziesii* and *Abies grandis* Stands in the Blue Mountains of Oregon and Washington" (M.A. thesis, University of Washington, 1994), pp. 21–33.

9. John D. Guthrie, "Blame It on the Indians: Forester Explodes Myth that Red Men Set Fires to Keep Forests Open" (John Day, Oreg.: Malheur National Forest, Supervisor's Office, 1933) (press release).

10. Hall, "Ecology of Natural Underburning in the Blue Mountains of Oregon," pp. 5–7.

11. See Nancy Langston, "The General Riot of the Natural Forest: Landscape Change in the Blue Mountains" (Ph.D. diss., University of Washington, 1994), pp. 31–58.

12. *Ibid.* See also Eastside Forests Scientific Society Panel (Dan Bottom, Sam Wright, Jim Bednarz, David Perry, Steve Beckwitt, Eric Beckwitt, James R. Karr, and Mark Henjum), *Interim Protection for Late-Successional Forests, Fisheries, and Watersheds, National Forests East of the Cascade Crest, Oregon and Washington: A Report to the United States Congress and the President,* eds. Ellen Chu and James R. Karr (Washington, D.C.: USDA Forest Service, 1994).

13. Donald Meinig, *The Great Columbia Plain: A Historical Geography, 1805–1910* (Seattle: University of Washington Press, 1968).

14. Langston, *Forest Dreams, Forest Nightmares*, pp. 42–85.

15. Press release, 12 September 1906 (Pendleton, Oreg.: Umatilla National Forest, Historical Files, Supervisor's Office).

16. Clary, *Timber and the Forest Service*, p. 16.

17. C. S. Judd, "Lectures on Timber Sales at the University of Washington, February 1911" (Forest Service Research Compilation Files, National Archives, Region 6, Entry 115, Box 136).

18. *Ibid.*

19. Henry Graves, *Annual Report of the Forester 1912* (Washington, D.C.: USDA Forest Service, 1912), p. 16.

20. Thorton T. Munger, "Basic Considerations in the Management of Ponderosa Pine Forests by the Maturity Selection System, 1936" (UNF Historical Files, Supervisor's Office, Pendleton, Oregon); C. J. Buck, regional forester, to Blue Mountains supervisors, 15 September 1936 (UNF Historical Files, Supervisor's Office, Pendleton, Oregon).

21. Harvests on private land were still substantial, especially along the Sumpter Valley Railway. See Jon Skovlin, "Fifty Years of Research Progress: A Historical Document on the Starkey Experimental Forest and Range" (Portland: Pacific northwest Research Station, General Technical Report PNW-GTR-266, 1991), p. 11.

22. For discussions of the Forest Service's attempts to sell timber before World War I, see Clary, *Timber and the Forest Service;* Cox, *This Well-Wooded Land;* Thomas B. Parry, Henry J. Vaux, and Nicholas Dennis, "Changing Conceptions of Sustained-Yield Policy on the National Forests," *Journal of Forestry* 81 (March 1983): 150–54. For a specific statement of national sales policy, see Graves, *Annual Report of the Forester 1912*. For the Blue Mountains, the forester R. M. Evans glumly noted in his 1912 report on the Wallowa Forest: "Owing to the inaccessibility of the government timber and to the large amount of privately owned timber surrounding it, there is no immediate prospect of a large sale." R. M. Evans, "General Silvical

Report Wallowa and Minam Forests" (Forest Service Research Compilation Files, National Archives, Region 6, Entry 115, Box 135, 1912).

23. USDA Forest Service, "Malheur River Working Circle Plan" (MNF Historical Files, Supervisor's Office, John Day, Oreg., n.d.) (internal evidence sets the date after 1926 and before 1929).

24. Assistant forester's letter, Sales, Sale Policy, 8 August 1922, cited in C. M. Granger, district forester, to Acting Forester Mr. Sherman, concerning the Baker Working Circle Plan, 1927 (WWNF Historical Files, Supervisor's Office, Baker, Oreg.).

25. USDA Forest Service, "Planning Report, Sales, Bear Valley Unit, Malheur National Forest 6/30/22" (MNF Historical Files, Supervisor's Office, John Day, Oreg.).

26. USDA Forest Service, "Malheur River Working Circle Plan."

27. E. A. Sherman, acting forester, to C. M. Granger, district forester, and John Kuhns, Whitman forest supervisor, 8 March 1927, concerning the Baker Working Circle Plan submitted on 14 February 1927 (Wallowa-Whitman National Forest Historical Files, Supervisor's Office, Baker, Oreg.).

28. For an excellent analysis of the postwar logging boom and its effects on federal forest policy, see Hirt, *A Conspiracy of Optimism*.

29. Charles Wilkinson, *Crossing the Next Meridian: Land, Water, and the Future of the West* (Washington, D.C.: Island Press, 1992).

30. Eric Pryne, "Unease over Logging to Control Fire," *Seattle Times*, 10 September 1994.

31. Steven Mealey, quoted in Eric Pryne, "Summer Forest Fires Spark Debate over Timber Policy," *Seattle Times*, 9 September 1994.

32. Larry LaRocco, quoted in Eric Pryne, "Unease over Logging to Control Fire."

33. Jay O'Laughlin, James G. MacCracken, David L. Adams, Stephen C. Bunting, Keith A. Blatner, and Charles E. Keegan III., *Forest Health Conditions in Idaho: Executive Summary*, (Moscow: Idaho Forest, Wildlife and Range Policy Analysis Group, 1993), p. 17, following the definition of forest health recommended by the Society of American Foresters in L. A. Norris, H. Cortner, M. R. Cutler, S. G. Haines, J. E. Hubbard, M. A. Kerrick, W. B. Kessler, J. C. Nelson, R. Stone, and J. M. Sweeney, *Sustaining Long-Term Forest Health and Productivity* (Bethesda: Society of American Foresters, 1993).

34. Agee, *Fire Ecology*.

35. Recent prescribed burns near Bend, Oregon, for example, consumed between 32 and 69 percent of the forest floor. After repeated moderate underburns in these ponderosa stands, growth slowed significantly. The duff layer in many places has also increased with fire suppression, forming an insulating mat over the soil. In the

Southwest, prescribed burns led to the ignition of heavy layers of duff that after burning formed an insulating ash cap, forcing heat into the soil, burning hot enough to kill small roots near the surface, which led to the death of 40 percent of the stand after three years. M. G. Harrington and S. S. Sackett, "Past and Present Fire Influences on Southwestern Ponderosa Pine Old Growth," in *Old Growth Forests in the Southwest and Rocky Mountain Regions* (Washington, D.C.: USDA Forest Service, General Technical Report GTR-RM-213, 1992).

36. S. S. Sackett and S. M. Haase, "Soil and Cambium Temperatures Associated with Prescribed Burning," *Natural Resource News* (La Grande, Oreg.: 1994).

37. Agee, *Fire Ecology*.

20 / Getting Out the Cut

A History of National Forest Management in the Northern Rockies

PAUL W. HIRT

National forest management in the northern Rockies is in crisis today: an environmental crisis, an economic crisis, and a crisis of confidence in the United States Forest Service (USFS) (fig. 20.1). The agency that was supposed to guarantee a steady, sustainable flow of timber from the national forests to help preserve the stability of timber-dependent rural communities has instead caused a crash in timber production that has brought with it unemployment, bitterness, and social conflict (fig. 20.2). The forest managers who were supposed to protect the health and ecological integrity of the land under their care have instead allowed widespread and severe environmental deterioration. For example, a report by a USFS hydrologist in 1994 indicated that as much as 60–70 percent of the watersheds in the region's national forests exceeded soil erosion tolerances (fig. 20.3).[1]

Congress told the Forest Service to provide for a balanced mix of uses, yet it stands widely accused from without and within of harboring a strong timber bias. A University of Montana Forestry School study of the Bitterroot National Forest concluded that "multiple use management, in fact, does not exist as the governing principle on the Bitterroot National Forest. . . . In a federal agency which measures success primarily by the quantity of timber produced weekly, monthly, and annually, the staff of the Bitterroot National Forest finds itself unable to change its course, to give anything but token recognition to related values, or to involve most of the local public in any way but as antagonists."[2]

Ironically, the foregoing quote is not from a contemporary source; it was

FIG. 20.1. Cartoon published in *Earth First! Journal*, September 22, 1990, p. 31.

written in 1970. The Forest Service's bias toward timber harvesting is a long-standing problem. Indeed, as long ago as 1955 USFS managers warned about overcutting of timber in the Lolo National Forest in western Montana: "As more of the privately owned timber lands are cut over, the local lumber industry is becoming more dependent on the national forest for its log supply. This situation is of concern to the Forest Service. . . . If milling capacity is expanded too far beyond the level of timber production which can be sustained, there must eventually be a contraction of the industry."[3] This report also acknowledged that by the early 1950s timber harvests in the Lolo already exceeded a sustainable yield.

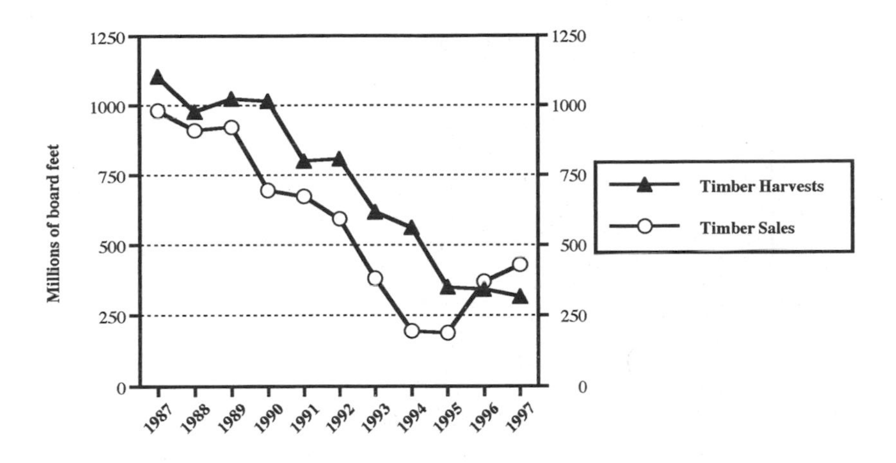

FIG. 20.2. Timber sales and timber harvests, national forests of Region 1 (northern Rockies), 1987–1997.

FIG. 20.3. Soil erosion caused by road construction and logging, upper Lolo Creek, Lolo National Forest, Montana, 1992.

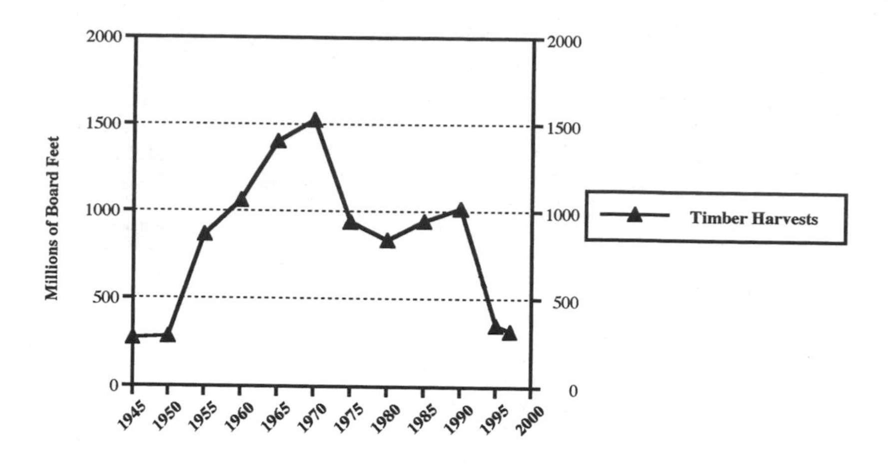

FIG. 20.4. Timber harvests, Region 1 (northern Rockies), 1945–1997.

Overcutting in the 1950s and 1960s did in fact contribute to a slow but steady decline in national forest timber harvests in the northern Rockies since 1970 (fig. 20.4). But the current crisis has a more immediate cause, too: during the 1980s, national forest managers labored under intense political pressure to raise the allowable cut at the very time they were struggling to resolve problems associated with three decades of overcutting. The Reagan administration and pro-industry members of Congress established specific timber harvest targets for the national forests in the federal budget process. The agency was required to meet these targets regardless of other needs or concerns and despite budget shortages for reforestation, soil-erosion control, and multiple-use management.[4] Although many foresters in the agency creatively invented technical justifications for the Reagan era timber targets, they also stepped up their warnings that those harvest levels were unsustainable. In a rather bold statement in 1989, the supervisors of Regions 1, 2, 3, and 4 (the intermountain West) sent a joint memo to the chief of the USFS saying that their 1980s timber targets were "unrealistic even with full funding."[5] The supervisors complained that Congress and the Reagan and Bush administrations continued to overemphasize "commodity" programs—such as timber production—at the expense of other resources, and in general, "national forest programs [did] not reflect the land stewardship values embodied in forest plans, nor . . . the values of many Forest Service employees and the public."[6]

The current national forest management crisis thus not only has substantial historical roots but was actually forewarned. How was it that professionals and politicians supposedly dedicated to sustained yield, multiple use, and environmental protection created exactly the opposite—a timber boom and bust, an obsession with commodity production to the detriment of other values, and widespread environmental degradation? The explanation lies in the political and economic framework within which the Forest Service, as a federal agency, has functioned. Encouraging the creation of wealth and providing social services are among the primary functions of government. These objectives are achieved in part by federal agencies' actions in controlling and allocating federal assets, as the Forest Service does with the national forests. Timber, range, recreation, water, wildlife, and even wilderness are marketable commodities, the use of which generates wealth or provides other benefits to users. These resources are managed primarily for economic objectives. Even the recently touted concept of "ecosystem management" is interpreted in the political arena as a tool for sustaining economic assets over time.[7]

The government motivates agencies to perform as desired by providing policy direction, legal authority, and financial resources. Of these, financial resources are perhaps the most important determinant of agency actions, because policy direction and legal authority cannot be exercised without money. The USFS essentially does what it is paid to do. A case in point is its 1994 *Eastside Forest Ecosystem Health Assessment* for the interior Northwest and the ambitious regional plan for ecosystem management that evolved from it. Funding to implement the plan has been the subject of intense political debate; the Clinton administration and Congress have agreed to fund only select portions of the overall management and restoration program. As a consequence, the agency finds itself unable to implement what it has the authority and will to accomplish—a lopsided budget results in lopsided management.

The Forest Service, however, is not simply a passive implementer of government directives. It is an organization with its own biases and a strong interest in its own growth, security, and autonomy. The agency can best serve these three fundamental organizational concerns by providing economic goods and services, because these build supportive constituencies that can apply pressure on the congressional committees that set the agency's budget. Agencies that provide goods are always better off than agencies that simply regulate or constrain behavior, because the former gain allies while the latter mainly generate antagonism. Thus, bureaucracies strongly favor production functions over regulatory functions—for their own good. Indeed, the USFS receives financial and organizational rewards from Congress, the executive branch, and its client groups for enabling the development of national forest resources. Powerful pro-development incentives are thus built into the Forest Service's working environment, which professionals in the agency, especially those in leadership positions, learn to adapt to. These sorts of development pressures moved the agency away from its early conservation orientation and toward an industrial model of maximized timber production, which ultimately led to the present forest crisis in the region. The following discussion explains how this scenario unfolded.

ESCALATING DEMAND FOR NATIONAL FOREST TIMBER AFTER WORLD WAR II

Prior to the mid-twentieth century, the USFS operated mainly as a custodian, engaged primarily in fire suppression, cooperative forestry, and education. Many people mistakenly believe this caretaker phase was deliberate and that the agency subsequently abandoned its former commitment to preserving forests. The USFS, however, has never been a preservation agency. It has always advocated production, but *responsible* and *sustainable* production was its goal. Following the Second World War, demand for national forest resources exploded, especially in the West. A rapid and unprecedented increase in a wide variety of uses of the national forests launched the USFS into a new era in its organizational history. The agency aided economic growth by providing wood, water, and forage for consumers and by providing hunting, fishing, camping, hiking, skiing, boating, berry-picking, and other nature experiences for outdoor-lovers. The more goods and services the USFS provided, the larger its work force and budget grew.

A variety of factors went into this rapid escalation of use of the national forests following World War II, including the postwar baby boom, the associated housing boom, the depletion of old-growth timber on private land, sustained economic growth, a westward shift in the national population, expansion of transportation systems, and an outdoor recreation boom. The northern Rockies were a major source of timber feeding the postwar housing boom. In the first year after the war, 83 new timber access roads were planned for the national forests in Montana and northern Idaho (U.S. Forest Service Region 1). Chambers of Commerce, logging companies, and investors saw an economic boom in the making and scrambled to get a piece of the action. Several hundred individuals and companies moved to the northern Rockies to set up logging operations. They found hundreds of existing operators already established, as well as an overworked and understaffed Forest Service trying desperately to train enough new "timber cruisers" to keep up with escalating demands for national forest timber sales. Although the Forest Service went to extraordinary lengths to accelerate logging on its land, it could not fully accommodate this flood of profit seekers.

Business investors, whose fiscal horizons extended only a fraction of the distance required to regrow a logged forest, felt frustration with the Forest Service, which seemed to be hoarding vast stretches of unused forest land. They often enlisted the aid of their congressmen to prod the agency to offer more timber sales. One disgruntled entrepreneur, Dwight Seymour of Seymour Brothers Lumber Company, called upon Montana's Senator James E. Murray for help in 1946. Seymour Brothers wanted a square mile or so of national forest timber to use as collateral for a loan to move its mill into the Swan Valley in west-central Montana: "As we explained before we cannot do a thing without having a source of timber supply guaranteed us prior to our plant construction, that would assure us of several years of continuous operation in order to warrant our investment. There are plenty of other people being serviced by the Forest Department in regards to this matter and we cannot understand why we are not given like consideration."[8] Senator Murray pressed Percy Hanson, the regional forester, to "get in touch with the Seymour Brothers and do everything you can toward meeting their requirements."[9]

The Forest Service responded in two predictable ways. First, it yielded to political pressure by adjusting a planned timber sale in Lolo Creek, about 60 miles from the Swan Valley, to make it more acceptable to Seymour Brothers, and it sent bid forms and a conciliatory letter to the loggers. Second, the agency then sent a lengthy letter to Murray explaining how sustained-yield and multiple-use policies required management considerations that private loggers often failed to understand or appreciate. Regional Forester Hanson pointed out that only about half of the trees in any timber sale area would be cut, "in order that there would be a reasonable stock of growing timber left both for future timber growth and to preserve the recreational values which are paramount in this area."[10] Selective cutting, rather than clear-cutting, was the dominant practice on national forest land in the northern Rockies at the time, in contrast to private industry timberland, which was mostly clear-cut. Seymour Brothers wanted a block of "mature" forest that the company could systematically log over a few years; then it would move on. This was not the Forest Service way—at least not at the time.

Agency limits on how much and how fast timber could be cut from the national forests grated on itinerant businessmen seeking to invest in the postwar timber boom. Moreover, very few foresters held office in Congress or in the administration, while quite a number of businessmen and businessmen's lawyers did. The preponderance of sympathy in the halls of power in Washington, D.C., leaned toward the economic entrepreneur. Although Dwight Seymour, for reasons unexplained in his correspondence with the senator, declined the Forest Service's offer of a sale of 8 million board feet of timber in Lolo Creek, his case is identical to dozens of others, and their combined impact on national forest management was devastating. Within a decade, migratory investors cashing in on a lucrative business climate stimulated by favorable government policies—and promoted by the agency's top-level bureaucrats—successfully quashed traditional Forest Service dedication to conservative timber production and multiple-use management.

MULTIPLE-USE MANAGEMENT

For the Forest Service to achieve multiple-use management in this climate, it had to attract multiple constituencies, not just the timber interests. Serving nontimber interests would enhance the agency's organizational resources and help guard against excessive pressures from the logging industry. As a result, the agency encouraged ranchers, miners, hunters, water developers, and the general recreating public to use forest resources too. The phrase "multiple use," which the agency brought consciously into parlance in the mid-1940s, reflects its effort to broaden its base of support.

The USFS made a strategic appeal to diverse and often opposing interest groups by promising to accommodate the demands of all of them. When hardly anyone was using the national forests, the agency found it relatively easy to promise to satisfy and harmonize everyone's demands for resources while protecting forest health. As uses increased, the promises became more difficult to fulfill, and conflicts among user groups escalated. The USFS then had to face the prospect of prioritizing uses—saying yes to one and no to another. It had to become regulatory, to ration resources rather than simply provide them. Rejecting the demands of an interest group with supporters in Congress (like Seymour Brothers Lumber Company) could have dangerous political ramifications.

How the USFS chose to respond to this novel situation in the 1950s is the key to understanding the forest management crisis of today. Essentially, forest managers dodged making the tough choices by adopting a strategy of expanding the resource pie. They proposed to *manufacture abundance* through intensive manipulation of the forest environment.

This response is reflected in the way in which managers of the Lolo National Forest dealt with the overcutting problem acknowledged in the 1955 report cited earlier. The allowable cut in the Lolo National Forest in the 1940s, based on sustained yield considerations, was 52 million board feet (mmbf) per year. Timber production in the forest greatly accelerated in the early 1950s, however, reaching 108 mmbf by 1955. This obviously posed a technical, if not an ethical, problem for the forest managers. And they had to resolve the problem while under continual pressure to further increase the harvest. As the 1955 Lolo report noted, "reduction of the timber cut to equal the sustained production of the land poses a serious problem for some communities and certain segments of the lumber industry."[11]

Caught between a rock and a hard place, what did the Lolo forest managers do? They decided to revise their assessment of how much logging could be sustained. Instead of reducing the harvest, they raised the allowable cut by adopting more optimistic assumptions about their ability to increase timber productivity through intensive management. As the report put it: "The allowable cut for the Lolo may be raised somewhat as a result of revision of timber management plans now in progress." This proved to be a vast understatement: the Lolo managers raised the ceiling on the allowable cut not once but three times in the next eight years, and not slightly but dramatically—trying to keep up with demand. The "allowable cut" in the Lolo National Forest nearly quadrupled from 52 mmbf in 1955 to 187 mmbf by 1963, and it stayed at that level until the 1980s. The Lolo was not unique; all the forests in the northern Rockies followed the same pattern, as did national forests all over the nation.[12]

INTENSIVE MANAGEMENT FOR "FULL USE"

Getting out the cut on the national forests of the northern Rockies in the 1940s, 1950s, and 1960s required a shift from conservative selective logging to large-scale clear-cutting (fig. 20.5). It was a change that produced the longest-running and bitterest public relations problem the USFS has ever faced. A public information flyer put out by the Northern Regional Office in 1962, titled "Timber Management *Facts*," addressed the growing debate over clear-cutting. Whereas the agency had strongly and consistently opposed clear-cutting during the first half of the twentieth century, it reversed itself in the 1950s and began arguing passionately that logging was in fact

FIG. 20.5. Clear-cutting in large blocks in the Lewis and Clark National Forest, 1962, from a photo featured in "Timber Management *Facts*," issued by the Northern Regional Office of the USFS.

the *essence* of forest conservation and that clear-cutting was not only economically preferable but biologically superior to selective harvesting.

Responding to the demand for rapidly increased harvests, the USFS generated a flood of scientific research defending clear-cutting and identifying other ways to enhance production. The scientific paradigm guiding such research reflected three important characteristics. First was what Donald Worster has called *economic thinking:* the view that nature is an economic asset, a set of commodities to be captured and marketed in the pursuit of wealth. Second was an *engineering mentality:* a view of nature as malleable physical material capable of being shaped in whatever fashion was needed to maximize its human utility. Third was *technological optimism:* the belief that for every problem there is a technological fix, and for every limitation, a technological enhancement.

These three ideological orientations, along with the gung-ho attitude of "maximization" that underlies them all, were widespread in American culture in the postwar era. This was not just an accident of history. Intensive management for maximum production served important social and political functions: it lessened popular fears of resource depletion with the promise of a future cornucopia; it infused the USFS with growing amounts of working capital and new personnel; it provided politicians an opportunity to bring development funds home to their constituents ("pork barrel" politics); and it offered to corporate America increasing access to federal resources while forestalling government rationing.

Forestalling government rationing was a particularly important motive behind this "expand the pie" ideology. Representative Harley O. Staggers (D-W.Va.) stated the underlying concern when he said during hearings on the 1960 Multiple-Use Sustained-Yield Act, "When the supply of natural resources and ability to produce fall to the point where government's chief function is the rationing of too few resources among too many people, democracy cannot survive."[13] Against the backdrop of such fears, support for maximum production became equated with support for liberty, democracy, and the "American way of life." In the postwar United States, virtually all economic growth was defined as good. A cult of prosperity captured the public imagination. To be a producer was to be a patriot. In this intellectual climate, the expanding business culture of the 1950s succeeded in redefining many key concepts, including "conservation." By the late 1950s the dominant definition of conservation—accepted even by the USFS—was "full use and development of all resources." Indicatively, a 1959 USFS regional development program went by the title "Full Use and Development of Montana's Timber Resources."[14] Ironically, those who continued to advocate conservative forest management were now accused of irresponsibly *withholding* resources from the public ("public" meaning lumber businesses in most cases). To achieve anything less than maximum levels of production was to stand in the way of progress.

The public eventually broke into factions over this redefinition of conservation, some adopting the new dogma and some remaining skeptical. The skeptics eventually turned against the USFS. By the 1960s, the ranks of the skeptics were expanding rapidly, spurred on by some stunning failures caused by the USFS in its efforts to manufacture abundance. The infamous Bitterroot National Forest clear-cutting and terracing experiment is one prominent example of such failures (fig. 20.6). But dissidents remained relatively marginal until the 1970s.

In seeking to maximize resource productivity, the USFS and the forestry profession as a whole began to ignore the distinction between theory and reality, to confuse what was theoretically possible with what was feasibly achievable. Aldo Leopold had warned the forestry profession of this potential pitfall in the 1940s when he complained that too many foresters were content to grow trees like cabbages. Leopold also complained that his colleagues were becoming too inured to biological violence as they tore apart natural ecosystems and tried to replace them with cellulose plantations.[15]

FIG. 20.6. Large clear-cuts and conversion of the forest to the single purpose of timber production through the costly practice of terracing entire mountainsides (shown here in the Bitterroot National Forest) in the 1960s and early 1970s led to national controversy over policies and practices of the U.S. Forest Service (photograph by Dale A. Burk).

But technological optimism reigned among foresters in the postwar decades. And with it came overinflated sustainable harvest calculations and extensive damage to ecosystems, of which forest managers were well aware. Here is an indicative quote from the Forest Service's northern regional forester written in 1956:

> Truck and tractor logging methods in steep country produce intolerable erosion and stream-clogging conditions. This has been discussed by forest officers for years and some efforts to reduce the damage have been made. Nevertheless, the damage continues and in many areas is growing worse as the size and power of logging equipment increases. It is apparent that we are not redeeming our responsibilities for the management of national forest land, for the prevention of damage to adjoining land and for the protection of downstream values far removed from the area logged.[16]

UNBALANCED BUDGETS AND THE TIMBER SUPPLY CRASH

Another serious problem for forest management, besides naive optimism, was the inconsistent and unbalanced funding for national forest management mentioned earlier. Congress and the executive branch repeatedly approved full USFS budget requests for timber sales and road construction while skimping on budget allocations for environmental protection and rehabilitation. Consequently, management was heavily skewed toward economic development. Resource protection fell by the wayside.[17] As a result of these and other problems, many timber inventories and management plans in the 1960s and 1970s warned about difficulties in achieving theoretical levels of control over timber supply and warned of future shortfalls. But rather than return to a more conservative orientation, the agency continued to propose ever more intensive management. Forest managers were honest enough to admit that things were not working out as planned, but not honest enough to admit that their whole approach was fundamentally flawed. In a very real sense, they maintained what I call a "conspiracy of optimism." By placing inordinate faith in their ability to expand the resource pie, responsible officials could defer corrective action—until a real crisis developed and avoidance was no longer possible.

The crisis in the northern Rockies national forests came in the late 1980s and early 1990s. Regional forester John Mumma, in charge of all the national forests of Region 1, had been striving for years to get his region to meet its congressionally mandated timber targets, but the forests kept falling short. Most of the forest supervisors under him began complaining that their Reagan-era timber targets were unachievable or that achieving them would violate a host of environmental laws designed to protect soil, water, wildlife, and scenic values and to ensure the sustainability of the timber program. Subject to increasing pressure from congressmen such as Idaho's Larry Craig, Mumma encouraged his foresters to cut as much as possible and then to document as thoroughly as possible the reasons why they could not meet their timber targets. The majority of reasons cited by forest supervisors in the Northern Region for their inability to meet timber targets were past mismanagement and flaws in agency data and assumptions. Other reasons included overcutting on adjacent private land, protection measures for endangered species, and "public pressures" to reduce the agency's timber emphasis.

Many forest supervisors admitted that their national forests had been overcut in previous decades and that reforestation had been inadequate. The supervisor of the Clearwater National Forest in northern Idaho wrote that his forest's "front country [was] exhausted," adding that "our currently developed lands are, in general, highly developed. The watershed and wildlife habitat conditions have been pushed to the limit by past activities. The same is true for the visual resource. Without what I believe to be substantial investments in major rehabilitation efforts in the very near future, I fear . . . we will be facing significant shortfalls below our historic trends until recovery occurs."[18] Other forest managers in the region echoed similar sentiments. For example, supervisor Orville Daniels of the Lolo National Forest published a "letter to concerned citizens" in 1991 explaining that the Lolo National Forest had been "intensively logged" during the 1960s and 1970s, and as a result, "many of these areas cannot be re-entered yet because more cutting would violate standards" established in the forest plan for minimum levels of environmental quality.[19]

Many forest plans contained exaggerated estimates of the amount of old-growth timber that remained in the forest. Lolo National Forest personnel discovered in 1990 and 1991 that 110,000 acres deemed "suitable" for harvesting in the 1986 forest plan had actually been logged already. They also discovered that an additional 280,000 scattered acres thought to be mature forest were actually covered by young growth where fires had burned between 1910 and 1935.[20] In their 1987 forest plan, managers of the Kootenai National Forest in the northwest corner of Montana apparently overestimated by 40 percent the amount of old growth available for harvesting and underestimated by 500 percent the areas of the forest that had been clear-cut.[21]

In monitoring the effects of timber harvesting on the Clearwater National Forest, managers discovered that they had overestimated their ability to mitigate impacts to soil, water, and fisheries. They found that even at a 30-percent reduction of the harvest level, they were "consistently bumping up against the watershed and fisheries thresholds established in the forest plan." They also discovered that meeting just the minimum environmental protection standards often rendered timber sales economically nonviable.[22] The Kootenai National Forest, which wanted to reduce its harvest target by approximately 30 percent in 1991, identified at least seven reasons for the reduction, three of which were traceable to overoptimistic assumptions: a slower than anticipated rate of recovery of wildlife habitat cover; fewer than expected improvements in efficiency for industrial utilization of logs; and a faster than anticipated rate of logging on adjacent private land.[23]

The methods and computer models used by the Forest Service to develop its forest plans often had built-in flaws that inflated allowable harvest levels. In a 1991 analysis of the failure of the Flathead National Forest in Montana to meet its timber targets, the U.S. General Accounting Office (GAO) found that the forest planners had improperly modified the computer model used to develop the forest plan in 1986 by arbitrarily assigning a minimum timber harvest target of 100 million board feet per year to meet the needs of the local timber industry. When this "constraint" of 100 mmbf was eliminated from the computer model, the computer indicated that 78 mmbf per year should be the harvest ceiling. Nevertheless, the planners stuck with the higher number. Later, while implementing the plan, the foresters found they could not achieve the higher harvest level and simultaneously meet minimum standards for environmental protection.[24] Clearwater and Lolo National Forest personnel similarly identified flaws in their computer modeling processes that overinflated their proposed harvest levels.[25]

Many national forests in north Idaho and western Montana have large areas in "checkerboard" patterns of landownership: every other square mile remains in private ownership. This is a legacy from the nineteenth century, when the federal government gave land grants to the Northern Pacific Railroad as an incentive to build a transcontinental line from St. Paul to Puget Sound. In land management planning, the agency makes assumptions about the rate and style of management on this intermingled private land. If the assumptions prove wrong, then the Forest Service faces unexpected changes in environmental conditions, to which it must respond.

Among the largest timber corporations in the West are Champion International and Plum Creek, both of which own substantial tracts of checkerboard land intermingled with national forest land in Montana, Idaho, and elsewhere. In the 1970s and 1980s, both companies began to dramatically accelerate the logging of their remaining timber in the northern Rockies, clear-cutting huge swaths, sometimes as large as a square mile, following the boundaries of the old railroad land grants (fig. 20.7). Champion logged what it could and then sold off most of its holdings in the 1990s.[26] The increased logging pushed soil, water, and wildlife protection standards in affected watersheds to the limit and beyond. The Kootenai, Clearwater, and Lolo National Forests cited this problem as a major reason for their reduced harvests, arguing that any additional logging on intermingled national forest land would exacerbate an already critical level of resource damage. In the case of the Lolo forest, a total of 400,000 acres of private checkerboard land experienced "higher than anticipated" rates of logging, prompting Lolo managers temporarily to prohibit cutting on 289,000 acres of its adjacent land.[27]

Spotted owls and Pacific salmon in the Northwest, grizzly bears and bull trout in the northern Rockies, goshawks and Mexican gray wolves in the

FIG. 20.7. An aerial view of checkerboard land. Trail Creek, near the Cabinet Wilderness in northwestern Montana, August 1992.

Southwest, red-cockaded woodpeckers in the Southeast—all are threatened or endangered species that are affecting timber harvesting or other commercial activities in the national forests. All were once relatively abundant (some, like the salmon, were superabundant), and all are now on the brink of extinction because of human-caused modifications of their habitat. Each species listed above, except salmon and trout, is dependent on old-growth forest habitat. The fish are affected by sedimentation and water temperature changes caused by logging—and of course they are adversely affected by dams. Traditional intensive silviculture called for the liquidation of all "overmature" forests and their replacement with fast-growing commercial timber plantations, leading to the current situation in which old growth has become scarce or fragmented or both. If the Forest Service had not subordinated other uses and resource values so completely to intensive timber production since World War II, many of these species might not have become endangered. Efforts to recover threatened and endangered species contribute to the other causes for reductions in timber harvests in the national forests.[28]

Finally, national forest managers often pointed to "changing public values" or "increasing public pressures" as one of their reasons for reducing timber harvests in the 1990s. Specifically, they meant pressures to keep roadless areas undeveloped and remaining old-growth forests intact. But conservationists had opposed the Forest Service's aggressive road construction and logging program for decades. These issues were not new. Public values did not suddenly change in the late 1980s either. Indeed, prior to the 1980s few Americans even knew that logging was allowed in national forests. What changed during the Reagan era was public *awareness* of the existence, extent, and impact of industrial logging in the national forests. Moreover, federal laws passed in the 1970s, such as the National Environmental Policy Act and the National Forest Management Act, required the Forest Service to analyze the environmental impacts of their proposed actions and solicit public input on their management plans. This resulted in citizens' becoming much more informed and involved in national forest decision-making by the 1980s. Forest managers have had to adapt.

In the midst of all this controversy and change, regional forester John Mumma told his northern Rockies forest supervisors in 1991 that if they had to choose between implementing environmental protection standards in their forest plans or implementing the timber harvest targets set by Congress, they should choose the former. Emboldened, the frustrated forest supervisors wrote a joint memo in September 1991 to Dale Robertson, chief of the Forest Service, asking for a reduction of about 30 percent in their timber sale target. Under pressure from pro-timber western states senators and representatives, Robertson compromised and instead promised Congress that the Northern Region would strive to meet at least 80 percent of its target. Indicative of the political pressure being applied, then-congressman Larry Craig had written to the chief in May 1991 saying, "Dale, I am very disappointed with the Forest Service's accomplishment and accountability for timber outputs in Idaho and the Nation as a whole. You have serious management problems that must be addressed. It is my hope you will move to assure targets are met and line officers are held accountable for targets."[29] The *New York Times* covered the story with the headline, "Forest Supervisors Say Politicians Are Asking Them to Cut Too Much."[30] Shortly thereafter, Robertson told Mumma that he, Mumma, would have to accept a transfer to an unspecified job in Washington, D.C., citing "poor performance ratings" as a reason. In fact, Mumma had met or exceeded all his targets for other renewable resources (recreation, fish and wildlife, watershed protection, etc.) and had received numerous awards over the years from the agency for outstanding service. In protest, Mumma chose to retire rather than accept transfer.

A nationwide storm of protest erupted against the "ouster" of Mumma, with supporters inside and outside the agency calling his forced retirement a "coup by hardliners" and "retrenchment" in the Forest Service.[31] At a press conference held in Missoula, Montana, after Mumma's ouster, four of the forest supervisors who had worked for Mumma expressed support for him, saying they would continue to implement his land stewardship policies and resist unsustainable timber harvest targets. The supervisor of the Clearwater National Forest, Win Green, flatly asserted, "Right now we have a timber target that is unrealistic and I don't hold my people accountable for something that is unattainable." Citing past overcutting, watershed damage, degradation of fish and wildlife habitat, and marred scenic views, Lolo National Forest supervisor Orville Daniels announced that he would allow a timber harvest level of only 50 percent of his assigned target.[32]

The media, which had been covering dissension within the Forest Service intermittently since 1989, recognized the controversy as seminal. *High Country News*, a biweekly newspaper covering environmental issues in the West, immediately featured a front-page spread on Mumma and also covered the congressional hearings subsequently held by the House Civil Service Subcommittee to determine whether civil service guidelines had been violated in the case.[33] At the hearings, the disgruntled head of the Forest Service's "whistle-blower" review program, John McCormick, testified that harassment and transfer of employees who bucked the timber targets was commonplace. The *New York Times* published an editorial by him in January 1992 (the month he retired), titled "Can't See the Forest for the Sleaze." In it, McCormick hammered his former employer: "The Forest Service simply does not tolerate freedom of dissent. . . . The agency has become comfortable with lying to the public, ignoring long-festering problems and serving the timber industry as Government agents of environmental destruction rather than environmental protection."[34] This highly visible episode was only one of a number of instances in which agency employees were transferred or otherwise punished for failing to support the timber targets.[35]

In an ironic postscript to the Mumma controversy, Robertson announced a few months later that he was adopting as official Forest Service policy a new approach called "ecosystem management." He explained: "We have been courting the ecosystem approach for 3 years and we like the relationship and results. Today, I am announcing the marriage and that the Forest Service is committed to using an ecological approach in the future management of the national forests and grasslands."[36] Again, many forest managers had trouble locating the reality behind the rhetoric. Lee Coonce, retiring supervisor of Oregon's Umpqua National Forest, wrote a blistering memo to the Pacific Northwest regional forester, John Lowe, in December 1992, expressing his frustration with the high timber targets assigned to his forest and the associated budget constraints that locked in funding for timber production but left nontimber programs out in the cold. "The facade is we're giving lip service to Ecosystem Management, operations and maintenance of recreation facilities, anadromous fisheries . . . etc., but in the budget realm it's business as usual and that spells timber production."[37] Despite encouraging signs like the revolt of conscience in Region 1 and the adoption of ecosystem management, it is not yet clear whether the agency will be able to successfully reform the practices and relationships that led to institutional failure in the first place.

CONCLUSION

The dramatic decline in national forest timber harvests in the 1990s—and the general decline since the 1970s in Region 1—is a legacy of organizational and political pressures to maximize timber production from the 1940s through the 1970s. The Forest Service allowed cutting beyond sustainable levels and at the expense of resource protection mandates. Pressures from industry and Congress created incentives for managers to make overoptimistic assumptions, to use faulty data bases and skewed planning procedures, and to place unjustified faith in intensive management solutions to environmental problems. Despite managerial warnings of failures and environmental deterioration, high-level bureaucratic and political efforts to maintain an optimistic view of timber production potential continued into the 1980s. Counterpressures to reform national forest management increased simultaneously, however. Amid the swelling controversy, ground-level managers of the national forests came to a growing sense by the end of the 1980s that their capabilities and the ecological capabilities of the national forests were overstretched and that dramatic change was both inevitable and necessary.

A particularly interesting aspect of this story is how decision-makers have claimed *not* to be responsible for problems that have resulted. In contrast, they are always willing to take credit when things go well. Why have decision-makers been able to dodge responsibility so effectively, even up to the present day? Because decision-making takes place in a complicated institutional environment that provides ample opportunity for people to duck or pass the buck. Foresters can blame vague or inconsistent policies, unbalanced funding, or unpredictable nature for management failures. Or they can blame a lack of information. Sometimes they simply blame their predecessors and claim to have reformed the flawed practices of the past. On the other hand, politicians can blame the other party or another branch of government or such things as the budget deficit for their fiscal irresponsibility. Industry can innocently point to public demand or international markets or tax policies as the determinant of their behavior. When cornucopian dreams fail to materialize, everyone involved can blame someone or something else for the failure. Everyone has what is referred to in politics as "plausible deniability."

This kind of evasion reflects what Patricia Limerick called "The Myth of Innocence" in her history of the American West.[38] Westerners never saw themselves as selfishly engaged in a process of exploiting a continent and dispossessing its native inhabitants, she said. Conquest was cloaked in wonderfully creative ideological justifications that left the victors innocent of blame for the seamier side of their actions. "Conquest" is in fact an apt metaphor for national forest management in the Northwest since World War Two. In a 40-year period, the timber industry, encouraged and subsidized by the federal government, consumed most of the marketable trees in the region's national forests and is now substantially "downsizing" its operations or abandoning the region. As forestry professor Richard Behan recently observed in an essay bluntly titled "Trashing the Northern Rockies": "We are witnessing the cutting-out and getting-out of the national forest system. . . . Our society, with the forest products industry as its agent, has treated the national forest timber resource as a capital asset. Within this decade [the 1990s], I believe, we will have succeeded in liquidating it."[39]

A potent irony in this story is that an agency of the federal government dedicated for a century to sustainable forestry actually facilitated a gigantic boom and debilitating bust—the very thing the agency was created to prevent. Perhaps the greatest irony, however, is that this liquidation of forest resources was accomplished under the guise of "conservation" by public servants claiming the best of intentions.

One of the functions of historical analysis is to illuminate contemporary social problems, to help us understand their genesis, and perhaps to suggest future courses of action. Looking forward from this legacy, there is reason to be both optimistic and pessimistic. On the bright side, clearly we have reached a watershed in national forest history. The ideology of intensive management for maximum production is under attack on all fronts and is widely repudiated even within the forestry profession now. Timber harvests for the next few decades are likely to remain at about one-third the average level of the past four decades—in some cases less—and that level should at least be sustainable. More "ologists" are being hired than ever before (biologists, ecologists, archaeologists, etc.), reducing the hegemony of traditionally trained, commodity-oriented foresters within the Forest Service. Indeed, President Clinton appointed a wildlife biologist, Jack Ward Thomas, as chief of the USFS during his first term, the first time in the history of the agency that anyone other than a forester or engineer had attained that position. Thomas's successor during Clinton's second term, Mike Dombeck, was trained as a fisheries biologist and has made "watershed health and restoration" a central policy emphasis of his administration—a real change from past emphases on the production of commodities.[40]

On the other hand, Congress continues to manipulate the agency through the budgeting process, approving lopsided funding and gutting the agency's ability to shift away from intensive commodity production and toward the new "ecosystem management" paradigm. One possibility for hope rests on how well we learn from past mistakes. But to learn, we must be very clear about what those mistakes were, and that is where the historian can play a role.

NOTES

Portions of this essay were previously published in Paul W. Hirt, *A Conspiracy of Optimism: Management of the National Forests since World War Two* (Lincoln: University of Nebraska Press, 1994).

1. J. Allen Isaacson, "Watersheds Overview: The Fish Are Seeing Red," in *Rocky Mountain Challenge: Fulfilling a New Mission in the US Forest Service* (Eugene, Oreg.: Association of Forest Service Employees for Environmental Ethics, 1994), p. 51.

2. "A University View of the Forest Service," prepared for the Committee on Interior and Insular Affairs, U.S. Senate, by a select committee of the University of Montana, 91st Cong, 2d Sess., Dec. 1, 1970, Senate Document no. 91–115, pp. 13–14.

3. U.S. Department of Agriculture [USDA], Forest Service, Lolo National Forest, *Report for 1955* (n.d.), p. 4.

4. Hirt, *A Conspiracy of Optimism*, pp. 271–78.

5. USDA Forest Service, *Sunbird Proceedings: Second National Forest Supervisors' Conference* (Tucson: November 13–16, 1989).

6. Forest Supervisors of Regions 1–4, "Feedback to the Chief," in *Sunbird Proceedings*.

7. In the "Executive Summary" of the Forest Service's *Eastside Forest Ecosystem Health Assessment*, the authors noted that in response to "a shift in public expectations" the Forest Service had adopted "a national strategy of ecosystem management to emphasize conservation of biodiversity, long-term productivity, and the capacity for sustained flows of renewable resources." Note the emphasis in the last two points on productivity and commodity flows. Richard Everett, Paul Hessburg, Mark Jensen, and Bernard Bormann, *Eastside Forest Ecosystem Health Assessment* (Portland, Oreg.: USDA Forest Service, Pacific Northwest Research Station, PNW-GTR-317, 1994), vol. 1 (Executive Summary), p. 3.

8. The information in this anecdote is drawn from a series of letters between a logging company, Senator James E. Murray (D-Mont.), and the Forest Service Regional Office in Missoula, Montana, located in the James E. Murray Papers, University of Montana Archives, Collection 91, Series I, Box 234. Quote is from a letter written by Dwight Seymour to Senator James E. Murray, 22 October 1946.

9. Quote is from a letter written by Senator James E. Murray to Percy D. Hanson, 14 November 1946.

10. L. A. Campbell (for P. D. Hanson) to Senator James E. Murray, 2 December 1946.

11. Lolo National Forest, *Report for 1955*, p. 5.

12. Hirt, *A Conspiracy of Optimism*, ch. 6.

13. Hearing on Multiple-Use, Sustained-Yield Act, H.R.10572, Subcommittee on Forests of the Committee on Agriculture, House of Representatives, 86th Cong., 2d Sess., 16 March 1960, p. 20.

14. USDA Forest Service, Northern Regional Office, "Full Use and Development of Montana's Timber Resources," 1959. Also published as Senate Doc. 9, 86th Cong, 1st Sess., 27 January 1959.

15. Aldo Leopold, "A Biotic View of Land," *Journal of Forestry* 5 (September 1939): 730.

16. Regional Forester Major Kelly, quoted by A. G. Lindh, "Watershed Management in Administration," in *Watershed Management Conference* (USDA Forest Service, Northern Region, 1956), p. 44.

17. For an analysis of the lack of balance in the Forest Service budget in the

1950s and 1960s, see Hirt, *A Conspiracy of Optimism*, chs. 9 and 10. For the 1970s and 1980s, see V. Alaric Sample, *The Impact of the Federal Budget Process on National Forest Planning* (New York: Greenwood Press, 1990), especially ch. 2.

18. Fred L. Trevey, Clearwater National Forest supervisor, to John Mumma, regional forester, "Timber Resource Strategy Update," 29 February 1990 (in the author's possession).

19. Orville Daniels, Lolo National Forest supervisor, "Letter to Concerned Citizens," 11 September 1991, p.1 (in the author's possession).

20. *Ibid.*

21. "Critics Say Forest Service Scapegoats Grizzlies," *High Country News* 24 (September 21, 1992); Liz Sedler et. al., "A Report on Kootenai National Forest Timber Inventory Data Used for Forest Plan Projections for Future Harvest" (prepared for Congress and Kootenai National Forest planners by the Inventory Inquiry Project, P.O. Box 1203, Sand Point, Idaho 83864, August 1992).

22. Trevey, "Timber Resource Strategy Update."

23. Robert Schrenk, Kootenai National Forest supervisor, to Liz Sedler, Inventory Inquiry Project, 31 August 1992 (in author's possession). See also Kootenai National Forest, "1991 Monitoring Report," p. 19.

24. Comptroller General, *Forest Service: The Flathead National Forest Cannot Meet Its Timber Goal* (Washington, D.C.: Government Printing Office, GAO Report no. RCED-91-124, 1991), p. 14.

25. Trevey, "Timber Resource Strategy Update"; Daniels, "Letter to Concerned Citizens."

26. For a book-length exposé of private forest liquidation in Montana in the 1980s, see Richard Manning, *Last Stand: Logging, Journalism, and the Case for Humility* (Salt Lake City: Peregrine Smith Books, 1991). Manning was a journalist for the *Missoulian* during this period, and his book is based on a series of investigative reports he wrote for the newspaper which won him a national journalism prize. See also Alan McQuillan, "Accelerated Cutting on Private Industrial Timberlands," *Missoula Independent*, 7 November 1991, pp. 19–20. McQuillan is a professor of forestry at the University of Montana who worked for the Forest Service and for Champion International in the late 1970s and early 1980s. See also the in-depth articles by Paul Koberstein, "Plum Creek Timber Leaves Its Mark on Montana" and "Private Forests Face Critical Log Shortages," *Oregonian*, 15 October 1990, special supplement: "Northwest Forests: Day of Reckoning," pp. 2–6; Sherry Devlin, "Eco-Battle Looms for Champion," *Missoulian*, 21 May 1992, pp. B-1, B-3. An article in *Forbes* magazine reported that Champion International accumulated a great deal of debt during an aggressive expansion program in the late 1970s and as a result had to radically accelerate logging on its vast holdings to service that debt. Jean A. Briggs, "Full

Speed Ahead, Damn the Recession!" *Forbes*, 5 March 1979, pp. 61, 64. On the Pacific coast aspect of this problem, see Casey Bukro, "Environmentalists Side with Loggers: Threat of Raiders Brings Foes Together," *Washington Post*, 29 May 1990, pp. E1, E6.

27. Trevey, "Timber Resource Strategy Update"; Daniels, "Letter to Concerned Citizens."

28. For a detailed, journalistic narrative of the biological, political, and social dimensions of the spotted owl controversy, see William Dietrich, *The Final Forest: The Battle for the Last Great Trees of the Pacific Northwest* (New York: Simon and Schuster, 1992), chs. 2, 4, and 14. Specifically regarding the Seattle Federal District Court injunction against logging in spotted owl habitat and its effect on timber harvests in the Northwest, see "Judge Dwyer Does It Again: Opinion Rebukes Forest Service for 'Remarkable Series of Violations,' " *Forest Watch* 11 (10) (May 1991): 10–12. For popular press coverage of the role of endangered species in the larger fight over the fate of remaining old growth on national forests, see Ted Gup, "Owl vs. Man," *Time* (June 25, 1990): 56–65; and Michael D. Lemonick, "Whose Woods Are These?" *Time* (December 9, 1991): 70–75.

29. Jeff Debonis, "Retrenchment in the Forest Service: Hardliners Oust Mumma in the Northern Rockies," *Inner Voice* 3 (5) (Fall 1991): 1–2.

30. Timothy Egan, "Forest Supervisors Say Politicians Are Asking Them to Cut Too Much," *New York Times*, 16 September 1991, pp. A1, A12. See also the thorough historical analysis by University of Montana professor Alan McQuillan, "Inside Mumma-Gate," *Missoula Independent*, 31 October 1991, pp. 18–19.

31. See articles in the *Missoulian:* "Mumma Fight Shifts to Congress," 7 September 1991, pp. A-1; "Dissenters' Fate Turns Focus on FS," 8 September 1991, pp. B-1; "Foresters Cry Coup in Mumma Ouster," 12 September 1991, pp. A-1; and "A Return to McCarthyism" (editorial), 30 December 1990. See also McQuillan, "Inside Mumma-Gate," pp. 18–19; Sherry Devlin, "Biologists' Union Throws Its Weight behind Mumma," *Missoulian*, 19 September 1991.

32. Debonis, "Retrenchment in the Forest Service," p. 1; "The Crumbling of the Coup," *Inner Voice* 3 (5) (Fall 1991): 3.

33. "Two Say Politics Rules Their Agencies," *High Country News* 23 (18) (October 7, 1991): 1, 10–13.

34. John McCormick, "Can't See the Forest for the Sleaze," *New York Times*, 29 January 1992, p. A21.

35. Speaking at a federal "whistleblowers" conference in Washington, D.C., in March 1992, Jeff Debonis of the Association of Forest Service Employees for Environmental Ethics (AFSEEE) observed: "We federal employees are here because it hurts. It hurts to work for once proud agencies that no longer meet the public trust. . . . These

agencies have literally been turned into instruments of mismanagement, instruments of political pork barrels, instruments of environmental destruction and, worst of all, instruments of repression against ethical employees." His statement, along with the statements of others on this topic, is reprinted in the proceedings of a conference sponsored by the Government Accountability Project and AFSEEE, "Protecting Integrity and Ethics: A Conference for Government Employees of Environmental, Wildlife, and Natural Resource Agencies" (Eugene, Oreg.: AFSEEE, March 1992). For congressional testimony on alleged harassment and repression of employees, see statements of Jeff DeBonis, Marynell Oechsner, and John McCormick before the Subcommittee on Environment, Energy, and Natural Resources of the House Committee on Government Operations, 102d Cong., 2d Sess., March 31, 1992, pp. 180–97. See also McCormick, "Can't See the Forest for the Sleaze"; Don Schwennesen, "Forest Biologist Cites Pressure to Sell," *Missoulian*, 31 March 1992, p. A-1; J. Todd Foster, "Critics Say Agency Is Eating Its Young," *High Country News* 25 (1) (January 25, 1993): 2; and J. Todd Foster, "A Combat Biologist Calls It Quits," *Inner Voice* 5 (1) (January/February 1993): 5.

36. F. Dale Robertson to regional foresters and station directors, re: ecosystem management of the national forests and grasslands (June 4, 1992), p. 1. For an academic analysis of the Ecosystem Management initiative, see Hanna J. Cortner and Margaret Ann Moote, "Ecosystem Management: It's Not Only about Getting the Science Right," *Inner Voice* 5 (1) (January/February 1993): 1, 6.

37. Lee F. Coonce to John Lowe, Umpqua National Forest, 14 December 1992, reprinted in *Inner Voice* 5 (1) (January/February 1993): 13.

38. Patricia Nelson Limerick, *The Legacy of Conquest: The Unbroken Past of the American West* (New York: W. W. Norton, 1987).

39. Richard Behan, "Trashing the Northern Rockies: A Legacy of Predatory Politics and Savage Capitalism," *Wild Forest Review* 1 (3) (February 1994): 23–25.

40. See http://www.fs.fed.us/intro/chiefbio.shtml.

21 / Changing Forests, Changing Needs

Using the Pacific Northwest's Westside Forests, Past and Present

THOMAS R. COX

So many things I "knew" as I grew up in central Oregon were not true! The state's pioneer settlers, I believed, had come out by covered wagon and, with a combination of perseverance and perspiration, had carved homes and farms from the unbroken forests that stretched from the Cascades to the sea. Even as an undergraduate biology major at Oregon State, when I really did know better, many of those of us from east of the mountains chided those from the Westside as products of "the great swamp" that, we insisted, extended from the mountains to the sea. And of course Governor Tom McCall and those Oregon "ungreeting cards" that appeared in the 1960s did nothing to correct that swampy image of the Pacific Northwest[1]—an image not all that different from my own view of the Westside as I grew up in the rain shadow on the east side of the Cascades.

Kenneth L. Gordon, professor of zoology at Oregon State and a pioneer in environmental education, disabused me of many of my illusions.[2] His marvelous, if fugitive, booklet *The Natural Areas of Oregon* detailed the varied ecological zones one passed through in moving inland from the Pacific shore to the eastern boundary of the state.[3] The Willamette Valley, I learned, had been a mosaic of woodlands, riparian strips, timbered north slopes of valley hills, and extensive open prairies on the edges of which the first white settlers made their homes. Western Washington had less extensive prairies—which helped to explain why most of the pioneers on the overland trail headed for the Willamette Valley—but it was equally varied.

William Bowen and others have carefully mapped these settlement patterns.[4] What they found was remarkably similar to patterns on the earlier frontiers of Ohio, Indiana, Illinois, and elsewhere. Pioneers settled where prairies and neighboring, relatively open woodlands met, in order to have a ready supply of fuelwood and building materials without having to clear away hosts of trees in order to get pasture and crop land.[5] Subsequent settlers from Kentucky, Missouri, and other Middle Border states found the woodlands in the hills around Oregon's great western valleys attractive sites for settlement, creating a population base that gave considerable strength to pro-Confederate sentiments in the state during the Civil War.[6]

What was not so clear until the work of Jerry Towle, Richard White, and Peter Boag was the extent to which these prairies and the neighboring woodlands were altered landscapes, shaped to a large extent by fires set by Native Americans seeking to improve their camas meadows and berry patches, to drive game, or to make hunting easier. To be sure, various pioneer observers left records graphically describing native use of fire, but few of us noticed these accounts. Even reports of the great Nestucca and Yaquina fires of the 1840s, both of which burned for more than a year (dying down in the winter but smoldering in roots and logs to flare up again in the spring, burning, before they were through, a good bit of the Coast Range from the Willamette Valley to the sea and many miles north to south), did nothing to change our views.[7]

Pioneers and Native Americans alike tended to avoid the Westside's dense coniferous lowland forests. Native Americans for the most part stayed close to the waterways that supplied much of their sustenance and offered relatively easy routes of communication. On the other hand, most early Anglo-American settlers avoided the forests in order to save scarce labor. They had little awareness of the limited value of the forestlands for agriculture. Folk wisdom from earlier frontiers held that forest bottomland made rich farmland. And the Jeffersonian ethos envisioned the frontiersman as one hacking farms from the forested wilderness and building a basis for a yeoman democracy in the process. Settlers brought this ideological baggage west with them, but even if they assumed the soil of the forested areas must be among the richest anywhere because it grew the biggest trees imaginable, they tended to settle on the Willamette Valley's prairies and surrounding, lightly forested foothills.[8]

In fact, Westside forestlands were not well suited for agriculture. Most of their soils were badly leached, gray-brown podzolic soils of little fertility—although, in addition to trees, they did support an ample growth of grass that would in time make the forest a valuable summer range for sheep and cattle.

Extensive areas in western Washington had been glacially scoured and were poor and gravelly.[9]

The grassland prairies, although somewhat leached, were decidedly superior, but only in the area's limited floodplains and deltas was truly deep, rich soil available. There the denseness of the vegetative cover and size of trees in the gallery forests caused settlers to shy away—at least until the arrival of latecomers such as the Danes who settled around Junction City, Oregon, cleared riparian bottomland (by then about all that was available to be taken up in the Willamette Valley), and prospered as a result. When a later generation moved onto the main forestlands around the 1920s, after they had been cleared (by loggers, not by farmers), people learned how poor the soil actually was. Many a family found itself barely able to eke out an existence on these "stump ranches." As Richard White put it, they were "poor men on poor lands," caught in a poverty trap.[10]

If I was slow to recognize the importance of Westside prairies and nearby open woodlands to early settlers (to say nothing of the reality of life in the cutover even as I had relatives living there), I was even slower to recognize the diversity of the dense Westside coniferous forests. Thornton Munger and other pioneer foresters, as well as forestry student friends, argued that spruce-hemlock stands were the natural climax vegetation of the area.[11] Douglas fir, they maintained, was a long-lived subclimax forest that moved in following fires, blowdowns, logging, and the like. This view justified the use of clear-cutting, or more accurately block cutting, in timber management. Under the spell of such common knowledge, I long continued to view forests west of the Cascades as an almost unbroken expanse of one or the other of these stands.

Although my Eastside chauvinism encouraged such views, my eyes and the evidence supplied by Kenneth Gordon should have told me otherwise. Ponderosa pine dominated in the Coburg Hills and in large stretches of Jackson and Josephine Counties, western white pine was common on various islands in Puget Sound, and lodgepole pine crowded areas along the coast (there so twisted by wind off the ocean as to justify the scientific name *Pinus contorta*). Elsewhere, western red cedar, Alaskan yellow cedar, and lowland white fir held sway. And wherever logging had disturbed the scene, there was the ubiquitous red alder. It was a far more varied forest than the simplistic Douglas fir–spruce–hemlock scenario would have it.[12]

But if people came to the Willamette Valley seeking farms, they came to the Puget Sound country seeking commercial opportunity. That opportunity, as often as not, meant tapping the riches of the surrounding forests. In the process, Dorothy Johansen noted some years ago, social and cultural norms were established that continue to distinguish Washington from Oregon. Under the circumstances, the term *Pacific Northwest* seems more an appellation of geographic convenience than a label for an area with a genuine regional character and identity.[13]

As California boomed with the gold rush and subsequent developments, western Washington became a major source of its building materials. A few large sawmills that shipped lumber by sea to California and other Pacific markets also appeared in Oregon. But with poorer ports, production in Oregon lagged further and further behind that in Washington. In contrast with its northern neighbor, however, Oregon had a plethora of small mills scattered through the valleys west of the Cascades. These supplied nearby farmers and the communities that supported them rather than the export trade.[14]

Although early Westside settlers gravitated to the prairies and neighboring open woodlands, it was the region's dense coniferous forests that drew awestruck comments. As one early settler, Michael Luark, put it, the trees grew "so thick, tall and straight that it must be seen to be believed." Charles Geyer, a pioneer German botanist who visited the area in 1844, was more matter-of-fact. "The pines . . . with which the country is overstocked, are of great size, 120 feet being about the average height; some of them measuring 220 feet, and more, with a proportionate diameter of trunk." But rather than seeing opportunity in such giants, Geyer saw barriers in the "mountains and ridges . . . bristling with impenetrable pine-forests." So isolated was the area by these and other natural barriers that Geyer found it hard to understand why settlers were moving there from the Mississippi Valley; settlement seemed practical only in the "beautiful and fertile, but narrow Wallamette [*sic*] valley (and a few still more limited localities)." Even these, Geyer was informed, were "subject to inundation almost every spring." He was sure that the area could never be incorporated into the United States.[15]

Although some sought from an early date to turn the forests to profit, and others sought simply to avoid them, appreciation of their beauty and magnificence did not have to await the emergence of twentieth-century environmentalism. Those who treasured the region's natural beauty and its handmaiden, outdoor recreation, appeared early in the Pacific Northwest. By the 1860s, Oregonians were going to the mountains during the summer and fall for hunting, fishing, berry-picking, and relaxation. Mount Hood was first scaled in the 1850s, Mount Rainier in 1870, and in 1887 a group of Portland-area residents formed the Oregon Alpine Club. The seashore drew people as well.

By the 1870s, northwesterners were vacationing in Seaside, Gearhart, Newport, and other coastal communities. At the same time, Leopold Samuel, Frances Fuller Victor, and other regional writers were turning out prose and poetry lauding the scenic beauty of the region.[16]

Yet save in connection with a few very special spots, there was little evidence of preservationism. Distance and terrain provided what protection most of the region's forests and scenery received. But that was enough. Throughout most of the nineteenth century, logging technology based on ox teams was capable only of tapping stands within about two miles of streams large enough to float logs to market. Human ingenuity devised ways of utilizing smaller and smaller streams, but the limit remained more or less constant. Maps of western Washington prepared by the U.S. Geological Survey in 1902 illustrate the result graphically. Along all the floatable streams, the timber was cut away, but farther back, dense coniferous stands still held sway. In Oregon, where the hills tended to be steeper and the streams more tumultuous, the impact of logging was even more limited.[17]

Near the turn of the century, steam donkeys—an invention of John Dolbeer of Eureka, California—began replacing ox teams in the woods. By using a steam engine attached to a windlass, loggers more than doubled the distance that logs could be pulled to the waterways. Furthermore, steam donkeys allowed logging in seasons when mud had previously made it impossible. But steam logging tended to leave terrible devastation behind, far worse than ox-team logging ever had. The resultant debris was not only aesthetically unpleasant but also a fire hazard. The region's era of great forest fires was at hand. With the massive fuel loads left on the ground by power logging, fires burned hotter and more destructively than before, consuming organic matter in the topsoil and crowning to destroy stands as the old ground fires seldom had.[18] Still, the basic pattern remained the same. Logging continued to be confined to the corridors along streams, and most of the region's forests remained untouched.

Railroads changed all this by opening vast stretches of previously inaccessible stands to loggers. The impact was greatest in western Washington, where the low-lying land between Puget Sound and Grays Harbor and southward toward the Columbia was quickly spanned by logging railroads. Logging proceeded apace, and Washington sprang into first place among the producing states. The builders of logging railroads even found ways of coping with Oregon's mountainous terrain, but Oregon continued to lag behind Washington until well into the twentieth century. Not until logging trucks replaced the railroad as the dominant means of log transport and thus opened stands that railroad loggers had never been able to reach did Oregon come to the fore in lumber production. Once trucks appeared, the state's economy became tied to the industry as never before.[19]

But the era that brought logging trucks also saw the emergence of automobile tourism, and that brought changes of its own. In 1920, Oregon's governor, Ben Olcott, was so appalled at the devastation left by logging operations along the new Cannon Beach–Seaside highway that he launched a campaign to save the state's roadside scenery. Others joined in, and strips of timbered land were soon set aside along major highways on both sides of the Cascade Range. Perhaps the most notable set-asides were along the McKenzie River highway running east from Eugene and in the Henry B. Van Duzer Forest Corridor along the Salmon River highway in the Coast Range. In Washington, Asahel Curtis and others launched a similar campaign to save timbered strips along the White River route to Mount Rainier National Park. Now forests were valued for their scenic attributes as well as the lumber they could provide, but it was clear that distance and terrain were no longer adequate to ensure preservation.[20]

William B. Greeley, chief of the Forest Service, was a prime mover in efforts to preserve forest scenery. Unlike Gifford Pinchot, who was one of his harshest critics, Greeley strongly supported programs to set aside roadside timber; he not only directed local Forest Service officials to cooperate in land exchange programs designed to protect corridors but also moved his agency into the construction of campgrounds, picnic areas, and similar facilities for lovers of the outdoors. Pinchot, with a much more materialistic yardstick for measuring forest values, criticized Greeley sharply. The new chief, undeterred, kept the Forest Service on a course that was to make it an ever more serious threat to the National Park Service, which liked to claim for itself the role of protector of scenery and provider of outdoor recreation. Moreover, in the years that followed, it was the Forest Service, not the Park Service, that launched the first federal programs of wilderness preservation.[21]

In the midst of all this, the forests themselves were changing. Although there were still old-growth stands of nearly every variety in the lowland areas, they were becoming more and more limited. Farsighted individuals began campaigns to save representative tracts before it was too late; the Washington Federation of Women's Clubs, for example, took steps that resulted in the preservation of the stand of old-growth Douglas fir southeast of Enumclaw that is now Federation Forest State Park. That such lowland forests were in danger was not just a figment of the overheated imaginations of nature lovers; pollen samples from sediment cores from Angle Lake, Washington, show a dramatic shift during the period from conifers to alder in the surrounding area, a sure sign of the changes that came in the wake of logging.[22]

During the twenties and thirties, most Westside logging was on private land, but in the 1950s a dramatic shift to the national forests commenced. Postwar demand was too great to be met from private stands alone, and prices pushed upward. Responding, the Forest Service shifted from its long-held custodial role to one focused on increased sales of timber, a shift that undermined the scenic and recreational programs Greeley had encouraged. These programs continued, but they no longer enjoyed as much support from the agency's upper echelons and increasingly had to be financed by users' fees.[23]

Private stands were affected too. Following World War II, the availability of cheap war-surplus trucks catapulted a number of new "gyppo" loggers onto the scene. No stand seemed too small or too remote to draw the attention of these small-time operators. Technological advances also added to the increased output of logs, and new mills sprang up throughout the region.[24]

At the same time, postwar prosperity, coupled with wartime savings, resulted in a sharp increase in tourism and outdoor recreation. A collision soon followed. Forest Service authorities removed key lowland areas from the Three Sisters Wilderness Area as a preliminary to timber sales there. Karl Onthank and a host of allies formed the Friends of the Three Sisters Wilderness in an unsuccessful effort to block the plan; in the process, they provided much of the initial impetus for the decade-long drive that was to culminate in 1964 in Congress's passing the Wilderness Act, which substituted much stronger statutory for administrative protection of wildlands. The struggle over the Three Sisters Wilderness was paralleled by similar fights between the Forest Service and preservationists in other quarters, especially over proposals for a North Cascades National Park.[25]

A basic difference in values was involved—as personal experience soon brought clearly home to me. Fresh out of college, I was teaching at a small high school in the Oregon Cascades. A college friend who had majored in forestry was attached to the local Forest Service ranger district. One day we heard that huge trout were biting at some small hike-in lakes nearby. We gathered our fishing gear and, together with the district ranger, headed up the trail toward the lakes. En route we passed through a stand of gigantic old-growth white fir that had been marked with the tell-tale blue paint signifying the trees to be cut following an upcoming timber sale.

"What a shame these trees are to be cut," I commented, "especially since white fir lumber is of so little value, a virtual drug on the market." The district ranger agreed but said production had to be maintained to meet the demand of local mills and the communities that depended on them—they had to "get out the cut," as the common phrase had it.

My college friend looked at the two of us in utter disbelief. "You can't mean that," he said. "Just look at this forest; the trees are old and in decline, the forest stagnant. It needs to be opened up to proper management to make it productive!" What he was espousing had nothing to do with being a handmaiden of industry or meeting politically (or socially) motivated production goals; rather, it sprang from a deeply held ideological position that blinded him to my aesthetic, nonmaterial arguments. Silently we trudged on toward the lakes.

What happened there in the high Cascades in the 1950s was a harbinger of much of what was to come in succeeding years. People with different perceptions of forest values increasingly talked past one another as they sought to debate policies and actions. Congress, state legislatures, advocacy groups, and the courts all added their voices to the efforts to shape the management and use of Westside forests—just as they did for forests elsewhere. As one forester working for a major timber firm commented to me in exasperation a few years ago, he had gone into his profession because he wanted to work in the out-of-doors, but things had reached a point where he found himself spending far more time in hearings and courtrooms than in the woods.

As Paul Hirt has shown, foresters sought to reconcile diverse demands on the forest by applying intensive, scientific management to the land. Greater production from a reduced acreage of commercial timberland would leave land free for other uses. But their faith was misplaced. Much of the science underlying forest management schemes was flawed—and even when it was not, bureaucratic, institutional, and political demands often overrode science in determining the Forest Service's policies.[26] The private sector fared no better. In the 1970s even some who would have preferred other approaches found their actions being shaped by economic demands generated by near-runaway inflation, rather than by the best scientific understanding of forest management that was then available.[27]

And the demands on the forests continued to mount. Log exports to Japan burgeoned in the 1960s, driving up stumpage prices and cutting into supplies to such an extent that they were now inadequate to meet domestic demand. Forest-dependent communities and smaller sawmills were especially hard hit, and protests mounted, making log exports the leading political issue in the Pacific Northwest for some two decades. In time, exports from federal forests were prohibited, although those from private and Washington state land were not. In the face of it all, mill closures continued, and communities such as Valsetz, in Oregon's Coast Range, became virtual ghost towns.[28]

Other contests over forest utilization followed. Especially noteworthy was that caused by the northern spotted owl. Old-growth forests, wildlife biologists held, were essential for the breeding, and thus the survival, of this bird. Since the owl was on the endangered species list, forest managers had a legal mandate to regulate logging so as to ensure its survival. In practice this would mean sharply reducing or even completely halting logging in most Westside old-growth forests. The marbled murrelet, which spends most of its time on rocky Pacific shores but nests in old-growth forests, provided a similar but less heralded case. The Forest Service sought to determine how much old-growth forest was necessary to maintain viable populations of these birds, but no one was satisfied with the agency's efforts. Environmentalists argued that they did not go far enough; woods workers and industry spokesmen argued that they were too restrictive. Jack Ward Thomas, a noted wildlife management specialist who in 1993 was named chief of the Forest Service by President Bill Clinton, settled on a sharp reduction in logging in federal old-growth forests while allowing some cutting to continue, but the compromise failed to satisfy the sharply divided forces that abounded both within and without the federal agency.[29]

The United States itself has changed in recent decades, making policies that might have been appropriate for one age unacceptable today. The country has grown more urban, and the postwar flight to the suburbs has now spread far beyond them into the very forests themselves. Increased affluence has led not only to more leisure time but also to a proliferation of snowmobiles, all-terrain vehicles, whitewater rafts, and backpackers. The forests are being used by more recreationists, in more ways than ever before. At the same time, new technologies—including such things as helicopter and balloon logging—have made even the smallest and most isolated stands accessible to loggers. Perhaps, under such circumstances, reconciliation of the conflicting claims on the forest is impossible—but perhaps not.

A letter I received not too long ago provides some grounds for hope. It came from a longtime friend from high school and college who had gone on to become a successful consulting forester. Together we had fished countless streams and lakes in the Oregon Cascades and hiked many a trail. Like numerous others, he was troubled by the recent course of events and the inability to communicate with the other side that now seems to afflict so many participants in debates over forest policy. He told me of the launching of his consulting business. "The best news about our new venture," he wrote

> is that we are demonstrating the leading edge of technology in "new forestry." Our harvest operations are best described as being kind and gentle with a minimum impact to the environment and the forest community. A typical job disturbs less than six percent of the soil surface area and removes 8 to 40 trees per acre, leaving on average 180 trees per acre to continue growing. Public acceptance of our work has been exceptional. 18 months after completion of logging, the general public would be hard put to tell where we had logged. See, things can change, and old dogs [and old foresters] can learn new tricks.[30]

My friend went on to discuss the status of the northern spotted owl and its habitat. I found his conclusion full of ecological wisdom: "When will we learn that management must be in balance for all species and stop this tragedy of single species tunnel vision?" His point seemed clear: we err whether the single species we seek to favor through management is the spotted owl for species protection, the Douglas fir for sawlogs, or something else for some other reason. What we should be doing is managing for healthy ecosystems.

Somehow, I find hope in hearing a forester argue thus. His tone suggests that communication and understanding among the forces contending over forest policies may be possible after all. But first, most of us will have to unlearn a lot of "truths" we have acquired over the years, just as I had to do when I went to college and began to find that my knowledge of the nature and early settlement of the Westside was badly flawed.

From the earliest days of Anglo-American settlement to today, residents of the Pacific Northwest's Westside have lived in an intimate, symbiotic relationship with the forests that cloak so much of the land. Over the years they have asked many different things of the forests, often—and increasingly in recent years—things that are mutually incompatible. This seems the result not so much of changing social values as of changes in population and technology that have extended human impact into previously isolated corners of the forest. Little is now beyond human influence.

Distance and terrain once provided protection for vast tracts of land and for the nonmaterial values they possessed. Now such protection must be extended through human wisdom, which must at the same time devise means of meeting the very real material demands on the forest that come with a growing regional and world population.[31]

Meeting the multifaceted challenge presented by the many, varied demands that we put on forests today will not be easy. But if we do as my forester friend suggests—manage for healthy ecosystems instead of single species or single uses—surely we will come closer to satisfying the often conflicting demands of species maintenance, of recreation in its many guises, and of commodity production (and its concomitant, job production) than will otherwise be the case. Since neither modern science nor capitalism is apt soon to go away, those who would protect the environment seem to have no practical choice save to turn them to their own purposes.

NOTES

1. Concerned about what population growth might do to quality of life in the state, McCall urged people to visit Oregon, "but for heaven's sake, don't move here to live." Echoing the sentiment, greeting card manufacturers sought humorous ways of making the state seem unattractive. "People don't tan in [western] Oregon," one read—"They rust." See Tom McCall with Steve Neal, *Tom McCall, Maverick* (Portland: Binford and Mort 1977), pp. 190–203.

2. Gordon's influence was primarily felt outside of university circles. Oregon State was the main institution preparing high-school biology teachers in Oregon, and all students in the program were required to take Gordon's three-term core sequence in the natural history of Oregon, which focused on what he called "the changing scene," that is, the environmental history of the state, although environmental history had not yet appeared on the academic scene as a discrete discipline.

3. Kenneth L. Gordon, *The Natural Areas of Oregon* (Corvallis: Oregon State College, 1953).

4. Peter G. Boag, *Environment and Experience: Settlement Culture in Nineteenth-Century Oregon* (Berkeley: University of California Press, 1992), pp. 43–57; William A. Bowen, *The Willamette Valley: Migration and Settlement on the Oregon Frontier* (Seattle: University of Washington Press, 1978), esp. chapters 5 and 6; Harlow Zinser Head, "The Oregon Donation Land Claims and Their Patterns" (Ph.D. diss., University of Oregon, 1971). See also Jesse S. Douglas, "Origins of the Population of Oregon in 1850," *Pacific Northwest Quarterly* 41 (1950): 95–108.

5. Boag, *Environment and Experience*, pp. 28–41 and *passim;* Edward E. Dale, "Wood and Water, Twin Problems of the Prairies," *Nebraska History* 29 (1948): 87–

104; Terry Jordan, "Between Forest and Prairie," *Agricultural History* 38 (1964): 205–16; Michael Williams, *Americans and Their Forests: A Historical Geography* (Cambridge: Cambridge University Press, 1989), pp. 130–33.

6. Dorothy O. Johansen and Charles M. Gates, *Empire of the Columbia: A History of the Pacific Northwest* (New York: Harper, 1957), pp. 251–53; Robert W. Johannsen, *Frontier Politics and the Sectional Conflict: The Pacific Northwest on the Eve of the Civil War* (Seattle: University of Washington Press, 1955), pp. 197–201, 209–10, 216–18; Alvin M. Josephy, Jr., *The Civil War in the American West* (New York: Alfred A. Knopf, 1991), pp. 263–64.

7. Jerry C. Towle, "Woodland in the Willamette Valley: An Historical Geography" (Ph.D. diss., University of Oregon, 1974); Jerry C. Towle, "Changing Geography of Willamette Valley Woodlands," *Oregon Historical Quarterly* 83 (1982): 66–87; Richard White, "Indian Land Use and Environmental Change in Island County, Washington: A Case Study," *Arizona and the West* 17 (1975): 327–38; Richard White, *Land Use, Environment, and Social Change: The Shaping of Island County, Washington* (Seattle: University of Washington Press, 1980), pp. 14–34; Boag, *Environment and Experience*, pp. 3–23; William G. Morris, "Forest Fires in Western Oregon and Western Washington," *Oregon Historical Quarterly* 35 (1934): 313–39; Thornton T. Munger, "Out of the Ashes of Nestucca," *American Forests* 50 (1944): 342–45, 366–68; Stephen J. Pyne, *Fire in America: A Cultural History of Wildland and Rural Fire* (Princeton, N.J.: Princeton University Press, 1982), pp. 336–37; Henry M. Gannett, *Forests of Oregon* (Washington, D.C.: Government Printing Office, United States Geological Survey, Professional Paper no. 4, 1902), pp. 11–13.

8. Robert Bunting, *The Pacific Raincoast: Environment and Culture in an American Eden, 1778–1900* (Lawrence: University Press of Kansas, 1997), pp. 74–77; Williams, *Americans and Their Forests*, p. 60; Boag, *Environment and Experience*, pp. 28–37. There has been considerable debate over the alleged connection between forest trees and soil fertility. For a brief summary, see Boag, *Environment and Experience*, pp. 169–70.

9. Arthur R. Kruckeberg, *The Natural History of the Puget Sound Country* (Seattle: University of Washington Press, 1991), pp. 4–33; Bowen, *Willamette Valley*, pp. 59–64.

10. White, *Land Use, Environment, and Social Change*, pp. 113–41.

11. Thornton T. Munger, "The Cycle from Douglas Fir to Hemlock," *Ecology* 21 (1940): 451–59; Thornton T. Munger, "A Look at Selective Cutting in Douglas-Fir," *Journal of Forestry* 48 (1950): 97–99. In some areas the climax trees would include western red or Alaskan yellow cedar in addition to or in place of spruce and/or hemlock. For an evaluation of the work and impact of Munger and his contemporaries, see Richard Allan Rajala, "Clearcutting the Pacific Coast: Production, Sci-

ence and Regulation in the Douglas Fir Forests of Canada and the United States, 1880–1965" (Ph.D. diss., York University, 1994), pp. 138–207.

12. For a fuller description of this diverse forest, see Richard M. Highsmith, Jr., *Atlas of the Pacific Northwest: Resources and Development* (Corvallis: Oregon State University Press, 4th ed., 1968), pp. 49–60.

13. Dorothy O. Johansen, "A Working Hypothesis for the Study of Migration," *Pacific Historical Review* 36 (1967): 1–12. See also Thomas R. Cox, *The Park Builders: A History of State Parks in the Pacific Northwest* (Seattle: University of Washington Press, 1988), pp. 167–69, 175–76.

14. Edmond S. Meany, Jr., "History of the Lumber Industry of the Pacific Northwest to 1917" (Ph.D. diss., Harvard University, 1935), pp. 80–83, 89–93.

15. Luark is quoted in Robert E. Ficken, *The Forested Land: A History of Lumbering in Western Washington* (Durham, N.C.: Forest History Society; Seattle: University of Washington Press, 1987), p. 4; Charles A. Geyer, "Notes on the Vegetation and General Character of the Missouri and Oregon Territories, Made during a Botanical Journey in the State of Missouri, and across the South Pass of the Rocky Mountains, to the Pacific during the Years 1843 and 1844," *London Journal of Botany* 5 (1846): 198–201, 522 (first quotation, p. 522; other quotations, p. 199). Geyer's "pines" were primarily Douglas fir, often referred to by Europeans and, later, in the lumber trade as Oregon pine.

16. Cox, *Park Builders*, pp. 5–11. See also Earl Pomeroy, *In Search of the Golden West: The Tourist in Western America* (New York: Alfred A. Knopf, 1957), pp. 114–18; Boag, *Environment and Experience*, pp. 74–83, 87–93, 142, 146–61.

17. Thomas R. Cox, "Trade, Development, and Environmental Change: The Utilization of North America's Pacific Coast Forests to 1914 and Its Consequences," in *Global Deforestation and the Nineteenth-Century World Economy*, eds. Richard P. Tucker and J. F. Richards (Durham, N.C.: Duke University Press, 1983), pp. 14–29, 180–84; George H. Plummer, F. G. Plummer, and J. H. Raveine, *Map of Washington Showing Classification of Lands* (Washington, D.C.: Government Printing Office, 1902). Unlike the forest maps prepared for other sectors of the Pacific Northwest, those for western Washington were never published with an accompanying report. Copies may be found in National Archives, Conservation Branch, Record Group 57.

18. Pyne, *Fire in America*, pp. 199–218, 327–45, esp. pp. 339–42; White, *Land Use, Environment, and Social Change*, pp. 87–91, 94–99, 105–12; Williams, *Americans and Their Forests*, pp. 300–302; Thomas R. Cox, *Mills and Markets: A History of the Pacific Coast Lumber Industry to 1900* (Seattle: University of Washington Press, 1974), pp. 227–33.

19. Ficken, *The Forested Land*, pp. 71–72; Cox, *Mills and Markets*, pp. 207–13;

Williams, *Americans and Their Forests*, pp. 318–20. The development of the logging truck and its impact still await thorough study.

20. Cox, *Park Builders*, pp. 32–78.

21. Harold K. Steen, *The U.S. Forest Service: A History* (Seattle: University of Washington Press, 1976), pp. 152–62; Craig W. Allin, *The Politics of Wilderness Preservation* (Westwood, Conn.: Greenwood Press, 1982), pp. 66–76, 94; Roderick Nash, *Wilderness and the American Mind* (New Haven: Yale University Press, rev. ed., 1973), p. 191. The Forest Service's interest in recreation during Greeley's tenure as chief is another topic that has received inadequate attention.

22. Cox, *Park Builders*, pp. 72–73; Kruckeberg, *Natural History*, pp. 116–93, esp. pp. 133–36, 268–69.

23. David A. Clary, *Timber and the Forest Service* (Lawrence: University Press of Kansas, 1986), pp. 112–25; Paul W. Hirt, *A Conspiracy of Optimism: Management of the National Forests since World War Two* (Lincoln: University of Nebraska Press, 1994), pp. 44–81; William G. Robbins, *Hard Times in Paradise: Coos Bay, Oregon, 1850–1986* (Seattle: University of Washington Press, 1988), pp. 132–36. See also Thomas R. Cox et al., *This Well-Wooded Land: Americans and Their Forests from Colonial Times to the Present* (Lincoln: University of Nebraska Press, 1985), pp. 238, 244–47.

24. Elmo Richardson, *BLM's Billion Dollar Checkerboard: Managing O&C Lands* (Santa Cruz, Calif.: Forest History Society, 1980), p. 113; Robbins, *Hard Times in Paradise*, pp. 110–15.

25. Cox et al., *This Well-Wooded Land*, pp. 240–44; Cox, *Park Builders*, pp. 140–43; Alan R. Sommerstrom, "Wild Land Preservation Crisis: The North Cascades Controversy" (Ph.D. diss., University of Washington, 1970). Cf. Dennis Roth, "The National Forests and the Campaign for Wilderness Legislation," *Journal of Forest History* 31 (1987): 112–25. Most studies of the passage of the Wilderness Bill emphasize events surrounding Dinosaur National Monument. A corrective is needed.

26. Hirt, *Conspiracy of Optimism*, pp. 54, 57, 59, 84–85, 169–70, 187, 192, and *passim*.

27. Robbins, *Hard Times in Paradise*, pp. 134–38.

28. For further discussion, see Thomas R. Cox, "The North American–Japanese Timber Trade: A Survey of Its Social, Economic, and Environmental Impact," in *World Deforestation in the Twentieth Century*, eds. John F. Richards and Richard P. Tucker (Durham, N.C.: Duke University Press, 1988), pp. 164–86.

29. Beverly A. Brown, *In Timber Country: Working People's Stories of Environmental Conflict and Urban Flight* (Philadelphia: Temple University Press, 1995), pp.

29–33, 277–81, and *passim;* Darrell Palmer to author, 17 December 1992. See also Jack Ward Thomas, "In My Opinion," *Sunday Oregonian*, 12 January 1997.

30. Palmer to author, 17 December 1992.

31. For further discussions, see John A. Zivnuska, *U.S. Timber Resources in a World Economy* (Washington, D.C.: Resources for the Future, 1967); *Environment and Trade: The Relations of International Trade and Environmental Policy*, eds. Seymour J. Rubin and Thomas R. Graham (Totowa, N.J.: Allanhead, Osmun, 1982).

PART VI / MINING

Mining joins farming, forestry, and fishing as one of the four mainstays of the Northwest's traditional, natural-resource-based economy. Without precious metals, the region's human history would have been very different. Mineral strikes in the nineteenth century led to the development of towns such as Lewiston, Boise, and Burke, Idaho. They strengthened the economies of existing regional market centers such as Walla Walla, Spokane, and Seattle, Washington. Montana's mining industry virtually owned state government until well after World War II.

Mineral wealth also had other influences on the Northwest. A rush of gold-seekers across the Yakama reservation in 1856 led to the Yakama War. Exploitation of wage workers and dangerous conditions in industrial mines and smelters led to some of the earliest union organizing, as well as some of the bloodiest labor confrontations in the Northwest. A lucky mineral strike in northern Idaho 100 years ago made May Arkwright Hutton a millionaire, giving her the time as well as the resources to advocate successfully for women's suffrage. Skills in mining, engineering, and explosives led Chinese immigrants into the industry—one of the few occupations available to them in the nineteenth century. By 1870 more than half of all miners in Idaho were Chinese.

In addition to the industry's contributions to the Northwest's economy and society, mining has imposed a wide array of social and environmental costs. Unlike farming, forestry, and fishing, mining harvests a nonrenewable resource. Consequently, the mining industry has spawned some of the most dramatic economic booms and busts in the region's history, wreaking havoc on communities dependent on it. It has wreaked similar havoc on local environments: air and water pollution and deep scars on the landscape attest to mining's continuing impact. The environmental impacts impose monetary costs, not least the companies' and taxpayers' share of the burden of cleaning up the 21-square-mile Superfund site in the Silver Valley of northern Idaho.

Both chapters in part six focus on the Silver Valley. Katherine Morrissey describes legal battles between mining companies and farmers who lived along the Coeur d'Alene River downstream from the mining district. Spring floods regularly washed mine tailings and other wastes downriver, leaving debris and poisons on meadows and farms. Morrissey offers a sensitive cultural analysis of the opposing interests: farmers who claimed the upstream mining companies were ruining their property, and mine owners who denied there was any damage or that it resulted from their activities. Like Zeisler-Vralsted, Morrissey focuses on cultural perceptions of the landscape; like Lang, she illuminates questions about people's "sense of place"; like Robbins, she offers readers a complex tale of contested perceptions.

Katherine Aiken's essay focuses on the town of Kellogg in the Silver Valley, examining legal battles surrounding air pollution from the ore smelters there. Her approach is political and legal rather than cultural. Although she examines a number of smelter smoke battles in other parts of the country, her story focuses on Idaho's Bunker Hill Company. She shows in careful detail how the mining company resisted government regulation and sidestepped liability for the pollution it generated—and was largely successful at these efforts prior to World War II, despite the proliferation of public health studies on the dangers of smelter smoke. Importantly, Aiken also demonstrates that pressures on mining companies to clean up smelter emissions did yield a slow but steady advance in pollution control technology. Finally, she shows how public perceptions of pollution changed after mid-century, leading to the strengthening of pollution control policies and making it increasingly difficult for Bunker Hill to continue emitting pollutants while avoiding liability.

22 / Mining, Environment, and Historical Change in the Inland Northwest

KATHERINE G. MORRISSEY

"Agriculture restores and beautifies, mining destroys and devastates." So asserted Isabella Bird while traveling in the late-nineteenth-century Rocky Mountain West. In her comparison, Bird reflected contemporary perspectives on the environmental impact of the economic use of natural resources. For those who valued land for its renewable productive purposes, farming clearly surpassed mining. Those who celebrated human control over nature, however, might have agreed with T. A. Rickard, who described an Idaho mine as "more than a hole in the ground; it is an expression of hope, initiative, energy, and accomplishment; it is the fine flower of industrial achievement."[1]

The dialectic opposition suggested by these two quotations has become an easy shorthand for interpreting environmental disputes. Historians, for example, in their descriptions of early-twentieth-century environmental conflicts, mark a divide between preservationists and conservationists, between ecocentric and technocentric thinking, between radical amateurs and professionals. A closer examination of such conflicts suggests that this oppositional narrative, although perhaps useful, often hides overlapping ideological concerns.[2]

For more than 150 years, mining has left distinctive imprints on the interior Pacific Northwest—on its environment, on the communities that have shared that environment, and on patterns of thought. One need only travel through the mining regions of western Montana, northern Idaho, and eastern Washington to see visible signs of mining operations: the tailings, smelter stacks, and open pits. Less visible to the eye are other alterations of the land, water, air, and wildlife. These material transformations mark ongoing ideological transformations and debates. The meanings of these visible and invisible marks on the landscape have been contested over time and continue to be disputed.[3] In the spring of 1996, for example, a dramatic Associated Press news photograph depicted the demolition of the enormous smokestacks at the abandoned Bunker Hill and Sullivan mines in northern Idaho. The accompanying story, reprinted in numerous newspapers, defined it as a historic moment of changed environmental understanding. Once the locale of mining wealth and western labor strife, the Bunker Hill area is now a Superfund site, slated for toxic cleanup. But such an interpretation did not go uncontested. Various residents, mining historians, and historic preservationists decried the loss of the symbolic stacks—whether as historical artifacts or as representations of a regional identity, a significant mining technology, or an economic livelihood.

Although often considered a recent phenomenon, public concern about the environmental impact of mining practices has long been an issue in Pacific Northwest communities. In legal battles, political disputes, scientific literature, and the public press, various groups have contested the "appropriate use" of the "natural resources" of the region. For inland Northwest residents involved in such ongoing struggles, the issues rarely fit into diametrically opposing camps. Even in the highly structured setting of a courtroom, where one might expect to find sharp divisions, the waters are often muddied. To consider the complex legacy of mining in the interior Pacific Northwest—the interplay between its material environmental and economic impacts and its cultural and ideological contexts—this chapter examines the stories involved in one early-twentieth-century contest.

In the fall of 1905, Elmer Doty, a northern Idaho homesteader, went to call on one of his neighbors, a family from the city of Portland, Oregon, who had rented farm and ranch land along the Coeur d'Alene River valley. As Doty surveyed his neighbor's property, he noted that "it was a shame to have such fine land ruined by lead poison." Portions of his own ranchland, he explained, had also been "entirely ruined by concentrates." According to Doty, these river ranches, some 50 miles downstream from the Coeur d'Alene mining district, showed the results of water pollution caused by the mining operations.[4]

The previous year, Doty had banded together with 65 of the other valley ranchers to bring a lawsuit for damages against the upstream mine companies—Bunker Hill and Sullivan Mining and Concentrating Company, Federal Mining and Smelting Company, Gold Hunter Mining and Smelting Companies, and Larson and Greenough. The ranchers claimed that the mine companies had polluted the river waters with lead and other poisonous materials, filled up the channels with tailings and debris, and contributed to spring flooding of their ranchland. The poisonous sediments left on their land when the floodwaters receded, and along the riverbanks throughout the year, had destroyed their vegetation and killed their livestock.

The neighbor listened intently to Doty's discussion of the environmental problems and the damage suit. When another visitor asked whether the ranchers had a chance to win the legal case, Doty asserted, "[We] certainly will, and the mining companies have admitted several times that there was damage beyond a doubt and now it is only a question of amounts they will have to settle." The optimistic Doty spoke from the perspective of his fellow ranchers and farmers: the mining companies had impaired the land's value for production, and therefore they should compensate the ranchers.[5]

The mine owners denied responsibility for any damages, and despite Doty's assertion, they also questioned whether such environmental alterations had occurred. Relying on their prior appropriation rights, they had reasonably used the water for beneficial purposes, they explained; they had even constructed dams and reservoirs on the south fork of the river to contain coarse tailings. The only effect their operations had on the river water, the owners asserted, was harmless, simply changing its color.[6] Indeed, the mining companies spoke with the conviction that their operations had contributed significantly to the development of the state's natural resources.

The neighbor who paid close attention to Doty's pronouncements, although not a party to the suit, had a vested interest in the case. But those interests were not entirely visible to Doty and the other river ranchers. The neighbor, who had moved into the community with his wife and two children that spring, was employed by the Thiel Detective Agency. The mine companies had hired the agency to investigate the ranching community and provide evidence to use in the ongoing litigation. The neighbor was a spy for the mine companies. From August 1905 to June 1908, Operative Number 14 wrote up a daily report on his activities and conversations for his employer.[7] This unnamed man offers an intriguing perspective on the environmental conflict. His almost three-year record provides a fascinating glimpse into the workings of a ranching community—by someone engaged in trying to disrupt it. In a case that hinged on perceptual distinctions between discolored and poisoned water and that spent rhetorical energy on questions of misrepresentation, the detective masquerading as a rancher only reinforces the apparent distortions.

The legal case, which would not come to trial until five years later, was not the first instance in which agriculturalists brought damage claims against mine companies; water contamination by industrial practices concerned a variety of water users throughout the nineteenth century. In the mining regions of Pennsylvania and California between the 1850s and 1880s, a flood of cases dealt with conflicts resulting from coal, quartz, and hydraulic mining and the use of stamp mills, concentrators, and smelters. Complaints included destruction of property, obstruction of waterways, and pollution of water, soil, and air. A series of California decisions particularly captured national legal, popular, and industrial attention and was cited by attorneys on both sides of the northern Idaho case. Starting in the 1870s, farmers in California's central valley had brought court cases against the hydraulic mining companies that were washing sediment and debris over their fields. Hydraulic mining—in which miners diverted streamwater through high-pressure hoses to wash gold-bearing rock gravel out of adjacent hillsides—caused a dramatic rearrangement of the landscape and used a tremendous amount of water. It also transported soil and sediments downstream, increased acidification of the water, and contributed to erosion and flooding. The California farmers succeeded not only in gaining financial awards for damages but also in ending hydraulic mining operations in the American and Yuba River watersheds.[8]

When Operative 14 entered the lower Coeur d'Alene River valley in eastern Kootenai County, he and his family moved into a settled northern Idaho community linked to the city of Spokane, Washington. Located in the Idaho Panhandle, the subregion stretched between Coeur d'Alene Lake and Indian reservation land, on the west, and the Bitterroot Mountains that formed the Montana-Idaho state border, on the east. Its established towns were strung along the transportation routes of the Oregon Railway and Navigation Company's steamer and railroad lines, which followed the river and its south fork eastward across the state. The towns served as market and communication centers for the larger region. Most of the residents engaged in economic activities closely tied to the environment—lumbering, ranching, farming, or mining. The ranchers and farmers held small homesteads, averaging 160 acres or less, acquired under homestead laws. They grew hay and oats along the fertile river property for their own use and for sale to the lumber camps and mining towns upstream. Although some of the lumber and mining companies were small-time operations run under local ownership, most were satellites of larger corporations with headquarters and stockholders located outside the region.[9]

The mining activities centered on the rich mineral lodes of Shoshone County, along the upper Coeur d'Alene River and its many tributaries. Discoveries of gold deposits along the north fork of the river during the early 1880s had drawn miners to the mountainous region that became known as the Coeur d'Alene district. Discoveries of high-grade silver and lead deposits along the south fork of the river attracted the attention of railroad developers, investors, and mine speculators in the late 1880s. This soon led to the development of the hard-rock industry, which was sustained by heavily capitalized corporations that employed hundreds of workers. By 1905, the mining operations had consolidated into several mining corporations—most of which were named in the damage suit.[10]

Along with material activities—staking claims, building rail lines, establishing mining operations—mining companies, railroad companies, civic boosters, and others created a literary portrait of the region. In their descriptions, the never-ending natural resources of northern Idaho offered great possibilities. As the *Coeur d'Alene Press* expressed it in 1892: "The people of Kootenai county are here for a purpose: Nature has placed before them the crude materials for building up a prosperous and wealthy country."[11] Promotional literature, including mining company prospectuses, civic booster literature, railroad pamphlets, and tourist guides, relied on exuberant prose and optimistic projections. Created to achieve economic ends, these forms of publication sought to provoke investment, settlement, tourism, or all three. Despite its hyperbole, promotional literature provides a window into the mind-set of those engaged in "settling" the region. Whether describing mountains, soils, or waters, they placed human activity between "nature" and "a prosperous wealthy country."[12]

Other local writers described a less felicitous relationship between human activity and nature. An 1895 report by botanist John B. Leiberg, who had been instructed by the Department of Agriculture to make a survey of the Coeur d'Alene Mountains "with special reference to the economic features of the flora," noted some detrimental impacts. A 10-year resident of northern Idaho, Leiberg described a changing environment and attributed the alterations to many factors, both "the operations of nature and those of man." As he explained the process, "the slow agencies of nature which are constantly destroying but as constantly replacing are augmented by the efforts of man, who, with all the means of destruction at hand, tears down the work of centuries, but gives no thought towards rebuilding of the fabric."

He blamed specific human actions—constructing dams, removing timber, eradicating beaver, overgrazing, setting intentional fires, and discharging mining slimes and wastes into streams—and enumerated the natural consequences—erosion, increased runoff and flooding, dense fogs, reduced botanical species diversity, and mineralized water.

A conservationist who advocated "proper use" and the need to "guard against waste," Leiberg railed against "the popular mind" and the northern Idaho belief in an inexhaustible supply of natural resources, especially timber; it was, he explained, "a most pernicious idea." Human nature, evident in the behavior of myopically self-interested settlers, miners, railroad operators, and loggers, led to carelessness, destruction, and hostility. Although they usually placed the blame elsewhere, other Kootenai and Shoshone County inhabitants besides Leiberg noted some of the human-related ecological changes, especially those that had a visible and financial impact on their own activities. Seasonal changes in water flow, especially the Coeur d'Alene River's susceptibility to overflowing its banks during heavy rains and spring runoffs, frustrated residents. During periods of high water, the river carried visible sediments and mine tailings farther downstream. It was in the spring of 1904, after floodwaters inundated river ranches near Cataldo and Dudley, Idaho, under as much as eight feet of water, that Elmer Doty and his neighbors brought their lawsuit against the mine companies.[13]

Mine owners and ranchers faced off in two court cases on the issue of damages. The initial case dragged on in the courts as mine company attorneys worked to create numerous delays, hoping that most of the ranchers would abandon the suit. Meanwhile, as the ranchers watched river waters swamp their meadows again in the spring of 1905, they brought a second case against the mine owners. They sought a permanent injunction against the mining companies to prevent further damage to their properties; they asked the court to shut down the mine operations. The injunction case came before Idaho federal district court judge James H. Beatty in 1906 and on appeal to circuit court in 1908. The original case finally came to trial in 1910. By the time the matter was decided, the testimony and printed materials ran to more than several thousand pages.[14]

In both cases, the mine owners, ranchers, and their lawyers wrangled over several key issues: Had the environment been changed detrimentally between 1901 and 1904, the period for which the ranchers sought damages? What were the causes of that change? Were the changes a result of the mining operations? Were the mine companies liable for any such damages?

Mine owners and ranchers drew on different visions of nature for the purposes of their arguments. Although both sides noted significant environmental changes over the previous 30 years, they clashed in their descriptions of the landscape prior to 1883. The ranchers' attorney focused on the richness of the natural environment, describing a valley of "luxuriant growth" with "grass as high as a horse's back;" it was "a hunter's paradise" where the "river abounded in trout and other food fish" and "the waters were clear and pure." The mine owners' attorney focused on the emptiness of an unused environment, describing the region before the 1880s as "vacant, unoccupied, unclaimed, unsurveyed."

They agreed that nature, by definition, constantly underwent change. As William Stoll, the rancher's attorney, phrased it: "The work of nature is not at rest, it is working all the time." And they agreed that human beings had an important role in regulating and controlling nature. Both sides noted significant environmental changes as a result of human actions, but they made distinctions between acceptable, appropriate changes and detrimental changes. The rich natural environment described by the ranchers had been, in their view, destroyed by the actions of the mine companies. The mine owners argued that their "lawful, safe, economical, healthful, and wholesome use of the waters" contributed to the productive wealth of the nation, developed the natural resources of the state, and supported more than 12,000 Shoshone County residents.[15]

Inside the courtroom, the mine owners' attorneys relied on establishing their legal rights to the water and on the contemporary judicial practice of evaluating natural resource use by its economic value. According to Idaho's prior-appropriation water laws, the mining companies had rights to the water because they had claimed its usage before the arrival of the homesteaders. The state constitution also gave mineral extraction the preferred right to the use of the waters of the state's streams and rivers in organized mining districts. The homesteaders had acknowledged these rights when they received the patents to their land. Since these water rights did not carry with them the right to destroy the property of other people, however, the focus of the case was on whether damage had occurred.[16]

Just as the mine owners' attorneys carefully denied any recognition of, or culpability for, changes in water quality, they also suggested that any detrimental changes resulted from other sources—from municipal sewage dumping or from mining activities in the region before their companies were incorporated. Perhaps, they even suggested, the changes attributed to mine tailings were the results of the ranchers' own actions; the visible deposits might be the markings of a grayish clay unearthed by the ranchers in digging ditches for their fields. Finally, not everyone considered mine tailings inherently dangerous or toxic. "The macadamized streets of Wallace, which are believed to be unsurpassed in the northwest," boasted one newspaper report, "are entirely made from the tailings from the concentrators," and one of the newest regional industries used the tailings to manufacture building blocks.

The ranchers, on the other hand, found the evidence of their eyes and experience utterly convincing. They saw the mine debris poisoning the water and land on which they, their domestic animals, and their crops relied. They called into question the generally accepted notion that the action of the stream dissipated any waste or toxic materials—"that running water purified itself through dilution." They compared the damage from the tainted Coeur d'Alene with the beneficial effects of freshwater flooding in from nearby Lake Killarney.

> the two waters would come together, the pure water from the hill side or from Lake Killarney in this direction and the polluted water from the river in this direction. And where the two bodies of water got together the farmers noted the clear line of demarcation. . . . The portion where the waters overflowed from Lake Killarney remained rich and productive and contained a luxuriant growth of grass and other vegetation. . . . And when you got over into the territory where the water was completely polluted you found no vegetation.

Such empirical evidence, verified by their understanding of the natural environment, supported their claims. Rancher Richard Sharpley, for example, "was satisfied beyond a doubt that this lead poison had killed many head of stock along this river and all the experts on earth would not convince him otherwise."[17]

The ranchers' attorneys, recognizing the two sides' sharp differences of view, often shifted the focus of the trials to visual evidence and perceptions. Initially they requested that the jury be allowed to visit the site in order to evaluate the damage for themselves. Their exhibits included photographs of the landscape and river as well as wooden boxes of crushed rock, numerous glass jars filled with water, dry tailings, and liquid tailings, and tin cylinders filled with settlings. Anyone who saw the markings of the mine tailings on the land, the damage to the soils, and the debris in the water, one rancher explained to Operative 14, would surely be convinced.[18]

Throughout the region, the ranchers' point of view found wide acceptance. In fact, the ranchers suffered economically because people believed their side of the controversy; their hay crops no longer found a ready market.

Aware of the damage suit and concerned about lead poisoning, buyers offered lower prices for river ranch crops and preferred to purchase hay and produce from Montana or Washington. Anxiety about water pollution also reached the region's urban areas. The citizens of Spokane, Washington, for example, feared that their water supply, drawn from the Spokane River, had been affected, because the river connected to the Coeur d'Alene River through Coeur d'Alene Lake. Spokane physicians, arguing about a newly proposed water filtration system, agreed that the mining companies did contaminate the waters of the Coeur d'Alene River but disagreed over whether or not the lead could travel as far as Spokane. To allay the anxiety and settle the questions, the Spokane County Medical Society requested an analysis of the city's water in 1906 to see whether it had been tainted.[19]

Regional cultural attitudes about mining, environment, and health can be read in contemporary newspaper reports and popular literature. Concern over the potentially toxic effects of mining on the air, water, and soil also resonated in regional tall tales. Consider one story published in Spokane as the tailings case wended its way through the courts. "Made an Odd Clean-up," set in an earlier time at a specific location, recounts the experiences of the teller, an assayer passing the winter months with the watchman at a closed silver mine. They take in an ill prospector, who gradually weakens and dies. By now they had determined that his suffering had been caused by his years as a California miner; he had absorbed the fumes of quicksilver and retained the metal in his system. The watchman, however, has a further fantastic theory—the miner's body, he surmises, would have attracted metals that had an affinity for the mercury, particularly as he continued to live in the mining landscape. The watchman's assumption was that minute amounts of various metals and minerals exist in the surrounding air, water, and soil. After much deliberation, they decide to cremate the body and assay the remains. Their financial gain: 41 ounces of pure gold and 18 ounces of silver. Juxtaposing the desire for wealth against the dangers to health, the tale relies on an experiential knowledge, one based on the watchman's—and the listener's—understanding of the interrelation of humans and the elements. Whether read as cautionary tales or as laughable absurdities, such stories were part of the cultural context within which the litigants presented their descriptions of alterations of the environment by mining practices.[20]

The mine owners, relying on the expertise of scientists and on legal rights, argued that their operations had not damaged the water and found the ranchers' accusations groundless. In accord with progressive-era authoritarian discourse and legal precedent in similar cases, both sides called on scientific experts to testify on their analyses of multiple soil and water samples—some of them gathered by Operative 14. While the ranchers' chemist found soluble sulphates of lead in their water samples—such sulphates were understood to be potentially toxic—chemists for the mine owners found "no trace of lead, zinc, copper or arsenic in any" of their water samples. Scientists from both sides also provided reports to the concerned Spokane physicians and citizens and published their findings in local papers. Chemist Charles M. Fassett, who testified for the mine owners, repeated his "no trace" findings. And while Professor Elmer Fulmer of Washington State College agreed that the city waters were unaffected by mining-induced problems, it was only because "the lead would be lost" in the lake and river before reaching the city.

In the face of strong regional acceptance of the ranchers' assessment, the mine owners' chemists continued to put forward other possible explanations. For example, although Fassett found no lead in the city waters, "sewage," he reported, "is another story." And in court the chemists even suggested that farming activities were inappropriate in the northern Idaho locales, declaring the river valley soils naturally unsuitable for agriculture and deficient in the most "important elements of plant food." The scientists also examined, and disagreed with, the work of their scientific rivals. "These results, so utterly at variance with the known facts and possibilities," declared Fassett of John S. Burd's chemical analysis, "stamp his work as unreliable, and, in my opinion, undeserving of any credence whatever."[21]

The affidavits and testimony in the trial papers are filled with strong assertions, accusations, and counteraccusations. Attempting to explain the discrepancy in soil sample findings, for example, mine witnesses claimed Carl Graf's ranch had been "salted" with tailings to provide a false reading. The appeals court judges, noting the "intense feelings," found the "exaggerations and misstatement of matters of fact . . . very gross." According to Operative 14's reports on conversations in the towns, fields, and homes along the Coeur d'Alene, these passionate beliefs, this incredulity regarding alternative views, and these accusations of illegal activities found expression outside the courtroom as well. In structuring their stories, in interpreting the landscape, and in perceiving their experiences, the residents of the Coeur d'Alene Valley consciously and unconsciously drew upon familiar modalities. Their variant descriptions of nature, for example, echo those found in other local publications, such as Chamber of Commerce literature. And the overblown language and narrative styles of the courtroom testimony and legal papers are shared by popular genres of exaggerated storytelling—both the tall tales and the promotional pamphlets.

When the injunction case came before Judge Beatty, he expressed concern over the "highly colored" affidavits and decided to visit the Coeur d'Alene River to assess the damage himself. Beatty, whose judicial decisions usually favored business interests, looked out over the land and river; he noted the "milky appearance" of the water and deposits that "had some appearance of the mining debris." He characterized the deposits and color change as "small, compared with the representations thereof made by the complainants," and concluded that there was "no justification" for the charges. He admonished the ranchers for their "wild assertions": "They cannot shelter themselves behind the flimsy veil that they believed them. . . . It must be concluded either that these complainants intended to deceive the court, or were themselves deceived by their own culpable negligence."[22]

Visual evidence continued to be important to the case when the ranchers appealed Beatty's decision to the U.S. Circuit Court of Appeals in San Francisco. The mine owners now turned to images as well, presenting photographs of the valley's agricultural land as evidence. They had hired T. N. Barnard, a prominent Wallace, Idaho, professional photographer, to take more than 100 photographs; Barnard also testified in the case. Although the judges acknowledged that "photographs are not very reliable, since the camera may be so placed as to give a deceptive impression," they found the preponderance of photographs showing "an unusual and vigorous growth of crops of various kinds" most convincing. Barnard, a former populist mayor of Wallace who had testified before a congressional investigating committee on behalf of striking miners during the 1890s, might have seemed an odd choice to present the mine owners' perspective. But much of the Barnard Studio's business relied on advertising and promotional work for the mining companies, and by the first decade of the 1900s, Barnard was also developing investments in mining deals.[23]

Photographs had been used as evidence to support particular uses of the environment for more than 50 years. Advocates for the establishment and preservation of national parks, for example, presented debating members of Congress with Yellowstone and Yosemite photographs taken by William Henry Jackson and Carleton Watkins. Under the employ of California mining companies, Watkins and J. A. Todd also created images that lawyers entered into evidence during the 1880s hydraulic disputes. Relying on the same techniques they used in their celebrated landscape images, the photographers combined art and science for commercial ends. In the courtroom, photographs carried cultural, not legal, weight as realistic depictions. By themselves, for example, photographs could not be entered into evidence;

someone had to testify about the images in order to lay a foundation for their admission. Invested with meanings through testimony, photographs were recognized by the courts as representations of a particular perspective.[24]

Lawyers, however, made effective use of cultural assumptions about the factual nature of photographic images. The testimony of the photographers guided the viewer's gaze to particular aspects of the images and asserted a visually based interpretation as an accurate depiction. While images of flooded fields and barren soil depicted the ranchers' perspective, the mine owners presented photographs of the agricultural land and crops of the valley. Once the photographs were entered as evidence, however, they could be reinterpreted by other viewers.

Both Beatty and the Circuit Court of Appeals judges denied the request for an injunction. Finding the evidence unsupportive of the damage claims, they argued that shutting down the mining operations would cause an unwarranted economic loss to the region. In their examination of the two positions before them, they sided with the mine owners, but in doing so they discounted the contradictory scientific data and drew on the same visual evidence upon which the ranchers had so confidently relied. Ruling in favor of the mine owners, Judge Beatty asserted that the "courts will endeavor to see that no man shall succeed through misrepresentation." Yet as Stoll, the ranchers' attorney, commented, "We don't all look through the same eyes, and we don't all understand a situation the same even after we see it."[25]

In their use of the Thiel detective, the mine owners did succeed, in part, through misrepresentation. Operative 14 played the role of a rancher while he was engaged for his economic livelihood as a detective. Representing mine owners' interests, he served as a business tool with which to uncover information for the legal case. Yet the detective serves, for us, as a reminder of the malleability of appearances and the impact of experience on perceptions.

In Operative 14's day-to-day life, his interests were more akin to those of the ranchers and farmers than those of the mine owners. As his reports reveal, he shared his neighbors' concerns about local politics, crop prices, and weather. Joining them on fishing trips, trading work, and fighting forest fires by their side, he also began to share their concerns regarding the impurities in the water. His early descriptions of the river, which "carries a strong showing of mine debris coloring," maintain the language of the mine owners. But over the years, his experiences begin to color his reports. During the spring floods of 1906, Operative 14 worried when the flooding of his low meadow ground also contaminated his well. He fretted over the health of his five-year-old son, who developed abscesses on his neck and legs; the doctor explained that the serious illness, which might leave the boy crippled for life, came from bathing in impure water. By the time of the spring flood of 1908, Operative 14 was using the language of the ranchers to describe the river: "The water is highly charged with dark blue lead from the mine debris from the mines on the river banks above here, and all the overflow is being carried back over the meadows by the raising current in the river. The river water can be seen at a glance as one is rowing over the meadows and one can tell to a nicety just where the lead water is circulating."[26]

The resolution of the northern Idaho controversy is undramatic but telling. As the legal issues dragged on in the courts, many of the ranchers involved, facing economic hardships—especially during the 1907 Panic—sold their ranches or lost their property to mortgagors. In 1904, the Coeur d'Alene mine owners, affiliated as a Mine Owners Association (MOA), had entered into an agreement with Washington Water Power Company (WWP) that would have an impact on the river ranches and the damage suits. Seeking a more permanent and reliable source of electricity, the MOA purchased the rights to Post Falls on the Spokane River, north of Coeur d'Alene Lake. Under an agreement with the mine owners, WWP agreed to supply electricity to the mining region and took over the development of the Post Falls site. The projected dam would raise the level of Coeur d'Alene Lake and back up the waters of the Coeur d'Alene River as well. Agents from WWP arrived in the valley to negotiate the purchase of the river ranch property that would be affected by the rising waters. The ranchers greeted them with suspicion; some "thought the Washington Water Power Co. formed an alliance [with the mining companies] to flood the river ranches and thus complicate the damage suits." But many agreed to sell their ranchland, especially after the unfavorable decisions in the injunction case.[27]

While the initial case remained in the courts, the mine owners considered the issue essentially settled with those 1906 and 1908 victories. The Thiel Agency ended its operation in Dudley in June 1908, and Operative 14 returned to the home office in Portland. Most of the initial complainants had moved from the area or had dropped out of the suit by the time the original case was settled in favor of the ranchers in 1910. By that time, a bitter Elmer Doty had given up on his property. "He said if all the judges were not ruled by money power he would have won suit against the mining co. long ago and they would have had to pay up for damage done or would have closed up the mines. Mr. Doty does not live on his ranch now as he says it is completely ruined and he has abandoned it and moved to Harrison." The ranchers' hollow victory was underscored by their award: the sum of $1.[28]

Although the legal case ended in 1910, the environmental conflict lingered on.[29] And so, too, did the mining debris. But the tailings and the tailings tales have shifted over the years. In the political arena, in the courtroom, and in the legal literature, decisions have refigured the issues. More recent scientific studies of acid drainage from lead mines and tailings piles, and specific studies of the Coeur d'Alene River, focus on the detrimental environmental legacy of the mine operations.[30] Environmental historians have also turned their attention to the significant impact of natural resource use—whether mining, farming, or ranching—on the environment and to cases like the northern Idaho tailings dispute.[31]

The controversies between ranchers and mine owners over the waters of the Coeur d'Alene River reveal some of the struggles over different visions of the environment, the region, and its future that underlay such conflicts at the turn of the century and still underlie them today. What is the appropriate use of natural resources? What distinctions should be made between "renewable" and "nonrenewable" resources? How should disputes over the use of the same resource be resolved? How should environmental change be regarded? How are negative environmental changes defined? How should future resource use be weighed against both immediate and long-term economic and social issues? These questions increasingly confronted Pacific northwesterners throughout the twentieth century.

The courtroom setting of this particular conflict influenced the rhetoric of the debate. Both sides used the legal arena to touch upon issues of rights, to present scientific data, to claim authenticity for their perspectives, and to accuse their opponents of misrepresentation. In such adversarial proceedings, the tendency is to place the struggle in sharp relief between two distinct groups with clear agendas. The ranchers and the mine owners did disagree on many issues, including the very existence of pollution, the appropriate ways in which to use and control nature, the value of their work for the community, and their visions of the future. Each side presented competing scientific and historical narratives along with visual evidence to support its view of mining landscapes.

Yet the ranchers and mine owners both used legal and scientific discourses to legitimate their vision; they shared rules of evidence and evaluation, a belief in technological progress, a confidence in human control of nature, and a profit mentality. They employed cultural and regional ways of communicating and thinking about these issues. Farmers, ranchers, mining engineers, scientists, and other community residents offered variant testimony arguing for the "public good." Their claims to authenticity rested on a common ground of experience and sensual knowledge, on a "naturalized" understanding of the natural world, even when their interpretations of that experience and knowledge clashed.

The ranchers and mine owners involved in the court cases shared more than the rhetoric of the debate; their cultural participation was connected to their material surroundings—the environment and economy of northern Idaho. Mutually dependent, they were intrinsically linked both to the market economy and the ecosystem to which they belonged. According to the scientific and legal criteria of the present, the mining and the ranching operations both contributed to environmental changes in the Coeur d'Alene River valley.

It might be tempting to tell this story with a clear set of heroes and villains. Certainly the characters fall into place in one type of narrative—the valiant homesteading families in the Coeur d'Alene River valley fighting the evil mine owners. But the muddied waters intrigue me. As viewers situated in the present, our interpretations of this story are similarly influenced by ideas about nature and human interactions with the environment, by legal definitions and rights, by scientific evidence, and by our perceptions. Mining creates visible and subterranean alterations to the environment; its imprints can be seen in the landscapes of nature as well as those of culture. The scars and tailings left by mining disputes offer valuable insights as we consider ways to deal with the region's present-day environmental issues.

NOTES

1. Isabella Bird, *A Lady's Life in the Rocky Mountains* (1879; reprint, Lincoln: University of Nebraska Press, 1960), p. 193; T. A. Rickard, *A History of American Mining* (New York: McGraw-Hill, 1932), p. 340. Other historians have also referred to these apt quotations; see Duane A. Smith, *Mining America: The Industry and the Environment 1800–1980* (Lawrence: University Press of Kansas, 1987), pp. xi, 61.

2. Samuel P. Hays, *Conservation and the Gospel of Efficiency: The Progressive Conservation Movement, 1890–1920* (Cambridge, Mass.: Harvard University Press, 1959); Stephen Fox, *The American Conservation Movement: John Muir and His Legacy*

(Madison: University of Wisconsin Press, 1985); Tim Bayliss-Smith and Susan Owens, "The Environmental Challenge," in *Human Geography: Society, Space, and Social Science*, eds. Derek Gregory, Ron Martin, and Graham Smith (Minneapolis: University of Minnesota Press, 1994), pp. 113–45; Cecelia Tichi, *Shifting Gears: Technology, Literature, Culture in Modernist America* (Chapel Hill: University of North Carolina Press, 1987).

3. Although environmental historians explore such alterations, they often employ a model of relations between humans and nature that, as William Cronon, among others, has pointed out, works "to elide social divisions and to downplay the impact of cultural differences." Furthermore, they frequently emphasize the material bases of change, especially market forces and industrialization, and tend to exclude or deemphasize the accompanying ideological and rhetorical struggles. William Cronon, "Modes of Prophecy and Production: Placing Nature in History," *Journal of American History* 76 (March 1990): 1122–31; Lewis Mumford, *Technics and Civilization* (New York: Harcourt, Brace, 1934); Richard V. Francaviglia, *Hard Places: Reading the Landscape of America's Historic Mining Districts* (Iowa City: University of Iowa Press, 1991); David Harvey, *Justice, Nature and the Geography of Difference* (Oxford: Blackwell, 1996), esp. pp. 117–204; Lynne Page Snyder, " 'The Death-Dealing Smog over Donora, Pennsylvania': Industrial Air Pollution, Public Health Policy, and the Politics of Expertise, 1948–1949," *Environmental History Review* 18 (Spring 1994): 117–39.

4. Thiel Reports, Stanly A. Easton Papers, MG 5, Special Collections, University of Idaho Library, Moscow, Idaho [hereinafter cited as UIdaho].

5. *Elmer Doty v. Bunker Hill and Sullivan Mining and Concentrating Company*, Boxes 31, 32, Civil Case File no. 309, U.S. Circuit Court, Northern Division (Moscow, Idaho), RG 21, National Archives and Records Center, Pacific Northwest Region (NARC-PNWR), Seattle, Wash. (hereinafter cited as *Doty v. Bunker Hill*, RG 21); Transcript of Record, *Doty v. Bunker Hill and Sullivan*, Bunker Hill and Sullivan Papers, MG 130, UIdaho. Complainants in the case, which was filed in November 1904, included at least seven women who owned title to valley lands. Elmer Doty, by all accounts, was a litigious individual; he frequently sought solutions to economic conflicts through the court system. See, e.g., *Doty v. Krutz*, 43 Pacific Reporter 17 (Washington Supreme Court, 1895). According to reports, other ranchers claimed that "Doty is always in some law-suit as a principal or witness. . . . They said he has had law-suits with nearly all of his neighbors." Thiel Reports, Easton Papers, November 6, 1905.

6. The emphasis on a change in color suggests the contemporary language of mining. A finding of mineral deposits was commonly referred to as finding "color"—

a reference drawn from gold prospectors, who frequently identified veins or pockets of gold by distinguishing the differences in hue on a rock face or in a streambed.

7. Transcript of Record, *Doty v. Bunker Hill*, MG 130; Thiel Reports, Easton Papers; R. L. Polk, *Idaho State Gazetteer and Business Directory . . . 1903–1904* (Portland: R. L. Polk and Co., 1903), vol. 2, p. 603; *Raymer's Dictionary of Spokane: An Encyclopaedic-Dictionary of the State of Washington, U.S.A., in General and the City of Spokane in Particular* (Spokane: Chas. D. Raymer and Co., 1906), p. 53. The mine companies, most of whom belonged to the Mine Owners' Association (MOA), were familiar with the use of detective agents for business purposes. The MOA had used the Pinkerton Detective Agency in its 1890s labor-capital struggles, and Bunker Hill and Sullivan had retained detectives in other legal disputes. During the years that Operative 14 lived in Dudley, Idaho, other Thiel and Pinkerton agents were employed in the Coeur d'Alene Valley mines to investigate labor relations. Clark C. Spence, *Mining Engineers and the American West: The Lace-Boot Brigade, 1849–1933* (New Haven: Yale University Press, 1970), pp. 205–6; Charles A. Siringo, *The Cowboy Detective: A True Story of Twenty-Two Years with a World-Famous Detective Agency* (New York: J. S. Olgivie Publishing Co., 1912); Detective Agency Reports, Box 75, Bunker Hill Mining Company Papers, MG 367, UIdaho.

8. *Woodruff v. North Bloomfield Gravel Manufacturing Co.*, 18 Federal Reporter 753 (U.S. Circuit Court for the District of California, 1884); Robert L. Kelley, *Gold vs. Grain: The Hydraulic Mining Controversy in California's Sacramento Valley* (Glendale, Calif.: Arthur H. Clark Co., 1959). The courts enjoined the hydraulic mining companies from dumping debris into the rivers, making their operations financially unprofitable. Hydraulic mining continued in other areas of California and the West, however. And by the turn of the century, the mining industry had mounted a successful political campaign to reopen hydraulic mining operations in central California. For a review of mining pollution legal contests, see Gordon Morris Bakken, "American Mining Law and the Environment: The Western Experience," *Western Legal History* 1 (Summer/Fall 1988): 211–36; Smith, *Mining America*. On contemporaneous water pollution issues, see Glenn Harris and Seth Wilson, "Water Pollution in the Adirondack Mountains: Scientific Research and Governmental Response, 1890–1930," *Environmental History Review* 17 (Winter 1993): 47–65. On sources, see Susan Chambers, "Western Natural Resources: Documenting the Struggle for Control," *Prologue: Quarterly of the National Archives* 21 (Fall 1989): 239–45; Waverly B. Lowell, "Where Have All the Flowers Gone? Early Environmental Litigation," *ibid.*: 247–55.

9. R. L. Polk, *Idaho State Gazetteer and Business Directory for 1905–1906* (Portland: R. L. Polk and Co., 1905); see also R. L. Polk directories for 1903–4 and

1908–9; James H. Hawley, *History of Idaho* (Chicago: S. J. Clarke, 1920); *An Illustrated History of North Idaho Embracing Nez Perce, Idaho, Latah, Kootenai, and Shoshone Counties, State of Idaho* ([Denver]: Western Historical Publishing, 1903), 2 vols. Until the 1880s, the ranchlands of Elmer Doty and the other litigants, located along the lower portion of the Coeur d'Alene River, had been included within the boundaries of the Coeur d'Alene reservation. Prompted by the mineral discoveries, the federal government held a series of councils in the 1870s and 1880s through which the Coeur d'Alenes ceded their rights to the northern portion of the reservation. See Jack Dozier, "History of the Coeur d'Alene Indians to 1900" (M.A. thesis, University of Idaho, 1961); U.S. Department of the Interior, *Annual Report of the Commissioner of Indian Affairs to the Secretary of the Interior for the Year 1889* (Washington, D.C.: Government Printing Office, 1889), p. 12; 50th Cong., 1st Sess., S. Ex. Doc. 76, 1887–88, p. 6; Katherine G. Morrissey, *Mental Territories: Mapping the Inland Empire* (Ithaca: Cornell University Press, 1997), pp. 67–92.

10. Rickard, *A History of American Mining*, pp. 318–40; Mark Wyman, *Hard Rock Epic: Western Miners and the Industrial Revolution, 1860–1910* (Berkeley: University of California Press, 1979); Patricia Hart and Ivar Nelson, *Mining Town: The Photographic Record of T. N. Barnard and Nellie Stockbridge from the Coeur d'Alenes* (Seattle: University of Washington Press and Boise: University of Idaho Press, 1984); John Fahey, *Hecla* (Seattle: University of Washington Press, 1989).

11. *Coeur d'Alene Press*, 20 February 1892, p. 4. See also Olin D. Wheeler, *Sketches of Wonderland: A Land of Rolling Plains, Boundless Grain Fields, Sculptured Lands, Alpine Lakes, Mining Camps, Indian Life, Dancing Rivers, Thriving Villages, Trackless Forests, Growing Cities, Lofty Mountains Penetrated by the Northern Pacific Railroad* (St. Paul: Northern Pacific Railroad; Chicago: Rand McNally, 1895), pp. 20–22; James L. Onderdonk, *Idaho: Facts and Statistics Concerning Its Mining, Farming, Stock-Raising, Lumbering, and Other Resources and Industries, Together with Notes on the Climate, Scenery, Game, and Mineral Springs. Information for the Home-Seeker, Capitalist, Prospector, and Traveler* (San Francisco: A. L. Bancroft and Co., 1885); F. L. Miller, *Guide to the Great Coeur d'Alene Gold Field* (1884).

12. Historians have begun to reexamine promotional literature, once denigrated as a source because of its "factual inaccuracies." See, e.g., David Hamer, *New Towns in the New World: Images and Perceptions of the Nineteenth-Century Urban Frontier* (New York: Columbia University Press, 1990); William Cronon, *Nature's Metropolis: Chicago and the Great West* (New York: Norton, 1991), pp. 41–46; David Wrobel, "Becoming Western: Regional Identity and America's Promised Land," *Origins* 12 (Fall 1996): 5–6; Morrissey, *Mental Territories*, pp. 129–42.

13. John B. Leiberg, *General Report on a Botanical Survey of the Coeur d'Alene Mountains in Idaho during the Summer of 1895* (Washington, D.C.: Government

Printing Office, Contributions from the U.S. Herbarium, vol. 5, no. 1, 1897), pp. 5, 64, 61, 63.

14. *Timothy McCarthy, William Raney and Elmer Doty v. Bunker Hill and Sullivan Mining and Concentrating Co.*, Boxes 193–96, Case no. 1397, U.S. Circuit Court of Appeals, Ninth Circuit (San Francisco, Calif.), RG 276, National Archives and Records Center, Pacific Central Region (NARC-PCR), San Bruno, Calif.; *McCarthy v. Bunker Hill and Sullivan*, 147 Federal Reporter 981 (U.S. Circuit Court, District of Idaho, 1906), *affirmed as modified*, 164 Federal Reporter 927 (U.S. Circuit Court of Appeals, Ninth Circuit, 1908); Transcript of Record, *McCarthy v. Bunker Hill and Sullivan*, Box 54, Bunker Hill Mining Company Papers, MG 367, UIdaho. The injunction case is more frequently cited in the legal literature. See David C. Frederick, "The Ninth Circuit and Natural-Resource Development in the Early Twentieth Century," *Western Legal History* 6 (Summer/Fall 1993): 183–215; David C. Frederick, *Rugged Justice: The Ninth Circuit Court of Appeal and the American West, 1891–1941* (Berkeley: University of California Press, 1991), pp. 116–17; Monique C. Lillard, "The Federal Court in Idaho, 1889–1907: The Appointment and Tenure of James H. Beatty, Idaho's First Federal District Court Judge," *Western Legal History* 2 (Winter/Spring 1989): 35–78; "Annotation: Pollution of Stream by Mining Operation," 39 *American Law Reports, Annotated* 891–914 (1925). The ranchers had a reasonable expectation that they would succeed in their legal actions. Bunker Hill and Sullivan and other mining companies had earlier settled claims with other river residents, and in 1904 the Idaho District Circuit Court had ruled in *Charles M. Brown et al. v. Federal Mining and Smelting Co. et al.* that the mining companies should pay for the damages they caused to river land by the building of a dam (Box 31, Civil Case no. 307, U.S. Circuit Court, Northern Division [Moscow, Idaho], RG 21, NARC-PNWR).

15. Transcript of Record, *Doty v. Bunker Hill*, MG 130; Transcript of Record, Appellants Brief, Brief of Appellees, *McCarthy v. Bunker Hill and Sullivan*, RG 276; 164 Fed. 927. These variant descriptions echo those found in other local publications, such as promotional pamphlets. On ideas about nature and the environment, see Hans Huth, *Nature and the American: Three Centuries of Changing Attitudes* (Lincoln: University of Nebraska Press, 1957); Roderick Nash, *Wilderness and the American Mind* (New Haven: Yale University Press, 1982); Lee Clark Mitchell, *Witnesses to a Vanishing America* (Princeton: Princeton University Press, 1981); Simon Schama, *Landscape and Memory* (New York: Vintage, 1995). For historical examinations of mining and the environment, see Smith, *Mining America*; Randall Rohe, "Man and the Land: Mining's Impact in the Far West," *Arizona and the West* 28 (Winter 1986): 299–338.

16. On the relevant mining and water laws, see Curtis H. Lindley, *A Treatise on*

the American Law Relating to Mines and Mineral Lands (San Francisco: Bancroft-Whitney Company, 2d ed., 2 vols., 1903, and 3d ed., 3 vols., 1914); R. S. Morrison and Emilio D. DeSoto, *Mining Rights on the Public Domain* (Denver: Smith-Brooks Printing Co., 12th ed., 1905), pp. 210–15; Orville E. Jackson, *Idaho Mining Rights: A Legal Guide for Miners, Prospectors, Locators, Investors, and All Others Interested in the Rich Mineral Lands of Idaho. A Compilation to Date of all the Laws of the United States and of the State of Idaho Relating to Mines and Mining, Rights of Way, Easements, Tunnel Rights, Mill Sites, Mining Contracts, Coal Laws, Etc.* (Boise: Author, 1899); Idaho Irrigation and Drainage Code Commission, *Laws of the State of Idaho Relative to Public Waters, Water Rights and Irrigation . . .* (Boise, 1915); Bakken, "American Mining Law and the Environment"; Gordon Morris Bakken, *The Development of Law on the Rocky Mountain Frontier, 1850–1912* (Westport: Greenwood Press, 1983). As a delegate to the Idaho constitutional convention, prior to his judicial appointment, James Beatty had argued against giving preference to agricultural use of water; Dennis C. Colson, *Idaho's Constitution: The Tie that Binds* (Moscow: University of Idaho Press, 1991), pp. 161–77.

17. *Spokesman-Review*, 25 September 1904; *Doty v. Bunker Hill*, RG 21; Thiel Reports, Easton Papers, 25 September 1906.

18. List of Exhibits, *McCarthy v. Bunker Hill and Sullivan*, RG 276; Thiel Reports, Easton Papers.

19. Thiel Reports, Easton Papers, 4 November 1905; *Spokesman-Review*, 5 August 1906.

20. Randall H. Kemp, "Made an Odd Clean-up," in *A Half-Breed Dance and Other Far Western Stories: Mining Camp, Indian, and Hudson's Bay Tales Based on the Experiences of the Author* (Spokane: Inland Printing Co., 1909), pp. 25–33. In other tales, bears ingest minerals from rocks and water; Kemp, "An Original Bear Story," *ibid.*, pp. 16–24.

21. Transcript of Record, *Doty v. Bunker Hill*, RG 21; 164 Fed. 935, 937; Affidavits of Elmer Fulmer and Charles M. Fassett, Bunker Hill Company Papers, MG 5075, UIdaho; *Spokesman-Review*, 12 August 1906; F. W. Bradley to Stanly Easton, 31 August 1906, Box 10, Bunker Hill Mining Company Papers, MG 367, UIdaho. In thinking about scientific practice in this context, especially the social construction of scientific knowledge, I have found useful the following: Stanley Aronowitz, *Science as Power: Discourse and Ideology in Modern Society* (Minneapolis: University of Minnesota Press, 1988); Timothy Lenoir, *Instituting Science: The Cultural Production of Scientific Disciplines* (Stanford, Calif.: Stanford University Press, 1997); and the Women and Scientific Literacy colloquium, University of Arizona, 1997–98.

22. Transcript of Record, *Doty v. Bunker Hill*, RG 21; *McCarthy v. Bunker Hill and Sullivan*, 147 Federal Reporter 981 (U.S. Circuit Court, District of Idaho, 1906),

affirmed as modified, 164 Federal Reporter 927 (U.S. Circuit Court of Appeals, Ninth Circuit, 1908); Thiel Reports, Easton Papers, 1 June 1906. Beatty, for example, had ruled in favor of the mine owners in the injunction cases brought during the Coeur d'Alene labor-capital struggles of the 1890s. The MOA supported Beatty when he ran for U.S. Senate in 1906, and when he resigned in 1907 the mine owners lobbied for a favorable pro-business replacement. Bradley to Easton, 5 July 1906, 30 January 1907, Box 10, Bunker Hill Mining Company Papers, MG 367, UIdaho; Easton to Bradley, 20 March 1907, *ibid*.

23. Transcript of Record, Brief of Appellees, *McCarthy v. Bunker Hill and Sullivan*, MG 367; *McCarthy v. Bunker Hill*, RG 276; Barnard-Stockbridge Collection, UIdaho. On Barnard, see Hart and Nelson, *Mining Town*, pp. 8–9; *Illustrated History of North Idaho*, pp. 1068–69.

24. J. A. Todd Evidence Photographs, *Woodruff v. North Bloomfield*, no. 2900, U.S. Circuit Court, District of California, 1882, 1883, California Historical Society, San Francisco, Calif.; Martha A. Sandweiss, "Views and Reviews: Western Art and Western History," in *Under an Open Sky: Rethinking America's Western Past*, eds. William Cronon, George Miles, and Jay Gitlin (New York: Norton, 1991), pp. 185–202; Howard Bossen, "A Tall Tale Retold: The Influence of the Photographs of William Henry Jackson on the Passage of the Yellowstone Park Act of 1872," *Studies in Visual Communication* 8 (Winter 1982): 98–109; Susan Sontag, *On Photography* (New York: Farrar, Straus and Giroux, 1977); Alan Trachtenberg, *Reading American Photographs: Images as History, Matthew Brady to Walker Evans* (New York: Hill and Wang, 1989), esp. pp. 144–63.

25. *McCarthy v. Bunker Hill and Sullivan*, 147 Federal Reporter 981 (U.S. Circuit Court, District of Idaho, 1906), *affirmed as modified*, 164 Federal Reporter 927 (U.S. Circuit Court of Appeals, Ninth Circuit, 1908), *certiorari denied*, 212 U.S. 583 (U.S. Supreme Court, 1909); Transcript of Record, *Doty v. Bunker Hill*, RG 21; Lillard, "The Federal Court in Idaho"; "Farmers Lose Tailings Case," *Spokesman-Review*, 14 August 1906. The dismissal of the scientific evidence was in keeping with contemporary Ninth Circuit practices and regional biases. Local stories about chemists and assayers suggest the ambivalence with which such industrial scientists were regarded. See, e.g., Kemp, "Wanted: A Partner," in *A Half-Breed Dance and Other Far Western Stories*, pp. 34–47.

26. Thiel Reports, Easton Papers, 21–24 April 1906; 10 August 1906; 22 April 1908. Physical experiences, such as those of Operative 14, helped to shape mental constructs, narratives, and perceptions. Ranchers and miners, for example, both assimilated knowledge of nature and environmental processes through work, through their bodily presence and involvement with natural resources. Mine owners and photographers, a step removed from such direct tangible involvement, maintained a

somewhat different understanding of environmental processes. All, nonetheless, mediated their engagement through technology and employed cultural abstractions, words and images, to communicate. In thinking about the implications of these relations I have found useful the following: Richard White, *The Organic Machine: The Remaking of the Columbia River* (New York: Hill and Wang, 1995); Thomas W. Laquer, "Bodies, Details, and the Humanitarian Narrative," in *The New Cultural History*, ed. Lynn Hunt (Berkeley: University of California Press, 1989), pp. 176–204; William Cronon, ed., *Uncommon Ground: Toward Reinventing Nature* (New York: Norton, 1995); Kim Barnes, *In the Wilderness: Coming of Age in Unknown Country* (New York: Anchor Books, 1996).

27. Not surprisingly, not all the river ranchers agreed to sell their property; some sued for damages. See *Gaskill v. Washington Water Power Co.*, 105 Pacific Reporter 51 (Idaho Supreme Court, 1909); A. G. *Kerns v. Washington Water Power Co.*, 135 Pacific Reporter 70 (Idaho Supreme Court, 1913); Thiel reports, Easton Papers, 5 November 1907. Washington Water Power was in the process of acquiring a monopoly over water power and transmission lines in the interior Pacific Northwest; *Report of the Commissioner of Corporations on Water-Power Development in the United States* (Washington, D.C.: Government Printing Office, 1912), pp. 110–11; Leiberg, *General Report*, p. 42.

28. Affidavits of Joseph La Rocque and Edwin L. Mottern, *Doty v. Bunker Hill*, RG 21; Thiel Reports, Easton Papers, 10 June 1908; Verdict, *Doty v. Bunker Hill*, RG 21; Easton to Bradley, 28 November 1908, Box 10, Bunker Hill Mining Company Papers, MG 367, UIdaho. The court verdict also required the mine owners to pay the ranchers' legal fees, but this part of the decision was overturned. See Plaintiff's Brief on Motion to Strike Judgement for Costs, 10 November 1910, *Doty v. Bunker Hill*, RG 21.

29. Disputes over Coeur d'Alene mining pollution—particularly caused by tailings debris and smelter smoke—have continued throughout the twentieth century; see Katherine Aiken's chapter in this volume. In 1983, after the Bunker Hill Mine and Smelter complex ended its operations, the Environmental Protection Agency named a 21-mile area along the river a Superfund Cleanup site, and the legal battle over who is responsible for the costs continues today. There is also a pending lawsuit brought by the Confederated Coeur d'Alene Tribes against both the federal government and the mining companies for the pollution of Coeur d'Alene Lake. Publications on the post-1910 disputes include Fred W. Rabe and David C. Flaherty, *The River of Gold and Green* (Moscow: Idaho Research Foundation, 1974); Nicholas A. Casner, "Toxic River: Politics and Coeur d'Alene Mining Pollution in the 1930s," *Idaho Yesterdays* 35 (Fall 1991): 2–19; Nicholas A. Casner, "Leaded Waters: A History of Mining Pollution on the Coeur d'Alene River in Idaho, 1900–1950"

(M.A. thesis, Boise State University, 1989); Inland Empire Public Lands Council, " 'Valley of Death': A Series from the *Coeur d'Alene Press* 1929–1930," *Transitions* 6 (July/August 1994): 25–63; Katherine G. Aiken, " 'Not Long Ago a Smoking Chimney Was a Sign of Prosperity': Corporate and Community Response to Pollution at the Bunker Hill Smelter in Kellogg, Idaho," *Environmental History Review* (Summer 1994): 67–85; Cassandra Tate, "American Dilemma of Jobs, Health in an Idaho Town," *Smithsonian* 12 (September 1981): 74–83. Specific cases include *Christ Luama v. Bunker Hill and Sullivan et. al.*, Case no. 6109; *Bunker Hill and Sullivan et al. v. Jacob Polak*, Case no. 4461, RG 276, NARC-PCR.

30. For examples, see M. M. Ellis, "Pollution of the Coeur d'Alene River and Adjacent Waters by Mine Wastes," M. M. Ellis papers, MG 5156, Special Collections and Archives, University of Idaho Library, Moscow, Idaho; Peter N. Norbeck, Leland L. Mink, and Roy E. Williams, "Ground water leaching of jig tailing deposits in the Coeur d'Alene District of northern Idaho," and Bryson D. Trexler, Jr., Dale R. Ralston, William Renison, and Roy E. Williams, "The Hydrology of an Acid Mine Problem," in *Water Resources Problems Relating to Mining,* eds. Richard F. Hadley and David T. Snow (Minneapolis: American Water Resources Association, 1974), pp. 149–57, 32–40; William S. Platts, Susan B. Martin, and Edward R. J. Primbs, *Water Quality in an Idaho Stream Degraded by Acid Mine Waters* (Ogden, Utah: Intermountain Forest and Range Experiment Station, General Technical Report INT-67, 1979). See also Sheila Ann Dean, "Acid Drainage from Abandoned Metal Mines in the Patagonia Mountains of Southern Arizona" (M.S. thesis, University of Arizona, 1982).

31. See, e.g., William K. Wyant, *Westward in Eden: The Public Lands and the Conservation Movement* (Berkeley: University of California Press, 1982), esp. pp. 161–92; Donald Worster, *Rivers of Empire: Water, Aridity, and the Growth of the American West* (New York: Pantheon, 1985); Arthur F. McEvoy, *The Fisherman's Problem: Ecology and Law in the California Fisheries, 1850–1980* (New York: Cambridge University Press, 1986).

23 / Western Smelters and the Problem of Smelter Smoke

KATHERINE AIKEN

When an official publication of the Bunker Hill Company opined in 1978 that "not long ago a smoking chimney was a sign of prosperity," it acknowledged that the image of smokestacks, once a potent symbol of the industrial landscape, had become an equally powerful symbol of pollution. The tallest smokestacks sprouted from metal smelters. Not coincidentally, smelters were also at the center of environmental litigation early in the twentieth century. An examination of public and governmental responses to smelter pollution provides a lens for exploring changing attitudes over time.

Smelter smoke carried particulate matter composed in part of heavy metals. This accounts for a significant characteristic of smelter smoke—it is clearly visible, so laypeople are conscious of its presence. But even its invisible contents caused problems: sulphur dioxide (SO_2) and other forms of sulphur that are frequent smelter by-products damage plant and animal life. Consequently, farmers often complained of smoke's effects on crops and livestock. By the beginning of the twentieth century, smelter smoke had become a target of widely publicized and lengthy lawsuits that pitted farmers against large metallurgical companies. The federal government—as a landowner itself, as the arbiter for disputes across state lines, and as the seat of bureaucratic oversight for both mining and agriculture—was also drawn into the fray.

In this essay I survey the smelter smoke problem in the United States in the first half of the twentieth century. Then, using the Bunker Hill Company as a case study, I examine how both perceptions regarding pollution and public policy directed toward ending it changed after World War II.

By the time construction of primary metals smelters in the United States shifted into high gear during the late nineteenth and early twentieth centuries, the industrial countries of western Europe had already experienced the impact of smelter pollution. Scientists and engineers had spent much time and money searching for a solution to the problem of smelter smoke. This research formed the basis for similar efforts in the United States. The smelter pollution problem in North America, however, soon overshadowed the European problem because the size of the smelters built in the Western Hemisphere dwarfed those in Europe. In 1899, all of the smelters operating in Germany's lower Harz Mountains treated 63,000 tons of ore. Within a few years, the Anaconda smelter in Montana (at the time the largest of its type in the world) handled more tonnage every week.[1]

A number of smelter smoke disputes in the early twentieth century highlighted the complexity of the problem. When mining and agriculture operated in close proximity, smelter smoke controversies often resulted. Small farmers confronted some of the West's most powerful industrialists; governmental entities were often drawn into the disputes. Mining and smeltermen throughout the country followed these controversies with considerable interest. And the controversies had a tremendous impact on western smelters and led to significant improvements in smelter technology to limit smoke pollution.

ROUND ONE

California

The great Selby smelting plant on San Francisco Bay had an annual production of gold, silver, copper, and lead of more than $50 million; its payroll was $300,000 a year. Ores from Nevada, Alaska, Canada, Mexico, and Central and South America were shipped to the smelter because its location on San Francisco Bay made it easily accessible. It became a cause célèbre in the smelting world in 1906 when a jury awarded C. B. Deming $200 in damages for lost livestock. Deming's case was only the first of 22 similar cases pending in the Solano County court system. On July 16, 1908, the court enjoined the Selby plant from emitting fumes between March 25 and November 15, when the prevailing winds carried the fumes onto the farmers' land. The California

Supreme Court affirmed the injunction on June 12, 1912. Nonetheless, farmers and the county were back in court repeatedly, arguing that the company was not in compliance. This was expensive for everyone. The Solano district attorney's time was taken up, and the county spent $25,000 and the company $50,000 in court costs and attorneys' fees. When the powerful Guggenheim family—which controlled most of the smelting industry in the United States through its American Smelting and Refining Company (ASARCO)—purchased the Selby plant and decided to construct a larger smelter, the stakes increased.

Both sides desired to halt the constant litigation. As a result, the parties agreed to an investigatory commission in 1913. The Selby Smoke Commission performed scientific research on smelter fumes and their impact under field conditions. It found that SO_2 damage to vegetation was negligible but that lead had affected a dozen horses. The following year, the State of California created a State Smelter Waste Commission. It was given power to call upon the California attorney general to close smelters found to be causing damage. When the commission issued its report in December, however, it did not even mention the Selby smelter, claiming that wet weather had precluded a viable study of that site.[2]

There is little doubt that a company as powerful as ASARCO was able to withstand scrutiny, even if the commissions established that it had caused some damage. The commissions served to convince the public that something was being done, and they gave the companies an opportunity to explore new methods of dealing with fumes while continuing production.

Utah

ASARCO also faced smelter smoke problems in Utah. Smelters in Murray and Bingham Junction were located along the Jordan River, a rich agricultural area. Complaints against smelters began around 1902; an investigation in 1903 found that farms lying in the path of the prevailing winds were most likely to be damaged by sulphur dioxide fumes. Farmers and representatives from ASARCO, Utah Consolidated Mining Company, and United States and Bingham Consolidated met in September 1904 to discuss the situation. It appeared that an amicable solution could be found, but in December 1904, three farmers received a judgment against Utah Consolidated for $65,539.60 for damage from smelter fumes. *Engineering and Mining Journal* noted that "the war between the farmers of the Salt Lake Valley and several smelting companies is on in earnest." On November 5, 1906, Judge John A. Marshall of the United States District Court in Salt Lake City enjoined the companies from smelting ores containing more than 10 percent sulphur. This injunction served as a wake-up call for the companies. Utah Consolidated and Bingham Consolidated decided to abandon their smelters and relocate. ASARCO set out to find a solution. By March 1907 ASARCO had agreed to use new technology to decrease pollution and to pay the farmers $60,000 to purchase smoke easements.

Despite these efforts, ASARCO continued to face legal difficulties. Eventually the court developed the so-called sea captain principle of smelter smoke management. Damage to crops appeared to be greatest in the morning when the plants were "breathing" and when winds were from a certain direction. Consequently, the smelters agreed to shut down their roasters during unfavorable weather periods and to take advantage of good weather, just as any sea captain would.

Under the threat of injunction, smelting companies made some concessions to farmers, although the companies were able to continue production. In both California and Utah, courts limited smelter operations based on climatological conditions, in the belief that weather was an important variable in smoke dispersal. This remained a central idea for smelter operators.[3]

Montana

Smoke complaints in Montana received a different response, perhaps because the metals industry was the primary economic activity there; the Anaconda company effectively controlled the state. In 1905, Fred J. Bliss and other farmers sued the Washoe and Anaconda Companies, claiming that the companies' smelters had damaged their crops. While Anaconda Copper Mining Company admitted that the farmers had sustained some damage and agreed to pay for it, the company argued that the farmers had enormously overestimated their losses. Judge Oliver T. Crane, master in chancery of the United States District Court, ruled that there was indeed some damage. He reasoned, however, that there would be much larger damage if the smelters were not allowed to operate. He noted that in fact the farmers' main markets were Butte and Anaconda and that if there was no mining and smelting, the farmers would have no market. He concluded that damages were only $350.

The farmers appealed to President Theodore Roosevelt, who had a reputation as a conservationist as well as a special place in his heart for Montana. Roosevelt directed experts from the agricultural and forestry departments to assist the farmers. In January 1909, a federal court granted Fred Bliss $450 a year, retroactive to 1904. The United States attorney general brought yet another suit; experts investigated yet again. The United States Circuit Court of Appeals sided with the companies, denying Bliss's request for a permanent injunction. Smelter operators throughout the country rejoiced in the victory and portrayed it as a commonsense approach to the economic realities of smelters.[4]

Tennessee and Georgia

While small farmers led the fight against smelter pollution in California, Utah, and Montana, state government assumed leadership in combating sulphur dioxide pollution from two copper smelters in Ducktown, Tennessee. When the smelters constructed taller smoke stacks, the smoke drifted into neighboring Georgia's broad-leafed forests. The State of Georgia responded by bringing suit against Tennessee to compel it to cancel the smelters' licenses. Because the case involved two states, it was heard directly by the United States Supreme Court. The federal government actively participated in the ensuing investigations that found definite smoke damage. In May 1907, the Supreme Court ruled that Georgia had the right to protect its forests and could force the closure of the Tennessee-based smelters. The Court rejected the companies' argument that Georgia had no right to bring suit where only individuals were concerned. The Ducktown experience thus marks a crucial shift: the Supreme Court held that states had the power to act to protect their resources. The decision also served as a warning to smelter companies that they could be held responsible for damage that occurred at a considerable geographic distance from their plants.[5]

Impacts of the Early Litigation

Although these smoke cases did not stop smelter production, the litigation and accompanying publicity did convince smelter companies to invest in scientific research seeking a solution to the problem. Smoke engineering research—combining chemistry, plant physiology and pathology, entomology, agriculture, veterinary medicine, and meteorology—became a sought-after specialty in metallurgical circles. ASARCO was a leader in funding research efforts. The company established a research center in Salt Lake City and developed several large experimental farms in order to study the effects of smelter smoke on agriculture. It also sponsored a large research laboratory at Stanford University under the direction of Robert E. Swain. The research led to several new technologies that came into use before World War I.

Baghouses were one approach. Simply put, smelter smoke and fumes were filtered through woolen or cotton bags designed to catch any nongaseous materials. The bags were suspended from the ceiling of a building—hence the term *baghouse*. Baghouses not only reduced lawsuits but also produced some return on the investment, because the metals caught in the bags could be resmelted. At the same time, baghouses improved working conditions inside smelting plants.

The first large-scale use of a baghouse took place at Denver's Globe Smelting Company in 1887, but the materials were too hot and the bags caught fire. Refinements soon resulted in practical applications of baghouse technology. The United Smelting, Refining, and Mining Company's plant at Midvale, Utah, completed in 1902, had the first truly practical baghouse. ASARCO used a baghouse to resolve the Utah smoke cases. Its Murray baghouse, costing $150,000 to construct, went into operation in July 1907. At that time it was the largest baghouse in the world: its 4,160 bags were each 30 feet long and 18 inches in diameter and used some 15 acres of material. The bags were housed in a building measuring 100 by 216 feet, and smoke from the smelter was forced into the baghouse with a fan capable of moving 250,000 cubic feet of gas per minute.[6] The baghouse allowed ASARCO to renegotiate its agreement with farmers; instead of being restricted to ores with 10 percent sulphur, ASARCO was able to smelt ores with a maximum of 25 percent sulphur and to mix high-sulphur-content ores with lower-grade ores when smelting. Baghouses allowed the companies to keep up the pace of their production.

The Tennessee cases led to the development of sulfuric acid plants, which converted sulphur oxide wastes into sulfuric acid. Enough sulphur was removed from smelter smoke this way that farmers no longer complained. Sulfuric acid plants were a practical solution to smelter smoke difficulties and often provided smelting companies with economic benefits as well, since sulfuric acid was a primary ingredient in fertilizers. The Tennessee Copper Company itself began to farm land near its smelter, both to show that the gases were no longer dangerous and to demonstrate the efficacy of sulfuric acid fertilizers.[7] Sulfuric acid plants remained the primary technological solution to SO_2 emissions into the 1980s.

Another alternative was an electrostatic precipitator, the product of chemist F. G. Cottrell's intensive research at the University of California. Powerful electrical discharges between electrodes placed in the flue attracted solids and the gaseous sulphur. This electrical precipitation process made gases diffuse more easily when they left the smokestacks, because they no longer carried solids and liquids. The Anaconda smelter added a large "Cottrell" in 1910 and limited smoke damages to neighboring farms.[8]

Finally, taller smokestacks were another technological solution to the problem of smelter smoke and fumes. If smoke left the smelter at a higher elevation, it would disperse over a wider geographical area, and the gases would be diluted by the time they reached the ground. Thus they would do minimal damage to vegetation and livestock. By 1917, the stack at Great Falls, Montana, was 506 feet high.[9] Companies studiously referred to these edifices as "stacks" rather than "smokestacks" in an effort to divert attention from the ubiquitous smelter smoke.

Smelter operators installed one or more of these technologies. By the start of World War I, the companies believed they had done what they could and adopted a tougher stance in combating damage claims. The importance of smelter products to the war effort played into company hands. The pro-business atmosphere in government during the 1920s was also very different from the conservation concerns that had been evident during the progressive era of the prewar period.

ROUND TWO

Decreases in metal production due to the Great Depression limited smelter smoke controversies during the 1930s. The tremendous demand for metals during World War II and general attention to the war also reduced the intensity of the smelter smoke controversy. After the war, however, the increased industrial production that was part of the postwar boom once again brought smelter pollution problems to the fore.

The Donora, Pennsylvania, disaster of October 1948 came to symbolize the new situation; it has become a legend in twentieth-century environmental history. The Donora Zinc Smelter, owned by the American Steel and Wire Company, was a World War I–era plant. From its construction, farmers had sued and the company had paid damages. On October 26, 1948, a smog enveloped the area; by October 30, 17 people had died and 5,910 were ill. A full-scale investigation followed.

Before the Donora smog, health officials had not considered air pollution to be an urgent issue. The scope of the disaster, however, forced the United States Public Health Service to take notice and led to a public health approach to air pollution. The health service began a program of air quality sampling designed to measure air quality against a norm, with an eye toward protecting the public. Donora thus marked the beginning of a shift from source-based research to health-effects research as the basis for public policy.[10] The effects of this paradigm shift were dramatic, and during the remainder of the century smelting companies found themselves increasingly under siege.

An early example of this new attitude was the passage in 1955 of Public Law 159, regulating air pollution. During the 1950s, the mining industry also addressed the pollution issue more fully, stating publicly that pollution control had economic benefits because it improved the environment, reduced maintenance costs, and led to the recovery of valuable metals. At the end of the decade, *Engineering and Mining Journal* featured a major article devoted to pollution control that offered a blueprint for the industry's response to pollution concerns. Associate editor James W. Franklin pointed out that whereas air pollution had cost each American $10 in 1949, the price was up to $65 in 1959. Consequently, he wrote, it was incumbent upon industry to make a case that it was working on the problem. Franklin noted that the metals industry was spending 5–10 percent of its capital investment (more than $300 million annually) on air pollution control, and he urged companies to publicize these efforts more aggressively. He also put forward another standard industry response—that there were other causes of air pollution, particularly automobile exhaust.[11] Metallurgical companies took Franklin's recommendations to heart and sought to publicize their concern with pollution and their considerable financial investment in solving pollution problems.

During the 1960s, environmental concerns came increasingly to the fore. A general reformist outlook accounts in part for this, as does scientific research proving the dangers of pollution. Metallurgical companies responded with the strategies that had served them well earlier in the century while relying on their considerable financial clout to protect them from interference. The balance of power was changing, however, as the Donora episode foretold. The post–World War II scenarios did not feature large companies against groups of small farmers, perhaps aided by government experts. Instead, the entire public was perceived to be endangered by the drive for large corporate profits. The federal government's role changed from one of arbitration and research to one of promulgation and enforcement of regulations.

The West became the primary battleground in the smelter smoke wars. Smelters in the region accounted for most of the country's primary metals production and most pollution as well. For example, by 1972 American smelters emitted 1,920,000 tons of sulphur oxides into the atmosphere each year, and 97 percent of these emissions came from western smelters.[12]

BUNKER HILL

Kellogg, Idaho's, Bunker Hill Company is a celebrated example of the recent history of the relationship among smelters, governments, and communities. The company's smelter was constructed in 1917. Company officials chose the Kellogg location, with its small population, partly to avoid the smoke litigation problems that had plagued the industry. The Bunker Hill smelter was state of the art and included both a baghouse and a Cottrell electrostatic precipitator system. Furthermore, the company entered into an agreement with the United States government that damage to trees in federally owned forests would not lead to litigation. Despite these precautions, complaints of smelter smoke damage began almost as soon as the smelter started operation. The addition of a zinc plant to the Bunker Hill complex in 1928 exacerbated the problem. The company's response followed the time-tested pattern of the rest of the industry: fight damage claims in court, seek technological solutions, try to assuage damages with financial offers—but continue production at any cost. By the 1970s, the Bunker Hill Company accounted for about 20 percent of the lead and zinc produced in the United States and 25 percent of the refined silver.[13]

As public awareness of environmental issues grew, the Bunker Hill Company stepped up expenditures for pollution control; during the 1960s, these expenditures totaled $8 million. The company also launched an advertising campaign highlighting "Our Clean Air Program." Company efforts paid off when the Idaho Board of Health and the Pacific Northwest Pollution Control Association both presented awards to Bunker Hill in recognition of its pollution control efforts. By November 1969, personnel director Ray Chapman was able to promise visiting state lawmakers and the North Idaho Chamber of Commerce, "I can tell you factually and truthfully that by 1971, the bulk of environmental problems in the Kellogg area attributed to Bunker Hill operations will be but a distasteful memory."[14]

The company's predictions proved overoptimistic. In 1970, Congress passed the Clean Air Act and established the Environmental Protection Agency (EPA). These developments marked the beginning of an increased federal involvement with Bunker Hill.

The company's response to the increased regulatory activity went through three phases. During the first, Bunker Hill often attempted to cooperate with government officials while it evaluated its options under the new laws. Between 1973 and 1977, company officials became less cooperative as they felt inundated by regulations. By 1978, Bunker Hill often refused to cooperate and became openly adversarial toward the regulatory agencies.

With the passage of the Clean Air Act in 1970, the Environmental Protection Agency and various Idaho regulatory agencies began to direct more attention toward Bunker Hill. A State Department of Health and Welfare letter to the company in June 1971 indicated that higher lead levels were measured in Kellogg than near an ASARCO plant in El Paso, Texas, that had been the subject of national attention. Consequently, the department launched studies of lead in the air, vegetation, and soil and implemented a screening program for residents. Bunker Hill public pronouncements denied any analogy with the El Paso scenario, but privately company officials were sufficiently worried to ask local physician Ronald K. Panke to conduct tests on area children. Although the results of these urinalyses indicated no elevated lead levels, the children's parents were not informed that the tests had been done.[15]

At the same time, the company was forced to delay the start-up of a new sulfuric acid plant when it could not market the acid. President Frank G. Woodruff wrote a letter to company employees explaining the situation and asking for patience in obtaining air quality improvement. The company also sought an Idaho Pollution Control Commission variance from state SO_2 requirements and constructed a $40,000 weather station designed to monitor atmospheric conditions and SO_2 levels. Even when the new sulfuric acid plant finally opened in August, problems continued to plague company officials. By October, the state filed a complaint against Bunker Hill for "not achieving the highest and best practical treatment and control."[16]

Despite these setbacks, management continued to insist that Bunker Hill was close to achieving its environmental goals. Houston-based Gulf Resources (Bunker Hill's parent company) claimed: "The enormous investment in pollution control equipment over the last several years has provided the company a viable smelter complex at a time when economic and environmental influences have brought about the closing of a number of other U.S. smelters."[17]

The company also embarked on a strategy of pitting state environmental agencies against the EPA. Federal law required states to develop State Implementation Plans (SIPs) that met federal air quality standards. Idaho's proposed SIP mandated 85-percent retention of SO_2.[18] In May 1972, the EPA disapproved the state plan and substituted its own 96.5-percent retention requirement. After hearing testimony, the state lowered its requirement to 82 percent and adopted the SIP as a formal regulation. The EPA took the matter under advisement.[19]

By 1973, environmental pressures converged on the Bunker Hill Company, and it became less cooperative. The company announced an Intermittent Curtailment System (ICS) to deal with SO_2 emissions not captured by the sulfuric acid plant. Reminiscent of the "sea captain" strategy developed earlir in Utah, the ICS required careful monitoring of weather and production conditions. When pollution levels threatened to exceed established standards, the company would cease operation. Although this was a considerable financial sacrifice, Bunker Hill claimed it was the only available option. Company officials believed that with their existing equipment they could meet the standard about 345 days a year. On the remaining days they would have to curtail operation.[20]

Disaster disrupted Bunker Hill plans on September 3, 1973, when a fire broke out in the smelter baghouse, the main emission control. Two of the seven sections of the baghouse suffered significant damage, including destruction of part of the roof. This created a difficult problem: without the roof there was no place to attach the cloth bags that caught the smelter emissions. The company decided to continue operations in order to take advantage of high metals prices and to use the remaining portion of the baghouse to control fumes. Although officials hurried to repair the damage, construction delays and the unavailability of replacement bags delayed their efforts. Baghouse section 7 was not replaced until November 3, 1973; section 6 remained inoperative until March 17, 1974. This situation placed added stress on the remaining sections, which had to compensate for the damaged areas; consequently, bags in these sections wore out more rapidly than usual.[21]

Despite this setback, the company remained optimistic at the end of 1973. Bunker Hill president James Halley wrote Gulf Resources president Woodruff that the company was "in the ball park with EPA" and was making progress in meeting regulatory requirements. Halley's statement that "Bunker Hill is and will continue to spend substantial sums of money to bring all facilities into compliance with meaningful standards and will contest any unreasonable application of the regulations" defined company policy for the remainder of the decade.[22]

The year 1974 was a pivotal one. In the wake of the baghouse fire, the airborne lead situation became more serious. Two Kellogg area youths were hospitalized with symptoms of lead poisoning. Dr. James A. Bax, director of the Idaho State Department of Health and Welfare, investigated and reported that average ambient lead levels in the Kellogg-Smelterville area had increased dramatically—from 3.9 micrograms per cubic meter in 1971 to 8.6 in 1972 and 13.2 in 1973. Bunker Hill's in-house publication claimed that the company was "completely unprepared when the State Department of Health and Welfare released the preliminary results of their blood lead level test of over 1000 children residing in Shoshone County." Blood samples from 1,047 children under the age of nine showed that 98 percent of the children living within a mile of the smelter had potentially dangerous amounts of lead in their blood—more than 40 micrograms of lead per 100 milliliters of blood. Forty-four children had levels high enough—above 80 micrograms—to be defined as lead poisoning.[23] The state sought to have the Centers for Disease Control launch a study of the lead situation. Local officials objected to state interference. In keeping with its strategy of limiting federal involvement, Bunker Hill offered to provide half the funds for a state study—the Shoshone County, Idaho, Lead Study—and the state accepted.[24]

At the same time, Bunker Hill began to lobby for a revision in the sulphur emissions compliance dates from the state. The company also stepped up control measures in Kellogg. It began to purchase homes near the smelter, demolish them, and revegetate the land; it also made free gravel and topsoil available to landowners wishing to replace lead-contaminated ground. The company's director of environmental control, Gene Baker, wrote to all area landowners warning them of the risks of pasturing livestock in the vicinity of Bunker Hill's plants. In January 1975, the company announced 26 short-term corrections designed to reduce smelter emissions and an additional $1.3 million in smelter modifications. That same month the company also reported that ambient lead levels had dropped from 29.8 micrograms in October 1973 to 20.4 micrograms in October 1974.[25]

The EPA expressed a lack of confidence in the Intermittent Control System (renamed the Supplemental Control System) for SO_2 emissions. Although state officials tried to help convince EPA to accept supplemental controls, state regulators also began to question company motives. When Bunker Hill attorney Bob Brown was quoted as saying that the company would use "every trick in the book to fight us on this matter," James Bax objected strenuously to Gulf Resources president Frank Woodruff. Bax complained of an "unreasonable lack of cooperation" and argued that since the state was trying to help Bunker Hill with EPA, it was entitled to some cooperation in return. Halley responded that his company was making an effort to improve things, but although they had promised to capture 85 percent of SO_2, the best they could do was 70 percent. The Supplemental Control System—with its suspension of smelting operations—was not economically viable when the difference between the emission target and the actual amount of SO_2 captured was so great.[26]

New smoke regulations were proposed and discussed at a public meeting in Kellogg on September 10. There were two potential approaches—cutbacks in production to begin meeting state standards by July 1975, and the installation of permanent control equipment to be in complete compliance by July 1977. An alternative proposal, favored by Bunker Hill, called for no interim deadlines but required the company to meet the July 1977 compliance date. President Halley warned the EPA that refusal to accept the state SO_2 plan would destroy mining in Idaho.[27]

The company did continue to make improvements, including a $360,000 expenditure for two hoods over the smelter blast furnaces. Nevertheless, in February 1975 the EPA disapproved that portion of Idaho's compliance schedule regulating SO_2 emissions from Bunker Hill.

Bunker Hill initiated a publicity campaign designed to portray the EPA as responsible for hamstringing legitimate company and state efforts to improve the environment. President Halley wrote James Bax, "I am personally troubled by the continued attempts of EPA to force us to experiment with unproven and in many cases completely untried systems suggested by their staff or their consultants." Halley urged Bax to stick with the state plan, claiming that further delay would prevent Bunker Hill from meeting the established state goals by the mid-1977 target date. Bunker Hill was convinced, Halley wrote, that no smelter in the United States could meet the EPA standard.[28]

While the EPA advocated sulphur burners or other permanent controls, Bunker Hill consistently claimed that these methods were untried and unproven. In their place, company consultants suggested the construction of a 715-foot stack at the lead smelter and a 610-foot stack at the zinc plant, designed to better disperse smelter emissions. Bunker Hill believed that the stacks would enable it to meet the 1977 state guidelines, and company officials claimed that EPA obduracy was delaying the beginning of stack construction and thus endangering the Kellogg economy. Gene Baker wrote to EPA hearing officer Theodore R. Rogowski that "the EPA's proposal is the only remaining barrier to the taking of final steps to ensure that the sulphur dioxide ambient air quality standards are met in Kellogg area without excessive curtailment of Bunker Hill's operations."[29]

During December 1975, the smelter was shut down seven days for supplemental control. These curtailments and others in January 1976 cost Bunker Hill over $3.5 million in lost revenues; employees lost 2,000 person-shifts of work. Gene Baker wrote to Dr. Lee Stokes that the shutdown had resulted in the furlough without pay of 150 workers. Consequently, these workers earned 25 percent less during the month of December. Baker warned, "In the future, there must be a better realization of the humanistic responsibility to individuals within the society the regulatory agencies were created to serve."[30] Many Kellogg residents echoed Baker's view.

The EPA, nevertheless, continued to argue that Bunker Hill could meet the 82-percent permanent control goal through the addition of new experimental equipment. EPA estimated that it would cost about $830,000 to install this equipment and about $550,000 to operate it. Halley's response was that the EPA was on record as saying that sulfuric acid plants were the "latest available technology" for sulphur dioxide control. Bunker Hill consultants had examined the company's sulfuric acid plants and found that they were operating at "peak efficiency." The EPA thus was asking for additional untried, experimental technology.

In a further initiative, Bunker Hill sought judicial review of the EPA SO_2 standard, contending that EPA had acted arbitrarily and capriciously in overturning the Idaho plan and promulgating its own regulation.[31] The company promised to carry its fight all the way to the United States Supreme Court. It also announced a delay in construction of the tall stacks—blaming EPA for creating a climate of uncertainty and jeopardizing the Kellogg area's economy and health. Bunker Hill officials touted the company's earnest efforts to meet EPA standards and claimed that the agency was "unreasonable."[32]

Bunker Hill faced a dilemma. A long delay in tall stack construction while the EPA ruling was in the courts would guarantee the company's inability to meet the state's 1977 SO_2 standards. Following a complete economic feasibility study and input from company lawyers, Bunker Hill announced on June 1, 1976, that it would go ahead with tall stack construction. Idaho governor Cecil Andrus praised the decision, and State Department of Health and Welfare officials deemed the company's plan acceptable. The EPA, on the other hand, continued to claim that Bunker Hill could capture more SO_2 through improved acid plant design. Rank-and-file State of Idaho employees also expressed doubts regarding the effectiveness of the tall stacks.[33]

Stack construction was estimated to cost $11 million; the stacks were scheduled for operation by July 1977, the original date for the SIP. *Engineering and Mining Journal* accurately described the Bunker Hill decision as a "gamble." According to the journal, Bunker Hill was gambling on "judicial reason" in applying the EPA standards. The company was also gambling that the $11-million expenditure would allow the smelter and zinc plant to continue operation long enough to recoup the costs.[34]

In July, the Ninth Circuit Court of Appeals held that EPA had not "exercised a reasoned discretion." The court refused to force Bunker Hill to meet the EPA standard and remanded the issue to the EPA for further consideration.[35]

This victory came amid increased difficulties on the environmental front. Although the preliminary report of the Shoshone County, Idaho, Lead Study had found "no permanent clinical impairments or illness," Billy and Marlene Yoss's three children and Edward and Janis Dennis's six children sued Bunker Hill for $20 million for health problems and loss of potential wages due to high blood levels of lead stemming from the 1973 baghouse fire. In August, the EPA proposed a new lead standard of 1.5 micrograms per cubic meter of air.[36]

By 1978, Bunker Hill had adopted a decidedly adversarial approach to both state and federal environmental regulators. The chief of the state's Bureau of Air Quality informed his superiors that "basically, the company has formally terminated our era of cooperation in SIP development." When the EPA issued an order requiring Bunker Hill to control emissions from the zinc fuming furnace, the company claimed that the technological changes EPA demanded were very expensive and would yield little result. James Halley argued that the central baghouse at the fuming plant represented all "reasonable precautions." The company opted to ignore the order. EPA filed a civil suit in U.S. District Court seeking injunctive relief and a civil penalty of $25,000 a day for every day the company failed to comply. Bunker Hill decided to continue operation and let the EPA press the suit.[37]

Bunker Hill officials voiced their opinion that the company was a victim of capricious regulations. The company reported that between 1968 and 1977, it had spent nearly 60 percent of its $37-million capital expenditures on pollution control, and it still wasn't enough. The company also argued that the proposed EPA ambient lead standards would force most American smelters to close.[38]

EPA criticism of the tall stacks and the entire Supplemental Control System continued, especially during January when a record-breaking cold wave hit and forced the company to shut down the smelter. President Howard pointed to the resulting damage to the smelter as proof that variances were needed for exceptional circumstances. EPA countered that the stacks and the Supplemental Control System simply were not effective. By June, however, Bunker Hill and EPA signed a document titled "Settlement Agreement and Interim Control of SO_2 Emissions." The agreement ended the SO_2 litigation filed in 1975 and represented another victory for the company, because it allowed the company to operate under the standards it had been meeting since the tall stacks were completed. In addition, the settlement included provisions for excess emissions during periods of start-up, shutdown, and equipment malfunctions. The company agreed to a research project to develop technology for permanent SO_2 control. The sulphur dioxide litigation was at least temporarily resolved.[39]

Company optimism that an end to environmental difficulties was at hand proved short-lived. In June, the U.S. Court of Appeals for the District of Columbia affirmed the EPA ambient air quality standard for lead of 1.5 micrograms. Bunker Hill officials claimed that the company would have had difficulty in meeting even the 5-microgram standard that the lead industry wanted. The Yoss case went to trial in Boise in January 1981, and after more than a month of testimony the two sides settled for $2 million. Most people in Kellogg viewed the decision as a "rip-off" of Bunker Hill. When the plaintiffs characterized Kellogg as an undesirable place to live, 4,337 people in northern Idaho signed a protest petition.[40]

Conflicts between the State of Idaho and the EPA also continued. Bunker Hill was convinced that the distance between Idaho and Washington, D.C., caused EPA officials to believe they could act without fear of oversight. For example, in early 1980 EPA administrator Douglas Costle announced that the agency was embarking on a two-year study of the lead problem, but Donald Dubois, EPA regional administrator, and Milton Klein, State of Idaho health and welfare administrator, demanded an immediate Bunker Hill compliance plan. Bunker Hill president J. W. Kendrick and Gene Baker traveled to Washington to meet with EPA officials; they argued that Bunker Hill was being separated from the rest of the lead industry. They "demanded that Costle evaluate the impact of the standard on the entire lead industry before singling out Bunker Hill." One Bunker Hill official asked Dubois, "Does the right know what the left hand is doing?" Gulf Resources CEO Robert Allen complained bitterly to Costle and sent blind carbon copies of his letter to Texas senator Lloyd Bentsen and Idaho senator James A. McClure and representative Steve Symms. In a letter to employees, Allen warned that if government regulation remained "unchecked," it "will surely result in loss of our economic freedom and when it does, loss of our political freedom will soon follow."[41] Many Idahoans shared Allen's viewpoint, fearing that their economic well-being was threatened by the uninformed attitudes of far-away bureaucrats.

And then, abruptly, Gulf Resources announced the closure of the Bunker Hill mine and smelter complex. Following almost 100 years of operation, the shutdown was completed in early 1982. More than 2,000 people were thrown out of work.[42] Gulf's reasons for the closure were low metal prices, a lack of available concentrates to purchase, and the burden EPA standards placed upon the company. Gulf Resources officials claimed that had they not attempted to meet the 1.5 microgram lead limit—a goal they were unable to achieve—they could have spent that money developing a lead-zinc mine that would have provided additional raw materials.

In the face of the shutdown, EPA agreed to five years of "business certainty to the company." But the regulatory concessions came too late. Forces put in motion during the 1970s had rapidly overtaken not only the Bunker Hill Company but also government regulatory agencies and area residents.

The EPA identified the 21 acres surrounding the smelter as a Superfund Cleanup site; the legal battle over who is responsible for clean-up costs continues. It was easy for people in Kellogg and for Gulf Resources to blame meddling government officials for the Bunker Hill closure. Clearly, regulators had often issued contradictory, constantly changing, and unclear messages to the company. Despite Bunker Hill's efforts to comply, to ignore, and to challenge the regulatory requirements, it was unable to keep pace.

CONCLUSION

The Bunker Hill experience raises as many questions as it answers. The closure of one of the nation's major lead-producing operations served no one's interest. There is uncertainty over whether or not the site can ever be cleaned up; progress in that direction has been slow. Perhaps the most notable event has been the removal of the tall stacks that symbolized Bunker Hill's environmental control efforts. No one knows what the long-term effects of Bunker Hill pollution will be on workers and community. Meanwhile, residents are left without an economic base and with a community that has been declared an environmental disaster area.

Similar scenarios have played out throughout the West as pollution cleanup costs have escalated.[43] The smelters at Selby, Anaconda, and Salt Lake are closed. Legal battles to determine who is responsible for cleaning up the residual pollution continue. Some western mines currently send their ores to Canada and Japan for smelting, raising the issue of whether or not Americans are exporting pollution to other countries. The EPA continues to be a target of considerable western opposition. The problem of finding a balance between the needs of the community for a clean environment and the needs of individuals to make a living has yet to be solved.

NOTES

1. Robert E. Swain, "Smoke and Fume Investigations: A Historical Review," *Industrial and Engineering Chemistry* 41 (November 1949): 2384–88. For a general discussion of the smelter pollution problem, see Duane A. Smith, *Mining America: The Industry and the Environment, 1800–1980* (Lawrence: University Press of Kansas, 1987), pp. 75, 94–99, 100–101.

2. Ligon Johnson, "History and Legal Phases of the Smelting and Smoke Problem—I," *Engineering and Mining Journal* 103 (May 19, 1917): 879–80; "Report of the Selby Smelting-Smoke Commission," *Engineering and Mining Journal* 98 (December 19, 1914): 1075–78; U.S. Bureau of Mines, *Report of the Selby Smelter Commission* (Washington, D.C.: Government Printing Office, Bulletin no. 98, 1914); *County of Solano v. Selby Smelting and Lead Company*, Superior Court of Solano County (August, 1909).

3. See Ligon Johnson, "History and Legal Phases of the Smelting-Smoke Problem—II," *Engineering and Mining Journal* 103 (May 26, 1917): 924–26; "Swain Reports on Smelter Smoke in Salt Lake Valley," *Engineering and Mining Journal* 111 (March 19, 1921): 519–20; *American Smelting and Refining Company v. Godfrey*, 158 Federal Reporter 225 (U.S. Circuit Court of Appeals, Eighth Circuit, 1907); *Anderson v. American Smelting and Refining Company*, 265 Federal Reporter 928 (U.S. District Court for the District of Utah, 1920); Walter R. Ingalls, "Lead and Copper Smelting at Salt Lake—I," *Engineering and Mining Journal* 84 (September 21, 1907): 527–31; Ernest E. Thum, "Smoke Litigation in Salt Lake Valley," *Chemical and Metallurgical Engineering* 22 (June 23, 1920): 1145–50.

4. *Bliss v. Anaconda Copper Mining Co.*, 167 Federal Reporter 342 (U.S. District Court for the District of Montana, 1909); J. K. Haywood, *Injury to Vegetation and Animal Life by Smelter Wastes* (Washington, D.C.: U.S Government Printing Office, 1910); Charles H. Fulton, *Metallurgical Smoke* (Washington, D.C.: U.S. Government Printing Office, Bureau of Mines Bulletin no. 84, 1915). For a discussion of Anaconda's dominating role in Montana history, see Michael P. Malone, Richard B. Roeder, and William Lang, *Montana: A History of Two Centuries* (Seattle: University

of Washington Press, rev. ed., 1991), pp. 231, 323. Sméltermen often referred to the farmers involved in these lawsuits pejoratively as "smoke farmers," to indicate that the farmers' main source of income was their lawsuits against smelters, not their agricultural pursuits.

5. Charles Baskerville, "Legal Status of Works Producing Noxious Gases," *Engineering and Mining Journal* 87 (May 1, 1909): 884–7; *Ducktown Sulphur, Copper and Iron Co. v. Barnes*, 60 Southwestern Reporter 593 (Tennessee Supreme Court 1900).

6. *Engineering and Mining Journal* 84 (July 20, 1907): 132.

7. Swain, "Smoke and Fume Investigations," p. 2385; W. H. Freeland and C. W. Renwick, "Smeltery Smoke as a Source of Sulfuric Acid," *Engineering and Mining Journal* 89 (May 23, 1910): 1116–20.

8. F. G. Cottrell, "Recent Progress in Electrical Smoke Precipitation," *Engineering and Mining Journal* 101 (February 26, 1916): 385–92; "The Research Corporation and the Cottrell Process, *Engineering and Mining Journal* 97 (May 30, 1914): 1107–9; Swain, "Smoke and Fume Investigations," p. 2385.

9. "Tall Chimneys in Metallurgical Plants," *Engineering and Mining Journal* 106 (July 27, 1918): 168–70.

10. James W. Franklin, "Air Pollution: Industry's Challenge," *Engineering and Mining Journal* 160 (July 1959): 68; Lynne Page Snyder, "The Death-Dealing Smog over Donora, Pennsylvania: Industrial Air Pollution, Public Health Policy, and the Politics of Expertise, 1948–1949," *Environmental History Review* 18 (Spring 1994): 117–38.

11. Franklin, "Air Pollution: Industry's Challenge."

12. H. R. Jones, *Pollution Control in the Nonferrous Metals Industry 1972* (Park Ridge, N.J.: Noyes Data Corp., 1972), p. 1.

13. Cassandra Tate, "American Dilemma of Jobs, Health in an Idaho Town," *Smithsonian* 12 (September 1981): 74–83. For a discussion of the Bunker Hill Company and lead pollution, see Katherine G. Aiken, " 'Not Long Ago a Smoking Chimney Was a Sign of Prosperity': Corporate and Community Response to Pollution at the Bunker Hill Smelter in Kellogg, Idaho," *Environmental History Review* 18 (Summer 1994): 67–86.

14. See "Smelter Smoke to Be Captured," *Kellogg Evening News*, 24 April 1970; "New Pollution Control Project On at Bunker Hill," *Kellogg Evening News*, 26 May 1970; "The Battle of Bunker Hill," *Pacific Building and Engineer* (August 21, 1970): 332–34; "End to Pollution Problem Expected by Bunker Hill," *Spokane Daily Chronicle*, 21 May 1971.

15. "Lead Testing Was 'Unreliable,' " *North Idaho Press*, 24 September 1981.

16. President Frank G. Woodruff, "Memo to All Employees," 28 January 1971, Bunker Hill Company Records, Manuscript Group no. 367, Special Collections,

University of Idaho Library (hereinafter cited as BHR); Bunker Hill News Release, 21 December 1971, BHR.

17. Bunker Hill Company news release, 15 November 1971, BHR.

18. The whole idea of mandated percentages was one that smelters had fought against for most of the century but now had given up on in principle and were fighting only in terms of specifics.

19. Bunker Hill newsletter, 22 January 1973, BHR.

20. "Intermittent Curtailment System," *Bunker Hill Reporter* (January–February 1973): 3.

21. Deposition of Gene Baker, Bunker Hill vice president for environmental affairs, 4 June 1979, pp. 105, 128, 130; "File, Note, Baghouse Fire at Bunker Hill," by Ian H. VonLindern, environmental engineer, Idaho State Department of Health and Welfare, 30 September 1974, *Yoss v. Bunker Hill Company*, U.S. District Court for the District of Idaho, no. CIV-77-2030 (hereinafter cited as Yoss case).

22. *Bunker Hill Reporter* (November–December, 1975): 1.

23. *Shoshone County Lead Project Report* (1975), pp. 92, 95.

24. James H. Halley to Dr. James A. Bax, director, Idaho State Department of Health and Welfare, 15 October 1974, including copy of "Protocol" for Shoshone County Lead Health Study, BHR; Tate, "American Dilemma," p. 77; Ian von Lindern and George Dekan to Jerry Cobb, Idaho State Department of Environmental and Community Services, 6 May 1974, Yoss case.

25. Bunker Hill news release, 27 January 1975, BHR.

26. James A. Bax to Frank Woodruff, 27 July 1974, Yoss case.

27. "New Smoke Regulations Proposed for Bunker Hill," *Kellogg Evening News*, 26 August 1974; *Engineering and Mining Journal* 176 (March 1975): 253.

28. Halley to Bax, 3 July 1975, BHR.

29. Gene Baker to Theodore Rogowski, EPA hearing officer, 12 August 1975, Yoss case.

30. Gene Baker to Lee Stokes, administrator, Environmental Services Division, Idaho State Department of Health and Welfare, 5 January 1976, Yoss case.

31. "Tall Stacks Deferred, EPA Ruling Appealed," *Bunker Hill Reporter* (November–December 1975): 1.

32. *Engineering and Mining Journal* 177 (January 1976): 148.

33. Privileged and Confidential Memorandum to Gene Baker from William F. Boyd and Edward Seger, 12 December 1975, BHR.

34. "Bunker Hill's Second Tall Stack: A Gamble on Reasonable SO_2 Rules," *Engineering and Mining Journal* 177 (September 1976): 23–4.

35. "Ninth District Court of Appeals Long Awaited Decision re SO_2," *Engineering and Mining Journal* 178 (September 1977): 304.

36. Memo, Bunker Hill president E. V. Howard to Frank Woodruff, Gulf Resource president, "B.H. 3rd Quarter Report," 23 October 1978, BHR.

37. Murray Michael, chief, Bureau of Air Quality, to Lee Stokes, Bob Olsen, Mike Christie, and John Stallings, 14 June 1978, "Bunker Hill Strategy Session, June 14, 1978," Yoss case; E. V. Howard to F. G. Woodruff, "Statements re: Fuming Plant Suit," 4 January 1977, BHR; Gulf Resources management meeting minutes, 27 February 1978, BHR.

38. E. V. Howard to R. H. Allen and F. G. Woodruff, 25 July 1978, "Smelter Profitability Data," BHR.

39. *Bunker Hill 1979 Annual Report to Employees*, BHR.

40. "Bunker Hill Agrees to Pay $2 Million," *Idaho Statesman*, 24 October 1981. *Statesman* reporters could find no one in Kellogg who agreed with the settlement: "Kellogg Folks Decry Bunker 'Rip Off,' " *Ibid*. Many people in Kellogg maintain that the children involved in the Yoss case suffered from neglect and a filthy home environment and that those who practiced basic hygiene were not leaded. Interview with Ray Chapman, former Bunker Hill public relations director, Kellogg, Idaho (8 June 1992).

41. D. R. Brandell to J. W. Kendrick, 25 February 1980, BHR; Robert Allen, CEO, to Douglas M. Costle, EPA administrator, 21 February 1980, BHR.

42. "Hunt for Bunker Hill Buyer Ends," *Idaho Statesman*, 11 November 1981.

43. See Bryce I. MacDonald and Moshe Weiss, "Impact of Environmental Control Expenditures on Copper, Lead, and Zinc Producers," *Mining Congress Journal* 64 (January 1978): 45–50.

Epilogue

Environmental History and Human Perception

WILLIAM DIETRICH

Environmental history is a relatively new way of interpreting history; its integration of natural and human history offers a different and revealing window on our past and our future. It is a splendid example of a new way of seeing: of the interconnection of thought, perception, and morality that becomes the stories we tell about ourselves and that reveal as much about the teller and his or her present as they do about the past. If there is a lesson in history it is that our vast, fascinating chronicle of human blunders—wars, hysteria, miscalculations, inquisitions, and genocide—should teach us the vital importance of perception, of seeing clearly, of honestly understanding ourselves and our fears and desires by understanding our troubled past. It is not that we humans cannot get what we want; rather it is that we are so clever we often achieve exactly what we wish for—which, because of the limits of our perception, often proves to be a disappointment or mistake of an entirely different sort.

To illustrate the limits of human perception that the environmental historian tries to overcome, consider the difficulties of observing what is arguably the most important object in Pacific Northwest history. You cannot see bacteria because they are microscopic, and yet the germ dictated the course of human development in this corner of the globe in two vital ways.

First, European germs largely wiped out the aboriginal population there. It was not the gun that conquered the Pacific Northwest for the newcomers, but disease. It eliminated upwards of 90 percent of many tribes. This was not only a tragedy for the Native Americans involved; it was, I suspect, a deep tragedy for those of us who came after. As a newspaper reporter I have had an opportunity to spend time with Native Americans at their traditional gatherings, and I often come away with mixed feelings: these are ceremonies and speeches that seem both intriguing and boring, wise and naive. Most of all, however, they seem profoundly *different*—a peek into a lost world where the inhabitants saw life and our planet in utterly different ways and had utterly different thoughts from those of us today. Native Americans lived in the Pacific Northwest at least 11,000 years before white pioneers arrived, and whatever wisdom they had accumulated about this place over that vast period of time was largely lost, wiped out in a holocaust of disease before they could influence and moderate the new culture.

The second importance of bacteria was the European development of a germ theory of disease, a medical breakthrough that inadvertently led to a global population explosion. Northwest history since Euro-American exploration is largely a history of waves of new immigration that is still going on today. British Columbia, Washington, and Oregon are each expected to gain nearly 100,000 more people a year over the next two decades. This growth is so diffuse, steady, and spread out that in many ways it is as invisible as microscopic bacteria—but it is this growth that has profoundly changed our environment so that the Northwest of the past is increasingly difficult even to imagine.

The limits to our perception are numerous. We can see only a small portion of the light spectrum. For every star our naked eye can see in the night sky, there are 100 million more in our own Milky Way galaxy alone. Our range of hearing is modest; a dog would consider our sense of smell pathetic. Similarly, most of what modern humans need to see in the environment is for all practical purposes invisible because of human limits in space and time. We do not notice the smothering of salmon spawning beds with silt from logging and agricultural erosion. We cannot see the way clearcutting disrupts the microscopic fungi vital to tree growth in northwestern soils. The invasion of alien plants that has transformed the region is unrecognized by most of us who assume that whatever plants we encounter have a long local ancestry. We do not see loss of water quality, cannot directly observe the horrific contamination of soil at the Hanford Nuclear Reservation, and are only dimly aware of the slow disappearance of hiking trails as forest-cutting extends into roadless areas. We mostly remain blind to creeping suburbanization. Certainly we cannot see loss of ozone and have difficulty detecting the signals of global warming. Our ancestors were naturalists by necessity, but most of us have lost any intimate connection to nature; it is a change that has crept up on us without conscious choice.

Similarly, our perception of time is faulty. Our "history" is like the history of an insect that lives only a few hours and measures existence by the events of a single day. If the history of the earth is compared to a 24-hour day, our species has inhabited the planet for only the last two seconds of its existence. Civilization has existed a fifth of a second, Europeans invaded the Pacific Northwest in the past two-hundredths of a second; the biggest dams and greatest conversion of land to logging, agriculture, and cities have lasted a thousandth of a second. We live in unique time: for the first time in the planet's history, a single species can change the entire earth, from modifying its atmosphere to triggering nuclear war. About 90 percent of the old-growth forest is gone, and more than 90 percent of the sage-steppe ecosystem. Half of the stream habitat has been lost and a third of stream miles permanently dammed. All this has taken generations of toil by the standards of conventional history—and a blink in time by natural history standards.

How can we overcome this national blindness? By environmental history. A popular song says, "You don't know what you've got till it's gone," and the sad truth might be that we do not really know what we've lost even *after* it's gone. We have no perception, no memory of the past. Unless it is taught.

History is that teaching, and its telling is a moral choice. The *lack* of its telling is also a moral choice: by ignoring our environmental past, we are choosing a deliberate blindness. Good history is not necessarily the history people want to hear. As a journalist I have sometimes been struck that what people want most is not "news" but "olds," stories that confirm the beliefs and prejudices they already have about the world. Write what they already think and they will call you a genius; propose something new and they will often react with confused doubt. Environmental historians may encounter the same reaction because they challenge our own myths, but by questioning our culture they ultimately serve to improve it. This is how humans learn.

Still, I should add that it is wrong simply to label our past errors as "mistakes." We are all prisoners of the assumptions and knowledge of our own time. What people strive for at any one time has a common sense to it, given the limits of their perceptions. So the point of education is to enable us not to become prisoners of narrow perception, not to become prisoners of history. For human culture, hindsight is a good pair of glasses: our telescope and microscope and spectrometer.

The history of the spotted owl fight, for example, makes no sense unless we understand that what happened was a shift in perception—a new way of looking at the same forest. What changed was not the woods but human perception of what kind of biology was decadent and what was valuable. Forestry scientists caution us to bring away from this story some humility. What we are certain of today will almost certainly be modified in 20 or 50 years.

Similar is the story of the Columbia River, a river ruled by changing human perceptions. The pioneers found a river that seemed rock-studded, erratic, and useless and so imagined a central Washington turned to Eden, dark cities turned to light. The fascinating point about the Columbia's story is that this imagination arguably outran human need; we built dams faster than we needed their power, and so development drove population growth, rather than the reverse. The transformation of the Columbia was a feat of human imagination, and because the pioneers had wiped out the aboriginal population with disease, they were essentially imagining on a blank slate; the historic native uses of the Columbia seemed hardly a factor for consideration. In contrast, we seem prisoners to our history today. We cannot imagine a Columbia different from the one our mothers and fathers created.

No one can deny the triumphs of Pacific Northwest development. It has resulted in a hundredfold increase in human population and perhaps a thousandfold increase in human wealth. We have just lived through the most incredible century in human history, of unprecedented violence and upheaval and economic growth and an explosion of human knowledge.

But we live in a hollow age, a spiritually confused one. Having satisfied many basic necessities, we are groping for meaning. Science has upset all our certainties, and history has overthrown all our smuggest beliefs. We are all in search of firm ground.

Human beings do have lives outside narrow material ones. We build churches, create art, and do crazy things for love. It was Ansel Adams who called national parks "vast areas devoted to the spirit." The only thing different today is that we are acutely aware that our parks don't seem so vast anymore. We look to the environment not just as a source of rest and recreation but as a source of wisdom, of perspective, of balance: of seeing, in ways we otherwise might be blind to.

In what directions, then, might environmental history point us? Perhaps it will help us to recognize truths about ourselves.

The naturalist E. O. Wilson points out that although the social insects such as ants and bees make up only a tiny fraction of insect species, they represent perhaps 80 percent of the insect biomass on earth. Similarly, we humans are successful social animals, and because of it we have come to dominate the planet. We are restlessly active at modifying the environment, because this has proven so successful for us in the past. We do not know yet if this tinkering will prove to be our ultimate salvation or our doom, but it certainly seems hardwired into what it means to be human.

On yesterday's field trip to the Palouse River,[1] we observed that children had used stones to try to dam or bridge the current. People like to manipulate things. We are also ambitious. We have recently lost a number of employees at the *Seattle Times* to software companies such as Microsoft in the Seattle suburbs, and those of us remaining are of course intensely curious about how our colleagues are doing in their new environment and, more importantly, how much money they are making. What is going on in Redmond today is a software gold rush, no different in its human motivation from the pioneer gold rushes that helped develop the early Northwest. Human character does not change much.

What this suggests, I think, is that humans—in our Western culture, at least—are unlikely to be content with passive protection of the environment. We are gardeners; we want to "improve" nature. What that implies is an environmental ethic that takes from our history the lesson that we are unlikely to be content with just preserving nature but rather must seize the opportunity to work with each other to nurture and rehabilitate it. It is the challenge of a hard task that truly excites us. Our relationship with the environment will be most successful when loggers and farmers and developers are enlisted in efforts to make a better future by pondering the triumphs and mistakes of the past. That enlistment could become one of the fruits of environmental history. And that is why the book you are holding is so important.

NOTE

1. Editors' note: This essay is based on extemporaneous remarks made by William Dietrich at the Northwest Environmental History Symposium, August 3, 1996.

Contributors

CARL ABBOTT is a professor of urban studies and planning at Portland State University. He is the author of numerous books and articles on urban and western history, including *The Metropolitan Frontier: Cities in the Modern American West* (1993) and *Political Terrain: Washington, D.C., from Tidewater Town to Global Metropolis* (1999). He is also the coauthor of *Planning a New West: The Columbia Gorge National Scenic Area* (1997).

KATHERINE AIKEN is an associate professor of history at the University of Idaho, where she teaches courses in twentieth-century United States history. She is the author of several articles dealing with the Bunker Hill Company and has also written *Harnessing the Power of Motherhood: The National Florence Crittenton Mission, 1883–1925* (1998).

MICHAEL C. BLUMM is a professor of law and codirector of the Northwest Water Law and Policy Project at Northwestern School of Law of Lewis and Clark College. He has taught and written extensively on natural resource law and policy for over 20 years. He teaches the only course in the country on Pacific salmon law.

THOMAS R. COX is a professor emeritus of history and Asian studies and an adjunct professor of American Indian studies at San Diego State University, where he taught for more than 30 years. He is the author of *Mills and Markets: A History of the Pacific Coast Lumber Trade to 1900* (1974) and *The Park Builders: A History of State Parks in the Pacific*

Northwest (1988). He coauthored *This Well-Wooded Land: Americans and Their Forests from Colonial Times to the Present* (1985).

WILLIAM DIETRICH is presently working as an author and freelance journalist after leaving the *Seattle Times* in 1997. He shared a Pulitzer Prize for coverage of the 1989 Exxon Valdez oil spill and is the author of two books on Northwest environmental issues: *The Final Forest* (1992) and *Northwest Passage* (1995). His first novel, *Ice Reich*, was published in October 1998.

KATHLEEN A. DWIRE has worked as a botanist for the Nature Conservancy, as a wetlands ecologist on contract to the U.S. Environmental Protection Agency, and as a range technician for the Bureau of Land Management. She is currently working on a Ph.D. in riparian ecology at Oregon State University, studying the interactions among hydrology, soils, and vegetation in montane meadows of the Blue Mountains, Oregon.

ERIC C. EWERT is a doctoral candidate and instructor at the University of Idaho. He brings a wide-ranging perspective to his essay, having lived in Alaska, Arizona, California, Nevada, Oregon, Utah, Washington, and Sonora, Mexico. His current research focuses on the implications of rapid demographic and economic change in the nonmetropolitan West.

MARK FIEGE is an assistant professor of history at Colorado State University. He is the author of *Irrigated Eden: The Making of an Agricultural Landscape in the American West* (1999).

DAN FLORES holds the A. B. Hammond Chair in Western History at the University of Montana, where his specialty is the environmental history of the American West. In 1997 he won a "Wrangler Award" for best article on the West from the National Cowboy Hall of Fame. He is the author of numerous books and articles on western history, most recently *Horizontal Yellow: Nature and History in the Near Southwest* (1999).

DALE D. GOBLE is a professor of law at the University of Idaho. His research and writing focus on the intersection of constitutional law, natural resource law, and history. He is the author of the usual "numerous articles."

KATHERINE HANSEN is a biogeographer and a professor of geography at Montana State University. Her research focuses on mountain environments and climatic and human impacts on vegetation.

PAUL W. HIRT is an associate professor of history at Washington State University specializing in environmental history and the history of the American West. He is the author of *A Conspiracy of Optimism: Management of the National Forests since World War Two* (1994) and editor of

Terra Pacifica: People and Place in the Northwest States and Western Canada (1998).

EUGENE S. HUNN is a professor of anthropology at the University of Washington. He is the author of *Tzeltal Folk Zoology: The Classification of Discontinuities in Nature* (1977) and *Nch'i-Wana "The Big River": Mid-Columbia Indians and Their Land* (1990), which grew out of a long-standing collaboration with James Selam, a Yakama elder of the John Day River people. In 1996 he returned to Mexico to initiate ethnobiological research in a Zapotec Indian village in Oaxaca. He has also done applied work on traditional Alaskan subsistence, emphasizing indigenous rights and comanagement strategies.

J. BOONE KAUFFMAN is an associate professor of ecology at Oregon State University. His primary research focuses on how natural and human-caused disturbances influence biological diversity and natural ecosystem functioning. His research sites include riparian/stream ecosystems in the Pacific Northwest and tropical rainforests undergoing rapid deforestation in Central and South America.

ARTHUR R. KRUCKEBERG, a Northwest naturalist for 40 years, is a professor emeritus of botany at the University of Washington. He is the author of *Gardening with Native Plants of the Pacific Northwest: An Illustrated Guide* (1982, 1995) and *The Natural History of Puget Sound Country* (1991). In 1976, he cofounded the Washington Native Plant Society.

WILLIAM L. LANG is a professor of history and director of the Center for Columbia River History at Portland State University. He is the author or editor of six books on Northwest history, including *Great River of the West: Essays on the Columbia River*, edited with Robert Carriker (1999).

NANCY LANGSTON, an environmental historian and ecologist, is an assistant professor of environmental studies at the University of Wisconsin at Madison. She is the author of *Forest Dreams, Forest Nightmares: The Paradox of Old Growth in the Inland West* (1997), which won the 1997 Forest History Society Weyerhauser prize. She is also the coauthor of a new ecology textbook, *Ecology*. Her current research examines the environmental history of northwestern riparian areas.

BRUCE A. McINTOSH is a research associate in the Department of Forest Resources at Oregon State University. He has conducted research and written several papers on riverine ecosystems and salmonid habitats in the Columbia River basin over the past nine years, focusing on historical changes in riparian and stream habitats.

ALAN G. MARSHALL is a professor of anthropology and social sciences at Lewis and Clark State College. He has worked with the Nez Perce people for the past 25 years. His academic interests lie in practice related to the symbolic dimensions of the natural environment and the material dimensions of culture. His most recent publication, coauthored with Kurt Torell, is "Socrates Meets Two Coyotes," a comparison of classical Greek and Nez Perce approaches to the problem of the one and the many.

PAUL S. MARTIN, an emeritus professor of geosciences at the Desert Laboratory of the University of Arizona, is a major contributor to theories explaining prehistoric animal extinctions over the last 50,000 years. Best known as coeditor of the path-breaking work *Quaternary Extinctions* (1984), he has also edited *Packrat Middens* and authored a book on the botany of Mexico's Rio Mayo watershed, *Gentry's Rio Mayo Plants* (1998).

CAROLYN MERCHANT is the Chancellor's Professor of Environmental History, Philosophy, and Ethics at the University of California, Berkeley. She is the author of *The Death of Nature: Women, Ecology, and the Scientific Revolution* (1980), *Ecological Revolutions: Nature, Gender, and Science in New England* (1989), and *Earthcare: Women and the Environment* (1996), as well as numerous articles on the history of science, environmental history, and women and the environment. She is also the editor of *Major Problems in American Environmental History* (1993) and *Green versus Gold: Sources in California's Environmental History* (1998).

KATHERINE G. MORRISSEY is an associate professor of history at the University of Arizona. She is the author of *Mental Territories: Mapping the Inland Empire* (1997) and a coauthor of *Washington: Images of a State's Heritage* (1988). Her essay in this collection is part of her current project, "The Nature of Conflicts," an interdisciplinary study of the environmental and cultural history of mining pollution in the twentieth-century North American West.

WILLIAM G. ROBBINS is Distinguished Professor of History at Oregon State University and the author of several books, including *Colony and Empire: The Capitalist Transformation of the American West* (1994) and *Landscapes of Promise: The Oregon Story, 1800–1940* (1997). He is currently working on a sequel to the Oregon book.

CHRISTINE R. SZUTER has conducted zooarchaeological research in the southwestern United States and northern Mexico for the past 20 years. She is the author of *Hunting by Prehistoric Horticulturists in the American Southwest* (1991) as well as numerous articles on the relationship of humans and animals with their environment. She is now director of the University of Arizona Press.

BARBARA LEIBHARDT WESTER is an attorney for the U.S. Environmental Protection Agency, where she specializes in tribal environmental issues. The essay in this collection is based on her dissertation, which won the American Society for Environmental History Rachel Carson dissertation prize in 1993.

WILLIAM WYCKOFF is a professor of geography at Montana State University. His research interests focus on settlement geography, cultural landscapes, and environmental history of the American West. He is the editor, with Lary Dilsaver, of *The Mountainous West: Explorations in Historical Geography* (1995).

DOROTHY ZEISLER-VRALSTED is a professor of history at the University of Wisconsin at La Crosse. She has worked with irrigation districts in Montana and Washington and has published several essays on water resources in the American West. She is working on a book examining agricultural and urban water use in the Yakima Valley, Washington.

Index